KT-237-292

216 998

Physiological Tests for Elite Athletes

Australian Sports Commission

Christopher John Gore, PhD
Australian Institute of Sport
Editor

Human Kinetics

Library of Congress Cataloging-in-Publication Data

Physiological tests for elite athletes / Australian Sports Commission ; Christopher J. Gore, editor.
p. cm.
Includes bibliographical references and index.
ISBN 0-7360-0326-6
1. Athletes--Medical examination. 2. Athletic ability--Testing. 3. Physical fitness--Testing. 4. Diagnosis, Laboratory. I. Gore, Christopher J., 1959- II. Australian Sports Commission.

RC1225 .P486 2000
612'.044'088796--dc21 99-088959

ISBN-10: 0-7360-0326-6
ISBN-13: 978-0-7360-0326-1

The following chapters were adapted with permission from the contributors: 3-6, 8-22, 24, and 26-29.

Acquisitions Editor: Loarn D. Robertson, PhD; **Developmental Editor:** Elaine Mustain; **Assistant Editors:** Derek Campbell, Amanda S. Ewing, Melissa Feld, Susan Hagan; **Copyeditor:** Joyce Sexton; **Proofreader:** Sarah Wiseman; **Indexer:** Sharon Duffy; **Permission Manager:** Heather Munson; **Graphic Designer:** Fred Starbird; **Graphic Artist:** Yvonne Griffith; **Photo Editor:** Clark Brooks; **Cover Designer:** Jack W. Davis; **Photographer (cover):** © James Startt; **Illustrator:** Mic Greenberg; **Printer:** Versa Press; **Binder:** Dekker Bookbinding

Printed in the United States of America 10 9 8 7 6 5 4

Human Kinetics
Web site: www.HumanKinetics.com

United States: Human Kinetics
P.O. Box 5076, Champaign, IL 61825-5076
800-747-4457
e-mail: humank@hkusa.com

Canada: Human Kinetics
475 Devonshire Road Unit 100, Windsor, ON N8Y 2L5
800-465-7301 (in Canada only)
e-mail: orders@hkcanada.com

Europe: Human Kinetics
107 Bradford Road, Stanningley, Leeds LS28 6AT, United Kingdom
+44 (0) 113 255 5665
e-mail: hk@hkeurope.com

Australia: Human Kinetics
57A Price Avenue, Lower Mitcham, South Australia 5062
08 8372 0999
e-mail: info@hkaustralia.com

New Zealand: Human Kinetics
Division of Sports Distributors NZ Ltd.
P.O. Box 300 226 Albany, North Shore City, Auckland
0064 9 448 1207
e-mail: info@humankinetics.co.nz

Contents

Contributors

Peter Abernethy, Department of Human Movement Studies at the University of Queensland

David Aitken, Queensland Academy of Sport

Peter Barnes, South Australian Sports Institute

Michael Blackburn, New South Wales Institute of Sport

Pitre Bourdon, South Australian Sports Institute

Andrea Buckeridge, Tennis Australia

Don Cameron, Australian Water Polo Incorporated

Kate Cameron, Australian Capital Territory Academy of Sport

Lindsay Carter, San Diego State University

Wayde Clews, Australian Water Polo Incorporated

David Cowley, Mater Misercordiae Adult Hospital

Neil Craig, South Australian Sports Institute

Bob Crudgington, Australian Institute of Sport

Richard Done, Commonwealth Bank Cricket Academy

Lindsay Ellis, Queensland Academy of Sport

Kieran Fallon, Australian Institute of Sport

Damian Farrow, University of Ballarat

James Finn, Northern Territory University

Danielle Fornasiero, Australian Institute of Sport

Peter Fricker, Australian Institute of Sport

Greg Gass, Griffith University

Paul Gastin, Victorian Institute of Sport

Ian Gillam, Exercise and Nutrition Consultant

Wayne Goldsmith, Moregold Sports Pty Ltd

Christopher Gore, Australian Institute of Sport

Simon Green, Queensland University of Technology

Allan Hahn, Australian Institute of Sport

Donna Harvey, Australian Institute of Sport

David Jenkins, University of Queensland

Deborah Kerr, Curtin University of Technology

Steve Lawrence, Western Australian Institute of Sport

Simon Locke, Queensland Academy of Sport

Peter Logan, Australian Institute of Sport

Katrina Lynch, University of Limerick

Mark McGrath, Victorian Institute of Sport

Craig Mansfield, Hawthorn Physiotherapy Clinic

Michael Marfell-Jones, Tai Poutini Polytechnic

David T. Martin, Australian Institute of Sport

Graeme Maw, Queensland Academy of Sport

Peter Morrow, Victorian Institute of Sport

Kevin Norton, University of South Australia

David Owies, Deakin University

Lachlan Penfold, Queensland Academy of Sport

Esa Peltola, Australian Institute of Sport

Rob Pickard, Triathlon Australia

Ted Polglaze, Tasmanian Institute of Sport

Frank Pyke, Victorian Institute of Sport

David Pyne, Australian Institute of Sport

Ann Quinn, Tennis Australia

Peter Reaburn, Central Queensland University

Aaron Russell, Deakin University

Rebecca Tanner, Australian Institute of Sport

Kaylene Saddington, Australian Institute of Sport

Bernard Savage, South Australian Sports Institute

Christer Skog, United States Ski Association

Darren Smith, Queensland Academy of Sport

Paul Smith, Queensland Academy of Sport

Tom Stanef, South Australian Sports Institute

Andrea Stapff, Western Australian Institute of Sport

Geoffrey Strauss, Curtin University of Technology

Richard Telford, Australian Institute of Sport

Douglas Tumilty, Australian Institute of Sport

Charlie Walsh, Australian Institute of Sport

Nancy Whittingham, University of South Australia

Greg Wilson, Rock Valley, New South Wales

Robert Withers, Flinders University of South Australia

Patricia Witt, Mater Misercordiae Adult Hospital

Sarah Woolford, South Australian Sports Institute

Tim Wrigley, Victoria University of Technology

Warren Young, University of Ballarat

Preface

At the Montreal Olympics (1976), Australia did not win a single gold medal; in an attempt to change this situation, the Australian government established a national system of state-based sport institutes. Just how successful these institutes have been in raising the level of athletic performance in the country is suggested by the fact that, in contrast to the 1976 Olympics, the Barcelona Olympics (1996) saw Australia place ninth among all nations. Part of this success can be traced to the standardized test procedures that were developed at the national sport institutes to allow comparison of results among laboratories. *Physiological Tests for Elite Athletes* contains the most current of these standardized physiological test procedures. Although other manuals of test protocols are in print, they generally address tests for gymnasium/health club clients or cardiac rehabilitation patients. This volume is unique in its focus on testing the elite athlete. If you work with elite athletes, or are a student who aspires to work with elite athletes, this manual is a comprehensive guide to the "how and why" of the principal physiological tests.

Physiological Tests for Elite Athletes provides test protocols for the physiological assessment of elite athletes in 17 different sports, as well as the rationale for these protocols. Although the protocols are used in Australia, they were developed from an understanding and scrutiny of the international literature related to athlete assessment. Thus the extensive normative data for the tests in this volume provide reference points for measuring elite athletes from any country. Sport scientists, academics, and students from many parts of the world may find the normative data particularly relevant because Australia has produced numerous world-class athletes. Readers will also find the reproducible forms for data collection and for preparticipation screening quite useful.

The book is divided into four sections. Part I (chapters 1 and 2) deals with the often overlooked issue of quality assurance in the exercise laboratory, as well as the importance of athlete preparation before testing. Part II (chapters 3-9) provides generalized test procedures for aerobic and anaerobic power, the lactate "threshold," anthropometry, blood sampling, and flexibility. These procedures, as well as those in the chapter on team sport testing, are extensively cross-referenced to the sport-specific chapters in part IV (chapters 14-30). The reader must keep this cross-referencing in mind, because many chapters are not complete unless read in conjunction with material elsewhere in the book. For example, the description of the 20 m multistage fitness test (used to estimate aerobic power) in the team sport chapter is referred to in the chapters on other sports that use this test: basketball, cricket, hockey, netball, rugby union, soccer, softball, and tennis.

Part III of the book (chapters 10-13) is an extensive examination of isometric, isokinetic, and isoinertial strength assessment. This is an area that requires further research and improved insights into mechanisms of acute and chronic adaptation to strength and power activity. Rather than providing definitive protocols, these chapters present a variety of approaches that may aid in the selection of appropriate test procedures.

Part IV (chapters 14-30) provides specific test protocols for 17 different sports, and many of these are sports at which Australia excels on the international stage—for example, cricket, cycling, golf, hockey, netball, rowing, rugby union, softball, swimming, tennis, triathlon, and water polo. Each chapter contains a rationale for the tests and a list of the necessary equipment and procedures; where possible, normative scores for each test are tabulated for national- and international-level athletes. Some of the chapters, such as the one on swimming (chapter 27), provide information regarding interpretation of test results. However, where normative data are provided, this volume does not cover how an athlete achieved that level of fitness. The test results provide a snapshot of the characteristics of elite athletes, but the coach—perhaps with the assistance of the physiologist—has to create the training regimen to achieve that level.

The overall approach of this book is to provide working procedures for testing athletes that have a sound theoretical basis and for which the precision of the technique is known. However, areas such as anaerobic capacity, lactate "threshold," and strength assessment involve mechanisms that are not well understood and are still heavily debated in the scientific literature. In these chapters, the attempt has been to provide procedures that may be useful stepping-stones until more refined tests are developed. For example, validation work is required for each of the tests of abdominal strength in the team sports chapter. Nevertheless, in situations in which the test is carefully administered and has good precision, repeat tests can be useful for tracking changes over time as a guide to the effectiveness of a training intervention.

The majority of chapters in this manual were written by sport scientists who spend every working day with athletes as their sole focus. Often their work extends to seven days a week, since that is the nature of elite sport. Most of the authors are not academics who theorize about what might help improve an athlete's performance. Rather, they interact closely with coaches and athletes to optimize sporting success. This book includes their collective insight and experience.

Acknowledgments

It would be remiss not to thank publicly the many people who contributed to this book. While there is always a danger that someone will be inadvertently forgotten, there are some that rate a special mention. First, the authors and coauthors of all chapters must be acknowledged for taking time to turn their experience into text. The many athletes whose normative data are contained in this book must also be thanked indirectly, since for ethical reasons none of them can be identified by name.

The support of the Australian Sports Commission (ASC) to publish two previous versions of this book in 1987 and 1991 has been instrumental in the genesis of this volume. Without the assistance of the ASC to elite sport science in Australia, this manual would not have been produced. In particular, the cooperation of Margaret Chalker of the Publications Section of the ASC should be recognized. Carol Miller, editor for the Publications Section of the ASC, put enormous effort into improving the original manuscript, and all of us who have been involved in the production of this text owe her a great debt of thanks. Dr. Ross Smith, Director of Sport Sciences at the Australian Institute of Sport, must also be thanked for allowing me to commence the manuscript in Chicago instead of Australia—telecommuting of the very long-range kind!

Finally, the excellent administrative assistance of Julie Hill over the three-year gestation of this book has been most appreciated.

Introduction

Frank Pyke

A national program of sport science servicing and research associated with the high-performance athlete requires testing protocols to be duplicated and comparable data obtained in different parts of the country. This in turn requires a high standard of quality control during both laboratory and field testing situations. Getting an accurate and precise result in a laboratory or field environment enables us to evaluate the appropriateness of coaching methods and to answer correctly the many research questions that arise as we attempt to optimize sport performance.

Before proceeding into the specifics of anthropometric and physiological testing in later sections of this manual, one must understand the potential benefits of a testing program and the criteria that should be applied when selecting tests.

Reasons for Testing

There are many reasons that testing is vital for sport science. The most important ones are discussed in the following list.

1. **Identify weaknesses.** The main purpose of testing is to establish where an individual's strengths and weaknesses lie. This involves identifying the major components of fitness in the sport and then conducting tests that measure those components. A training program that is geared toward the development of the individual can then be prescribed.

2. **Monitor progress.** By repeating appropriate tests at regular intervals, the coach can obtain a guide to the effectiveness of the prescribed training program. A "one-shot" testing experience provides very little benefit either for the athlete or the coach and is strongly discouraged.

3. **Provide feedback.** The feedback of a specific test score often provides the incentive for an athlete to improve in a particular area, as he or she knows that the test will be repeated at a later date. The opportunity for members of a training squad to periodically compare their performances on objective and relevant tests is a useful motivational tool to encourage them to strive for improvements.

4. **Educate coaches and athletes.** A testing program can provide coaches and athletes with a better understanding of the demands of the sport and the attributes that are required to be successful. This facilitates systematic planning of athlete development programs.

5. **Predict performance potential.** Many nations have experienced some success at identifying individuals who may be suitable for sports that depend on certain anthropometric characteristics and physiological capacities.

These characteristics and capacities can be identified only through appropriate and accurate testing.

Criteria for Selecting Tests

There are a number of criteria that need to be considered when selecting tests that are appropriate to the circumstances in a particular sport.

- **Relevance.** In order to obtain valid results, it is important for athletes to respond positively to fitness tests. Their immediate recognition of the relevance of the particular test to their sport is therefore critical. The test should be selected in accordance with the known energy requirements of the sport. For example, a pursuit cyclist is involved in a maximal effort for four to four and one-half minutes. Maximal explosive and sustained power using both aerobic and anaerobic energy pathways should be measured. In addition, some assessment of mechanical efficiency and the blood lactate transition threshold would be appropriate. By the same token, a field hockey player requires both speed and endurance with energy supplied in both an intermittent and continuous manner. Hence, a short sprint (to assess both acceleration and speed), a series of interval sprints (to determine a fatigue index), and a longer run such as a 20-meter shuttle run to exhaustion (to estimate maximum aerobic power) would provide a useful guide to the major fitness components for field hockey.

• **Specificity.** Fitness tests should address the performance capabilities of the muscle groups and muscle fiber types actually involved in the sport. For example, strength testing should relate not only to the specific muscle groups involved in the activity but also to the patterns and speeds of movement followed. Also, distance runs or treadmill tests should be used for measuring endurance in running sports, and cycle, kayak, swimming, and rowing ergometers should be used for assessing explosive and sustained power in the specific sports for which they were designed. Field tests on the road or the water are ultimately the most specific, if the athlete is able to move freely, unhindered by testing apparatus, but only if environmental influences such as wind, tide, or temperature are not too extreme.

• **Practicality.** The concept of practicality is also an important consideration when choosing which tests to administer. Factors such as the location and availability of both the subjects and the venue as well as the duration and cost of the tests need to be taken into account when determining the suitability of a test battery.

• **Validity.** An appropriate fitness test should measure what it claims to measure. That is, the test should be valid. The degree of validity can be gauged from close inspection of its content. For example, if a test is measuring aerobic fitness, it should be of sufficient duration to test the power of this energy system. A test lasting longer than 5 minutes places sufficient emphasis on the aerobic energy pathways for it to be called an aerobic fitness test. Tests of shorter duration, while still placing stress on the aerobic system, require a significant anaerobic energy contribution and therefore lack sufficient content validity to be called aerobic fitness tests.

• **Accuracy.** An appropriate fitness test should also be accurate when compared to a criterion method. For example, the high correlation that has been established between performance on the 20-meter shuttle-run test and maximal aerobic power as measured in the laboratory ensures that the 20-meter shuttle run is an acceptable measure of aerobic running power. That is, the accuracy of the 20-meter shuttle run as a "field" measure of running endurance fitness has been established.

Quality Control in Administering Tests

Once a test has been selected it must be conducted in such a way as to obtain valid and useful information in an ethically acceptable manner.

Standardizing Test Conditions

It is important for the coach and the athlete to be aware that test results are affected greatly by the conditions under which they were obtained. In the laboratory, measurements can be made that minimize many of the variables that may affect performance on competition day. However, even in the laboratory there will be small changes in test results as a consequence of both day-to-day biological variability and small variations in the calibration and use of the measurement equipment. To reduce these variations, on each test occasion the tester should allow the same warm-up, the same order of tests, the same recovery period between tests, and as far as possible ensure that the environmental conditions of heat, humidity, and air movement are similar. The athlete should be tested at approximately the same time of day and be in a similar fluid and nutritional state. It is also important for the athlete to be adequately rested and not be suffering from injury or illness. After careful instruction by the testing staff, it is the responsibility of athletes to present themselves for fitness testing in such a way that they are capable of giving a performance that reflects their peak physiological status at that time. Otherwise, changes in fitness test scores may be attributed to a host of factors that have nothing to do with the prescribed training method.

While quality control is a basic practice in clinical chemistry, it has not been as rigorously pursued in the exercise sciences. Such quality-control procedures should involve control of instruments, including calibration of ergometers, analyzers, and recorders; control of subjects before, during, and following a test; and control of the knowledge of both the scientist and the technician via continuing education programs. It should be mandatory to implement quality-control procedures.

Even with optimal quality control, there will be small errors associated with laboratory measures that are conducted on athletes. Simple statistics can be used to quantify these small measurement errors and also can be used to determine the probability that any measured change is a consequence of the training program and not merely a result of poor control of subject preparation or of calibration procedures (Pederson and Gore 1996). Quality-control statistics are founded on an understanding of precision and reliability, and underpin any interpretations that can be made of results.

- **Precision.** It must be possible to obtain the same result on a test on two or more separate trials or occasions. However, for any test, the repeated measurements will usually vary (to a smaller or larger degree) in an unpredictable fashion. The size of that variability determines the level of *precision* and is a characteristic of a particular measurer using a particular measurement procedure on a particular variable. High precision corresponds to low variability in successive measurements and is the aim of a competent sport scientist. A statistical procedure called the Technical Error of Measurement (TEM) can be used to quantify the precision of a measure (Pederson and Gore 1996). For example, the TEM for a laboratory measure of maximum aerobic power is approximately 3%. This TEM value allows the scientist to determine the probability that any measured change is a real consequence of the training program and not merely an artifact of poor subject preparation or poor calibration procedures.

- **Reliability.** Another property of a measurement that is based on the concept of repeated measurement of the same subjects is *reliability*. Reliability is the consistency of successive measurements. Although it is true that a similar statement can be made about precision, the distinguishing feature of reliability is that it depends on the variability between subjects as well as the variability within subjects. This concept provides additional information to the TEM. Pederson and Gore (1996) refer the reader interested in these statistics to a comprehensive treatment of the topic.

- **Interpretable Results.** Each test should be described to the athlete before it is conducted. There is a much greater chance of the athlete giving maximum effort if he or she understands the reason for the test, its relevance to performance, and the physical commitment required. The test results should be returned promptly and then interpreted in a way that the coach and athlete easily understand. This is part of the educational value of a good testing program. In reporting results, it is useful to have the individual's score on each test, the group average, and the rank of the individual in the group. In addition, the test result sheet should include the precision (TEM) of the measurement so that the likelihood of a real physiological change can be distinguished from combined biological variation and technical measurement error. Finally, a comment on the exposed strengths and weaknesses can then provide the coach with useful guidelines for an appropriate training prescription.

Ethical Considerations

Before administering a test to an athlete, you should thoroughly explain its purpose, clearly enunciate the risks associated with being involved, and give assurance that the results will be kept confidential. It is imperative to have athletes sign an informed consent form indicating that they are fully aware of these matters before proceeding with the test(s). This is done to respect the athlete's human rights.

Concluding Comment

The test protocols contained in this manual should be reviewed regularly as a means of further improving their value to a particular sport. In this context, the creativity of sport scientists and coaches needs to be encouraged within a continual learning environment.

Meanwhile, I recommend this manual to you in the hope that it will provide a sound basis for implementing high-quality laboratory and field testing practices.

References

Pederson DG and Gore CJ (1996). Anthropometry measurement error. In Norton KI and Olds T (Eds.) *Anthropometrica*. Sydney, University of New South Wales Press, 77-96.

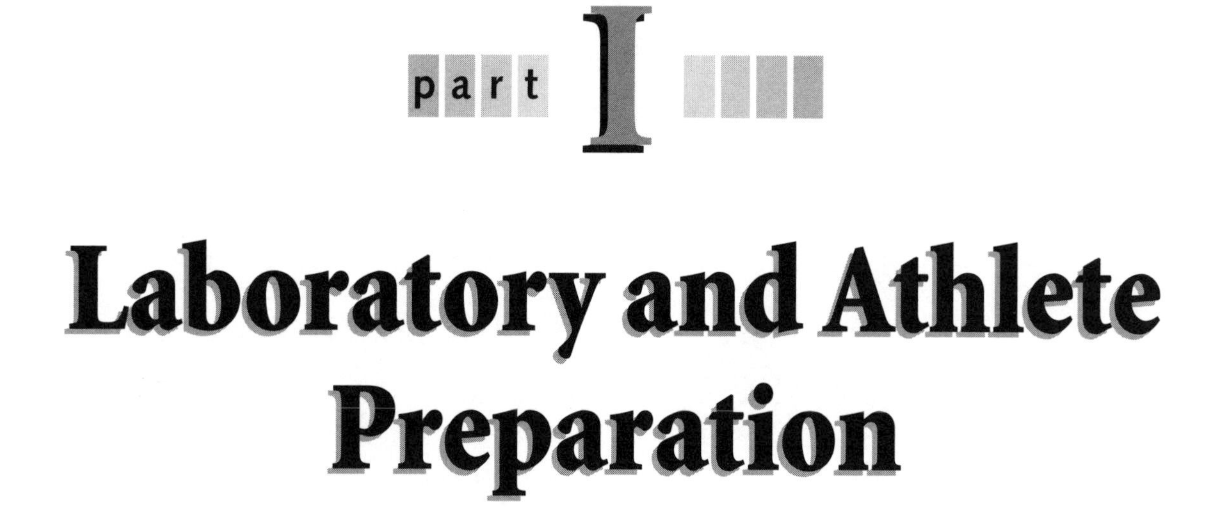

part I

Laboratory and Athlete Preparation

chapter 1

Quality Assurance in Exercise Physiology Laboratories

Christopher Gore

Quality assurance, commonplace in many areas of manufacturing and service industries, can be defined as "all those planned and systematic actions necessary to provide adequate confidence that a product or service will satisfy given requirements for quality" (National Association of Testing Authorities [NATA] 1989). In sport science, quality assurance allows the coach and athlete to be confident that the correct protocol has been used; that the equipment employed in data collection has been properly calibrated; and that test results have been collected in a standard format that readily identifies all the test details, the scientist, and the size of the uncertainty associated with the tests. Quality assurance of a service also means that the results are returned to the athlete or coach within two days and in a form that can be readily understood. If for any reason the test has to be repeated at another location, or several months (or even years) later, the data record sheet should contain enough information about all pertinent factors to allow replication.

Quality assurance in an exercise science laboratory is difficult for at least three reasons:

1. Test procedures may change as techniques are refined and new approaches are implemented.
2. Some athlete testing may be more experimental than routine.
3. The athlete's characteristics (for instance, motivation, training on the day before testing, or even muscle glycogen levels) may vary from day to day; that is, measurement accuracy is compromised by the changes that can occur in athletes.

For reasons such as these, quality assurance should be routine in all sport science laboratories that test athletes. With adequate documentation, it is easier to identify whether a test used an outdated or a current protocol or to identify an aberrant score—for example, because an athlete appeared for testing after an unusually hard training session on the previous day.

Because exercise science is dynamic and constantly evolving, the need to develop and evaluate new protocols will always exist. The aim of quality assurance is neither to stifle creativity nor to make all laboratories clones of one another. Instead, well-implemented quality assurance means that data from different locations can be pooled and users can have confidence in the test results. For example, if a national sporting association stipulates that all athletes from that sport should be tested with specific protocols, quality-assurance procedures enable sport scientists to follow those protocols exactly and to report results in a uniform format, regardless of the city or town in which the athletes are tested.

Quantifying the uncertainty of athlete tests has become an important component of quality assurance in exercise physiology laboratories. Tests and retests can be conducted on a representative subpopulation of athletes to identify the precision or uncertainty of a specific procedure. These trials can be used to calculate a statistic called the technical error of measurement (TEM), which identifies possible error due to the equipment and to biological variation of athletes (see "Technical Error of Measurement" on page 83). In addition, this statistic can be used to generate 68% or 95% confidence intervals of a real change (Pederson and Gore 1996). No longer does an exercise physiologist have to guess whether a change from, say, 62 to 63 $ml \cdot kg^{-1} \cdot min^{-1}$ is meaningful or just measurement artifact. The TEM provides a statistical basis for making such decisions. In addition, TEM data provide an objective method to evaluate the competency of a sport science laboratory. To provide a quality service to athletes and coaches, it is not considered sufficient to merely possess a pair of skinfold calipers, a gas analysis system, and a treadmill. Sport scientists must prove, for example, that they can use skinfold calipers with a TEM of better than 2% and that they can measure maximum oxygen consumption to better than 3% (see table 1.1).

Assessing Quality Assurance

Despite the benefits of quality assurance, some form of evaluation is needed to maintain the quality system at a high level. Without audits of the effectiveness of the quality-assurance system, procedures are likely to degenerate to those that are "most expedient" rather than "best practice."

In General

Worldwide, the effectiveness of quality-assurance systems in manufacturing and service industries is measured against the International Organization for Standardization document, ISO 9001/2/3 series (ISO 1994). Companies actively pursue this certification because it improves the quality and marketability of their product or service and the competitiveness of their organization. A company seeking ISO 9001 certification must prepare extensive documentation showing how it fulfills each of the 20 clauses of the ISO standard, and must submit to an external assessment of how well it implements its documented procedures. The company must also complete and maintain records of regular internal audits of its quality-assurance procedures and submit to an external review every six months and a full reassessment every two to three years. The ISO 9001 certification has prestige and credibility because it is a truly international standard.

Table 1.1 Target Technical Error of Measurement (TEM) Data

Measurement	Units of raw data	Target TEM
Anthropometry		
$\Sigma 7$ or $\Sigma 8$ skinfold sites	mm	<2%
Field testing		
Shuttle run	$ml \cdot kg^{-1} \cdot min^{-1}$	<3%
Vertical jump	cm	<5%
Sprint tests—20 m sprint	s	<1.5%
505 agility	s	<3.5%
Repco cycle ergometry—10 s work	W	<3%
Oxygen consumption (at $\dot{V}O_2max$)		
$\dot{V}O_2$	L/min	<3%
Ventilation	L/min	<5%
Respiratory exchange ratio		<5%
Heart rate	Beats/min	<2%
Blood analysis—lactate (and associated measures at "anaerobic threshold")		
[La]	mmol/L	<15%
Power output	W	<3%
$\dot{V}O_2$	L/min	<5%
Heart rate	Beats/min	<3%

These targets were developed from data acquired from the first five Australian laboratories that achieved Phase 1 (Precision) certification.

In Exercise Physiology Laboratories

Quality assurance in laboratories is well established in areas such as pathology and hematology, materials testing, and metrology. For example, if one receives a cholesterol result from a doctor, it will have come from a calibrated analyzer, using documented procedures with the cholesterol value printed on the report actually pertaining to the sample that was submitted for analysis. Many government laboratories must, by law, have quality-assurance certification.

Since exercise physiology laboratories often have fewer than six staff and cannot afford to have a full-time quality-assurance officer, the ISO 9001 standard is considered to be excessive for them. An appropriate and functional certification scheme that is within the constraints of time and resources of such laboratories has been developed by the Laboratory Standards Assistance Scheme (LSAS) of the Australian Sports Commission.

While the "essence" of the ISO 9001 standard was retained, only relevant sections of the 20 clauses are assessed for exercise physiology laboratories. Though some have argued that a watered-down version of the ISO 9001 document is worthless because the standard is already minimalist, those exercise physiologists who have implemented quality assurance are adamant that the LSAS scheme has resulted in quantifiable improvements to the assessment of athletes in Australia. These scientists are more confident than before about the accuracy and interpretation of results that they return to coaches and athletes.

This chapter outlines some of the key elements that have contributed to implementing successful quality assurance in Australian exercise physiology laboratories. The process continues to be refined over time because quality assurance is, by definition, a process of ongoing improvement. It is also worth noting that quality assurance in Australian exercise physiology laboratories has been implemented gradually after being initiated by the sport scientists themselves. It is not merely an "examination" of a laboratory, but a collaborative effort to improve standards of athlete exercise testing. It is the responsibility of the LSAS to provide advice to applicant laboratories on how to rectify any perceived deficiencies observed during assessment of a laboratory.

Australian Certification Model

This section summarizes the certification rationale and mechanism that the LSAS has developed.

Why Certify Laboratories?

The primary aim of certification is to enable comparison of the results of athlete testing from different laboratories.

The secondary aims of certification are to

- establish greater confidence in exercise physiology services among coaches and athletes,
- encourage sport scientists to exchange information,
- develop a national database of accredited laboratories and encourage national sporting organizations and coaches to use it,
- develop a database of appropriate measurement "error" tolerances for commonly conducted tests, and
- develop a database of test results on elite athletes by pooling results from all accredited laboratories.

In the first six years of the certification model in Australia, several of these aims have been achieved. Particularly useful has been the improved collaboration among sport scientists who, by sharing their knowledge, have gained a collective information base far in excess of what an individual could achieve. Error tolerances are now well established (see table 1.2), and more rigorous calibration requirements have achieved the primary aim for several of the common athlete tests.

What Is Certificated?

Certification is granted for demonstrated competence in the most common groups, or "streams," of physiology tests conducted on athletes. The streams, with examples of tests within each, are the following:

- Anthropometry (e.g., skinfolds, height, and mass)
- Field testing (e.g., 20 m shuttle run, sprint tests, vertical jump, and agility runs)

Table 1.2 Technical Error of Measurement (TEM) Data and 95% Confidence Intervals

Test	No. of *staff or no. †labs for mean	No. of subjects at each lab	Grand mean	TEM range	$\bar{X}$TEM	95% CI for a real change
Anthropometry						
Σ8 skinfolds (mm)	*13	~20	75.7	0.6–3.3	1.2	±3.4
Blood analysis						
Lactate at anaerobic threshold (mmol/L)	†3	~10	3.45	0.41–0.49	0.46	±1.30
Cycle ergometry						
Peak power—10 s test (W)	†3	~10	1135	29–53	38	±107
Field testing						
20 m shuttle run—predicted $\dot{V}O_2max$ ($ml \cdot kg^{-1} \cdot min^{-1}$) test (W)	†4	~10	49.5	0.9–2.0	1.4	±4.0
Vertical jump (cm)	†5	~12	50.5	1.0–5.4	2.2	±6.2
20 m sprint (s)	†4	~10	3.32	0.02–0.08	0.05	±0.14
505 agility test (s)	†4	~10	2.80	0.05–0.91	0.17	±0.48
Oxygen consumption						
$\dot{V}O_2max$ (L/min)	†5	~12	4.45	0.06–0.14	0.10	±0.28
Ventilation at $\dot{V}O_2max$ (L/min)	†5	~12	164.9	3.7–9.9	7.3	±20.6
Respiratory exchange ratio at $\dot{V}O_2max$	†5	~12	1.15	0.01–0.06	0.03	±0.08
Heart rate at $\dot{V}O_2max$ (beats/min)	†5	~12	193	1.5–3.7	2.7	±7.6

These results were developed from data acquired from the first five Australian laboratories that achieved Phase 1 (Precision) certification. Units for the mean, TEM range, $\bar{X}$ TEM, and 95% confidence interval are each in the same units as those of the respective tests.

- Oxygen consumption (e.g., $\dot{V}O_2max$ and maximum accumulated oxygen deficit)
- Ergometry (e.g., performance tests on calibrated cycle ergometers, rowing ergometers, and treadmills)
- Blood analysis (e.g., measurement of blood lactate, blood gases, ferritin, and hemoglobin).

Note that a laboratory that is certified in all five streams offers a greater range, not necessarily higher quality, than a laboratory certified in only one or two streams. The quality of measurements from any certified laboratory meets similar criteria based on TEM data and calibration records.

How Are Laboratories Certified?

As with the ISO 9001 certification model, the basis of certification is an on-site assessment by three expert peers to "critically evaluate all aspects of the laboratory management, staff, facilities and operations likely to affect the reliability of its test results" (NATA 1989). In Australia, a report on the visit, together with the TEM data for each laboratory, is evaluated by the National Sports Science Certification Committee (figure 1.1), convened by the Australian Sports Commission and the LSAS. Other countries interested in replicating (or adapting) this model should select

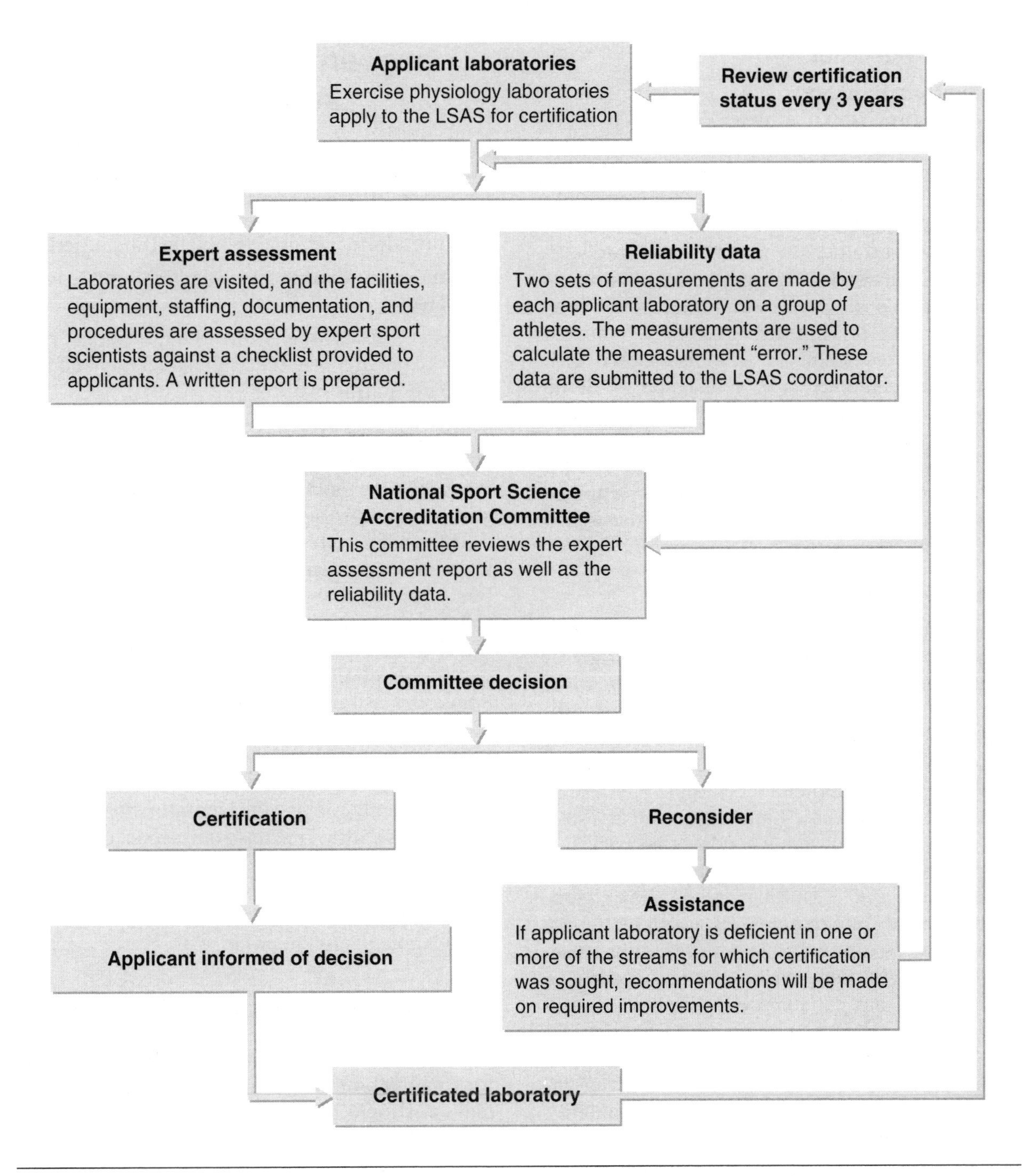

Figure 1.1 Flow chart of the certification process for Australian exercise physiology laboratories.

a committee composed of both senior sport physiologists who are well published in the international literature, and senior scientists with at least five years of experience testing state- or international-level athletes. In Australia, this mix of "academic" and "practical" scientists has provided a counterbalance between what is ideal and what is achievable in testing athletes and working with their coaches.

After successfully completing a rigorous certification process, laboratories receive a formal certificate that indicates the test areas in which they are highly proficient. This will include up to five of the common measurement streams such as anthropometry, field testing, or oxygen consumption. The certificate also indicates the period for which a laboratory is certificated.

There are two phases of certification in the Australian model, and both phases require on-site assessment by expert peers.

Phase 1—Precision

The first phase required laboratories to demonstrate the precision of athlete test results and to establish the measurement "error" associated with these test results. Equipment limitations combined with biological, day-to-day changes in the performance of athletes mean that all measures have an "uncertainty" or "error" associated with them. Thus, instead of reporting a sum of skinfolds as 65 mm, it is reported to an athlete and coach as 65 ± 3 mm. The expression "± 3 mm" indicates the level of uncertainty or TEM associated with the measure. Similarly, a $\dot{V}O_2max$ result would be reported as 70 ± 2 ml · kg^{-1} · min^{-1} rather than 70.2 ml · kg^{-1} · min^{-1}. Note that this is exactly the same kind of measurement error that is associated with any scientific measurement; for instance, cholesterol results are usually reported as a value with a measurement error range of ±0.1 mmol/L.

Phase 2—Accuracy

In the second phase of the certification scheme, laboratories must demonstrate the accuracy of measures; that is, they must show that the measures not only have a small level of uncertainty but also are accurate when compared with a criterion (or "first principles") measure. For example, an athlete in a maintenance period of training might have his or her $\dot{V}O_2max$ measured as 73 ± 2 ml · kg^{-1} · min^{-1}, but one week later be tested at another laboratory as part of a training camp and be measured as 68 ± 2 ml · kg^{-1} · min^{-1}. Which laboratory is right (or accurate)? Establishing the accuracy of measures will help interpret differences in measurements between laboratories, and to this end the LSAS has collaborated with others to develop devices such as a dynamic calibration rig for skinfold calipers (Carlyon et al. 1996), a gas analysis calibrator that can simulate the ventilation pattern and expired gas fractions of an elite athlete (Gore et al. 1997), and a torque meter to calibrate kayak ergometers. Accuracy is essential if a national database of athlete test results is to be established and if meaningful comparisons between laboratories are to be made.

Certification in either Phase 1 or Phase 2 is granted for a period of three years and reviewed every three years thereafter. Recertification requires a laboratory to collect new TEM data to demonstrate competence in all techniques of athlete testing, as well as to undergo an on-site assessment of the laboratory quality systems by three expert peers.

Implementing Quality Assurance

Implementing quality assurance in an exercise physiology laboratory requires a functional quality system that documents

- the laboratory quality policies and practices;
- the calibrations, test procedures, and reports; and
- procedures for ensuring the traceability of all documents (see section on document control later in this chapter).

These are the elements that the LSAS has extracted from the 20 clauses in the ISO 9001 standard as those most pertinent to small-scale sport science laboratories.

The quality system is documented and recorded in a handbook (hereafter referred to as the laboratory manual) that should contain, but is not restricted to, a number of key elements as listed in the next section. While some items such as the quality policy may take up only one page, other items such as test procedures may take up dozens of pages. One important aspect of the laboratory manual is that it must be truly relevant and useful to the laboratory that developed it. There is little value in merely copying the laboratory manual from another institution, because one aspect of the peer assessment of a laboratory is to verify that the documented procedures—procedures intended to improve the precision of the test results—are actually implemented in practice.

The Laboratory Manual

There is not one right way to construct a laboratory manual, but regardless of its structure, the manual should be useful. In essence, the aim of the quality system is to make the operation of the laboratory more efficient and to provide traceability of results. For example, if a sport scientist changes jobs, that person's documentation should be sufficient to enable another scientist to replicate the protocols he or she was using. Ideally, the records of previous athlete tests should contain sufficient information to identify exactly which test protocol was used, how equipment was calibrated, and the error associated with that test. Furthermore, it should be possible to trace results of an individual athlete across many years and identify whether different protocols were used during that time period. The quality system with the associated laboratory manual is merely a more formal organization of the Post-it notes and dog-eared pages that sport scientists often use to cope with the multiple demands of their work.

A convenient method of organizing a laboratory manual is to have sections on

- laboratory quality policies and records,
- test procedures/calibrations, and
- work instructions.

This organization allows for good flexibility in preparation, approval, use, and amendments.

Laboratory Quality Policies and Records

The Policy Statement for Quality Assurance should indicate the overall commitment of the laboratory management and staff to implement and maintain quality assurance in all aspects of their work. Full support for quality assurance is essential from the highest level of management because such a system requires resources to implement and maintain.

Records of Staff Qualifications and Training

Quality assurance depends on adequate training of staff, and organizations assure a high level of competence by encouraging continuous development of skills. It is often useful to include job descriptions of all staff in this section of the laboratory manual. A diagram of the organizational structure should indicate who has overall and day-to-day responsibility for quality assurance within the laboratory. Key among the roles of this person(s) are maintaining document control (as discussed in the section on document control) and calibration records, writing test procedures, and ensuring that staff are adequately trained in quality issues. This person is also responsible for identifying deficiencies in the quality systems and developing procedures for review and continuous improvement of the quality systems.

Equipment Inventory and Manuals

When equipment needs to be ordered, reordered, or repaired, a comprehensive inventory that contains records of the supplier, the approved service agent (including address, contact numbers, and names), and item reorder numbers can be useful. This section may also include equipment maintenance schedules.

Equipment operating manuals are usually kept with each piece of equipment for ready reference to operating and troubleshooting procedures. A quality system should maintain records of the location and status of operating manuals.

Principles for Technical Error of Measurement Calculations

The first principles of TEM calculations should be in the manual for new or current staff to review as required. The raw TEM data should also be kept in the laboratory manual. For detailed information on TEM, see page 83 and Pederson and Gore (1996).

Schedules for collection of TEM data at regular intervals should also be prepared and implemented to ensure that the precision estimates are relevant to the current staff and equipment. As a guide to the uncertainty of common athlete tests, table 1.2 (page 6) presents TEM data that have been acquired from five certificated exercise physiology laboratories.

Consent Forms

The laboratory manual should include sample consent forms. Informed consent is one of the guiding principles of the Declaration of Helsinki for Research Involving Human Beings (*Journal of Applied Physiology* 1996). Routine monitoring of athletes also falls under the intent of this declaration, and therefore consent must be obtained from all athletes prior to testing.

Document Control

While it is important to document calibration procedures, test protocols, and data record sheets (as discussed in the next section), this process will quickly deteriorate without a system of document control. In order for these documents to be effective, master control sheets for calibration procedures, test protocols, and data record sheets must be prepared to enable ready identification of the current protocols for any sport and also to identify redundant protocols and the dates for which these protocols were effective. Obsolete documents should be removed from use but kept on record. The master control sheets should include the signature of the senior sport scientist or quality-control manager who authorized implementation of new protocols or procedures.

Ideally the process for developing, authorizing, issuing, reviewing, and updating all new documents should itself be documented. Overall, this approach is referred to as a quality document register for document control.

The status of and authority for issuing new documents can often be incorporated in the header or footer of a document. For example, a new cleanup procedure for rubber mouthpieces authorized by Joan Joyce Smith in June 1999, to replace the January 1996 protocol of Adam Brown, could be indicated as follows: (Status: CLEANUP_0699 replaces CLEANUP_0196 Authorized by: JJS Date Effective: June 20 1999).

Test Procedures and Calibrations

The most widely used aspect of the laboratory manual is the test and calibration procedures. Adequate records of the protocols are essential to allow replication of tests irrespective of new staff, while documented calibration protocols and logs facilitate the maintenance of equipment in optimal condition.

Calibration Procedures and Logs

Most equipment can be calibrated against external standards. The procedures for calibration should be documented, and the instructions should include the calibration tolerances as well as the corrective action to be taken if calibrations are outside the tolerances. Sample calibration logs should be included in the laboratory manual to indicate the frequency of calibration, and it may be useful to attach a sticker to the equipment to indicate its calibration status. Commonly, calibration logs are attached to or kept near the corresponding pieces of equipment. These logs allow one to readily identify departures from calibration over time or even a sudden loss of calibration as a consequence of mechanical or electronic failure.

In some instances, such as for barometers, calibration may have to be conducted by a metrology laboratory (ISO, 1990). Such an organization should have ISO certification and will issue calibration documents showing the traceability of the equipment that was used for calibration. (Note that calibrations from metrology laboratories can be expensive.)

In addition to keeping calibration logs for individual equipment, it is good practice to prepare a complete laboratory schedule, as a chart or on computer, that reminds laboratory staff on a weekly, monthly, or yearly basis when equipment is due for recalibration. Quality-assurance software packages for this purpose are also commercially available.

Copies of any interlaboratory quality-assurance schemes in which the laboratory participates, such as blood quality-assurance schemes, should also be kept in this section of the laboratory manual.

Test Protocols

The methods used for athlete assessment should be adequately documented to allow for replication at a later date and to allow traceability of altered protocols. If the test procedure departs from the standard protocol, this should be noted on the results sheet so that possible reasons for unusual results can be traced and understood. If the protocol is taken from a published article, a copy of the article should be included in the laboratory manual. Even if the protocol is a modification of a published article, it is relevant to include a copy of the article to allow easy reference to the source material.

While some laboratories may choose to have very extensive test protocols, others may write procedures that contain only the essential information that a competent scientist could follow.

If particular methods of data smoothing or analysis are used, explanations of these must be included with the test protocol, because a different interpretation of the data is likely if different methods of analysis are employed.

Data Record Sheets and Reports to Coaches and Athletes

While allowing for flexibility according to coaches' requirements, standardized templates for results sheets and reports should identify fundamental information, including

- the athlete's name and identification code,
- the test date,
- the laboratory name, address, and telephone number,
- the test method and equipment (i.e., a few words or code to identify this),
- the signature or identification of the sport scientist who conducted or supervised the test, and
- relevant TEM data for the test method.

Work Instructions

To facilitate day-to-day use of the documented calibration procedures and logs, equipment operating manuals, and test protocols, it is frequently relevant to have summary work instructions in close proximity to the operator. For example, a plastic-laminated summary of the calibration procedure for an oxygen consumption system might be posted on the side of the machine; wash-up instructions for respiratory mouthpieces might be attached to the wall next to the appropriate sink; and a summary of the workloads and sampling times for a progressive lactate profile test might be available in a folder next to the ergometer. A master copy of all work summary instructions should be located in the laboratory manual.

chapter 2

Pretest Preparation

Peter Fricker and Kieran Fallon

Preparation for testing the physiological function of athletes involves numerous considerations. This chapter provides a framework within which the laboratory can document its own pretest procedures. This documentation can be available for ready reference by staff involved in the physiological assessment of athletes.

The chapter includes samples of the following:

- A pretest checklist covering general considerations for the athlete
- A pretest checklist covering general considerations for the laboratory
- A pretest diet
- A pretest questionnaire
- A medical examination form
- A pretest health questionnaire
- An informed-consent document
- A form explaining physiological assessment procedures

These forms are intended as examples, but readers may copy and use them if they appear suitable as presented here. Each laboratory will have its own modifications to make to these materials. Furthermore, readers should pay attention to the special considerations recommended for the sport-specific protocols presented in later chapters of this manual.

Pretest Preparation Checklist

The Athlete

- ☐ Familiarize the athlete fully with all test procedures before formal testing. This should occur on a day prior to testing.
- ☐ Obtain the athlete's informed consent and ensure that he or she receives medical clearance where appropriate.
- ☐ Ensure that the athlete has completed a medical examination and health questionnaire.
- ☐ While many protocols recommend no exercise on the day of the test, and preferably none on the previous day, this is impractical when testing high-performance athletes. With such athletes there should be no unaccustomed exercise on the previous day, and the lead-up to testing should be individually standardized.
- ☐ Athletes should maintain a normal diet in the days leading up to physiological assessment and should not eat for two hours before the test. Figure 2.1 presents an example of a diet that would be appropriate for the athlete weighing about 155 lb (70 kg) for the three days leading up to testing and on the day of testing.
- ☐ On a test day, athletes should avoid smoking; drinking alcohol, tea, or coffee; and taking any substances that are known to affect or may be suspected of affecting human physiological functions.
- ☐ Athletes should wear light and comfortable clothing.
- ☐ On the day of testing, fully explain to the athlete the test procedures and objectives.
- ☐ Ensure that any other pretest conditions, as specified in the sport-specific guidelines, have been implemented.
- ☐ The athlete should be in good health and fully recovered from previous injuries on the day of the testing.
- ☐ Have the athlete fill out the Pretest Questionnaire (pages 16-18) on the day of the test.

The Laboratory

- ☐ The testing environment is well ventilated.
- ☐ Temperature is controlled and maintained between 18° and 23° C
- ☐ Relative humidity is less than 70%.
- ☐ All conditions are recorded for every test.
- ☐ All equipment being used in a test procedure is calibrated against appropriate known standards, and the calibration checks are recorded.
- ☐ Time of day for testing is standardized as much as possible and recorded.
- ☐ Test is ordered to minimize, as much as possible, interference with subsequent test performances.
- ☐ All tests are conducted in quiet and reassuring environments to keep the athlete's anxiety to a minimum.
- ☐ Testing staff are well trained and are able to recognize potentially dangerous situations and prevent them from occurring.
- ☐ The laboratory has immediate access to first aid facilities and prompt access to medical assistance.
- ☐ Telephone numbers and addresses of hospitals and doctors are listed and placed by the telephone in the laboratory.

Sample Pretest Diet

Days before test	Meal or snack				
	Breakfast	**Morning snack**	**Lunch**	**Afternoon tea**	**Dinner**
3	Cereal (2 c) + skim milk (1 c) Strawberries (1/2 c) Toast (1–2 slices) + jam/honey 1 fruit juice	1 banana	1 roll + ham or chicken or tuna or reduced-fat cheese + salad (no butter) 1 carton low-fat fruit yogurt or fresh fruit	English muffin (1) + honey (2 tsp)	Chicken (breast, no skin) + 2 c rice + 2 stir-fry vegetables Fruit salad or low-fat custard
2	Cereal (2 c) + skim milk (1 c) Peaches (1/2 c) Toast (1–2 slices) + jam/honey 1 fruit juice	1 cereal bar (4 fruits)	1 roll or Lebanese roll + ham or chicken or tuna or reduced-fat cheese + salad (no butter) 1 carton low-fat fruit yogurt or fresh fruit	Rice cakes (2) + spread + margarine (1 tsp)	Pasta + meat sauce (2–3 cups) Green salad Canned fruit in natural juice
1	Cereal (2 c) + skim milk (1 c) English muffin (1 whole) + margarine (2 tsp) + spread 1 fruit juice	2 rice cakes or crackers + jam/honey	2 rice cakes or crackers + jam/honey	Fruit loaf (2 slices) + 1 tbsp jam	Pasta (2–3 c) + chicken + tomato-based sauce + vegetables Green salad Fresh fruit salad
Testing day	7:00 AM test: 5:30 AM–6:00 AM 2–3 toast + honey Fruit juice	2:00 PM test: Have breakfast and snack as for previous days. Early lunch as for pregame—2–3 h before: Canned spaghetti on toast or banana or salad sandwiches or roll or pasta + plain tomato-based sauce		4:00 PM test: Have breakfast and snack as for previous days. Early lunch as for pregame—2–3 h before: Canned spaghetti on toast or banana or salad sandwiches or roll or pasta + plain tomato-based sauce	

Abbreviations: c = cup(s), tsp = teaspoon, tbsp = tablespoon.

To meet fluid replacements: you must drink 50–60 ml of fluid/kg body weight—equivalent to 3–4 liters of fluid per day as water, noncaffeinated drinks, sports drinks (you may have other drinks such as tea or coffee but meet your fluid requirements first).

Figure 2.1 Although your athletes may not wish to follow precisely the sample diet offered here, you should ensure that any substitutions are nutritionally equivalent.

The Importance of the Pretest Questionnaire

Many fitness traits commonly quantified in the laboratory can be influenced by variables such as diet, fatigue, medications, illness, injury, and environmental conditions. Because a primary objective in testing is to assess fitness in a controlled environment, a standardized pretesting protocol is strongly encouraged. For reliable testing, athletes who present themselves for testing sessions should be in a similar state with regard to nutrition and fatigue for every testing session.

In addition, it can be useful for an athlete to fill out a pretest questionnaire so that many of the variables that could potentially influence testing results are documented. This information can be particularly useful when data are retrospectively evaluated and unique results examined. The following pretest questionnaire was designed to gather information quickly from athletes immediately prior to testing that could potentially influence fitness test results.

Pretest Questionnaire

Personal

Name: ______________________ Test date: ____________

Address: ______________________ Date of birth: ____________

Diet

1. Evaluate your diet over the last two days. ☐ POOR ☐ OK ☐ GOOD ☐ EXCELLENT
2. How many hours ago did you eat your last meal? ____________
3. What foods did you eat at your last meal? ____________

   ____

   ____
4. Record foods eaten over the last 24 h.

Meal	**Foods (including drinks)**	**Portion size (i.e., cup, grams)**
Breakfast	____________	____________
	____________	____________
	____________	____________
Snack	____________	____________
	____________	____________
	____________	____________
Lunch	____________	____________
	____________	____________
	____________	____________
Snack	____________	____________
	____________	____________
	____________	____________
Dinner	____________	____________
	____________	____________
	____________	____________

Environment

1. Have you been training in hot conditions over the last two weeks? ☐ NO ☐ YES

   If yes, provide details:

   ____

   ____

2. Have you been training or sleeping at altitude over the last two weeks? ☐ NO ☐ YES

If yes, provide details:

____

____

Illness

1. Are you currently suffering from any type of illness? ☐ NO ☐ YES

If yes, provide details (type, severity):

____

____

2. Have you had any type of illness or health problem for the last two weeks? ☐ NO ☐ YES

If yes, provide details (type, severity):

____

____

Injury

1. Do you currently have any injuries? ☐ NO ☐ YES

If yes, provide details (type, severity):

____

____

2. Have you had any injuries for the last two weeks? ☐ NO ☐ YES

If yes, provide details (type, severity):

____

____

Medication/Supplements

1. Are you currently taking any medication? ☐ NO ☐ YES

If yes, provide details (type, severity):

____

____

2. Have you taken any medication over the last two weeks? ☐ NO ☐ YES

If yes, provide details (type, severity):

____

____

Please list any supplements you are currently taking.

Supplement ______________________________ Daily dose ______________

(continued)

Motivation

1. Evaluate your motivation for training today. ☐ POOR ☐ OK ☐ GOOD ☐ EXCELLENT
2. Evaluate your motivation for testing today. ☐ POOR ☐ OK ☐ GOOD ☐ EXCELLENT

Training

1. Evaluate your last week of physical training. ☐ EASY ☐ MODERATE ☐ HARD ☐ VERY HARD
2. How fatigued are you today? (0 = not at all; 5 = extremely)

   0 1 2 3 4 5
3. How many hours ago did you last exercise? ________________
4. Describe your last three training sessions.

Time	**Training session**	**Difficulty (easy, moderate, hard)**
Today	____________________	____________________
	____________________	____________________
	____________________	____________________
Yesterday	____________________	____________________
	____________________	____________________
	____________________	____________________
Two days ago	____________________	____________________
	____________________	____________________
	____________________	____________________

Travel

1. Have you had to travel over the last seven days? ☐ NO ☐ YES

   If yes, provide details (e.g., plane, car; duration of trip):

   ____

   ____

Miscellaneous

Please provide any additional information that you believe may influence your fitness test results.

____

____

For Women Only—Menstruation

1. Please indicate your current menstrual status:
   ☐ No menstruation ☐ Irregular menstruation ☐ Regular menstruation
2. How many days since your last menstruation? ______________
3. Do you currently take an oral contraceptive? ☐ NO ☐ YES

Medical Examination Form

This form is to be completed by a physician.

Name of athlete: ______________________________ Date: ______________
Surname Given names

Sport: ______________________________ Event: ______________________________

Sex: ______________ Age: ______________ Date of birth: ______________________

Mass (kg): ______________ Σ skinfolds (mm): ______________ (at n = _____ sites*)

____

* Actual sites must be recorded, for example triceps, subscapular, biceps, mid-abdominal, supraspinale, midthigh, medial calf.

List any medication being taken currently and/or taken during the last year: ______________

____

____

____

____

Comments: ____

____

____

____

____

Ears, Eyes, Nose, and Throat

Relevant comments on history (e.g., fractures, surgery): ______________________

____

____

____

____

____

____

____

(continued)

	Left		Right	
	Normal	Abnormal	Normal	Abnormal
Eyes: General	______	______	______	______
visual fields	______	______	______	______
visual acuity	______	______	______	______
Ears	______	______	______	______
Nose	______	______	______	______
Throat	______	______	______	______
Lymph nodes (neck)	______	______	______	______
Teeth	______	______	______	______

If abnormal findings are present, please specify: ______________________________

____

____

____

____

____

____

Respiratory

Relevant comments on history (e.g., asthma): ______________________________

____

	Normal	Abnormal
Breath sounds	______	______
Chest wall movements	______	______
Air entry	______	______

If abnormal findings are present, please specify: ______________________________

____

____

____

____

____

Cardiovascular

Relevant comments on history: ____

____

Recorded heart rate: ____

Recorded blood pressure: ____

	Normal	Abnormal
Heart sounds	_______	_______
Heart size	_______	_______
Fundoscopy	_______	_______
Bruits (abdominal/carotid)	_______	_______
ECG (optional)	_______	_______

If abnormal findings are present, please specify: ____

____

____

____

____

____

Abdominal

Relevant comments on history (e.g., appendix surgery): ____

____

	Normal	Abnormal
Liver	_______	_______
Spleen	_______	_______
Kidney	_______	_______
Abdominal wall	_______	_______
Testes	_______	_______

Other findings (e.g., hernia):

____

____

____

____

(continued)

If abnormal findings are present, please specify: ______

______

______

______

______

______

______

Central Nervous System

Relevant comments on history: ______

______

	Normal	Abnormal
Reflexes	______	______
Proprioception	______	______
Balance (Romberg's test)	______	______
Peripheral sensation	______	______
Gait	______	______
Strength (gross with manual resistance)	______	______

Other tests: ______

______

______

______

______

If abnormal findings are present, please specify: ______

______

______

______

______

______

______

Musculoskeletal System

Relevant comments on history: ______________________________

____

	Normal	Abnormal
Spine		
Cervical	_______	_______
Thoracic	_______	_______
Lumbar	_______	_______
Sacral (sacroiliac)	_______	_______
Coccyx	_______	_______

If abnormal findings are present, please specify: ______________________________

____

____

____

____

____

____

Joints

Relevant comments on history: ______________________________

____

Please examine for pathology, deformity, range of motion, swelling, stability, tenderness of each joint:

	Left		Right	
	Normal	Abnormal	Normal	Abnormal
Shoulders	_______	_______	_______	_______
Elbows	_______	_______	_______	_______
Wrists	_______	_______	_______	_______
Hands	_______	_______	_______	_______
Hips	_______	_______	_______	_______
Knees*	_______	_______	_______	_______
Ankles	_______	_______	_______	_______
Feet	_______	_______	_______	_______

*A detailed knee examination should be carried out for athletes participating in sports in which the knee joint is of major importance or for those with specific knee complaints/injuries. Refer to "Knee Examination (Optional)," page 26.

(continued)

If abnormal findings are present, please specify: ____________________

____

____

____

____

____

____

____

For Sports Utilizing Running as a Portion of Training

Alignment: Specify the degree of varus and valgus for the normal as well as abnormal alignment:

		Normal	Abnormal
Patella:	Q angle	______	______
Knee:	Varus	______	______
	Valgus	______	______
	Recurvatum	______	______
Tibia:	Varum	______	______
	Valgum	______	______
	Torsion (medial)	______	______
	(lateral)	______	______
Foot:	Pronated	______	______
	Supinated (cavus)	______	______
	Leg length (true)	______	______

If abnormal findings are present, please specify: ____________________

____

____

____

____

____

____

____

Laboratory Studies

(Only if indicated on the basis of history and physical examination)

Date of examination: ______________

Urinalysis: Specific gravity
Albumin
Glucose
Blood
Microscopic

Hematology: Hemoglobin
Serum ferritin

Blood chemistry: Biochemical screen, e.g., CK, AST, ALT, LD, SAP, Alb, Glob, Na, K, Creat, urea, Glu, etc. (optional)

Respiratory: FEV_1/FVC
(pre- and post-bronchodilator)
Peak flow rate

Other tests recommended (e.g., bone scans, CT scans): ______________________________

____

____

____

____

If abnormal findings are present, please specify: ______________________________

____

____

____

____

____

____

Additional Comments

____

____

____

Physician's name: ______________________________

Signature: ______________________________

Knee Examination (Optional)

History: ____

____

____

____

____

Nature of knee complaint:

Symptoms:	☐ acute	☐ chronic	☐ intermittent
Swelling:	☐ acute	☐ chronic	☐ intermittent
Instability:	☐ giving way	☐ locking	
Pain:	☐ activity related	☐ postactivity	☐ night pain
	☐ sitting	☐ squatting	☐ kneeling
	☐ up stairs	☐ down stairs	☐ running
	☐ cutting	☐ stopping	☐ jumping

Physical Examination:

	Left	Right
Leg length:	_______ cm	_______ cm
Thigh circumference (4.0 cm above patella):	_______ cm	_______ cm

	Left Yes	Left No	Right Yes	Right No
Swelling: Extra-articular	☐	☐	☐	☐
Intra-articular	☐	☐	☐	☐
Joint line tenderness	☐	☐	☐	☐
Ligamentous tenderness	☐	☐	☐	☐
Patellar tenderness	☐	☐	☐	☐
Popliteal space: Swelling	☐	☐	☐	☐
Tenderness	☐	☐	☐	☐
Tibial tubercle	☐	☐	☐	☐
Ligamentous laxity:				
Forced varus 0° flexion	☐	☐	☐	☐
Forced varus 30° flexion	☐	☐	☐	☐
Forced valgus 0° flexion	☐	☐	☐	☐
Forced valgus 30° flexion	☐	☐	☐	☐

	Left		Right	
	Negative	Positive	Negative	Positive
Lachmann's test	________	________	________	________
McMurray's test	________	________	________	________

		Left	Right
Standing alignment:	Genu varus	________ cm	________ cm
	Genu valgus	________ cm	________ cm
	Genu recurvatum	________ cm	________ cm

Other Comments

____

____

____

____

____

____

____

____

____

____

____

____

____

____

____

Pretest Health Questionnaire

This form is to be completed by the athlete (parent or guardian) before visiting the medical practitioner.

Personal

Date of examination: ____________ Sport: ______________________________

Event: ____

Name of athlete: ____
Surname Given names

Sex: ______________ Age: ______________ Date of birth: ______________________

Adddress (home): ______________________________ Postal code: __________

Telephone (home): (___) ______________ Telephone (business): (___) ______________

Medicare number: ______________________ Private insurance number: ______________

Family Doctor

Name: ____

Address: ______________________________ Postal code: __________

Telephone: (___) ____________ Date of last medical examination: ______________

Family Dentist

Name: ____

Address: ______________________________ Postal code: __________

Telephone: (___) ____________ Date of last dental examination: ______________

Emergency

In case of emergency, please notify:

Name: ______________________________ Relationship: __________

Address: ______________________________ Postal code: __________

Telephone: (___) ____________

Family History

Please identify any health problems that have occurred in your immediate family. ______________

____

Has someone in your family died suddenly (before the age of 50 years)? ☐ NO ☐ YES

Do you know why? ____

____

Is there any family history of:

- ☐ High blood pressure
- ☐ Heart problems
- ☐ Cancer or tumor
- ☐ Migraine headache
- ☐ Emotional problems
- ☐ Diabetes
- ☐ Bowel disorder
- ☐ Problem with pregnancy
- ☐ Genetic disorder (e.g., hemophilia, Marfan syndrome)
- ☐ Other—please name: ______________________
- ☐ Anemia
- ☐ Epilepsy
- ☐ Arthritis
- ☐ Kidney/Bladder disorder
- ☐ Stomach disorder
- ☐ Allergies/Asthma

Your Recent Condition

I have recently experienced or at present experience:

- ☐ Allergies
- ☐ Difficulties with eye(s) or vision
- ☐ Difficulties with nose or throat
- ☐ Problems with hearing
- ☐ Headaches, dizziness, weakness, fainting, a problem with coordination or balance
- ☐ Numbness in a part of the body
- ☐ Tendency to shake or tremble
- ☐ Cough, shortness of breath, chest pain, dizziness, or palpitations (awareness of rapid heartbeat) with exercise
- ☐ Poor appetite, vomiting, abdominal pain, abnormal bowel habits (diarrhea, bleeding, etc.)
- ☐ Symptoms referable to the muscle, bones, or joints—that is, stiffness, swelling, pain
- ☐ Problems with the skin such as sores, rashes, itchy or burning sensations, etc.
- ☐ Menstrual irregularities or problems (period pains, missed periods, heavy bleeding, etc.)
- ☐ Other symptoms—please list: ______________________
- ☐ Has your weight changed in the past year? If so, gain or loss? _____ kg increase _____ decrease _____
- ☐ Have you changed in height in the past year? If so, how much? _____ cm increase _____ decrease _____

Past Illness or Medical Problems

I have had, or been told I have, or consulted a physician for:

- ☐ Heart disease (rheumatic fever)
- ☐ High blood pressure

(continued)

☐ Diabetes, goiter, or any disease of the glands (e.g., mononucleosis)

☐ Tuberculosis, asthma, or any lung disease or respiratory disorder

☐ Hepatitis or jaundice

☐ Epilepsy

☐ Nervous disorder or any disease of the brain or nervous system

☐ Disease of the blood (e.g., anemia), easy bruising or bleeding tendency (e.g., hemophilia)

☐ Varicose veins, phlebitis, hemorrhoids

☐ Ulcers or disease of the stomach, intestine, liver, or gallbladder

☐ Sugar, albumin, or blood in the urine or disease of the kidneys or genitourinary organs

☐ Arthritis, rheumatism, or injury or disease of the bone, peripheral joints, back, or spine (including the neck)

☐ Hernia or disease of the muscle

☐ Cancer, tumor, or growth of any kind

☐ Heatstroke (failure of the body's heat-regulating system resulting in body temperature above 40.5° C/105° F)

☐ Heat problems or disorder (specify in the Medical Chart provided, next page)

☐ Head injury (including concussion) causing severe dizziness, loss of memory, vomiting, or loss of consciousness or requiring medical attention or hospitalization

☐ Psychological or psychiatric disorders

☐ Injury (e.g., fracture) to head, neck, ribs, lumbar spine, or sacroiliac joints

☐ Injury (e.g., dislocation, fracture) to shoulder(s), arm(s), elbow(s), wrist(s), or hand(s), which incapacitated me for ______ week(s)

☐ Pain in the back, very seldom/frequently/only after vigorous exercise (circle the appropriate frequency)

☐ Injury to hip(s), knee(s), ankle(s), or foot (feet), which incapacitated me for ______ week(s)

☐ Surgical procedure (please specify in the Medical Chart provided, next page)

☐ Particular rehabilitation/physiotherapy programs being undertaken—please specify: ______

Drugs, Food Supplements, and Miscellaneous Agents

☐ I am taking nutrient supplements at present (e.g., vitamins, iron, calcium).

☐ I am taking stimulants (benzedrine, amphetamine, etc.).

☐ I am taking anabolic agents (e.g., steroids, growth hormone).

☐ I am taking sleeping pills.

☐ I am taking an oral contraceptive (pill).

☐ I am taking some nonprescription drugs not listed above (please specify in the Medical Chart provided, next page).

☐ I smoke. Amount per day: ______________

☐ I drink alcoholic beverages. Amount per week (e.g., five beers per week): ______________

☐ I have been advised for some medical reasons not to participate in certain sport(s) for a period of time (please provide details).

☐ I wear spectacles/glasses for sports.

☐ I wear contact lenses for sports.

Medical Chart

Please give relevant details of those illnesses, health problems, or drugs that apply to you.

Nature of illness, medical problems, or drugs	Date (year)	Details (e.g., duration, period of any hospitalization, name and address of physician or hospital)

I (print name) ______________________ have given true and complete information to the best of my knowledge, and hereby give permission for transmission of the results of this medical examination to my sport governing body (print name) ______________________

Athlete signature: ______________________ Date: ______________

Parent/Guardian name (required if age under 16): ______________________

Parent/Guardian signature: ______________________

Consent Form

(sample)

I (print name) ______________________________ consent to participating in this physiological assessment on the following terms:

1. I have read the Explanation of Physiological Assessment Procedures attached and understand what I will be required to do.
2. I understand that I will be undertaking physical exercise at or near the extent of my capacity and that there is possible risk in the physical exercise at that level, that is, episodes of transient light-headedness, fainting, abnormal blood pressure, chest discomfort, and nausea.
3. I understand that this may occur though the staff in this laboratory will take all proper care in the conduct of the assessment, and I will fully assume that risk.
4. I understand that I can withdraw my consent, freely and without prejudice, at any time before, during, or after testing.
5. I have told the person conducting the assessment about any illness or physical defect I have that may contribute to the level of that risk.
6. I understand that the information obtained from the test will be treated confidentially with my right to privacy assured. However, the information may be used for statistical or scientific reasons with privacy retained. (If I am an athlete, I may have made special arrangements about treatment of individual data with my coach or manager. I understand these arrangements.)
7. I release this laboratory and its employees from any liability for any injury or illness that I may suffer while undertaking the assessment, or subsequently occurring in connection with the assessment, or that is to any extent contributed to by it.
8. I will indemnify this laboratory with respect to any liability it may incur in relation to any other person in connection with the assessment.
9. I hereby agree that I will present myself for testing in a suitable condition having abided by the requirements for diet and activity prescribed for me by laboratory staff.

Participant signature: ______________________________ Date: ________________

Parent/Guardian name (required if age under 16): ______________________________

Parent/Guardian signature: ______________________________ Date: ________________

Witness name: ______________________________

Witness signature: ______________________________ Date: ________________

Explanation of Physiological Assessment Procedures

Please note: The tests you will be participating in are marked with a circle.

1. **Blood Test.** The test for routine blood measures involves taking a small sample of blood from a vein in the forearm. If you have any concerns, please ask the person taking the sample to fully explain the procedure to you.
2. **Body Composition.** This involves simple measurement of body height, weight, and body fat levels. Body fat is estimated using standard skinfold thicknesses as measured by handheld calipers.
3. **Submaximal Aerobic Power Test.** This test of "cardiovascular fitness" or "aerobic power" involves cycling at low to moderate intensity. The athlete pedals at a light rate (25 W) to begin, and this is increased in small steps (by 25 W) every minute until the heart rate reaches approximately 75% of the age-predicted maximum, which in your case is ________________________________ beats/min.
4. **Maximal Aerobic Power Tests.** Maximal aerobic power can be measured in the laboratory in a number of different ways including the bicycle, treadmill, rowing, kayaking, and arm-cranking ergometers. In each case the athlete will have his or her ventilation and heart rate continually monitored by the appropriate electronic equipment. Tests are usually incremental; that is, the initial workload is fairly light and then increases gradually through to such a level that the athlete cannot sustain the work rate.
5. **Anaerobic Tests—Cycle.** These tests, usually conducted on the bicycle ergometer, assess anaerobic power and capacity, that is, "outright speed" and "sustained speed." The first of these is assessed by a 10-second "all-out" or greatest possible effort and the latter by an "all-out" 30- to 60-second effort. The former normally presents no problems; however, a 30- to 60-second test is very exhausting, and some individuals may feel ill afterward.
6. **Anaerobic Test—Treadmill.** Athletes are required to run to exhaustion (approximately 1-2 minutes) on a fast-moving treadmill on a moderate gradient. This test requires a high level of skill and fitness and is used only with trained runners.
7. **Anaerobic Test—Vertical Jump.** To assess explosive lower-body or leg power, a standing single vertical jump is used. This simple test requires the athlete to jump as high as possible from a standing position.
8. **Anaerobic Test—Upper Body.** To assess upper-body power, a 10-second "all-out" or greatest possible effort test on the arm-cranking ergometer is used. This requires the athlete to exert the greatest possible effort on the arm-cranking ergometer.
9. **Strength Tests.** Strength is routinely measured using an isokinetic dynamometer, which measures the force produced by various major muscle groups of the body. This test involves moderate and greatest possible effort during various body movements; for example, knee extension and flexion, and shoulder extension and flexion. Some strength tests may be conducted using free weights or using the athlete's own body weight. If there is a history of joint instability or muscle strain, strength tests should not be performed unless special arrangements are made.
10. **Flexibility Test.** General lower back/hamstring flexibility is assessed approximately with the standard sit-and-reach test. The athlete is required, in a seated position, to reach as far as possible. Athletes with a history of lower back or hamstring problems may have difficulty in adequately performing this test.

    More specific tests of shoulder, hip, knee, and ankle joint flexibility may also be administered by a physiotherapist. Any previous injury or pre-existing instability of a joint should be mentioned to the tester before the test is conducted.
11. **Environment Chamber—Acclimatization to Heat and Humidity.** This procedure involves having the athlete exercise in a sealed room in which the temperature and humidity can be manipulated using a computerized system. You will be exercising at a moderate level at a temperature of ____ °C and a humidity of ____ % for a period of ____ minutes.

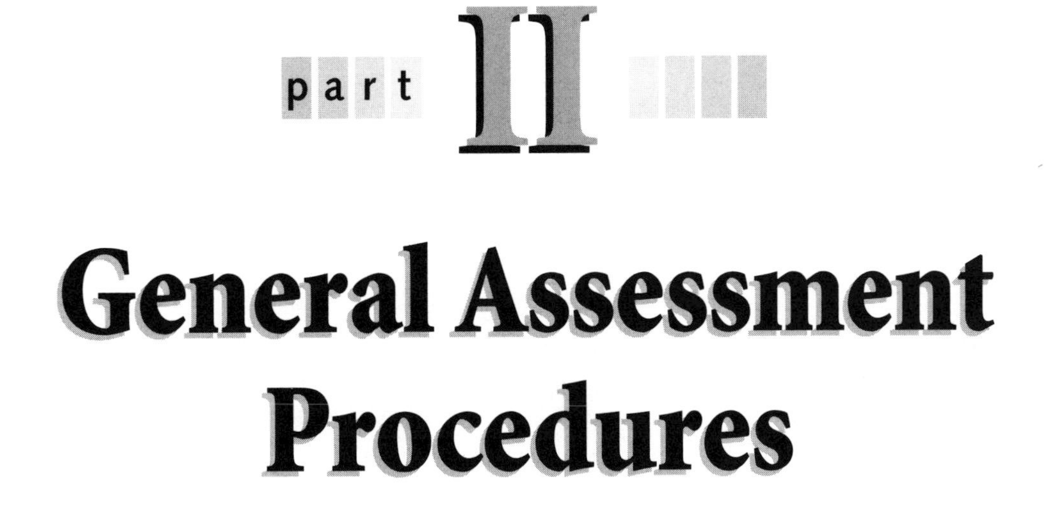

part II

General Assessment Procedures

chapter 3

Estimation of Peak Power and Anaerobic Capacity of Athletes

James Finn, Paul Gastin, Robert Withers, and Simon Green

Coaches of high-performance athletes are often interested in standard laboratory measures that can be used to indicate the potential for peak power and exercise at intensities greater than $\dot{V}O_2$max (supramaximal). A wide range of tests has been outlined by Vandewalle et al. (1987) and more recently by Bouchard et al. (1991). However, more valid anaerobic tests have resulted from recent research that examined the metabolic processes involved in short-duration, high-intensity performance. A discussion of these processes and a rationale for using the maximum accumulated oxygen deficit (MAOD) as an estimation of anaerobic capacity appear later in this chapter. For some athletes, measurement of peak power and anaerobic capacity may occur during the same test.

Our aim is not to present detailed protocols for assessing peak power and MAOD, as the reader may find these protocols in the chapters for sports in which they have particular value. Rather our intention is to highlight issues that one needs to attend to in the estimation of peak power and MAOD of athletes.

Preliminary Issues

If peak power and MAOD of athletes are to be of value to sport scientists, we need to appreciate not only what these variables measure but also the assumptions, advantages, and limitations inherent in such measurements.

- **Peak power** is the highest power output achieved in a short high-intensity sprint test. It is averaged over 1 s and is measured in watts or watts per kilogram.
- **The MAOD** is the difference between the oxygen equivalent of the work performed and the oxygen consumed during an anaerobic capacity test.
- **A basic assumption made in using the MAOD** is that the mechanical efficiency of supramaximal work is identical to that for submaximal work.
- **Advantages of using the MAOD** are the following:
    1. The method has face validity because it factors out the significant aerobic contribution to short-duration, maximal-intensity exercise.
    2. The method has been validated empirically by comparisons with biochemical measurements.
- **A limitation of the MAOD** is that extrapolation of the linear relationship between $\dot{V}O_2$ and submaximal work intensities may not yield a valid estimate of the O_2 requirement of supramaximal work (Green et al. 1996).

Calculating the Maximum Accumulated Oxygen Deficit

The technique for estimating the MAOD relies on extrapolation of the linear relationship between submaximal work intensity (such as power output, or running or swimming speed) and oxygen consumption (Medbø et al. 1988). Once this relationship has been established, subsequent work intensities can be quantified in terms of their oxygen consumption equivalents. Figure 3.1 illustrates an extrapolation technique commonly used to estimate oxygen consumption equivalents at work intensities above the $\dot{V}O_2$max. Steady-state oxygen consumption values are required, but there appears to be some discrepancy concerning the relative intensity this should encompass. Poole et al. (1990) suggested that the work intensity for this phase must not require greater than 79% of $\dot{V}O_2$max. This suggestion was subsequently supported by Green and Dawson (1995), who demonstrated that the slope of the regression line above the anaerobic threshold was 14% greater than that below. However, recent work at the South Australian Sports Institute (SASI) with elite cyclists (Craig et al. 1995; Woolford et al. 1999) has shown no significant difference between anaerobic capacities calculated using values below and above the anaerobic threshold (30-91% $\dot{V}O_2$max). There is therefore no evidence to preclude those $\dot{V}O_2$ values above the anaerobic threshold for trained athletes, provided that the $\dot{V}O_2$ has not begun to plateau as in the final stages of a $\dot{V}O_2$max test.

The slope and intercept of the regression line indicate the mechanical efficiency of the subject. It is important to establish these on an individual basis, since there are variations in slope of ~16% in running (Medbø et al. 1988) and ~20% in cycling (Withers et al. 1991). Many earlier tests suffered from a lack of validity because they assumed an overall mechanical efficiency for cycling of 22.5% (Hermansen and Medbø 1984).

Recommendations

The limitation of using the MAOD to estimate anaerobic capacity is that the validity of doing so is based on the unverified assumption that mechanical efficiency is independent of work intensity. One must acknowledge that since this assumption is not verified, it is possible that extrapolation of the linear relationship between $\dot{V}O_2$ and submaximal work intensities will not yield a valid estimate of the O_2 requirement of supramaximal work. With use of an air-braked ergometer, the regression of oxygen consumption on work intensity is determined at cadences less than those used in the peak power and anaerobic capacity tests. The energy demand during heavy exercise may therefore be underestimated. This could be due to recruitment of muscle groups to stabilize the upper body (Åstrand et al. 1986), the necessity to overcome joint resistance and connective tissue elasticity (Whipp and Wasserman 1969), and sarcomere shortening exceeding its most efficient rate for power production (Goldspink 1978). While the mechanical efficiency of supramaximal exercise is unknown (Bangsbo et al. 1990; Saltin 1990), some investigators have suggested that it may be as low as half that of submaximal work (Gladden and Welch 1978). Oxygen deficit values may be underestimated at supramaximal exercise intensities; but because the advantages of the MAOD outweigh the limitations, it remains the most promising technique for estimating anaerobic capacity (Hermansen and Medbø 1984; Saltin 1990).

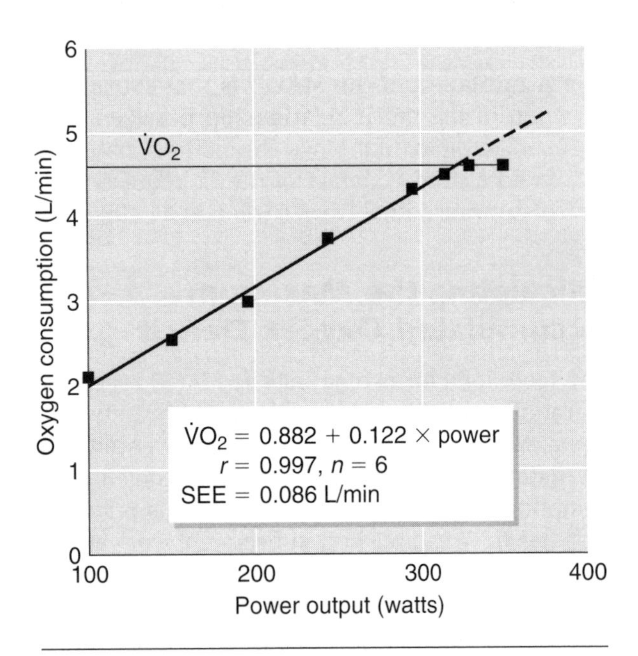

Figure 3.1 The extrapolation technique. Note that values at or near the $\dot{V}O_2$max have not been used in constructing the regression line.

Some of the MAOD is actually derived from stored oxygen, and Abler et al. (1986) believe that this quantity needs to be known with some accuracy. It is hypothesized that ~8-10% of the oxygen deficit is stored as oxygen in hemoglobin and myoglobin, dissolved in body fluids and initially present in the lungs. It represents ~5-6 ml/kg (Saltin 1990). Although one can argue that this stored oxygen should be subtracted from the calculated oxygen deficit, few reports in the literature acknowledge that this has been done. Some investigators have found that the MAOD was slightly reduced following hypoxia and cited this as a demonstration that there was a small contribution from stored oxygen (Di Prampero et al. 1983). Overestimation of the MAOD resulting from ignoring the contribution of stored oxygen may actually be counterbalanced by underestimation of the MAOD resulting from ignoring the reputed lower mechanical efficiency at maximal power outputs (Withers et al. 1993). However, in the absence of consensus on this issue, it is recommended the MAOD be reported unadjusted for stored oxygen because this is the most widely used practice.

Laboratory Environment and Subject Preparation

To prepare the laboratory environment and the subjects to be tested, follow the instructions in chapter 2.

Equipment Checklist

The accuracy of ergometers needs to be verified throughout the range of measurement using a dynamic calibration rig (Woods et al. 1994). Determination of the MAOD relies on accurately measuring the power output or treadmill speed and elevation.

Data collection and processing software is available for ergometers. One of the features of these programs is that they enable the power output to be integrated over 1 s periods. Programs include those developed by the Flinders University of South Australia (Finn et al. 1996), the SASI, Lode BV (Groningen, The Netherlands), and Sports Medicine Industries, Inc. (St. Cloud, Minnesota).

The following equipment is required:

- ☐ Barometer
- ☐ Thermometer
- ☐ Weighing scale
- ☐ Sling psychrometer or equivalent device
- ☐ Sport-specific ergometer or treadmill
- ☐ Cadence and work monitor units preferably interfaced with a computer
- ☐ Optional data collection and processing software to expedite the procedure
- ☐ Heart rate recording unit (suitable units include electrocardiogram chart recorders and Polar Electro Sport Testers)
- ☐ Gas collection and analysis system according to Laboratory Standards Assistance Scheme recommendation (see chapter 8)

PROTOCOLS: Peak Power and Anaerobic Capacity

To calculate the MAOD, one must first determine the relationship between power output or treadmill speed and oxygen consumption for each subject. Because of large interindividual variability, a common mechanical efficiency cannot be assumed.

Protocol for Testing Oxygen Consumption at Submaximal Power Outputs or Treadmill Speeds

1. Barometric pressure, temperature, and humidity are measured and recorded.

2. The subject's mass is measured and recorded along with any other information pertinent to the individual's sport. For example, frontal surface area is important in cycling, and this can be calculated using a multiple regression equation (McLean 1993).

3. The subject performs a mild warm-up. As this graded test begins with low power outputs or treadmill speeds, stretching and easy locomotor activity are all that are required.

4. Four to eight power outputs or treadmill speeds spanning a range up to 80-90% of the subject's $\dot{V}O_2max$ are needed. Steady-state oxygen consumption must be achieved at all intensities used. The maximal intensity that can be used depends on the caliber of the athlete.

5. The average power output or treadmill speed for the final 2 min of each 4 min period is recorded. With use of ergometers, this may be derived from the work performed in the final 2 min. It does not matter whether or not the subject rests between these work bouts if each work bout does not exceed 4 min. The length of work bouts will need to be extended if steady state is not achieved in 4 min. If a continuous protocol is used, it is not recommended that work bouts exceed 8 min because this may result in a greater slope for the $\dot{V}O_2$-power regression (Green and Dawson 1996). Two minutes of exercising before sampling at each power output or treadmill speed should ensure that a steady state of oxygen consumption has been reached in elite athletes. Alternatively, curve-fitting analysis of breath-by-breath data refines the procedure of steady-state determination and may reduce the duration of each exercise period (Gastin and Lawson 1994a).

6. The cool-down consists of exercising at low intensity until the heart rate recovers to below 120 beats/min.

Considerations:

- Oxygen consumption at low power outputs or treadmill speeds should be checked for linearity, as it is often elevated as a result of factors other than the exercise intensity.
- If rest periods are required, subjects should be encouraged to make the best use of them. Stretching and hydration are important. Exercising at low power outputs or treadmill speeds after high-intensity exercise will assist recovery.
- Controlling environmental conditions and using a fan for body cooling are important to prevent oxygen drift associated with elevated body temperatures.

Protocol for Peak Power and Anaerobic Capacity Tests

1. Barometric pressure, temperature, and humidity are measured and recorded.

2. The subject's mass is measured and recorded along with other information pertinent to that person's sport.

3. The subject performs his or her own warm-up, specific to that person's sport and level of competition.

4. The subject is informed of the following:

- Test duration
- The need to keep a good seal on the mouthpiece
- The need to complete as much work as possible during the test period if exercising on an ergometer
- The need to start with the preferred pedal 30° above horizontal and to remain seated throughout the test if cycling

5. The test starts as soon as practicable after the finish of the warm-up.

6. Heart rate is recorded continuously and $\dot{V}O_2$ is measured during the test. Power output can be recorded every second if the ergometer is interfaced with a computer.

7. The cool-down consists of maintaining a low power output or treadmill speed until the heart rate recovers to below 120 beats/min.

Considerations:

- The validity of all maximal testing procedures is dependent on the motivation of the subject. Strong verbal encouragement may be warranted.

Computations

Sport-specific air-braked ergometers are commonly used, and the procedures involved are outlined next. For runners, the quantification of power output from treadmills is problematic. Peak power may be estimated only if a nonmotorized treadmill is used. Power is calculated as the product of the instantaneous force registered on a load cell, which is tethered to the athlete, and the treadmill belt velocity (Cheetham et al. 1985; Lakomy 1984, 1987). Only anaerobic capacity may be estimated on a motorized treadmill. The need to quantify power output may be negated if the running efficiency and anaerobic capacity tests are performed at the same treadmill elevation (usually 5-10%), so that only speed is increased for the maximal effort. Such constant-intensity tests should exhaust the athlete in 2-4 min.

The data sheet on page 42 can be used to summarize the following computations:

- **Computation #1:** The Repco work monitor unit (see pages 129, 136) will display power (W) and record total work performed (kJ) for the anaerobic capacity test. If the unit is not interfaced with a computer, the maximal deflection of the needle on the power scale needs to be noted as the peak power. The total work performed is displayed on the digital readout. When using an Exertech work monitor unit, press the button marked "Hold" the moment the test ceases. This terminates the accumulation of work displayed on the digital readout meter. Once this is recorded, release this button and depress the "Watts" key to display peak power. The total work performed is converted to mean power (1W = 1J/s) using the following equation:

$$\frac{\text{Work (kJ)} \times 1000}{\text{Time (s)}}$$

- **Computation #2:** One cannot rely upon work monitor units indefinitely to accurately measure the power output of an ergometer. Users can correct inaccuracy by determining the difference between the actual and displayed power outputs with a dynamic calibration rig. Accurate measurement of power output is critical to reporting peak power and calculating oxygen equivalents. If dynamic calibration is not possible, there is some limitation to the accuracy of test results. After dynamic calibration, peak and mean power outputs can be adjusted. Determining the relationship between actual and displayed power outputs involves a polynomial equation. On different occasions we have found that either a third- or fourth-order expression provides the line of best fit. A fourth-order polynomial equation takes the form:

$$y = k_1 + k_2x - k_3x^2 + k_4x^3 - k_5x^4$$

where:

y = corrected power output

x = uncorrected power output

k_1, k_2, k_3, k_4 = constants from polynomial regression

Statistical computer software packages such as Statview or Statistica provide the line of best fit and thus simplify this process.

• **Computation #3:** The following equation is used to correct the peak and mean power outputs from the air-braked ergometer for barometric pressure, temperature, and humidity:

where:

P_2 = barometric pressure during test (mm Hg)

P_1 = barometric pressure during ergometer calibration (mm Hg)

RH_2 = relative humidity during test (decimal)

RH_1 = relative humidity during ergometer calibration (decimal)

$ppWV_2$ = partial pressure of water vapor during test (mm Hg)

$ppWV_1$ = partial pressure of water vapor during ergometer calibration (mm Hg)

T_2 = temperature during test (K)

T_1 = temperature during ergometer calibration (K)

• **Computation #4:** If the work monitor unit is interfaced with a computer, the following equation can be used to calculate the fatigue index when power outputs have been corrected for calibration and ambient conditions (temperature, pressure, and humidity):

$$\frac{5\ \text{s Maximal Power} - 5\ \text{s Minimum Power}}{5\ \text{s Maximal Power}} \times 100$$

• **The subject's peak power is expressed in watts and watts per kilogram.** It is equal to the highest power output recorded in 1 s and is corrected for both the work monitor calibration and ambient conditions. The corrected power outputs should be recorded on the data sheet.

• **The 4-8 submaximal power outputs also need to be measured rather than estimated.** If the work monitor unit is not interfaced with a computer, the equation in Computation #1 can be used to convert total work to mean power for each power output. It is important not to forget to correct these mean powers for the work monitor calibration and environmental conditions. The oxygen consumption for each of the 4-8 submaximal power outputs is then regressed on the corrected power outputs so that the oxygen equivalent can be predicted according to the linear relationship:

$$y = a + bx$$

where:

y = oxygen equivalent

x = power output or treadmill speed

b = slope of regression line

a = intercept of regression line

The oxygen equivalent is then estimated for the anaerobic capacity test. Subtracting the measured oxygen consumption from the estimated oxygen equivalent yields the oxygen deficit for the anaerobic capacity test.

• **The subject's anaerobic capacity or MAOD is expressed as L O_2 or ml/kg O_2.**

Expected Test Scores

There is considerable intersubject variation in the values reported for peak power and anaerobic capacity. It is not possible to produce a summary of pooled data for peak power and anaerobic capacity because

- not all investigators reported standard deviations, and
- there were problems with classifying athletes according to event and competitive level.

Table 3.1 provides a rough range for expected MAOD values for trained and untrained males. The values reported for peak power (table 3.2) and anaerobic capacity (table 3.3) are more detailed and should be useful for benchmark comparisons. The following have not been cited in tables 3.2 and 3.3:

- Tests that assumed the same mechanical efficiency for all the subjects rather than measuring it individually

Data Sheet—Peak Power and Anaerobic Capacity

Name ______________________ Mass kg ______ Time ________

Date ________ Other information ______________________

____

____

Temperature ____________ Pressure mm Hg ________ Humidity % __________

Peak Power (Not for Motorized Treadmills)

Peak 1 s power = maximal anaerobic power:

_______ W _______ W/kg

If computer recording of power output was used:

Peak 5 s power: _______ W _______ W/kg

Final 5 s power: _______ W _______ W/kg

Fatigue index: _______ %

Fatigue Index = ([Peak 5 s power – Final 5 s power]/Peak 5 s power) × 100

Efficiency

Power output or treadmill speed-$\dot{V}O_2$ relationship:

1. _______ W or km/h _______ L/min
2. _______ W or km/h _______ L/min
3. _______ W or km/h _______ L/min
4. _______ W or km/h _______ L/min
5. _______ W or km/h _______ L/min
6. _______ W or km/h _______ L/min
7. _______ W or km/h _______ L/min
8. _______ W or km/h _______ L/min

Regression equation:

$\dot{V}O_2$ (L/min) = _______ + (_______ × Power output or Treadmill speed)

Total work (ergometers only): _______ kJ _______ J/kg

Mean power or treadmill speed: _______ W or km/h Duration: ________ min

Using regression equation, O_2 equivalent of mean power or treadmill speed:

_______ L/min. For the duration of this test the O_2 equivalent = _______ L.

Maximal Accumulated Oxygen Deficit (MAOD)

O_2 equivalent (_______ L) – O_2 consumption (_______ L) = _______ L

MAOD = anaerobic capacity = ________ L _______ ml/kg

- Tests shorter than 60 and 120 s, since these are the minimum duration for all-out and constant intensity tests, respectively
- Tests that pooled endurance- and sprint-trained athletes, since one of the purposes for measuring the MAOD is to be able to discriminate between these groups
- Data for untrained persons, since these data are similar to those of endurance-trained athletes

One can see from this information that

- sprint-trained athletes incur greater MAODs than endurance-trained athletes and
- adult males have greater MAODs than adult females.

Testing Peak Power and Anaerobic Capacity: Metabolic Processes

At the onset of high-intensity exercise, the limited stores of oxygen in muscle myoglobin are inadequate to resynthesize the adenosine triphosphate (ATP) required. Although abundant oxygen can be delivered, maximizing the rate at which additional oxygen can be supplied to the active muscle mass takes some minutes; and even then, the power output of the muscles requires more oxygen for ATP resynthesis than can be supplied by the cardiovascular system. The rate at which energy can be turned over is known as the *power* of the energy system, whereas the total amount of energy

Table 3.1 Range of Expected MAOD Values for Trained and Untrained Males

Untrained and endurance athletes		Sprint athletes	
L	ml/kg	L	ml/kg
2.5-5	40-65	4.5-7.5	55-100

Table 3.2 Peak 1 s Power of Trained Athletes When Cycling

Reference	n	Age (yr)	Training status	Mass (kg)	Peak power (W)	Peak power (W/kg)
Serresse et al. (1989)	8	29	ET	65	674	10.4
	10F	27	ET	53	441	8.3
	8	22	ET	70	752	10.7
	3	16	ET	64	733	11.5
	9F	15	ET	50	442	8.8
	7	16	ST	61	845	13.9
	3F	16	ST	56	566	10.1
	8	23	ST	69	928	13.5
	9	19	ST	57	590	10.4
Withers et al. (1993)	12	25	ET	73	899	12.3
Gastin and Lawson (1994a)	8	26	ET	73	822	11.2
	6	25	ST	81	1113	13.7

ET = endurance trained, ST = sprint trained, F = female.

Table 3.3 Anaerobic Capacity of Trained Athletes as Measured by the MAOD

Reference	n	Age (yr)	Training status	Weight (kg)	O_2 deficit (L; ml/kg)	Exercise mode	Duration (s)	Intensity
Hermansen and Medbø (1984)	6	25	ET	70	2.9; 42	Treadmill	55	Constant
	6	22	ST	75	4.2; 56	Treadmill	58	Constant
Medbø et al. (1988)	3	24	MDT	73	5.5; 75	Treadmill	170	Constant
	1	26	ST	79	6.5; 82	Treadmill	120	Constant
Graham and McLellan (1989)	4	22	ET	73	4.4; 61	Cycle	145	Constant
Medbø and Sejersted (1985)	6	25	ET	70	2.9; 41	Treadmill	60	Constant
	6	25	ST	75	4.2; 56	Treadmill	60	Constant
Medbø and Burgers (1990)	6	27	ET	69	4.4; 64	Treadmill	120–180	Constant
	8	24	ST	76	6.4; 84	Treadmill	120–180	Constant
	5	35	ST	79	4.6; 58	Treadmill	120–180	Constant
	7F	29	ST	66	2.8; 43	Treadmill	120–180	Constant
Scott et al. (1991)	4	21	ET	67	3.8; 57	Treadmill	120–180	Constant
	5	22	MDT	70	5.2; 74	Treadmill	120–180	Constant
	3	19	ST	74	5.8; 78	Treadmill	120–180	Constant
Withers et al. (1991)	6	25	ET	70	3.7; 53	Cycle	60	All-out
	6	25	ET	70	3.4; 49	Cycle	90	All-out
Olesen (1992)	5	20	ST	75	7.5; 100	Treadmill	160	Constant
Perez-Landaluce et al. (1992)	16		ET		3.3; 51	Cycle		Constant
Bangsbo et al. (1993)	14	25	MDT	70	3.6; 52	Treadmill	181	Constant
	3	23	ST	80	4.5; 57	Cycle	179	Constant
Withers et al. (1993)	12	25	ET	73	3.8; 52	Cycle	60, 75, or 90	All-out
Gastin and Lawson (1994a)	8	26	ET	73	3.8; 52	Cycle	90	All-out
	6	25	ST	81	4.8; 60	Cycle	90	All-out
Weyland et al. (1994)	13	28	MDT	70	47	Treadmill	120–240	Constant
	12F	29	MDT	52	38	Treadmill	120–240	Constant
	9	20	ST	72	55	Treadmill	120–240	Constant
	7F	20	ST	57	45	Treadmill	120–240	Constant
Craig et al. (1995)	12	20	ET	73	4.5; 62	Cycle	300	Paced
	6	20	ST	79	5.3; 67	Cycle	70	Paced
Green and Dawson (1996)	7	26	ET	76	4.3; 56	Cycle	173	Constant

ET = endurance trained, ST = sprint trained, MDT = middle distance trained, F = female.

turned over is known as the *capacity* of the energy system. Since oxygen cannot be delivered at sufficient rate during high-intensity exercise (even though abundant supplies are available), the aerobic (oxygen) system is a low-power, high-capacity one.

High-intensity exercise is possible because although resynthesis of ATP via oxidative phosphorylation may be the most efficient means of energy turnover in humans, it is not the only mechanism available. Adenosine triphosphate may also be resynthesized, at a great rate although in limited quantity, without the presence of oxygen via anaerobic energy supply systems. In contrast to the aerobic energy system, these are high-power, low-capacity systems.

Anaerobic energy supply systems operate within the cytosol of the cell. The first of these, known as the alactic system, provides energy for ATP resynthesis by splitting molecules of creatine phosphate (CP). Because muscle CP stores are limited, the alactic system can contribute for up to only 10 s of maximal exercise.

Subsequent resynthesis of ATP is achieved by the second anaerobic energy system, known as the glycogen-to-lactic acid system, which utilizes the metabolic pathway of anaerobic glycolysis. As oxygen is not involved, lactic acid is produced as a by-product. Lactic acid production is related to fatigue as muscle pH is lowered when the concentration of hydrogen ions exceeds the buffering capacity. A reduction in pH inhibits metabolic reactions because ionization of an enzyme's amino acid side chains alters its shape, and therefore its ability to catalyze reactions. The glycogen-to-lactic acid system contributes the most to the total energy output from around 10 s until approximately 60 s, after which the aerobic energy system becomes the major contributor for the resynthesis of ATP (Medbø and Tabata 1989; Withers et al. 1991).

It is important to realize that energy delivery is achieved through a sequentially overlapping involvement of all three energy systems. No one system exclusively provides the energy turnover; relative contributions vary over time in a coordinated metabolic response to the demands of exercise. Figure 3.2 illustrates the relative energy system contribution to the total energy supply for any given duration of maximal exercise. Data from 16 studies using the oxygen deficit method, including 52 maximal exercise efforts during cycling, running, and swim bench ergometry, have been used in developing the figure. Percentage anaerobic contribution has been divided into estimates of alactic and lactic components based on anaerobic ATP turnover rates (Bangsbo 1998).

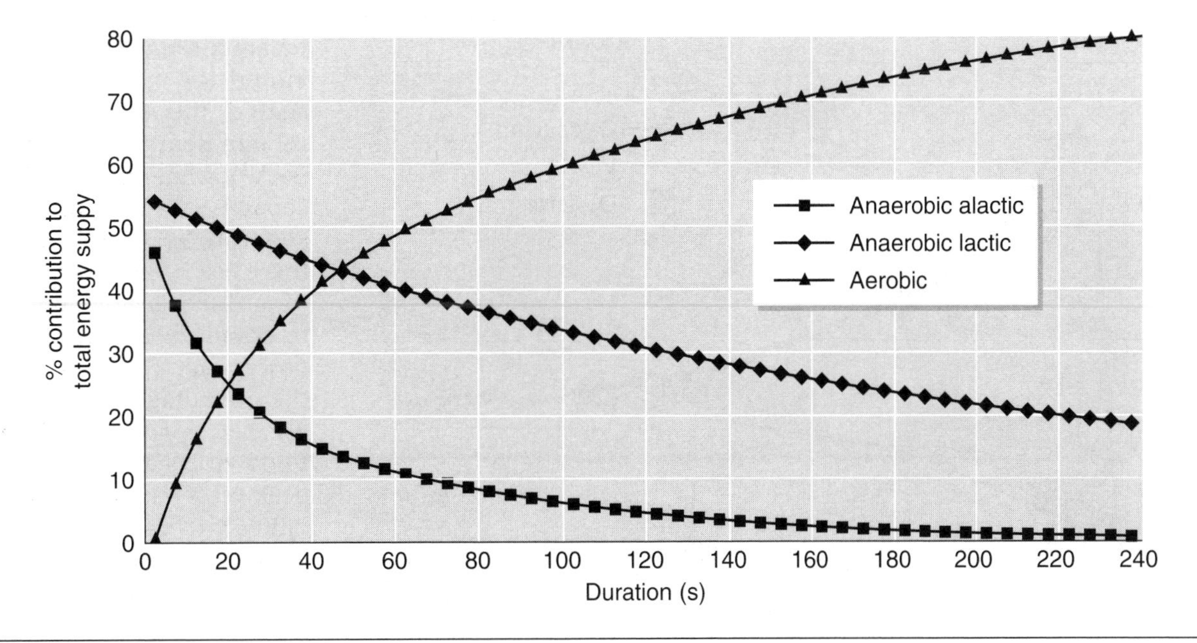

Figure 3.2 Percentage energy system contribution to the total energy supply for any given duration of maximal exercise. (This figure graphically represents information often presented in tabular form for given exercise durations; it does not illustrate the time course of each of the energy systems.)

Developed by Gastin (previously unpublished) from data in the literature. Data from Bangsbo et al. 1993; Craig et al. 1995; Gastin and Lawson 1994a, 1994b; Gastin et al. 1995; Green et al. 1996; Medbø and Sejersted 1985; Medbø and Tabata 1989; Morton and Gastin 1997; Nummela and Rusko 1995; Olesen et al. 1994; Ramsbottom et al. 1994; Spencer et al. 1996; Withers et al. 1991, 1993.

During high-intensity exercise, when the required rate of ATP resynthesis in skeletal muscle cannot be met by the low-power aerobic system, high energy turnover is still possible, as all of the energy systems contribute to the energy turnover. The aerobic energy shortfall, apparent in the difference between the amount of oxygen theoretically required to produce such power outputs and the amount actually consumed during high-intensity exercise, can be met by the previously described anaerobic energy supply systems. This shortfall can be expressed as oxygen equivalents and is known as the oxygen deficit (figure 3.3). The anaerobic capacity, or total amount of energy available from nonaerobic sources, is analogous to the MAOD. The methodology for the measurement of maximal aerobic power is well established. However, despite a strong theoretical basis, use of a standardized protocol involving the oxygen deficit to quantify anaerobic metabolism is not universal.

Peak Power and Anaerobic Capacity Tests

Traditionally, peak power has been expressed as the highest power output averaged over periods between 1 and 5 s, and anaerobic capacity as the work performed in a short high-intensity sprint test. Perhaps the most widely used of the ergo-

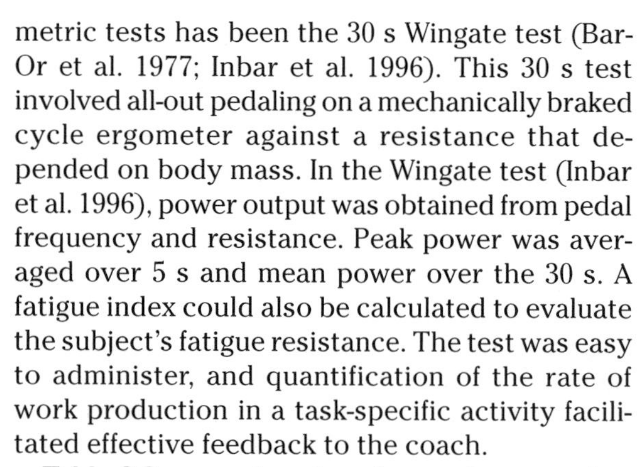

metric tests has been the 30 s Wingate test (Bar-Or et al. 1977; Inbar et al. 1996). This 30 s test involved all-out pedaling on a mechanically braked cycle ergometer against a resistance that depended on body mass. In the Wingate test (Inbar et al. 1996), power output was obtained from pedal frequency and resistance. Peak power was averaged over 5 s and mean power over the 30 s. A fatigue index could also be calculated to evaluate the subject's fatigue resistance. The test was easy to administer, and quantification of the rate of work production in a task-specific activity facilitated effective feedback to the coach.

Table 3.2 presents values for peak power during cycling as reported in the literature. Comparisons are sometimes difficult owing to variation in the sampling period used, so only studies using a 1 s period are cited. A sampling period of 1 s is desirable because of the transient nature of the peak. Averaging power output over a longer duration simply reduces the measure attained. Variation in the ergometer used also makes comparison difficult. Many of the investigations reported in table 3.1 employed mechanically braked ergometers. Boulay et al. (1985) used an electromechanically braked ergometer, whereas Telford (1982), Withers et al. (1991, 1993), and Craig et al. (1995) chose to use air-braked ergometers.

Finally, comparisons are made difficult by variation in the test protocols used. As the subject's mechanical efficiency (used in calculations to estimate the MAOD) is determined for cycling while seated, the subject must remain seated when the peak power and anaerobic capacity are measured. If separate tests are used, there has been a tendency toward allowing the subject to stand during the peak power test. Subjects gain the advantage of using a greater portion of their body mass to generate peak power. It is therefore difficult to compare their peak power with that of subjects whose peak power and anaerobic capacity were estimated in the same test.

While measurement of peak power is reasonably straightforward, the prob-

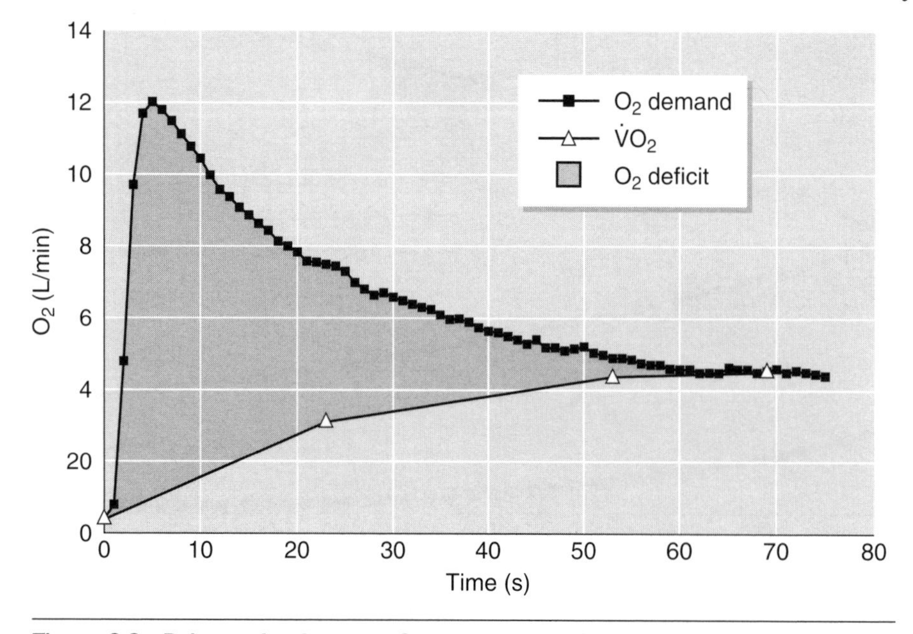

Figure 3.3 Relationship between O_2 consumption, O_2 demand (oxygen equivalents), and O_2 deficit at any point in time during maximal-intensity cycling. (An "all-out" protocol was used with an air-braked cycle ergometer. It is clear that 60 s was required to exhaust the anaerobic capacity of these endurance-trained cyclists [Finn et al. 1996].)

lem with treating the work performed as equivalent to the anaerobic capacity is that this fails to factor out the considerable aerobic contribution, which ranges from 28% in a 30 s to 64% in a 90 s all-out test (Withers et al. 1991). One cannot assume that the aerobic contribution to a 30 s test is negligible. In many publications, mean power has been called anaerobic capacity. It is not surprising that the originators of this test say that equating mean power with anaerobic capacity is based on an unproven assumption and that they prefer not to do this (Inbar et al. 1996). Variation between subjects in the aerobic contribution to work performed for any given test duration requires gas analysis to quantify anaerobic metabolism.

Test Validity

Although most anaerobic capacity tests appear reliable, it is their validity that is questionable. There is no "gold standard" or acceptable criterion for what should be measured. The following have been advanced as measures of anaerobic capacity: mean power, total work, oxygen debt, peak muscle or blood lactate, and MAOD. The disadvantages of mean power and total work as indicators of anaerobic capacity were outlined in the preceding paragraph. The validity of oxygen debt measurements is dubious because factors other than lactate elevate postexercise oxygen consumption, and the reproducibility of measurement is low. Peak blood lactate occurs after and not during exercise. The extent to which lactate is oxidized to pyruvate in muscle after exercise is not known accurately. This lactate oxidation occurs before an equilibrium between muscle and blood lactate is reached. Thus anaerobic glycolysis cannot be quantified from postexercise blood lactate.

The most promising measure of anaerobic capacity is the MAOD because it isolates the anaerobic contribution to performance in brief maximal-intensity tests. Comparisons with biochemical measurements further validate the use of the MAOD in quantifying the anaerobic capacity. Bangsbo et al. (1990) demonstrated in one-legged exhausting exercise that the oxygen deficit was equivalent to estimations of the energy yield based on ATP and CP reductions and lactate production in the active muscle. Dividing the oxygen deficit by 22.4 normalizes liters of gas to moles using Avogadro's principle. This can then be compared with the changes in intramuscular ATP, CP, and lactate concentrations via needle biopsy. Such validation provides powerful support for the use of the MAOD. The work of Bangsbo et al. (1990) was subsequently augmented by Withers et al. (1991), who used the assumptions of Medbø et al. (1988). Recent reviews of the various methods for quantifying anaerobic capacity have been published by Green and Dawson (1993), Gastin (1994), and Foster et al. (1995).

Constant-Intensity, All-Out, and Paced Protocols

Anaerobic capacity tests have traditionally used either constant-intensity or all-out protocols. In the former, the subject exercises at a constant intensity (determined during several pretest trials) that is sufficient to cause exhaustion in a given time period. All-out tests require subjects to exert their maximal effort for the duration of the test rather than pacing themselves. The minimum duration to exhaust anaerobic capacity is 120 s for a constant-intensity test (Medbø et al. 1988) and is 60 s for an all-out test (Withers et al. 1991, 1993). Anaerobic capacity tests should therefore equal or exceed 60 s in duration. The advantages of all-out tests over constant-intensity protocols are that they do not require pretest trials to determine exercise intensity, are of shorter duration, and have a definite end point. Furthermore, peak power may often be measured in the same test when an all-out protocol is employed.

Figure 3.4 illustrates the time course of the relative contribution of the three energy systems to

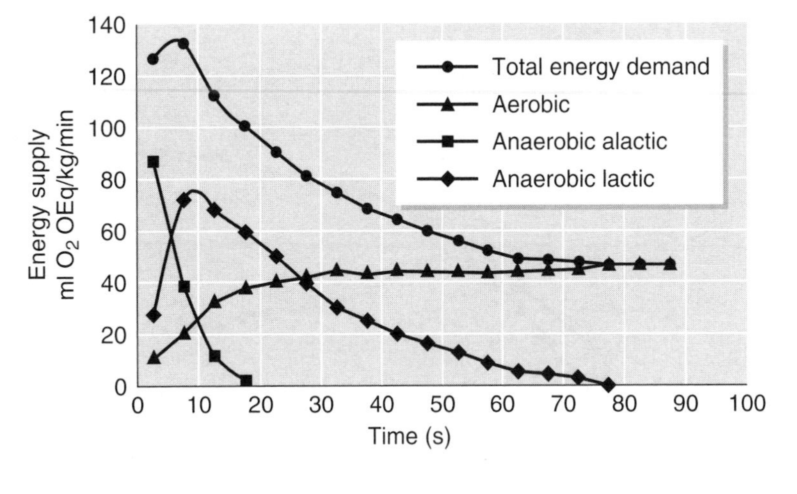

Figure 3.4 The relative contribution of the three energy systems to the total energy supply at any point in time during 90 s all-out exercise. Modified from Gastin et al. (1995).

the total energy supply during a 90 s all-out effort. Estimated oxygen demand of the power output and actual oxygen consumption were taken from Gastin et al. (1995). Alactic and lactic components were estimated from the anaerobic contribution as given by the calculated oxygen deficit. The early and relatively quick response of the oxygen consumption is important, and it somewhat contradicts the common view that the aerobic system is slow to respond to the demands of high-intensity exercise. It is also apparent that exhaustion is associated with a close matching of the power output/oxygen demand with the actual oxygen consumption. This is consistent with the concept of MAOD and the existence of an anaerobic capacity.

A constant-intensity test to exhaustion is an alternative to an all-out test (figure 3.5). While no difference between MAODs achieved during all-out and constant-intensity tests has been demonstrated (Gastin et al. 1995), event specificity appears to be an influencing factor. A paced maximal effort, which has the advantage of event specificity, employs a protocol in which the subject goes out hard at the start and then settles into a pace that can be maintained for the remainder of the test, such that the maximum amount of work is achieved. Foley et al. (1991) reported no significant differences in oxygen deficit between a constant-intensity and a competition-specific time-trial test protocol (paced) using well-trained cyclists. The oxygen deficits of sprint and endurance cyclists were found to be greatest in 70 s and 300 s tests, respectively, with use of a paced protocol (Craig et al. 1995). Furthermore, estimation of the MAOD in cyclists depends on the cadence used to establish the relationship between their power output and oxygen consumption. Sprint and endurance cyclists compete in the 120-130 and 90-100 rpm ranges, respectively. It is therefore recommended that the duration of the anaerobic capacity test, and the cadences used in anaerobic capacity tests to establish the power output and oxygen consumption relationship, mimic those of the athlete's event.

Whether peak power and anaerobic capacity are measured in the same all-out or paced test depends on the duration of the test, which needs to exceed 60 s for the anaerobic capacity measure. In longer events, subjects may struggle to complete the test. When oxygen deficit is calculated for each second of the test, negative oxygen deficits may be recorded during the later periods because of a delay in matching the $\dot{V}O_2$ with a declining power output, which is likely to be compounded by a reduction in efficiency. This is less of an issue with air-braked ergometers. The resistance decreases as the subject fatigues, thus facilitating test continuity. Gastin et al. (1991) outlined a procedure whereby variable resistance loading can be used with friction-braked ergometers. They showed that by reducing the resistance during a 60 s all-out effort, reliable values for peak power and mean power could be obtained in the one test. The decision whether separate tests are required for peak power and anaerobic capacity must be weighed by the facility and should also take into account the motivation of the subject.

Apart from sport-specific ergometers, treadmills need to be considered. They are the most appro-

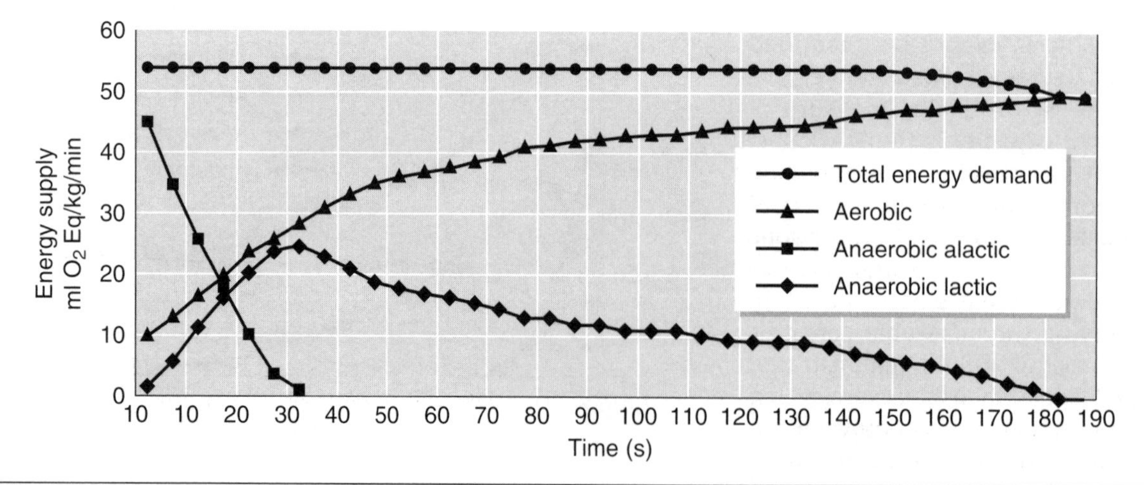

Figure 3.5 The relative contribution of the three energy systems to the total energy supply at any point in time during constant-intensity exercise at 110% $\dot{V}O_2$max.

Modified from Gastin et al. (1995).

priate tool for estimating MAOD in runners. A treadmill protocol would be expected to yield a greater MAOD for a runner than a cycle test because of specificity and the reputedly greater exercising muscle mass. Although a grade of 10% is normally used, Olesen (1992) found a significantly greater MAOD in sprint-trained runners when the treadmill inclination was increased to 15%. The main shortcoming of the treadmill is in quantifying the subject's power output. Accurate measurement of power output is critical to calculating oxygen demand and therefore the MAOD. A way to overcome this inadequacy is to perform the efficiency runs at the same treadmill elevation as in the anaerobic capacity test, increasing only the speed from the submaximal to the maximal effort and therefore negating the need to quantify power. The decrease in mechanical efficiency at high running speed is on the same order as the limitations inherent in the cycle ergometer tests. Although constant-load tests are more common on treadmills, Lakomy (1984, 1987) and Cheetham et al. (1985) offer some insights concerning an all-out test on a treadmill.

The following recommendations apply to conducting peak power and anaerobic capacity tests for runners:

- Treadmills are the ergometer of choice, but peak power for running can be measured only on nonmotorized treadmills.
- Efficiency and anaerobic capacity tests are best conducted on separate days—especially for 200-400 m runners who lack an endurance base.
- A lower treadmill grade relates better to track running yet requires high maximal velocities and the use of a safety harness.

chapter 4

Blood Lactate Transition Thresholds: Concepts and Controversies

Pitre Bourdon

Blood lactate accumulation during incremental exercise tests is a measure commonly used to evaluate the effects of training, to set training intensities, and to predict performance. Typically this is done through the determination of deflection points or transition thresholds on the blood lactate versus workload curve. Although the concept of blood lactate transition thresholds has been developing for over 60 years, there is still much controversy both about the explanation of these phenomena and about the methods that should be employed to identify them. In fact, points of contention probably outnumber points of agreement—which can frustrate those searching for practical applications of these concepts (Thoden 1991). In their review, Jones and Ehrsour (1982) suggest that many studies on the blood lactate response to exercise offer valid information that has applications to sport; however, use of these concepts requires precise description of workloads and increment duration, as well as carefully defined calculation techniques.

The popularity of the use of blood lactate-related thresholds as performance indicators has increased dramatically over the past 10-15 years, with many exercise science laboratories around the world now routinely measuring various blood lactate transition thresholds as an integral component of the physiological assessment of endurance athletes. Probable reasons for this burgeoning popularity are the following:

- The predictive and evaluative power associated with the lactate response to exercise
- The development of automated lactate analyzers that offer ease of sampling and improved accuracy
- The reliability of such measures under standardized conditions
- Increased levels of coach education and understanding of such modern training methodologies

While lactate testing has proven useful for evaluating endurance performance, prescribing exercise intensities, monitoring training adaptations, and subsequently enhancing performance, there are many different approaches to test methodologies, data analysis, and interpretation. This chapter presents an overview of the major concepts and controversies relating to the practical or applied implications of the blood lactate response to exercise as they relate to endurance exercise. A review of the more theoretical and mechanistic considerations pertinent to this issue is beyond the scope of the chapter.

Reasons for Blood Lactate Testing

Measurement of the blood lactate response to exercise in conjunction with heart rate, oxygen consumption ($\dot{V}O_2$), and workload is often a part of the routine physiological assessment of the high-performance athlete. There are three main reasons for these measures:

- They serve as indicators of training adaptation.
- They correlate with endurance performance.
- They may indicate optimal training stimuli.

As outlined in the three sections that follow, numerous studies give strong support to the evaluative and predictive power of the blood lactate response to exercise, suggesting that it can serve well as a monitoring test for endurance performers.

Indicator of Training Adaptation

In the past, endurance training/detraining studies commonly used changes in maximal oxygen consumption ($\dot{V}O_2$max) to indicate alterations in the capacity to perform endurance exercise (Coyle et al. 1986; Daniels et al. 1978). However, in recent years, research has suggested that the blood lactate response to training adapts to a greater degree than does $\dot{V}O_2$max (Denis et al. 1984; Hurley et al. 1984; Katch et al. 1978; Sjodin et al. 1982). In particular, blood lactate transition thresholds have been shown to be more sensitive indicators of training adaptations (Denis et al. 1982; Gollnick et al. 1986; Sjodin et al. 1982). This is especially true in highly trained athletes who may show little or no change in $\dot{V}O_2$max but significant changes in endurance performance (Coyle et al. 1986; Daniels et al. 1978; Foster et al. 1982). Table 4.1 and figure 4.1 provide evidence supporting this fact by showing changes in various physiological parameters in a national-level middle-distance runner across two years of his training history. Furthermore, numerous studies have demonstrated that blood lactate-related thresholds can increase with training beyond the point where $\dot{V}O_2$max fails to increase (Davis et al. 1979; Denis

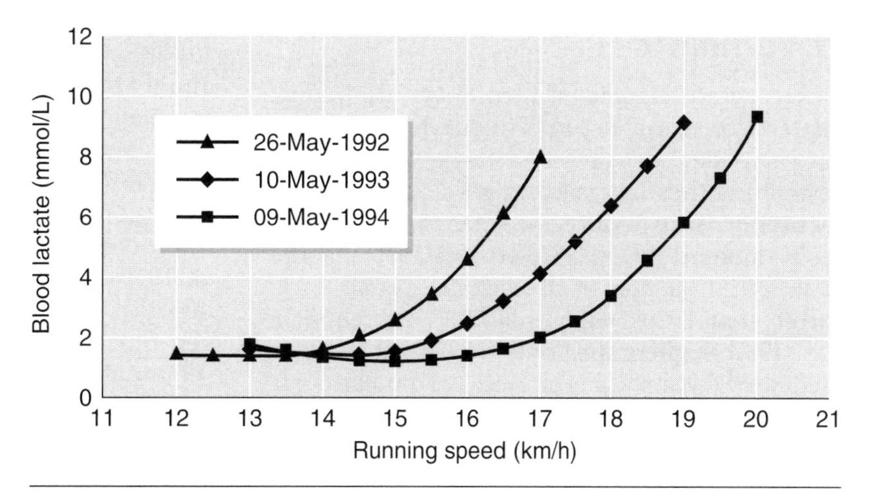

Figure 4.1 Blood lactate response curves monitored over two seasons. Subject is a male middle-distance runner.

Table 4.1 Blood Lactate Transition Threshold Changes Monitored Over Two Seasons on a Male Middle Distance Runner

Date	Threshold name	Blood lactate (mmol/L)	Running speed (km/h)	Heart rate (HR) (beats/min)	Percent HRmax (%)	$\dot{V}O_2$max (L/min)	Percent $\dot{V}O_2$max (%)	800 m time (min:s)
5/26/92	LT	1.41	13.0	166	83.4	3.61	69.4	
	AT	3.02	15.2	186	93.5	4.04	77.7	
	Max	8.18	17.0	199	100.0	5.21	100.0	1:51.9
5/10/93	LT	1.4	14.7	163	81.5	3.67	66.2	
	AT	3.84	16.8	185	92.5	4.39	79.2	
	Max	8.34	19.0	200	100.0	5.54	100.0	1:50.1
5/9/94	LT	1.36	15.8	162	85.3	3.99	73.2	
	AT	3.79	18.2	179	94.2	4.61	84.6	
	Max	9.25	20.0	190	100.0	5.45	100.0	1:48.1

LT and AT = lactate threshold and anaerobic threshold, respectively.

et al. 1982; Foster et al. 1995; Hurley et al. 1984; Jacobs 1986; Karlsson et al. 1972; MacRae et al. 1992; Poole and Gaesser 1985; Poole et al. 1990; Sjodin et al. 1982; Svedenhag and Sjodin 1985; Tanaka et al. 1986; Weltman et al. 1992; Weltman 1995). This would suggest that $\dot{V}O_2max$ and the ability to perform at prescribed submaximal intensities are limited by different mechanisms (Gollnick and Saltin 1982; Karlsson and Jacobs 1982; Saltin and Rowell 1980; Weltman 1995). A further extension of this concept would suggest the need for specific training programs for the body's central (heart and lungs) and peripheral (skeletal muscles) components of endurance performance (Gollnick 1982; Saltin et al. 1976).

Correlation With Endurance Performance

Blood lactate-associated thresholds are highly related to performance in various types of endurance activities. Many researchers have in fact suggested that these parameters are a better indicator of endurance performance than the traditional "gold standard" $\dot{V}O_2max$ (Allen et al. 1985; Bishop et al. 1998; Conconni et al. 1982; Craig et al. 1993; Farrell et al. 1979; Foster et al. 1995; Fohrenbach et al. 1987; Hagberg and Coyle 1983; Ivy et al. 1981; Jacobs 1986; Kumagai et al. 1982; LaFontaine et al. 1981; Londeree and Ames 1975; Olbrecht et al. 1985; Sjodin and Svedenhag 1985; Weltman et al. 1987; Weltman 1995).

Optimal Training Stimulus

Accumulated data suggest that the various blood lactate-related thresholds may provide the best indexes of exercise intensity by which to prescribe guidelines for training (Coen et al. 1991; Craig 1987; Fohrenbach et al. 1987; Heck et al. 1985; Jacobs 1986; Kindermann et al. 1979; Olbrecht et al. 1985; Schnabel et al. 1982; Sjodin et al. 1982; Stegmann and Kindermann 1982; Weltman et al. 1992; Weltman 1995; Yoshida, Suda, and Takeuchi 1982). This is of particular interest to coaches, since these parameters can potentially provide them with a means to optimize training intensity and help prevent overreaching and overtraining.

Blood Lactate Response to Exercise: Concepts and Controversies

Although the blood lactate response to exercise is used widely to control and monitor training programs, many factors in addition to training adaptation can affect the blood lactate response. Therefore one needs to consider the following points when collecting, analyzing, and interpreting blood lactate measurements.

Terminology

One problem in understanding and interpreting the available literature about discontinuity in the blood lactate response to exercise is the plethora of terms used to describe similar phenomena. These terms include lactate threshold (Allen et al. 1985; Beaver et al. 1985; Craig 1987; Ivy et al. 1981; Tanaka et al. 1985; Weltman et al. 1990; Yoshida et al. 1987), aerobic threshold (Aunola and Rusko 1986; Skinner and McLellan 1980), anaerobic threshold (Aunola and Rusko 1986; Heck et al. 1985; Skinner and McLellan 1980), individual anaerobic threshold (McLellan et al. 1991; Stegmann et al. 1981; Stegmann and Kindermann 1982), aerobic-anaerobic threshold (Kindermann et al. 1979), onset of blood lactate accumulation (Karlsson and Jacobs 1982; Sjodin et al. 1981), onset of plasma lactate accumulation (Farrell et al. 1979), lactate turnpoint (Davis et al. 1983), maximal lactate steady state (Beneke and Petelin von Duvillard 1996; Heck et al. 1985) and Dmax (Cheng et al. 1992). Without doubt there are some other relevant terms, but clearly this is an extensive and complicated topic of discussion.

The situation is even more complicated in that some researchers have used the same term that another investigator used but to refer to a different phenomenon. For example, the term "lactate threshold" has been defined as the highest $\dot{V}O_2$ (workload) attained during an incremental work task not associated with an increase in blood lactate concentration above the resting level (Beaver et al. 1985; Ivy et al. 1980; Tanaka et al. 1985; Weltman et al. 1990; Yoshida et al. 1987); as the workload corresponding to a lactate concentration that is 1.0 mmol/L above the baseline (Coyle et al. 1983); or as the highest exercise intensity that elicits a blood lactate concentration of 2.5 mmol/L after 10 min of steady-state exercise (Allen et al. 1985). Table 4.2 presents a number of commonly used terms for the blood lactate response and their corresponding definitions. Irrespective of the name assigned to an assessment technique, the user must have a clear understanding of the protocol required to detect the blood-lactate-related threshold.

Definitions

A number of researchers have independently suggested that there are at least two apparent

Table 4.2 Sample Terminology Used to Classify Changes in the Blood Lactate Response to Progressive Exercise

Lactate threshold (LT): First workload at which there is a sustained increase in blood lactate concentration above resting values	Anaerobic Threshold (AT): The workload causing a rapid rise in blood lactate indicating the upper limit of equilibrium between lactate production and clearance
Aerobic threshold (Kindermann et al. 1979)—fixed 2.0 mmol/L value	**Aerobic–anaerobic threshold** (Mader et al. 1976), **onset of blood lactate accumulation (OBLA)** (Sjodin and Jacobs 1981), **4 mmol/L threshold** (Heck et al. 1985)—fixed 4 mmol/L value
Lactate threshold (ADAPT 1995)—workload preceding a 0.4 mmol/L rise in blood lactate above the baseline	**Anaerobic threshold** (ADAPT 1995)—modified Dmax utilizing LT instead of first workload as start point
Lactate threshold (Beaver et al. 1985)—point of deflection in the log [blood lactate] versus log $\dot{V}O_2$ transformation	**Anaerobic threshold** (Kindermann et al. 1979)— steep part of exponential increase in lactate concentration, approximately 4 mmol/L
Lactate threshold (Ivy et al. 1980)—workload preceding nonlinear rise in blood lactate during progressive work	**Dmax** (Cheng et al. 1992)—point on curve at maximal distance from line connecting starting and finishing workloads
Maximal steady state (LaFontaine et al. 1981)—fixed 2.2 mmol/L value	**Individual anaerobic threshold** (Keul et al. 1979)—a fixed slope point on the lactate power curve whose tangent is equal to 51°
Onset of plasma lactate accumulation (Farrell et al. 1979)—exercise intensity that elicited a blood lactate concentration 1.0 mmol/L greater than baseline	**Individual anaerobic threshold** (Stegmann et al. 1981)—based on a model to define workload at maximal lactate steady state (rate of diffusion in equilibrium with rate of elimination)
	Lactate threshold (Coyle et al. 1984)—Nonlinear increase of at least 1 mmol/L
	Maximal steady-state workload (MSSW) (Borch et al. 1993)—fixed 3 mmol/L value

discontinuities or thresholds in the blood lactate response to incremental exercise that may serve as general concepts for many of the terms proposed by other researchers (Heck et al. 1985; Kindermann et al. 1979; Skinner and McLellan 1980). The first of these is associated with the first workload at which there is a sustained increase in blood lactate above resting levels. This point is generally consistent with blood lactate concentrations of less than 2.0 mmol/L. The second of these discontinuities is marked by a very rapid rise in blood lactate concentration. This point is representative of a shift from oxidative to partly anaerobic energy metabolism during constant workloads, and it refers to the upper limit of blood lactate concentration indicating an equilibrium between lactate production and lactate elimination (i.e., maximum lactate steady state) (Beneke and Petelin von Duvillard 1996). This second point is generally associated with blood lactate concentrations between 2.5 and 5.5 mmol/L. While the present author acknowledges arguments concerning the correct nomenclature for these two discontinuities or thresholds, as well as the problems of whether or not true thresholds actually exist, this chapter will use the terms "lactate threshold" (LT) and "anaerobic threshold" (AT), respectively, to describe the first and second thresholds. Figure 4.2 is a visual representation of the LT and AT in relation to their respective positions on the blood lactate-exercise response curve as interpreted in the context of this chapter.

Categories of Blood Lactate Results

Among the many terms and definitions used for blood lactate transition thresholds, most can be

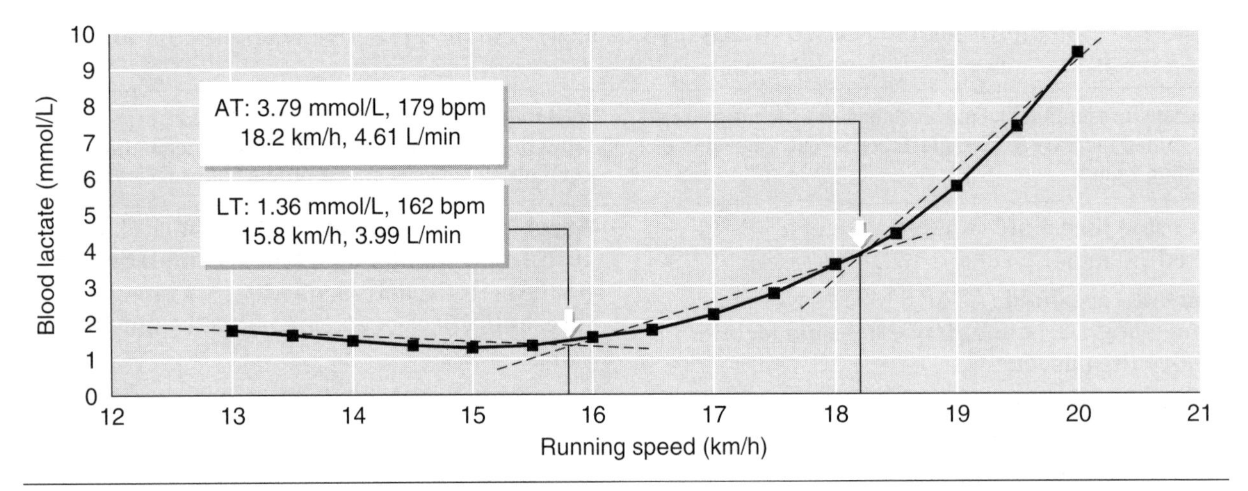

Figure 4.2 A lactate response curve identifying the lactate threshold and anaerobic threshold. Subject is a male middle-distance runner.

categorized into one of two broad classifications: (1) fixed blood lactate concentrations and (2) individualized lactate and anaerobic thresholds.

Fixed Blood Lactate Concentrations

As a strategy for minimizing the problems of biological noise associated with detecting inflections in the blood lactate response curve, fixed blood lactate concentrations of 2.0 mmol/L (Kindermann et al. 1979), 2.2 mmol/L (LaFontaine et al. 1981), 2.5 mmol/L (Allen et al. 1985; Foster et al. 1995; Hagberg 1986), 3.0 mmol/L (Borch et al. 1993), and 4.0 mmol/L (Heck et al. 1985; Mader et al. 1976; Mader and Heck 1986; Sjodin and Jacobs 1981) have all been used. The actual workload associated with fixed blood lactate concentrations is determined by interpolation from visual plots of workload versus blood lactate as illustrated in figure 4.3. Fixed blood lactate concentrations are, however, strongly influenced by an athlete's nutritional and training/recovery state (Dotan et al. 1989; Hughes et al. 1982; Ivy et al. 1981; Jacobs 1981; Maassen and Busse 1989; Yoshida 1984a), and care must be taken to control for such factors when testing an athlete.

Individualized Lactate and Anaerobic Thresholds

Stegmann et al. (1981) reported that steady-state blood lactate concentrations can vary greatly among individuals. On the basis of this fact, in combination with arguments founded on the diffusion of lactate from active muscle to blood, they proposed the concept of individualizing blood lactate threshold determinations. Numerous others have since proposed methodologies such as log transformations, rates of metabolite accumulation, tangential methods, and even subjective assessments to determine individualized

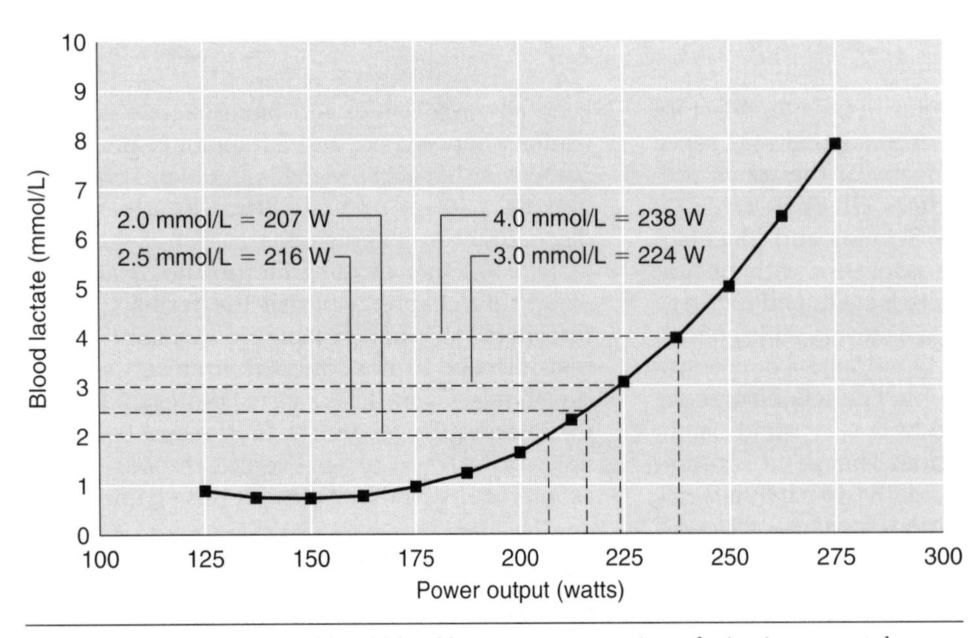

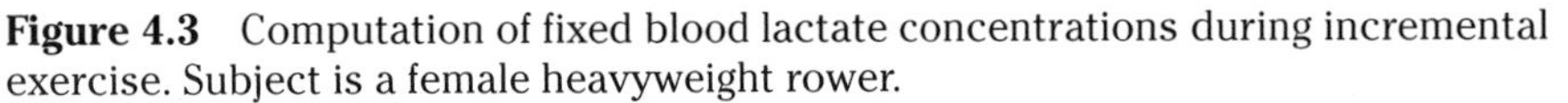
Figure 4.3 Computation of fixed blood lactate concentrations during incremental exercise. Subject is a female heavyweight rower.

LT (Coyle et al. 1984; Beaver et al. 1985; ADAPT 1995) and/or AT intensities (ADAPT 1995; Cheng et al. 1992; Keul et al. 1979; Stegmann et al. 1981). Figure 4.4 is a schematic presentation of seven methods commonly employed to determine blood lactate transition thresholds.

Since the terms and their corresponding definitions affect the interpretation of blood lactate response to exercise, before selecting a methodology the user must have a clear understanding of the protocol required to assess and detect the response. One must also make a decision as to the appropriateness of the methodology for evaluating and/or prescribing training or performance. Currently most laboratories in Australia dealing with high-performance athletes use the ADAPT (1995) program to determine LT and AT. The technical error of measurement (TEM) (see pages 61 and 83) for LT and AT using ADAPT has been shown to have good precision for athletes tested at the South Australian Sports Institute in the endurance sports of cycling, running, and rowing.

Test Protocols

Protocol-related factors such as the sampling site, workload duration, continuous versus discontinuous exercise bouts, and choice of ergometer can all affect

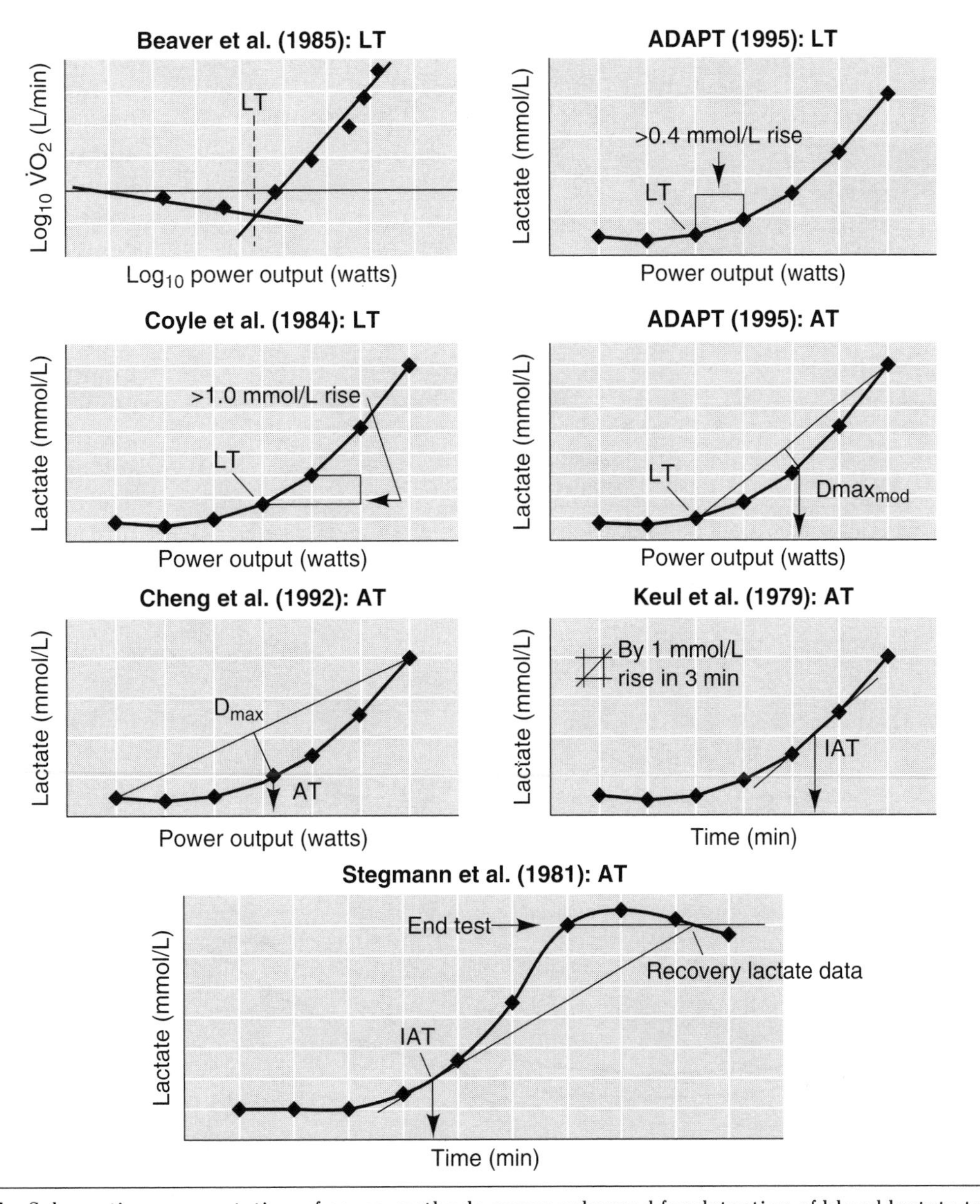

Figure 4.4 Schematic representation of seven methods commonly used for detection of blood lactate transitions.

Adapted, by permission, from D. Bishop, D.G. Jenkins, and L.T. MacKinnon, 1998, "The relationship between plasma lactate parameters, W peak and 1-h cycling performance in women," *Medicine and Science in Sport and Exercise* 30 (8): 1270-1275.

the measurement of blood lactate response to incremental exercise. Therefore, one must consider such factors carefully when establishing a test protocol.

Blood Collection, Handling, and Analysis Techniques

The blood vessel sampling site has been shown to affect measurement of the relationship between exercise intensity and blood lactate concentration (Foxdal et al. 1990, 1991; Welsman 1992; Yoshida, Takeuchi, and Suda 1982). Differing samples include arterial blood, venous blood, capillary blood, and arterialized venous blood (the procedures for each site are described in chapter 6). The blood lactate transition thresholds determined from differing sample sites are not comparable unless one has applied a correction factor. Samples taken from the same vessel type but from different locations on the body—for example, capillary samples drawn from the fingertip and earlobe—are, however, comparable (author's own observations).

Blood handling practices prior to analysis for lactate concentration can also affect measurement of the relationship between exercise intensity and blood lactate concentration. Such practices include untreated whole blood, plasma, serum, deproteinized blood, and lysed whole blood. Whole-blood lactate concentrations, for example, have been reported to be between 10% and 30% lower than plasma values (Telford 1984; Welsman 1992).

Once the blood has been sampled and treated, the way in which it is analyzed for lactate concentration can also have a significant effect on the result obtained. Various brands of automated blood lactate analyzers use different assaying techniques to determine blood lactate concentration. Figures 4.5 and 4.6 and table 4.3 demon-

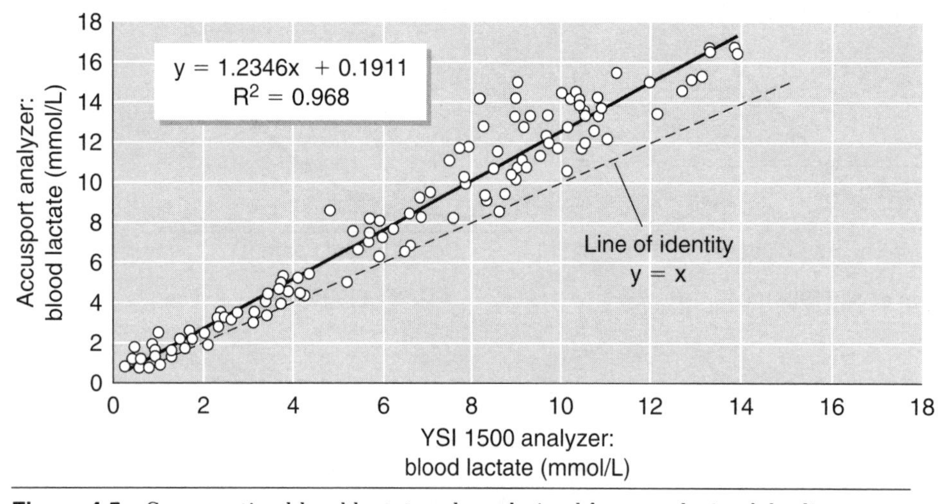

Figure 4.5 Comparative blood lactate values derived from analysis of duplicate samples analyzed in different automated blood lactate analyzers. YSI = Yellow Springs Instruments; Accusport = Boehringer Mannheim.

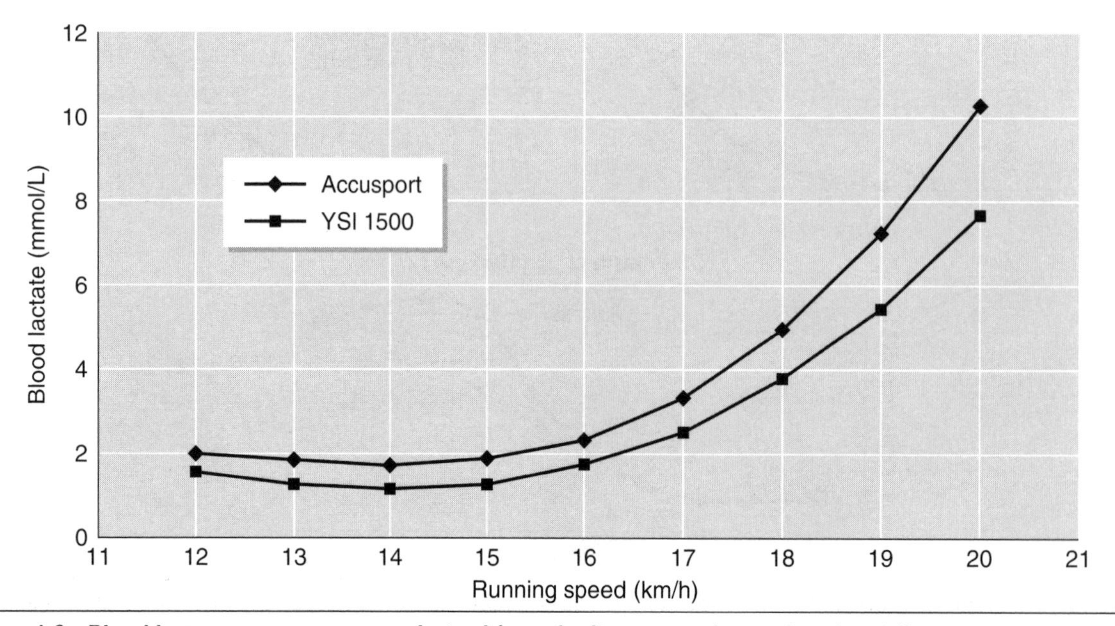

Figure 4.6 Blood lactate response curves derived from duplicate samples analyzed in different automated blood lactate analyzers. YSI = Yellow Springs Instruments; Accusport = Boehringer Mannheim. Subject is a junior male middle-distance runner.

Table 4.3 Comparison of Select Blood Lactate Transition Thresholds Determined From Duplicate Samples Analysed With Different Automated Blood Lactate Analysers

Blood lactate threshold	Blood lactate analyzer	Blood lactate (mmol/L)	Running speed (km/h)	$\dot{V}O_2$max (L/min)	HR (beats/min)
LT—ADAPT	YSI 1500	1.3	15.0	3.31	155
	Accusport	1.9	15.0	3.31	155
AT—ADAPT	YSI 1500	3.2	17.6	3.94	172
	Accusport	4.4	17.7	3.97	173
2.0 mM	YSI 1500	2.0	16.4	3.65	163
	Accusport	2.0	15.3	3.38	157
4.0 mM	YSI 1500	4.0	18.2	4.09	177
	Accusport	4.0	17.5	3.92	171

Subject is a junior male middle-distance runner.

YSI = Yellow Springs Instruments, Accusport = Boehringer Mannheim.

strate the difference in blood lactate concentration values for duplicate whole-blood samples between two commonly used automated lactate analyzers. As these data show, the results from these two analyzers should not be compared unless an appropriate correction factor has been applied. These data also highlight a potential problem in using fixed blood lactate thresholds, since one can see that the threshold intensities derived from the different analyzers differ markedly for the fixed values calculated from the same samples.

Within a laboratory it is therefore important to standardize blood sampling site, treatment procedures, and assay methodology for lactate determinations in order to allow direct and meaningful comparisons between results collected under test-retest conditions.

Workload and Rest Duration

Another factor that affects the blood lactate-exercise response curve is the workload duration. Numerous studies have demonstrated that the longer the workload duration, the lower the AT values (Foxdal et al. 1996; Freund et al. 1989; Heck et al. 1985; McLellan 1987; Rusko et al. 1986; Yoshida 1984b). Most exercise physiology laboratories around the world generally use incremental-intensity exercise tests with work durations between 3 and 5 min. These durations are considered adequate for measurements of $\dot{V}O_2$ and heart rate, because these variables are usually in steady state within these time frames for trained individuals. However, these durations seem less adequate when the aim of the test is to determine the highest exercise intensity corresponding to a maximal lactate steady-state concentration or AT. It is thus possible that the use of exercise durations shorter than 5 min may overpredict the AT exercise intensity. A number of recent studies have indicated that exercise periods of at least 5-7 min may be required to attain steady-state blood lactate concentrations and therefore allow an accurate determination of AT (Foxdal et al. 1996; Foxdal et al. 1994; Heck et al. 1985; LaFontaine et al. 1981; Oyono-Enguelle et al. 1990; Rieu et al. 1989; Stegmann and Kindermann 1982).

Figure 4.7 and table 4.4 clearly demonstrate that varying workload duration leads to different blood lactate response curves, which in turn affects the calculation of the exercise intensity corresponding to a specific blood lactate threshold. The selection of a workload duration should depend largely upon the reason for conducting the test. If one is testing to prescribe exercise intensities for endurance training, it is necessary to remember that such training generally takes place under steady-state conditions. Thus a test protocol using a number of exercise bouts of at least 5 min will provide more valid data for prescription of training intensities (author's own observations).

In addition, the intrastage rest interval can affect the determination of blood lactate transition thresholds. With longer breaks between work bouts in discontinuous protocols, such as interruptions in the rowing and kayaking protocols to allow blood sampling, blood lactate thresholds tend to move to higher work intensities (Heck et al. 1985; Foster et al. 1995). Therefore

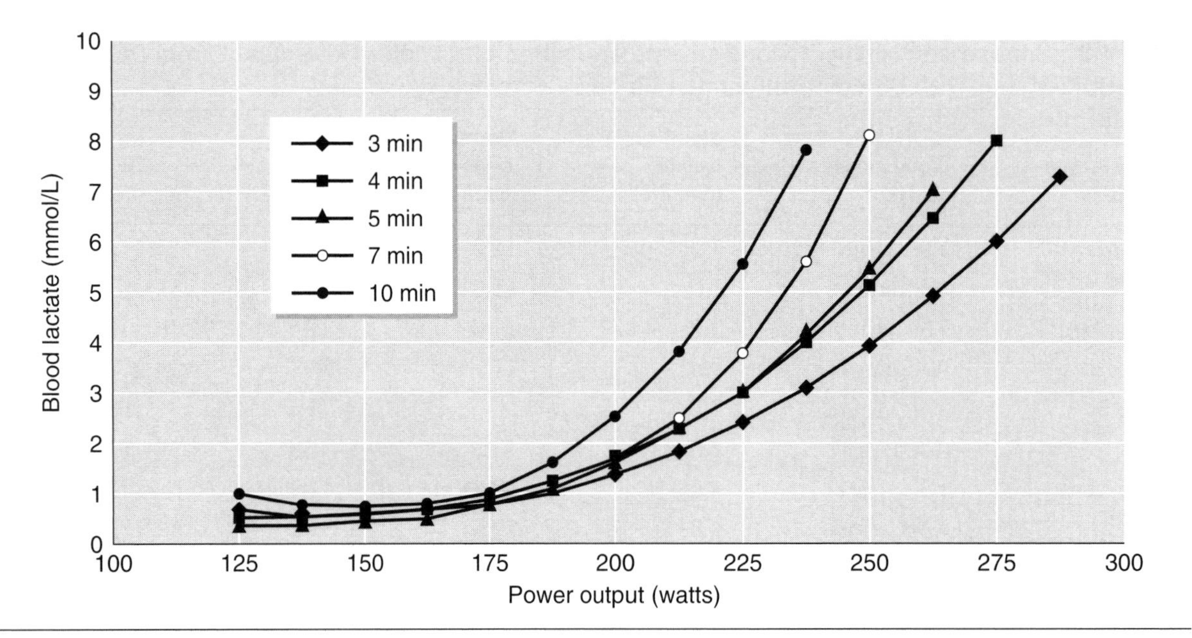

Figure 4.7 Effect of workload duration on the blood lactate response curve. Subject is a heavyweight female rower.

Table 4.4 Comparison of Select Blood Lactate Transition Thresholds Determined From Incremental Blood Lactate Tests With Varying Work Duration

Blood lactate threshold	Increment duration (min)	Blood lactate (mmol/L)	Power output (W)	$\dot{V}O_2max$ (L/min)	Percent $\dot{V}O_2max$ (%)	HR (beats/min)	Percent HRmax (%)
LT—ADAPT	3	0.72	172.4	2.83	70.8	166	81.8
	4	1.02	174.4	2.92	73.1	169	83.3
	5	0.75	170.6	2.84	71.2	168	82.8
	7	0.87	167.5	2.83	71.0	167	82.3
	10	0.96	169.9	2.85	71.4	170	83.7
AT—ADAPT	3	2.26	221.8	3.41	85.4	185	91.1
	4	2.96	223.9	3.62	90.7	188	92.6
	5	2.79	220.7	3.52	88.1	188	92.6
	7	2.68	209.1	3.42	85.6	187	92.1
	10	2.55	200.6	3.31	82.9	185	91.1
2.0 mmol/L	3	2.0	216.4	3.35	83.8	183	90.1
	4	2.0	206.8	3.38	84.6	181	89.2
	5	2.0	207.8	3.34	83.6	183	90.1
	7	2.0	200.7	3.34	83.7	181	89.2
	10	2.0	193.7	3.21	80.3	181	89.2
4.0 mmol/L	3	4.0	250.5	3.75	93.9	197	97.0
	4	4.0	237.4	3.82	95.7	193	95.1
	5	4.0	235.9	3.72	93.1	195	96.1
	7	4.0	223.8	3.67	92.0	191	94.1
	10	4.0	214.2	3.52	88.1	191	94.1
MLSS		2.4	199.9	3.29	82.4	183	90.1

Subject is a female heavyweight rower.

any intervals between work stages should be kept as brief and standardized as possible, with 1 min being viewed as a possible maximum.

Also, with many of the methodologies outlined earlier, mathematical manipulations are performed to best fit the data curve and generate the blood lactate transition thresholds (ADAPT 1995; Beaver et al. 1985; Cheng et al. 1992; Foster et al. 1995; Keul et al. 1979; Stegmann et al. 1981). Such manipulations require a minimum number of workloads; five to eight workloads are generally sufficient. Most investigators agree that the work intensity increments should be relatively small to allow more precise determination of the transition thresholds.

Within sport science laboratories, the decision regarding workload duration, intensity increase, interval between workloads, and number of workloads has generally been a matter of convenience and logistics (Foster et al. 1995). However, the evidence presented here suggests that longer work stages (>5 min) are definitely preferable over shorter stages, that breaks between stages should be as short as possible (<1 min), and that there should be at least five work stages with relatively small increases in intensity.

Ergometer Type

The choice of test ergometer will also influence the intensity at which a blood lactate threshold will occur. In assessment of high-performance athletes, the choice of ergometers should be dictated by the principle of training specificity (Craig 1987; Jacobs 1986; Withers et al. 1981). For example, cycle ergometers should be used to test cyclists, treadmills for runners, and rowing ergometers for rowers. This factor has major implications for triathletes, as here the need is to assess the blood lactate response to exercise for each sporting discipline. Table 4.5 supports this concept in presenting blood lactate transition threshold data for each discipline generated on three male triathletes.

In addition, one may also need to consider event specificity within a sport. For example, varying cadence can affect the work output at various blood lactate thresholds during cycling ergometry (Craig et al. 1993, 1995; Woolford et al. 1999). This point is relevant in the assessment of road cyclists whose preferred cadence tends to be 90-100 rpm. On the other hand, track endurance cyclists tend to pedal at cadences around 110-120 rpm, suggesting that their protocol needs to be modified from that of road cyclists (Craig 1987; Craig et al. 1995).

As with all testing equipment, ergometers should be regularly calibrated to minimize potential errors from one test to another. Research conducted by Heck et al. (1985) shows a distinct difference in the lactate curves with the use of identical loading procedures but different treadmills.

Table 4.5 Mean Blood Lactate Transition Threshold Values From Three Male Triathletes

Blood lactate threshold	Event discipline	Blood lactate (mmol/L)	Work intensity*	$\dot{V}O_2$max (L/min)	Percent $\dot{V}O_2$max (%)	HR (beats/min)	Percent HRmax (%)
LT—ADAPT	Swim	1.59	1.136	n/a	n/a	143	80.5
	Ride	1.01	213.2	3.08	59.2	133	72.3
	Run	1.41	14.0	3.69	68.1	156	81.0
AT—ADAPT	Swim	3.60	1.270	n/a	n/a	163	91.6
	Ride	3.09	315.7	4.24	81.5	161	87.6
	Run	3.17	16.7	4.51	83.2	178	92.7
2.0 mmol/L	Swim	2.00	1.179	n/a	n/a	149	83.7
	Ride	2.00	280.2	3.84	73.8	151	82.3
	Run	2.00	15.3	4.08	75.3	167	86.5
4.0 mmol/L	Swim	4.00	1.286	n/a	n/a	165	92.6
	Ride	4.00	337.9	4.49	86.4	167	90.9
	Run	4.00	17.5	4.75	87.7	184	95.8

* Swim intensity: m/s; ride intensity: W; run intensity: km/h; n/a = not available.

The calculated threshold values are reported for each discipline.

Environmental and Experimental Conditions

The blood lactate response to exercise and derived transition thresholds have been shown to be reproducible under standardized conditions (Heitkamp et al. 1991; Jacobs 1986; Karlsson and Jacobs 1982). Table 4.6 offers further support in showing TEM values for blood lactate transition threshold values generated in the South Australian Sports Institute's exercise physiology laboratory. A number of factors have been identified that affect the blood lactate response to exercise, however. These factors generally fall into one of the following categories:

- **Environmental factors.** High ambient temperatures have been shown to increase blood lactate concentrations both at rest and during exercise (Fink et al. 1975; MacDougall et al. 1974). In contrast, during exercise in the cold, LT has been shown to occur at higher workloads than in normothermic environments (Flore et al. 1992; Blomstrand et al. 1984). Blood lactate concentrations during submaximal workloads are also greater during acute exposure to altitude than at sea level (Weltman 1995). Such findings are potentially of major importance when one is taking blood lactate measures in the field.
- **Subject attributes.** Blood lactate concentration is decreased at any given workload when muscle glycogen stores are depleted (Dotan et al. 1989; Hughes et al. 1982; Ivy et al. 1981; Jacobs 1981; Maassen and Busse 1989; Yoshida 1984a). Muscle glycogen depletion can result from either dietary (Jacobs 1981; Greenhaff et al. 1987; Yoshida 1986) or exercise (Genovely and Stanford 1982; Jacobs 1981) manipulations. Since a glycogen-depleted state leads to decreased lactate production and there is a resultant shift to the right of the blood lactate-performance curve, however, neither the shape nor the slope of the curve is changed (Frohlich et al. 1989). Therefore, in a glycogen-depleted state, endurance capacity is overestimated in the determination of fixed blood lactate thresholds, while individually based thresholds calculated from the individual shape of the curve are hardly changed at all (Frohlich et al. 1989; Urhausen and Kinderman 1992). Obviously, one must consider the implications of depleted muscle glycogen stores when evaluating the blood lactate response to exercise, particularly if fixed blood lactate thresholds are to be calculated. Standardization of pretest diet and exercise patterns is therefore recommended.

Other personal factors such as caffeine ingestion (Gaesser and Rich 1985), extreme deviations from normal acid-base balance (i.e., following bicarbonate ingestion) (Davies et al. 1986; Kowalchuk et al. 1984), motivation, and anemia (Celsing and Ekblom 1986) also have been shown to affect the blood lactate response to exercise. Therefore it is highly advisable to keep a detailed record of an athlete's pretest condition and, where possible, to standardize these conditions for future test sessions.

Relationship of Lactate Concentrations in Muscle and Blood

A strong correlation between blood and muscle lactate concentrations exists during exercise (Foster et al. 1995; Jacobs 1986; Karlsson and Jacobs 1982). However, it is erroneous to interpret blood lactate accumulation as wholly reflective of muscle lactate production. Blood lactate concentration depends on the existence of a net positive gradient for lactate between muscle and blood, and is affected by dilution in the body water; by removal by organs such as the liver, heart, and inactive skeletal muscle; and by the temporal lag before lactate produced in the muscle appears in the blood (Foster et al. 1995; Jacobs 1986; Weltman 1995). Jorfeldt et al. (1978) reported that the rate of lactate release from the muscle to blood is partially dependent upon concentration, with maximal release occurring in the range of 4 mmol/kg wet weight of muscle. Therefore, at high muscle lactate concentrations there may be a significant time lag before lactate equilibrates with the blood. This is supported by a number of studies that have simultaneously measured lactate concentrations in the muscle and blood. Rusko et al. (1986) employed work bouts of about 5 min duration and demonstrated a consistent relationship between the two. Jacobs (1981) reported that the blood lactate concentration progressively underestimated the muscle lactate concentration with use of shorter work durations (<4 min). In contrast, in a rapidly incremented exercise test (1 min stages) as employed by Green et al. (1983), the difference between the two lactate concentrations was rather large.

Thus, blood lactate concentration may be thought of as an effective shadow of muscle lactate concentration (Foster et al. 1995). However, this relationship is influenced by the duration of the exercise stages, with longer work durations (at least 5 min) providing a more consistent

Table 4.6 Technical Error of Measurement (TEM) Values for Blood Lactate Transition Thresholds

Test	TEM	TEM%	95% CI	Range
Peak values				
Test duration (min)	0.99	3.9	2.80	25.0–33.5
Heart rate (beats/min)	1.5	0.8	4.2	180–205
$\dot{V}O_2$ (L/min)	0.042	1.02	0.12	3.183–5.649
Lactate threshold				
Blood lactate (mmol)	0.18	19.43	0.51	0.33–1.47
Power output (W)	5.2	3.3	14.6	122.0–204
Heart rate (beats/min)	2.1	1.4	6.0	132–168
%HRmax	1.3	1.6	3.6	71.1–84.5
$\dot{V}O_2$ (L/min)	0.061	2.19	0.17	2.18–3.51
%$\dot{V}O_2$max	1.2	1.7	3.3	61.0–78.8
2 mmol/L				
Power output (W)	4.8	2.4	13.4	150.1–271.8
Heart rate (beats/min)	2.0	1.2	5.6	153–181
%HRmax	0.7	0.8	1.9	83.1–91.6
$\dot{V}O_2$ (L/min)	0.047	1.43	0.13	2.650–4.409
%$\dot{V}O_2$max	0.7	0.9	2.0	72.4–89.4
Anaerobic threshold				
Blood lactate (mmol)	0.39	10.23	1.10	2.75–4.77
Power output (W)	3.4	1.5	9.7	166.2–303.5
Heart rate (beats/min)	1.9	1.0	5.3	158–190
%HRmax	0.9	1.0	2.7	87.8–95.3
$\dot{V}O_2$ (L/min)	0.045	1.22	0.13	2.808–4.811
%$\dot{V}O_2$max	0.8	0.9	2.3	84.4–94
4 mmol/L				
Power output (W)	4.9	2.2	13.8	169.4–306.1
Heart rate (beats/min)	1.8	1.0	5.0	163–189
%HRmax	0.7	0.8	2.1	89.0–95.8
$\dot{V}O_2$ (L/min)	0.043	1.18	0.12	2.937–4.879
%$\dot{V}O_2$max	0.8	0.9	2.3	83.2–95.0

Calculated from rowing data, n = 10 athletes.

indication of the relationship between muscle and blood lactate concentrations (author's own findings).

Laboratory Versus Field Testing

The use of laboratory-derived values to assess and regulate exercise intensity in the field is open to question. For example, does rowing an ergometer in the laboratory require the same technique as rowing a racing shell on water? How do you account for the lack of environmental factors (e.g., wind resistance in cyclists, water resistance/drag in swimming and kayaking) in the laboratory? Does the difference in running surface (i.e., treadmill belt vs. grass) affect the blood lactate-exercise response curve? Further research will be needed to provide the answer to such applied questions. Numerous studies have, however, investigated the heart rate-blood lactate relationship in both the laboratory and field environments. Nonsignificant differences between these test environments have been reported for running (Brettoni et al. 1989; Coen et al. 1991), rowing (Lormes et al. 1987), and cycling (Brettoni et al. 1989; Foster et al. 1993). Similar applied research conducted at the South Australian Sports Institute, comparing the blood lactate-heart rate response of kayakers on water and in the laboratory, supported these findings. This latter study showed no significant difference between the heart rates corresponding to the calculated LT and AT (ADAPT 1995) for either environment. Table 4.7 presents the results of this research. But these findings are not unequivocal, as Schmidt et al. (1984) reported that heart rates for cross-country skiers were 16 beats/min higher in the laboratory than in the field.

Compared with laboratory investigations, field-based tests monitoring the blood lactate response to exercise have the following potential advantages:

- They are extremely specific to the actual athletic performance.
- They have more practical significance to the athlete and coach.
- They enable the investigator to test more than one subject at a time.

Possible disadvantages of field-based tests are as follows:

- Environmental conditions (for example, ambient temperature, wind speed and direction, surface conditions) cannot be controlled.
- The number of physiological parameters that can be monitored is limited.

Conversely, laboratory-based tests have the following advantages:

Table 4.7 Comparative Heart Rates at the LT and AT for 16 Kayakers

	Lactate threshold heart rates		Anaerobic threshold heart rates		Test maximum heart rates	
Athlete	**Field (beats/min)**	**Lab (beats/min)**	**Field (beats/min)**	**Lab (beats/min)**	**Field (beats/min)**	**Lab (beats/min)**
Females						
DA	149	148	165	164	180	179
RJ	156	157	176	180	189	197
JP	153	151	177	178	190	194
DB	144	141	171	168	187	189
LL1	149	147	175	175	189	190
LL2	148	147	176	174	194	191
LL3	145	145	176	176	189	189
Males						
BF	155	161	178	176	191	195
BM	156	153	173	169	189	190
BS	144	143	167	168	185	189
JB	139	140	166	163	181	178
SA	142	138	165	165	186	186
MD	148	145	169	167	185	184
DF	154	154	185	182	194	193
DG	157	154	178	179	196	197
TD	141	143	170	172	200	201
SD	152	153	184	183	206	207
DS	154	150	180	180	198	197
Mean	149.2	148.3	173.9	173.3	190.5	191.4
SD	5.7	6.3	6.1	6.5	6.6	7.2

Values were determined from blood lactate response curves conducted both in the laboratory and in the field. No significant difference was found between any of the three heart rate groups.

- They are performed in a controlled, stable environment.
- Physiological data acquisition is easier.
- There is greater accuracy in the control of workloads.

The potential disadvantages of laboratory-based tests include the following:

- Specific technique may be questionable due to ergometer design.
- Generally only one athlete can be tested at a time because of equipment limitations.

In deciding which environment is most suitable for athlete assessment, the coach and sport scientist need to collaborate and to have a clear understanding of the test's purpose and desired outcomes. Furthermore, if the aim is to make direct comparisons between laboratory- and field-based test protocols to determine the blood lactate response to exercise, it is advisable that the protocols mimic each other as closely as possible.

Practical Applications of the Blood Lactate Response to Exercise Data

The way blood lactate levels change with increased exercise intensity can provide valuable information about how the athlete is adapting to training. The following sections describe how this information helps in the evaluation of the athlete's endurance adaptations and training status.

Optimal Blood Lactate Concentration for Training

Several groups have proposed that blood lactate transition thresholds be monitored regularly during training in order to provide information relative to training intensity (Coen et al. 1991; Fohrenbach et al. 1987; Keith et al. 1992; Kindermann et al. 1979; Madsen and Lohberg 1987). However, the question coaches and athletes most often ask is, "What is the best blood lactate threshold(s) to train at?" At present, there is a lack of scientific consensus about the best way to use blood lactate concentrations in the design and modification of training programs. In the case of endurance training, researchers have reported optimal blood lactate levels for training between 1.3 and 10.0 mmol/L (Fohrenbach et al. 1987; Hagberg and Coyle 1983; Heck et al. 1985; Katch et al. 1978; Mader et al. 1976; Madsen and Lohberg 1987; Sjodin and Jacobs 1981; Stegmann and Kindermann 1982). Most recent research, however, favors the use of individually derived LT and AT values over fixed blood lactate concentrations (Beneke 1995; Bishop et al. 1998; Cheng et al. 1992; Coen et al. 1991; Keith et al. 1992; McLellan et al. 1991; Urhausen and Kindermann 1992). The question of optimal training intensities will remain unanswered until we have more controlled training studies with athletes.

Among sport scientists currently working with endurance sports in Australia there exists a general consensus on the classification of training intensities. However, the nomenclature used to describe these classifications varies widely among sports (refer to chapters on cycling [chapter 17], rowing [chapter 21], swimming [chapter 27], and triathlon [chapter 29]). This can cause confusion among coaches as well as sport scientists who are trying to draw reference across sports. Generally, though, work intensities are divided into six categories based on identification of the LT and AT on the lactate-workload curve. Table 4.8 defines and provides some approximate guidelines for each of these training zones. Training performed below LT is regarded as recovery or very low intensity aerobic; that performed between LT and AT is divided into two halves, with the lower half termed "extensive aerobic" and the upper half "intensive aerobic." Anaerobic threshold work (i.e., work performed at AT) constitutes a category by itself. Training performed above AT is classified as "$\dot{V}O_2$max" work. The sixth training zone comprises "anaerobic" work. This category is not based on the lactate curve, but includes all efforts performed at or above specific race pace. Note that there are areas of overlap within this table, as its purpose is to provide general, not individual, guidelines defining six commonly used international training classifications.

Interpretation of Shifts in Blood Lactate Response Curves

The relationship between blood lactate and increasing exercise intensity can also provide useful information for the evaluation of adaptations to endurance training (Denis et al. 1984; Hurley et al. 1984; Katch et al. 1978; Sjodin et al. 1982). Any analysis or interpretation of the blood lactate response curve is best performed at the individual level, since values can be affected by the various factors already discussed in connection with experimental

Table 4.8 Classification of Training Zones as a Function of Lactate Threshold and Anaerobic Threshold

Description of training zone	Blood lactate threshold relationship	Percent HRmax (%)	Blood lactate (mmol/L)	Perceived exertion
Recovery (U3, E1)	<LT	<75	<2.0	Easy
Extensive aerobic (U2, E2a)	LT to LT+ ([AT – LT]/2)	75–84	1.0–3.0	Comfortable
Intensive aerobic (U1, E2b)	AT to AT– ([AT – LT]/2)	82–89	1.5–4.0	Uncomfortable
Threshold (AT, E3)	AT	89–93	2.5–5.5	Stressful
$\dot{V}O_2max$ (Transport, E4)	>AT	>92	>5.0	Very stressful
Anaerobic	Maximal	n/a	>7.0	Maximal

conditions. Even after as many as possible of these confounding factors have been controlled for, each athlete's results should be analyzed and interpreted separately. This is not to discount the value of intersubject or group comparisons.

There are basically two ways of assessing the blood lactate-intensity curves for improvement, maintenance, or degradation of fitness. These involve (1) subjective evaluation of the graphical and raw data and (2) objective or statistical evaluation of the data.

Graphical overlays of the individual athlete test-retest blood lactate profiles, with a subjective assessment of any curve shifts, are the most commonly employed method of determining the extent of any adaptation. Some of the most common changes in the blood lactate-intensity curve and their interpretations (Madsen and Lohberg 1987; Pyne 1989; Weltman 1995) are as follows:

- A shift in the curve down and/or to the right is indicative of increases in the event-specific aerobic capacity of the athlete. This is expressed by the athlete's ability to exercise at a greater intensity for a given blood lactate level or to express lower blood lactate levels for the same intensity (see figure 4.8, graphs A and B).

- A shift in the curve up and/or to the left indicates a deterioration in the event-specific aerobic capacity of the athlete. This is expressed by the athlete's exercising at a lesser intensity for a given blood lactate level or expressing a higher blood lactate level for the same intensity (see figure 4.8, graphs C and D).

Changes in specific blood lactate transition thresholds values (i.e., LT and AT) can also serve as distinct indicators of change in an athlete's training status. Increases in the intensity at LT reflect an improvement in base aerobic condition. This adaptation presents itself as a shift to the right and/or downward in the initial three to four workloads. This is thought to be the result of delayed blood lactate production due to increased fat oxidation and enhanced aerobic mechanisms. Similarly, increases in exercise intensity at AT, represented by graphical shifts down and/or to the right, may be indicative of an improvement in higher-level aerobic endurance. Possible causes may be improved lactate clearance or acid buffering.

Determining TEM (or precision) values for the various blood lactate transition thresholds allows for a more meaningful interpretation of these points of reference. These values represent a quantitative means of assessing significant changes in the threshold-related variables (see table 4.6, page 61). They also give the scientist the ability to account for methodological and tester-influenced errors (refer to page 83).

More objective means of interpreting the significance of curve shifts also exist. These usually involve the use of least squares linear regression analysis and post hoc statistical procedures. Such analytical methods, however, should not override the need for subjective interpretations, as they simply serve to present the data in a way that aids in interpretation. Objective methods tend to be very rigid and as such do not fully account for the

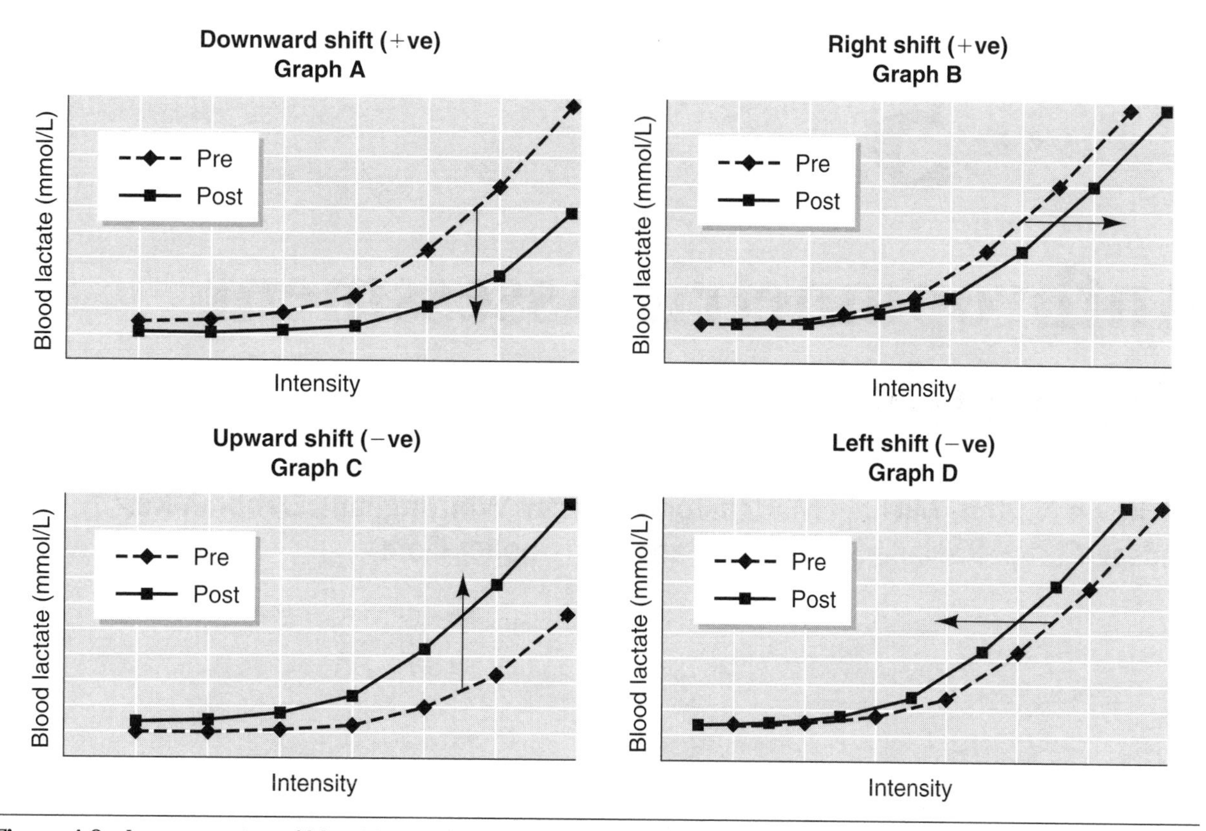

Figure 4.8 Interpretation of blood lactate-intensity curve shifts.

individual nature of the data. The tester must always take into account the raw data together with information relating to the athlete's state at the time of the test. This should involve asking the coach and athlete questions about current training and medical and psychosocial status. Such a process allows the application of the "art of the science" and often leads to more meaningful interpretation.

Summary

Discontinuities—or "thresholds," as they are commonly termed—in the blood lactate response to exercise may be used as indexes of endurance capacity and markers for training prescription. Irrespective of methodological variations, there are two main areas of discontinuity in the blood lactate-intensity curve. The first of these discontinuities, LT, is associated with the first workload at which there is a sustained increase in blood lactate above resting levels. The second, AT, is marked by a very rapid rise in blood lactate concentration. This point represents a shift from oxidative to partly anaerobic energy metabolism during constant workloads, and it refers to the upper limit of blood lactate concentration indicating an equilibrium between lactate production and lactate elimination.

A number of factors (e.g., work duration; calculation method; technique associated with blood collection, handling, and analysis; ergometer type; ambient environment; pretest exercise and diet) can all modify the blood lactate-intensity relationship and its dependent thresholds. Therefore it is necessary to take due care to control for these factors if meaningful data are to be collected. As identified throughout this chapter, blood lactate transition thresholds based on fixed blood lactate concentrations are generally affected more by such factors. The recommendation therefore is to use individually based methods of determining LT and AT.

The aim of this chapter has been to provide a brief overview of the major concepts and controversies associated with the use of the blood lactate accumulation to evaluate responses to exercise. The discussion is by no means exhaustive: the frequency with which research related to this area is appearing in the sport science literature makes comprehensiveness extremely difficult, yet attests to the growing acceptance of blood lactate and its associated thresholds as valuable diagnostic, interpretive, and predictive tools.

chapter 5

Anthropometric Assessment Protocols

Kevin Norton, Michael Marfell-Jones, Nancy Whittingham, Deborah Kerr, Lindsay Carter, Kaylene Saddington, and Christopher Gore

Anthropometry, like any other area of scientific measurement, depends upon adherence to the particular requirements involved in the standards of measurement as determined by international standards bodies. The anthropometric standards body is the International Society for the Advancement of Kinanthropometry (ISAK), and it is this organization's standards that have been adopted for the purpose of this manual.

The standards of measurement relate to the following:

- The observance of landmarks in the determination of measurement sites
- Adherence to standard procedures when using equipment
- The continuous calibration of equipment

For readers interested in information regarding measurements not included in this volume, Norton et al. (1996) provide additional technical advice. The descriptions of anthropometric sites and measurement procedures in this chapter are from Norton et al. (1996), reproduced with kind permission of the University of New South Wales Press.

Preliminary Matters

A number of preliminary steps should be taken to ensure the comfort of the subject and to facilitate ease of measurement. For example, ensure that the physical space is large enough to allow the measurer to move freely around the subject. An environmentally controlled room is also desirable. Appreciate that all people have an area around their body known as a "personal space" and that when this area is invaded they feel uncomfortable or threatened. This is particularly true for the front of a person, and this is why most measurements are taken from the side or from behind. It is also important to see that the room is not cluttered with equipment or furniture, that there is plenty of light, and that subjects are made aware of what parts of their body will be touched and measured as you proceed.

The Subject

Throughout the marking and measurement session, the subject stands relaxed, arms comfortably to the side, allowing the measurer to move around and manipulate the equipment (figure 5.1).

So that measurements can be made as quickly and efficiently as possible, subjects should be asked to present themselves in minimal clothing. Swimsuits (two-piece suits for women) are ideal for ease of access to all measurement sites.

Subjects must be informed which measurements are to be taken and must complete a consent form as part of the preliminaries of the test period.

Data Collection

When possible, it is advisable for a recorder to enter data. Ideally the recorder will be knowledge-

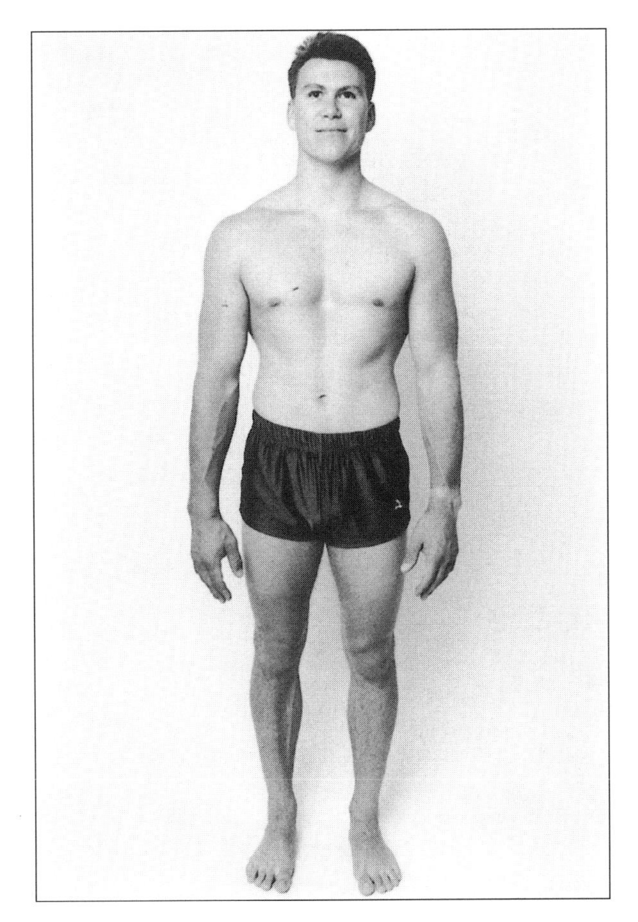

Figure 5.1 Subject's stance.

able in the skill of measurement and so will be able to assist the measurer.

Despite careful attention to the standards, it is still possible for errors to occur in the recording of data—because of poor pronunciation by the measurer, inattention on the part of the recorder, or the recorder's failure to follow the steps designed to eliminate such errors. Ideally, data collection should involve one measurer and one recorder.

The measurer should use single numbers only—for example, reading 29.16 as "two, nine, point one, six." The recorder repeats the value while recording it so that the measurer can do an immediate check. Remember that the measurer and the recorder work as a team, and it is the responsibility of the recorder to help the measurer wherever necessary.

The recorder must be particular in forming the written numbers, using the European 7 and clearly distinguishing between the 6 and the zero. The recorder should also avoid erasing a number to correct a recording error. It is better to cross out the error and rewrite the whole value.

In some cases, measurements may be repeated or even taken a third time. In the first case, the average value is used; in the second case, the median value is used.

Stature

There are three general techniques for measuring stature: free standing, stretch, and recumbent. The latter may be used for infants or adults unable to stand and will not be considered here. The other two methods will give slightly different values, because stretch height removes the minor effect of daily gravitational compression.

Height exhibits diurnal variation; generally, subjects are taller in the morning and shorter in the evening. Repeated measures should be taken as near as possible to the same time of day as the original measurement.

Equipment

In the laboratory, a stadiometer should be mounted on a wall and used in conjunction with a right-angled head board that is at least 6 cm wide and that can be placed firmly on the subject's head and fixed to the stadiometer. The floor surface must be hard and level.

The stadiometer should have a minimum range of measurement of 600 mm to 2100 mm. The accuracy of measurement required is 0.1 cm. The stadiometer should be checked every month, or after 30 subjects, against a standard height.

In the field, when a stadiometer is not available, a girth tape fixed to a wall, checked for height and vertical positioning, is used in conjunction with a head board. As a "last-resort" method, a piece of paper taped to a wall may be used to identify the height, using a head board. Assessment of the height can then be completed using a steel tape. This method is not acceptable in a laboratory.

Method

Both the free-standing and the stretch-stature methods require the subject to stand with the feet together and the heels, buttocks, and upper part of the back touching the scale. The head when placed in the Frankfort plane need not be touching the scale.

The *Frankfort plane* is achieved when the orbitale is in the horizontal plane with the tragion. The *orbitale* is the lower edge of the eye socket. The *tragion* is the notch superior to the flap of the ear. When these are aligned, the vertex is the highest point on the skull (figure 5.2).

In the stretch-stature technique, the measurer places the hands along the jaw of the subject with

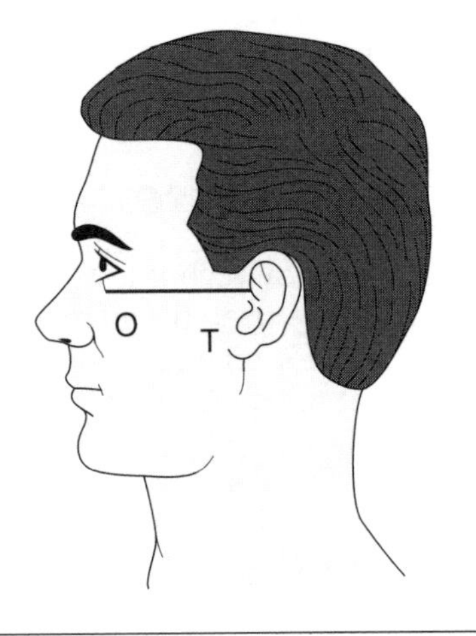

Figure 5.2 The Frankfort plane (O = orbitale, T = tragion).

Reprinted, by permission, from C. Gore and D. Edwards, 1992, *Australian fitness norms: A manual for fitness assessors* (Adelaide: The Health Development Foundation, South Australia), 7.

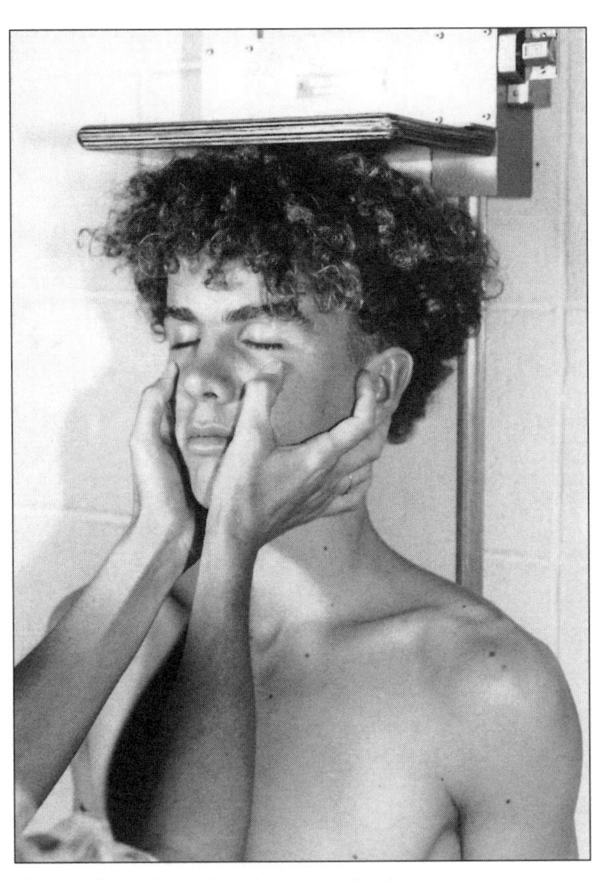

Figure 5.3 Stretch-stature method.

the fingers reaching to the mastoid processes. The subject is instructed to take and hold a deep breath, and while keeping the head in the Frankfort plane the measurer applies firm upward lift through the mastoid processes. The recorder places the head board firmly down on the vertex, crushing the hair as much as possible. The recorder further assists by watching that the feet do not come off the floor and that the position of the head is maintained in the Frankfort plane (figure 5.3).

Body Mass

The instrument of choice for measuring body mass is a beam balance accurate to the nearest 0.1 kg. In the field situation, it may be necessary to use a spring balance with accuracy to the nearest 0.5 kg.

The use of electronic digital scales is becoming more general, and the accuracy of this type of scale is equal to that of the beam balance (given maintenance of calibration of both machines). Calibration masses totaling 100 kg are required as standard equipment.

To measure nude body mass, first weigh the clothing that the subject will wear during measurement and then subtract this from the weight. Generally the body mass in minimal clothing is sufficiently accurate.

Body mass exhibits diurnal variation; the most stable values are those obtained routinely in the morning 12 h after any food, and after voiding.

The subject stands still without support, with the weight evenly on both feet over the center of the scales.

Landmarks

Accurately identifying landmarks is one of the most critical aspects of good anthropometry. A sound knowledge of anatomy is essential. Landmarks are anchor or reference points and must be located in the same way each time. You can then be confident that from one measurement session to the next you will be very precise in your location of these sites.

General Method to Locate Landmarks

Landmarks are identifiable skeletal points that generally lie close to the body's surface and are the "markers" identifying the exact location of the measurement site. All landmarks are found by palpation. For the comfort of the subject, the measurer's fingernails should be kept trimmed.

The landmark is identified with the left hand, usually with the thumb. The site is released to remove any distortion of the skin, then is relocated and marked with the thumbnail. Using a fine-tipped felt pen, the site is marked directly over the landmark. The pen mark is then checked to ensure that there has been no displacement of skin relative to the underlying bone.

The landmarks described here are those required for the measurements included in this chapter. Readers will need to consult other references for landmarks required for additional measurements. All landmarks are identified and marked before any measurements are made. The order of their identification is as listed in the next section.

Definition and Location of Specific Landmarks

Note that some of the marks are short lines, while others are x's. Actually, all the landmarks are points, but in getting to these points, it is often useful/essential to draw short vertical or horizontal lines as reference marks.

PROTOCOLS: Landmarks

Acromiale

Definition: The point at the most lateral, superior border of the acromion process and which is midway between the anterior and posterior borders of the deltoid muscle when viewed from the side (figure 5.4).

Location: Standing behind and on the right-hand side of the subject, palpate along the spine of the scapula to the corner of the acromion. This represents the start of the lateral border, which usually runs anteriorly, slightly superiorly, and medially.

Apply the straight edge of a pencil to the lateral aspect of the acromion to confirm the location of the border. The landmark is a point on the most lateral and superior part of the border that is adjudged to be in the mid-deltoid position when viewed from the side.

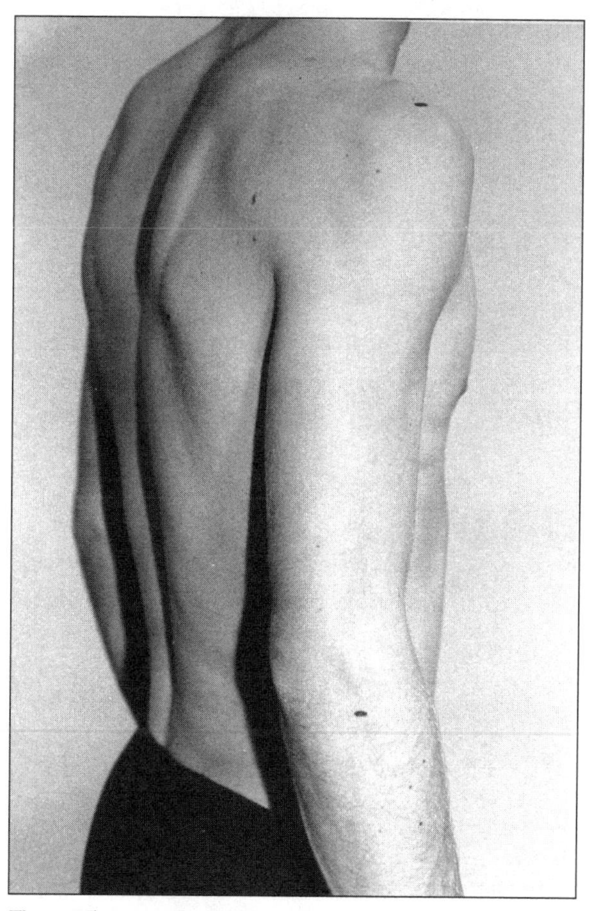

Figure 5.4 Acromiale and radiale landmarks.

Radiale

Definition: The point at the upper and lateral border of the head of the radius (figure 5.4).

Location: Palpate downward into the lateral dimple of the right elbow. It should be possible to feel the space between the capitulum of the humerus and the head of the radius. Slight rotation of the forearm is felt as rotation of the head of the radius.

The most superior lateral border of the radius is marked.

Mid-Acromiale-Radiale

Definition: The point equidistant from acromiale and radiale (figure 5.5).

Location: Measure the linear distance between acromiale and radiale with the arm relaxed and extended by the side. Place a small horizontal mark at the level of the midpoint between these two landmarks. Project this mark around to the posterior and anterior surfaces of the arm as a horizontal line. This is shown for the anterior surface (biceps skinfold site) in figure 5.5.

This is required for locating the triceps and biceps skinfold sites. When one is marking the sites

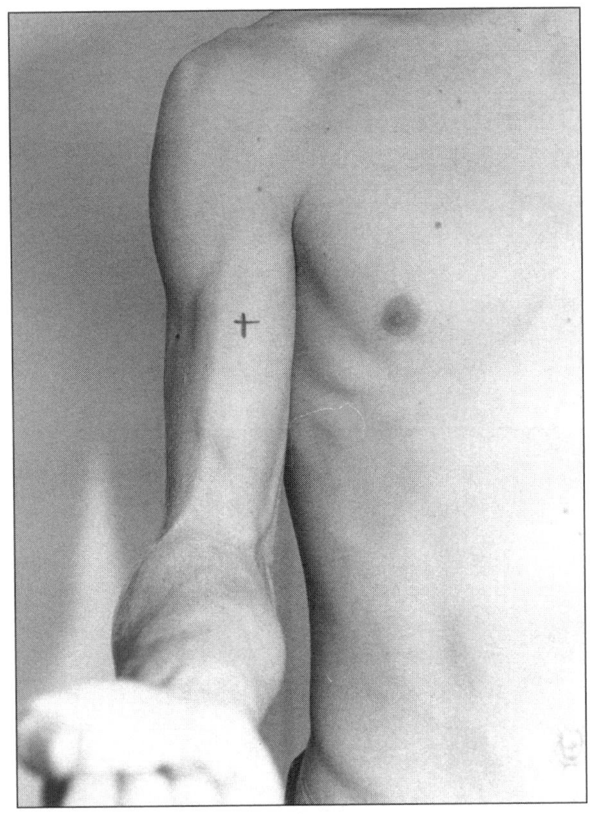

Figure 5.5 Mid-acromiale-radiale landmark.

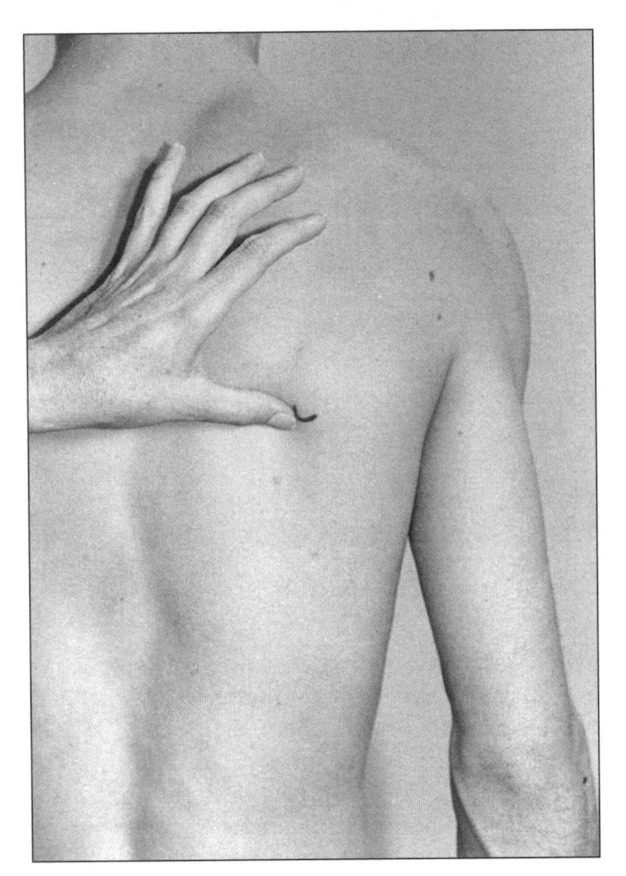

Figure 5.6 Subscapulare landmark.

for triceps and biceps skinfolds, the subject must assume the anatomical position. The triceps skinfold is taken over the most posterior part of the triceps, and the biceps skinfold is taken over the most anterior part of the biceps when viewed from the side (at the marked mid-acromiale-radiale level).

Subscapulare

Definition: The undermost tip of the inferior angle of the scapula (figure 5.6).

Location: Palpate the inferior angle of the scapula with the left thumb. If there is difficulty locating the inferior angle of the scapula, the subject should slowly reach behind the back with the right arm. The inferior angle of the scapula should be felt continuously as the subject's hand is again placed by the side of the body. A final check of this landmark should be made with the subject's hand by his or her side in a relaxed position (i.e., the subject in figure 5.6).

Xiphoidale

Definition: The xiphoidale is found at the lower extremity of the sternum. The landmark is the tip of the xiphion (figure 5.7).

Location: Locate the xiphoidale by palpating in the medial direction of the left or right costal arch toward the sternum. These arches form the infrasternal angle and articulate at the xiphisternal joint.

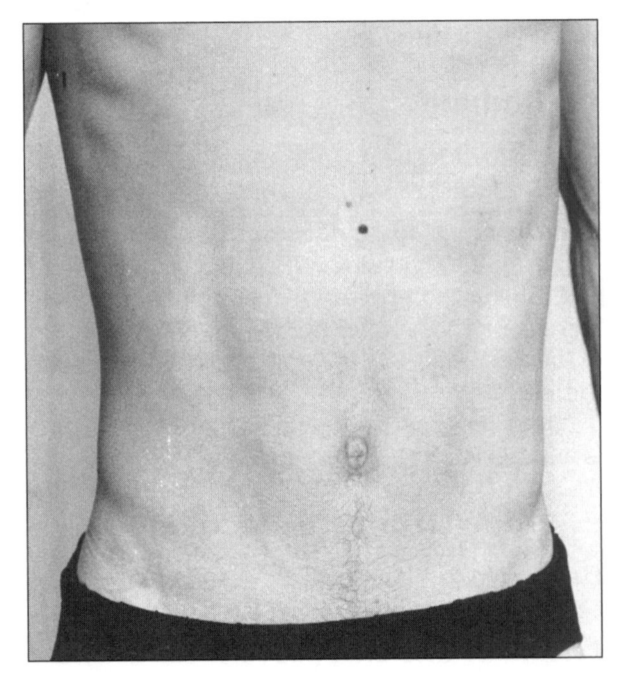

Figure 5.7 Xiphoidale landmark.

Iliospinale

Definition: The most inferior aspect or undermost tip of the anterior superior iliac spine (figure 5.8).

Location: Palpate the superior aspect of the ilium and follow anteriorly along the crest until it seems that the bony structure "disappears." The landmark is the underpoint of the edge where the bone "disappears."

If there is difficulty locating the landmark, the subject should lift the heel of the right foot and rotate the femur outward. Because the sartorius muscle originates at the site of the iliospinale, this movement of the femur enables palpation of the muscle and tracing to its origin.

Once the landmark is identified, it is marked using the standard procedures, as the subject stands with the feet together and the weight evenly distributed.

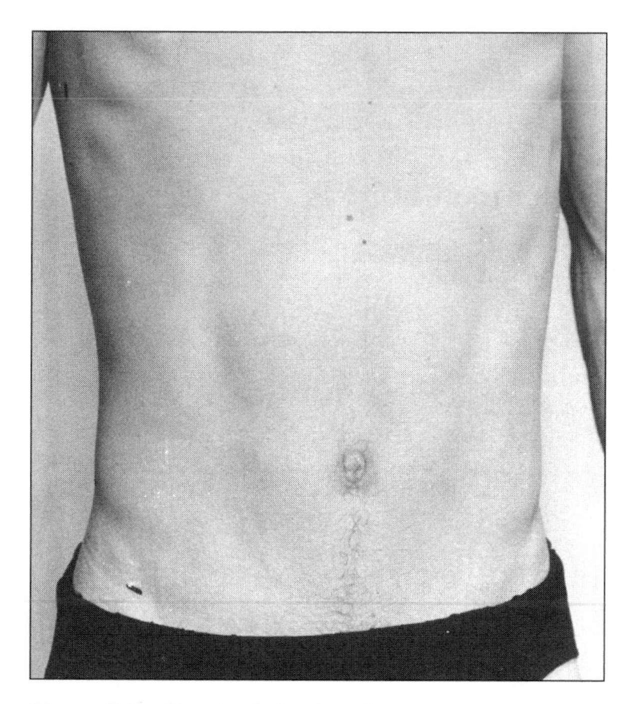

Figure 5.8 Iliospinale landmark.

Ilio-Axilla

Definition: The imaginary line joining the observed midpoint of the armpit with the most lateral superior edge of the ilium (iliocristale; figure 5.9).

Location: With the subject's arm abducted to the horizontal and the trunk erect, locate the most lateral superior edge of the ilium using the right hand. The left hand is used to stabilize the body by providing resistance on the left side of the pelvis as the right hand is used to palpate the ilium.

The landmark is made at the identified edge of the ilium in line with the imaginary line to the midpoint of the axilla.

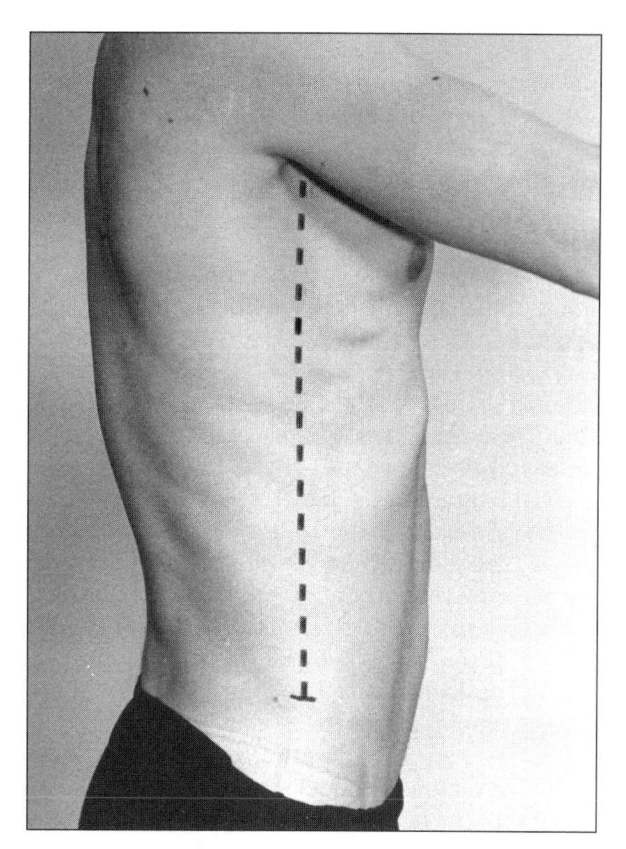

Figure 5.9 Ilio-axilla and iliocristale landmarks.

Skinfolds

Skinfolds are the most commonly measured anthropometric sites. They provide a good indication of the level of fatness located over the body in the subcutaneous storage areas. They can also be used to track changes in peripheral fat stores over time and with intervention programs. However, they are the most difficult measurements to take with accuracy and precision. Careful attention to the following guidelines and considerable practice are the best ways to improve your measurement skills.

Equipment and Measurement Technique

The Harpenden caliper has been used as the criterion instrument by ISAK. These calipers have a compression of 10 g/mm^2, are calibrated to 40 mm in 0.2 mm divisions, and may be accurately interpolated to the nearest 0.1 mm. The instructions in this chapter are for the Harpenden caliper. It is appropriate to apply skinfold data to a regression equation only if one has used the same calipers as those with which the regression equation was derived.

An anthropometric tape is required to locate a number of skinfold sites. The recommended tape is a Lufkin Executive Diameter tape (W606PM). If any other type of tape is used, it should be nonextensible and flexible, should be no wider than 7 mm, and should have a stub (blank area) of at least 3 cm before the zero line.

The skinfold is raised by the left hand, which is positioned so that the thumb points downward and the back of the hand is in full view of the measurer. If you are left hand dominant it is possible to use your right hand to raise the skinfold, but it is critical that you place the caliper faces in precisely the same spot as a right-hand-dominant person would, as described here.

To raise the skinfold, the measurer uses the thumb and index finger of the left hand to pick up a fold that contains a double layer of both adipose tissue and skin. In order to eliminate muscle, the finger and thumb roll the fold slightly, thereby also ensuring that there is a sufficiently large grasp of the fold. Note: It is important to hold the fold throughout the measurement procedure.

The near edges of the finger and thumb, in line with and straddling the landmark, raise the fold in the direction specified for each site. The fingers pull the fold slightly away from the body; this gives the calipers a better grip on the fold and reduces the chance of their slipping off.

Hold the calipers in the right hand with the fingers operating the movable arm. A full sweep of the needle is 20 mm, and this is reflected on the small scale on the caliper face. Before using the caliper, make sure that the needle is on zero. Rotation of the outer ring of the caliper is used to adjust the position of the caliper dial directly under the needle.

The calipers are applied to the fold so that there is 1 cm between the near edge of the fingers and the nearest edge of the caliper face. The caliper needs to be placed high enough that it is not in contact with the nonfold surface, which might grip and exaggerate the size of the fold, but not so close to the top of the fold that it might slip off. It is important to remember that the calipers are always applied at right angles to the fold.

The reading of the dial to 0.1 mm is made 2 s after the complete release of the caliper trigger. In the case of large skinfolds, the needle may still be moving at this point. The measurement is nevertheless recorded at this time.

Specific Skinfolds

In general, all measurements are made on the right side of the body. In some studies, however, bilateral body changes or development may be of interest, in which case one will take measurements on both sides of the body.

PROTOCOLS: Skinfolds

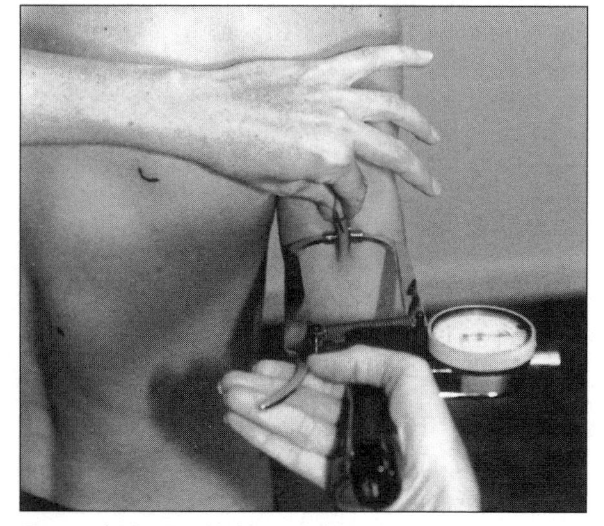

Figure 5.10 Triceps skinfold.

Triceps

This skinfold is raised with the left thumb and index finger on the marked posterior mid-acromiale-radiale line. The fold is parallel to the line of the upper arm (figure 5.10).

Subscapular

This skinfold is raised with the left thumb and index finger at the marked site 2 cm along a line running laterally and obliquely downward from the subscapulare landmark at an approximate 45° angle as determined by the natural fold lines of the skin (figure 5.11).

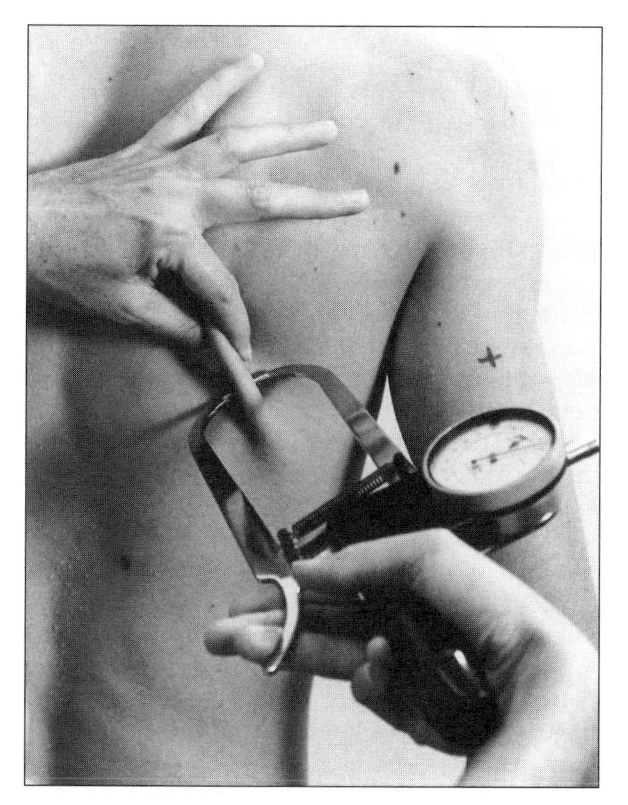

Figure 5.11 Subscapular skinfold.

Biceps

This skinfold is raised with the left thumb and index finger on the marked mid-acromiale-radiale line so that the fold runs parallel to the axis of the upper arm (figure 5.12).

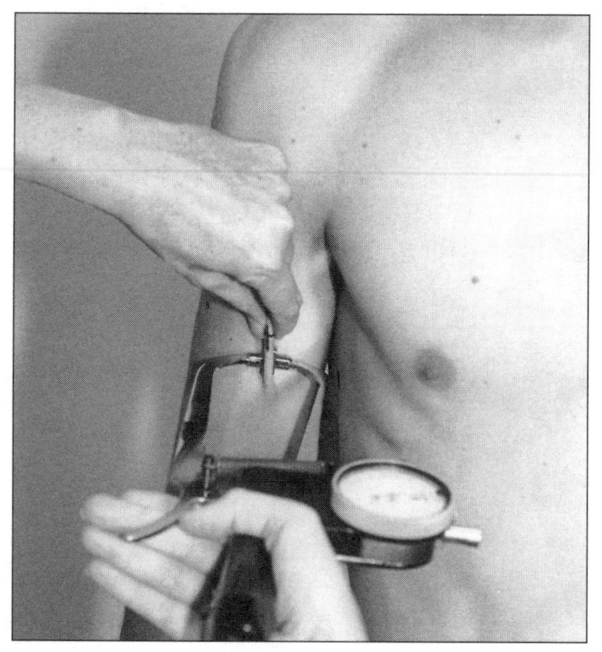

Figure 5.12 Biceps skinfold.

Iliac Crest

This skinfold is raised immediately superior to the iliac crest on the ilio-axilla line. The subject abducts the right arm to the horizontal. Align the fingers of your left hand on the iliocristale landmark, and exert pressure inward so that the fingers roll over the iliac crest. Substitute the left thumb for these fingers, and relocate your index finger a sufficient distance superior to the thumb that the grasp becomes the skinfold to be measured. The fold runs slightly downward to the medial aspect of the body (figure 5.13).

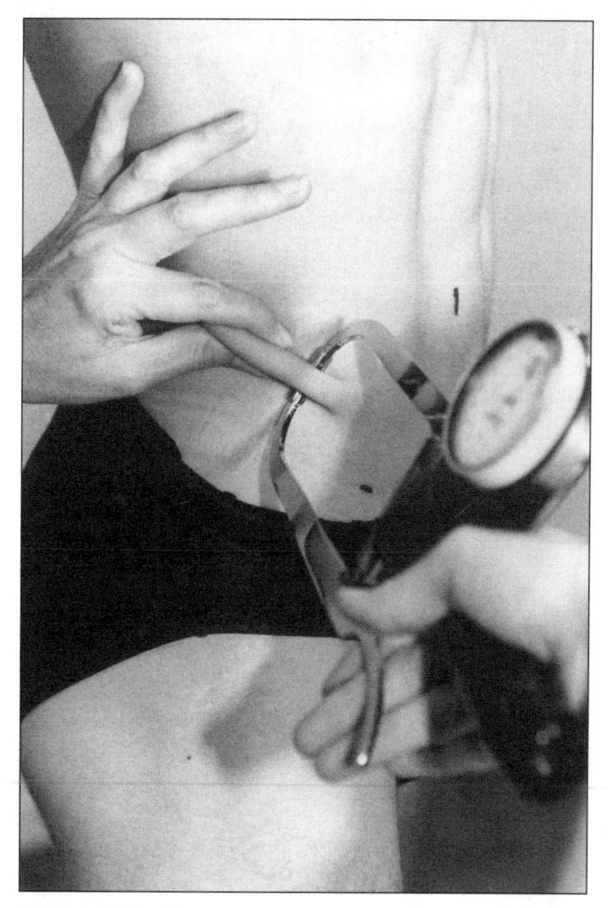

Figure 5.13 Iliac crest skinfold.

Supraspinale

This fold is raised at the point where the line from the iliospinale landmark to the anterior axillary border intersects at the horizontal level of the superior border of the ilium. This is about 5-7 cm above the iliospinale, depending on the size of the subject. The fold runs medially downward at about a 45° angle (figure 5.14).

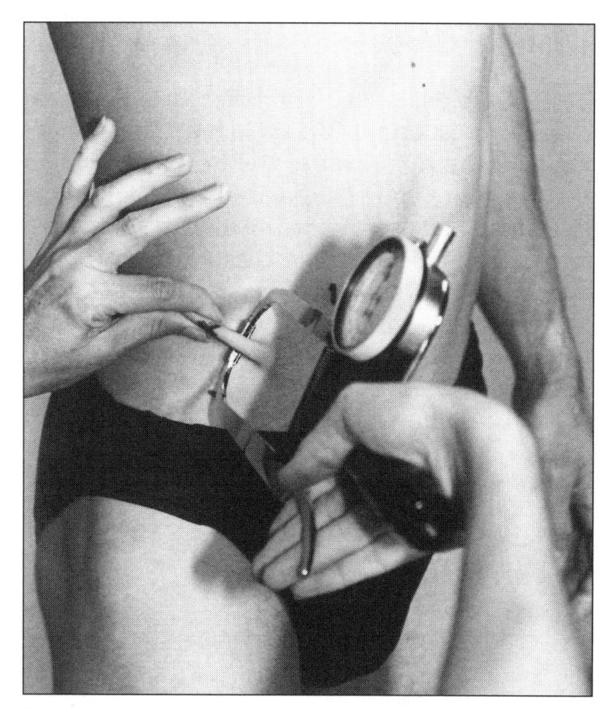

Figure 5.14 Supraspinale skinfold.

Abdominal

This is a vertical fold raised 5 cm from the right-hand side of the omphalion. The *omphalion* is the midpoint of the navel (figure 5.15). Note: It is important not to place the calipers or fingers inside the navel.

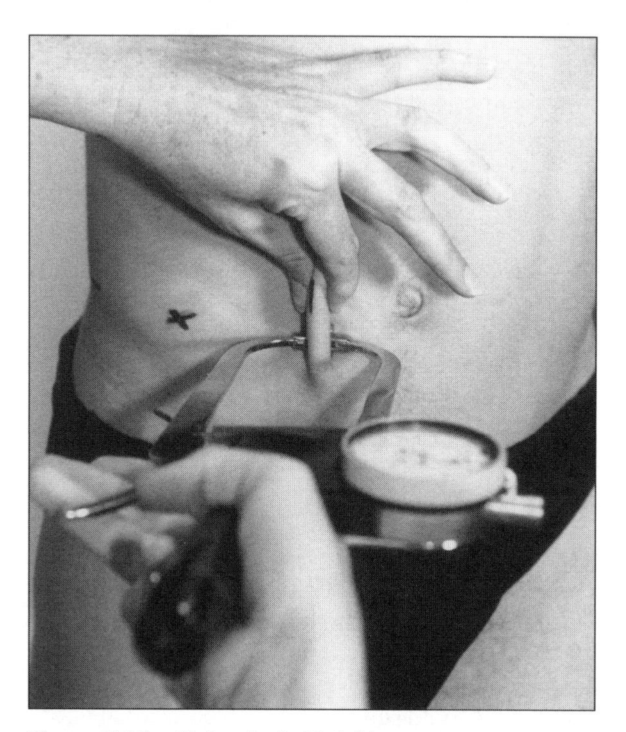

Figure 5.15 Abdominal skinfold.

Front Thigh

The measurer stands facing the right side of the subject on the lateral side of the thigh. The subject's knee is bent at right angles by having the subject place the right foot on a box or by being seated. The site is marked parallel to the long axis of the femur at the midpoint of the distance between the inguinal fold and the superior border of the patella (while the leg is bent at 90°). The skinfold measurement can be taken with the leg straight and resting on a box or with the knee bent. If the fold is difficult to raise, the subject may be asked to extend the knee joint slightly by moving the foot forward to relieve the tension of the skin. If the difficulty remains, the subject may assist by lifting the underside of the thigh to relieve the tension of the skin. As a last resort for subjects with particularly tight skinfolds, a recorder (standing at the medial aspect of the subject's thigh) can assist by raising the fold using two hands so that there is about 6 cm between the fingers of the right hand raising the fold (at the correct anatomical landmark) and the left hand, which raises a distal fold. The calipers are then located in between the recorder's hands, 1 cm from the thumb and forefinger of the recorder's right hand (figure 5.16).

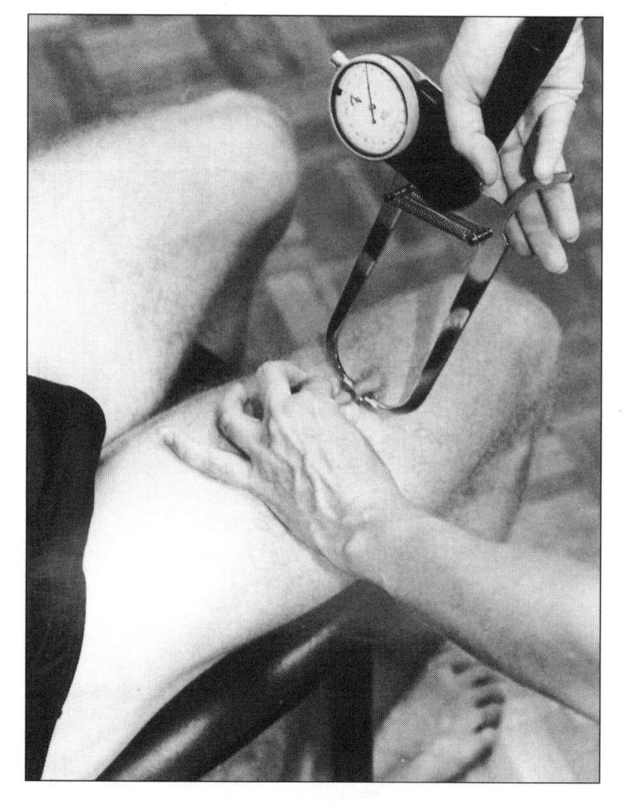

Figure 5.16 Front thigh skinfold.

Medial Calf

The subject either is standing with one foot on a box or is seated; the knee is flexed at 90°, and the calf is relaxed. The vertical fold is raised on the medial aspect of the calf at a level where it has maximal circumference. The maximal circumference will be determined during measurement of girths, and this level must be marked on the medial aspect of the calf during this process (figure 5.17).

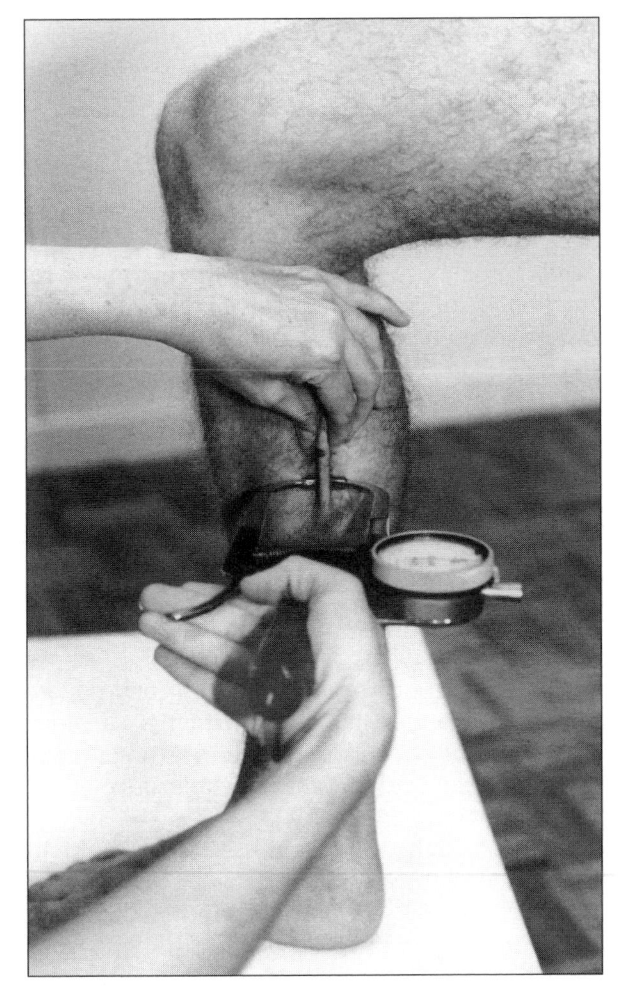

Figure 5.17 Medial calf.

Mid-Axilla

This is a vertical fold raised in the mid-axillary line at the level of the xiphoid process (Norton et al. 1996). Although this skinfold may not be routinely used, it illustrates the fact that less standard skinfolds may be effective provided that they can be exactly located in terms of landmarks (figure 5.18).

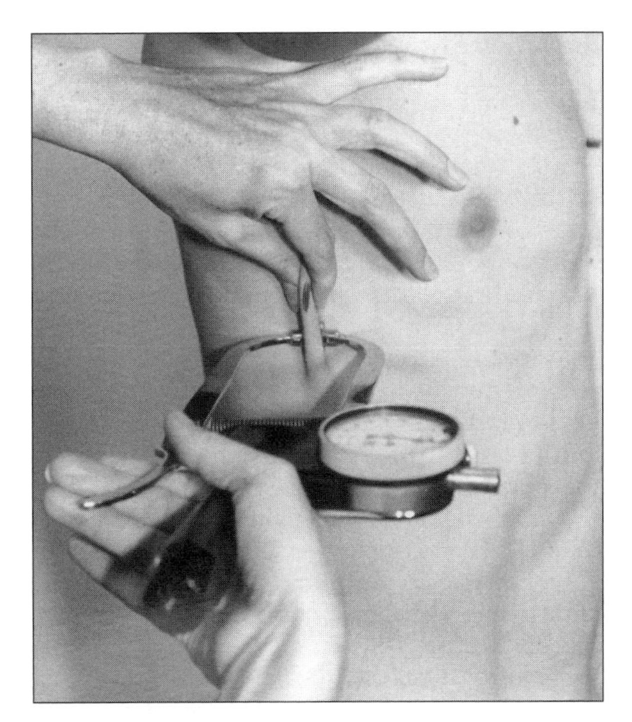

Figure 5.18 Mid-axilla skinfold.

Pectoral/Juxta-Nipple (Men Only)

The skinfold is an oblique fold located 5 cm from the right nipple in a line running from the nipple to the left acromion process (figure 5.19). This site was established by Withers et al. (1987) as the best single predictor of total body fat in the Australian and state-level athletes they tested.

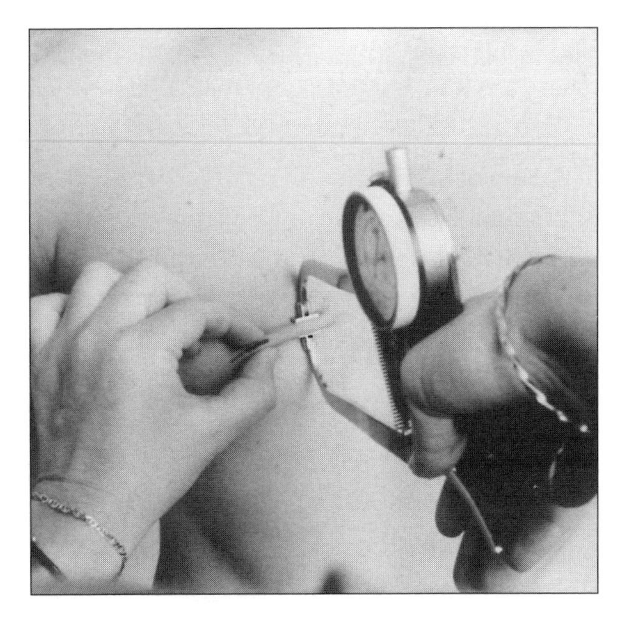

Figure 5.19 Pectoral/juxta-nipple skinfold.

Sum of Skinfolds

Rather than adopt the doubly indirect method of estimating percent body fat from body density estimates that are in turn derived from skinfold measurements, it is now common practice to monitor body mass and sums of skinfolds of an athlete through various training phases. The normative data presented in this chapter (see tables 5.1 and 5.2) are those collected on elite South Australian athletes and are to be considered only as a guide to optimal values. One can anticipate that as more data are collected and pooled, these skinfold sums will be modified accordingly.

A number of skinfold sums can be obtained from the 10 sites listed in the previous section. The sites selected will depend upon the needs of the anthropometrist, coach, and athlete. However, the fundamental principle of the Laboratory Standards Assistance Scheme (LSAS) is to use skinfolds that are in agreement with those recommended by ISAK. Since July 1996 the following seven sites have been used on all athletes:

- Triceps
- Subscapular
- Biceps
- Supraspinale
- Abdominal
- Front thigh
- Medial calf

Collectively, these seven sites are known as the sum of seven skinfolds and are recommended by the LSAS as the standard to be used for men and women athletes. All sites meet ISAK specifications.

Historically some athletes have also been profiled using skinfold totals from 8, 9, or 10 sites. These are the variations from the recommended 7 sites:

1. **Sum of 8 skinfolds A**
    - Sum of 7 skinfolds + iliac crest
    - Suitable for male and female athletes
2. **Sum of 8 skinfolds B**
    - Sum of 7 skinfolds + axilla
    - Suitable for male and female athletes
3. **Sum of 9 skinfolds**
    - Sum of 7 skinfolds + iliac crest + axilla
    - Suitable for male and female athletes
4. **Sum of 10 skinfolds**
    - Sum of 7 skinfolds + iliac crest + axilla + pectoral
    - Suitable for male athletes only
    - Allows all subsets of skinfold sums to be derived and population-specific body fat equations to be calculated, given the limitations of this method (e.g., Withers et al. 1987)

Girths

Girth measurements are a useful part of anthropometric profiling. They provide information about changes in segment/trunk volumes and, when combined with skinfold measurements, can also indicate relative changes in body composition. For example, increased limb girths without increased overlying skinfold thicknesses indicate lean tissue increases.

Equipment and Method

Measurement requires a flexible steel tape calibrated in centimeters with millimeter gradations. The tape needs to be enclosed in a case with automatic retraction. It should be 1.5 to 2.0 m long with a stub (blank area) of about 7 cm before the zero mark.

The Lufkin (W606PM) is the preferred tape. Fiberglass tapes require regular calibration, as they may stretch over time. Although constant-tension tapes may be available, one can achieve equal reliability with practice using the Lufkin tape.

The cross-hand technique is used for measuring all girths; and the reading is taken from the tape where, for easier viewing, the zero is located more lateral than medial on the subject.

In measuring girths, hold the tape at right angles to the limb or body segment being measured; the tension of the tape must be constant. To achieve constant tension, ensure that there is no indentation of the skin and that the tape holds its place at the designated landmark.

To position the tape, hold the case in the right hand and the stub in the left. Facing the body part to be measured, pass the stub end around the back of the limb and take hold of the stub with your right hand, which then holds both the stub and the casing. At this point your left hand is free to place the tape at the correct level.

Apply sufficient tension to the tape with the right hand to hold it at that position while you reach with your left hand underneath the casing to take hold of the stub again with your left hand. The tape is now around the part to be measured.

The middle fingers of both hands are free to exactly locate the tape at the landmark for measurement and to orient the tape so that you can easily read the zero. The juxtaposition of the tape ensures contiguity of the two parts of the tape from which the girth is determined.

Table 5.1 Normative Data for the Sum of Seven Skinfolds for National-Level Australian Female Athletes

Sport		Sum 7 skinfolds (mm)			
		Mean	SD	Range	Count
Athletics	SASI jumps	61.1	12.7	41.7–72.8	4
	SASI throws	95.3	49.4	53.0–203.7	9
	SASI race walking	76.8	14.0	62.8–90.8	2
	SASI sprint	60.3	11.9	45.1–83.9	7
	SASI middle distance	59.2	19.6	37.4–110.6	20
	SASI long distance	51.3	8.8	40.4–68.3	6
	SASI heptathlon	75.0	11.4	63.6–86.3	2
Badminton	SASI senior	89.2	14.1	62.1–120.0	34
	National junior	101.2	24.4	60.8–146.4	17
Basketball	SASI U20	90.0	16.2	62.8–135.3	64
	National senior	70.5	21.8	45.5–98.6	3
Cricket	National senior	90.8	19.7	55.9–141.1	27
Cycling, road	National senior	61.9	12.0	33.8–89.5	32
Diving	SA junior squad	80.1	23.8	40.6–119.2	7
Gymnastics	SASI elite	37.9	6.1	27.4–57.6	68
Hockey	SASI senior	87.4	18.5	48.1–140.3	57
	SASI talent	86.2	23.7	49.3–195.8	92
Kayaking	SASI senior	83.1	15.6	65.0–112.7	20
Lacrosse	SASI senior	89.3	16.6	67.5–132.8	17
	National senior	98.9	35.9	68.1–212.2	12
Netball	SA senior	83.4	17.3	51.5–124.0	33
	SA U21	97.9	29.3	56.6–168.4	31
	SA U19	92.6	24.9	49.4–187.4	48
	SA U17	87.6	18.9	49.7–132.1	51
Orienteering	SASI squad	89.0	16.0	55.5–132.1	21
Rowing	SASI lightweight	73.4	13.4	55.5–105.2	24
	SASI heavyweight	87.5	17.8	60.7–119.4	30
Shooting	SASI senior	137.9	2.1	135.2–140.3	3
Speed skating	SASI senior	73.3	11.9	58.6–87.7	3
Softball	SASI senior	99.1	21.5	56.2–153.5	51
	SASI U19	100.4	23.9	56.8–156.8	39
	SASI U16	94.8	18.9	62.9–157.2	37

(continued)

Table 5.1 *(continued)*

Sport		Sum 7 skinfolds (mm) Mean	SD	Range	Count
Squash	SASI squad	114.3	30.2	80.4–187.8	10
Swimming	SASI senior	80.4	18.6	45.3–133.1	63
Table tennis	National junior	109.2	24.1	74.8–149.5	11
Tennis	SASI junior	90.5	24.1	53.5–151.0	23
Triathlon	SASI senior	73.1	11.9	52.4–91.9	9
Volleyball	SASI senior	90.5	25.1	35.8–147.1	29
Water polo	SASI senior	124.3	42.7	58.1–220.6	21
Weight lifting	SASI squad	80.5	12.4	58.2–95.4	6

1. Body composition results for female squads 1987-1992.

2. Sum of seven skinfolds = triceps, subscapular, biceps, supraspinale, abdominal, front thigh, medial calf.

3. SA = South Australia, SASI = South Australian Sports Institute.

Reprinted, by permission, from S. Woolford, P. Bourdon, N. Craig, and T. Stanef, 1993, "Body composition and its effects on athletic performance," *Sports Coach* 16 (4): 24-30. Courtesy The South Australian Sports Institute.

Table 5.2 Normative Data for the Sum of Seven Skinfolds for National-Level Australian Male Athletes

Sport		Sum 7 skinfolds (mm) Mean	SD	Range	Count
Athletics	Decathlon	37.7	2.50	35.8–42.7	6
	Javelin	62.8	1.91	61.4–64.1	2
	Jumps	37.3	4.75	32.7–42.2	3
	Long distance	50.1	11.80	37.3–71.0	6
	Middle distance	42.0	11.53	28.3–79.1	24
	Pole vault	43.2	6.24	37.2–55.1	6
	Sprints	46.8	5.39	39.6–52.6	4
Australian rules badminton	National U17	67.1	13.88	41.9–94.5	20
	Senior	59.6	30.70	32.6–201.3	46
	National junior	53.4	10.37	37.1–72.3	21
Basketball	Junior U20	62.8	19.18	36.7–145.0	68
Boxing	Senior	62.3	16.82	38.6–100.6	13
	Junior	45.4	10.96	35.7–63.2	8
Cricket	Shield	70.6	29.37	33.0–193.0	46
	SA U19	61.2	17.71	36.0–101.7	11
	AIS academy	62.2	19.11	33.2–120.5	96
	Australian squad	69.1	21.26	46.6–121.0	23

Sport		Sum 7 skinfolds (mm)			
		Mean	SD	Range	Count
Cycling	Junior	46.2	8.00	30.7–59.9	21
	Nat jnr track	46.9	5.62	38.3–60.5	19
	Nat snr road	51.2	9.78	37.7–76.9	24
	Nat snr track sprint	54.9	10.60	36.7–71.5	13
	Nat snr track endurance	46.1	8.53	29.7–72.5	46
Diving	SA junior squad	69.0	30.41	39.1–117.7	6
Gymnastics	Elite	38.7	7.14	24.3–61.8	63
Hockey	SA U21 squad	52.6	15.46	34.4–96.6	22
	Talent	56.8	17.30	34.6–105.2	113
Jockey	Talent	41.4	8.29	26.5–59.7	28
Kayaking	National U18 squad	52.5	11.44	42.2–83.9	14
	Senior	51.6	12.39	33.2–79.7	52
Orienteering	Squad	48.7	6.96	35.6–60.8	16
Rowing	Lightweight	41.6	7.07	31.4–64.2	34
	Heavyweight	59.7	17.60	35.1–109.9	30
Rugby union	Senior	89.1	37.09	40.1–196.2	101
	Talent	81.9	36.20	40.7–197.2	46
Shooting	National senior	77.8	25.15	55.8–119.1	6
Soccer	SA U15 squad	56.9	17.34	36.9–101.2	24
	Talent U18	53.8	10.18	33.6–75.4	28
Speed skating	Senior	49.4	4.02	44.9–55.5	6
Squash	Squad	56.0	9.22	41.4–74.2	23
Swimming	Senior	51.9	10.93	32.4–91.9	59
	Junior camp U13	59.0	16.40	40.0–89.6	23
	Junior elite U15	41.4	6.35	34.1–50.8	6
	Club	52.8	14.46	33.3–90.2	27
Table tennis	National junior	76.2	17.50	41.4–106.3	20
Tennis	Junior	49.8	13.92	34.0–97.6	33
Triathlon	Senior	45.5	7.84	36.2–64.8	12
Volleyball	SASI senior	49.3	11.99	32.5–70.6	16
	SA U19 squad	57.1	23.34	31.3–108.8	16
	SA U17 squad	55.2	18.35	36.5–111.9	16
Weight lifting	SASI squad	74.9	34.45	33.9–190.2	47
Water polo	SASI senior	54.9	10.68	38.3–77.9	12
	SASI junior	66.7	42.56	37.6–203.7	13

1. Body composition results for male squads 1987-1992.

2. Sum of seven skinfolds = triceps, subscapular, biceps, supraspinale, abdominal, front thigh, medial calf.

3. SA = South Australia, SASI = South Australian Sports Institute.

Reprinted, by permission, from S. Woolford, P. Bourdon, N. Craig, and T. Stanef, 1993, "Body composition and its effects on athletic performance," *Sports Coach* 16 (4): 24-30. Courtesy The South Australian Sports Institute.

For reading the tape, the measurer's eyes must be at the same level as the tape to avoid any parallax error.

Technique

All measurements are made on the right side of the body.

PROTOCOLS: Girths

Arm Girth—Relaxed

The subject stands with the arm relaxed and hanging to the side. The girth is measured at the mid-acromiale-radiale level (figure 5.20).

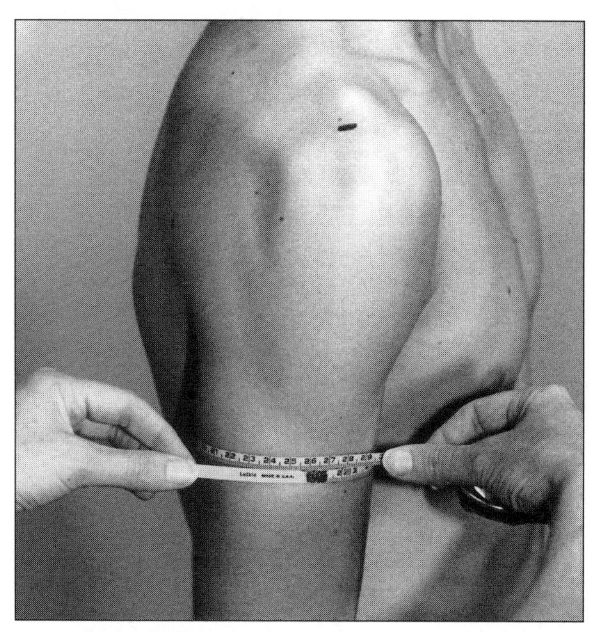

Figure 5.20 Relaxed arm girth.

Arm Girth—Flexed and Tensed

This is the maximum circumference of the right upper arm, which is raised laterally to the horizontal with the forearm at about 45° to the upper arm. The measurer stands to the rear of the subject and with the tape in position asks the subject to partially flex the biceps to identify the point where the girth will be maximal. Loosen the tension on the casing end and then instruct the subject: "Clench your fist, bring your hand toward your shoulder, keeping your elbow at 45°, and fully tense the biceps and hold it." Make the measurement while the subject is doing this (figure 5.21).

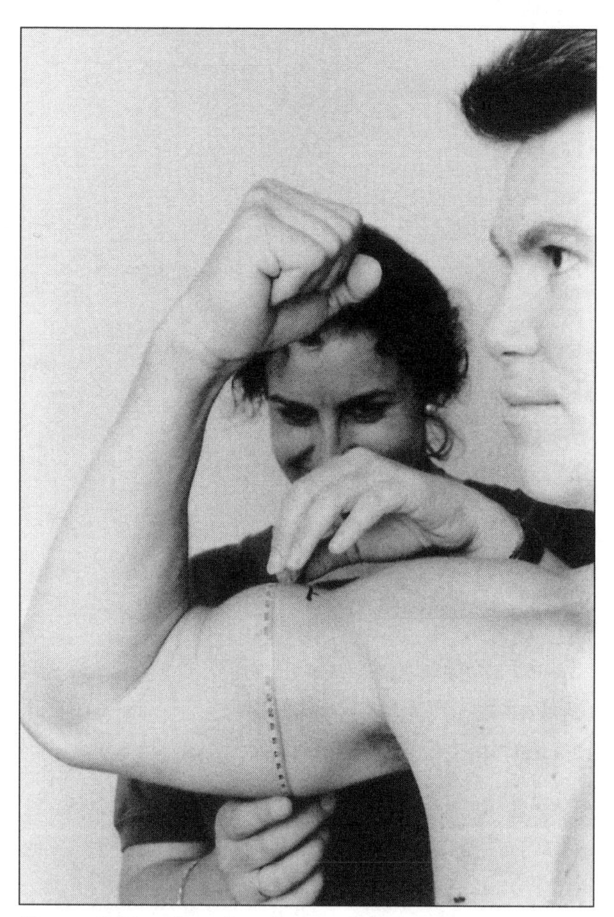

Figure 5.21 Flexed and tensed arm girth.

Waist Girth—Minimal

The location of this measurement is at the level of the trunk where the girth is minimal; this is the location where there is a noticeable indentation of the trunk when viewed from the front. If there is no such indentation, the measurement is made at the level that is midway between the lowest rib (laterally) and the iliocristale landmark (figure 5.22).

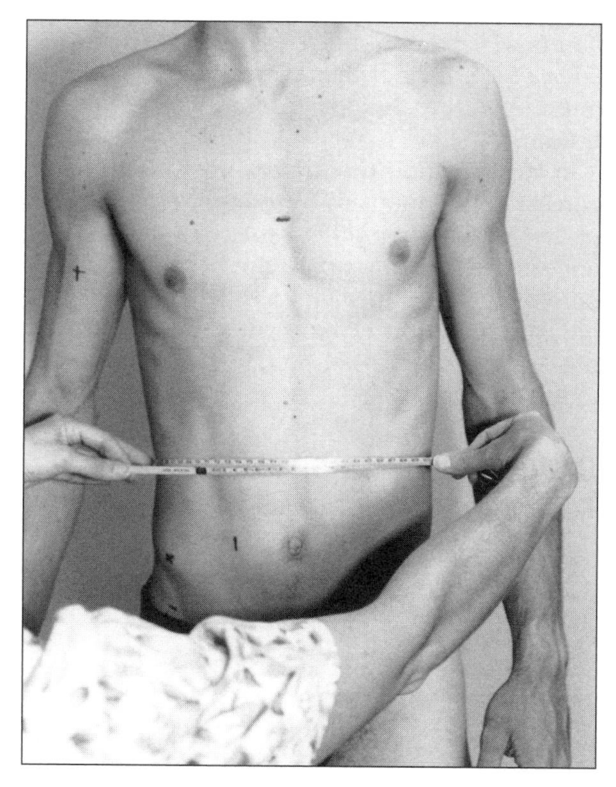

Figure 5.22 Waist girth.

Gluteal Girth—Maximal

Sometimes referred to as the hip girth, this measurement is made at the level where there is greatest protuberance of the gluteals. The subject must stand with both feet together. The measurer stands at the subject's side (figure 5.23).

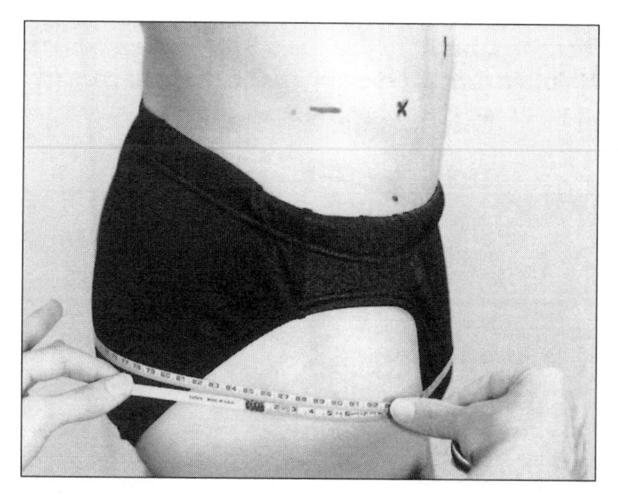

Figure 5.23 Gluteal girth.

Calf Girth—Maximal

The subject stands facing away from the measurer in an elevated position (e.g., on a box or low table) with the weight equally distributed on both feet. The elevated position will make it easier for the measurer to align his or her eyes with the tape. Place the tape around the calf in the prescribed manner. Find the maximal girth by using the middle fingers to manipulate the position of the tape in a series of up-or-down measurements to identify the maximal girth. Mark this point medially in preparation for skinfold assessment (figure 5.24).

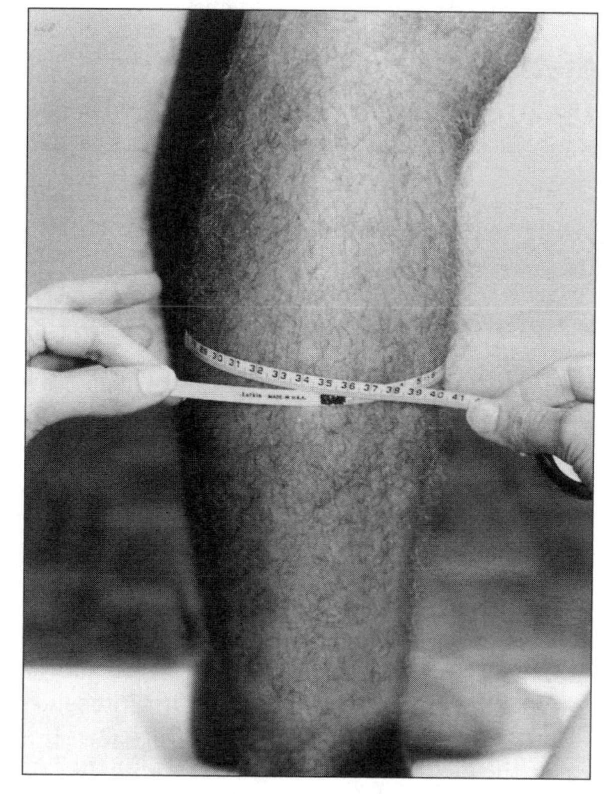

Figure 5.24 Calf girth.

Limb Breadths (Widths)

It is often forgotten that bone, like other tissues, can increase and decrease in size and density over time. Factors such as exercise training and nutrition may impact on these changes. The bony breadths in this chapter represent two sites that have been commonly measured in athlete assessments.

Equipment

The Mitutoyo adapted engineers' caliper is the preferred equipment for these measurements. These are vernier calipers to which longer arms have been added; the longer arms can encompass the bi-epicondylar widths of the femur and humerus.

Alternatives to the Mitutoyo are the large sliding caliper that is part of the Siber-Hegner anthropometer and the Harpenden bone calipers. The Harpenden bone calipers are easy to use, but the measurement scale may be less accurate than with the Mitutoyo, especially if the arms become loose. The Siber-Hegner caliper is more cumbersome to handle over relatively small breadths.

Method

All the equipment mentioned here is held in the same way. The calipers lie on the backs of the hands while the thumbs rest against the inside edge of the caliper arms and the extended index fingers lie along the outside edges of the arms.

In this position the fingers are able to exert considerable pressure to reduce the thickness of any underlying soft tissue, and the middle fingers are free to palpate the bony landmarks on which the caliper faces will be placed. The measurements are made with the calipers in place, with the pressure maintained along the index fingers.

Definitions and Specific Techniques

All measurements are made on the right side of the body.

PROTOCOLS: Limb Breadths

Humerus Breadth (Width)

Definition: The distance between the medial and lateral epicondyles of the humerus when the arm is raised forward to the horizontal and the forearm is flexed at right angles to the upper arm.

Location: With the calipers gripped correctly, use your middle fingers to palpate the epicondyles of the humerus, starting proximal to the sites. The bony points first felt are the epicondyles. The calipers are placed directly on the epicondyles so that the arms of the calipers point upward at a 45° angle to the horizontal plane. Maintain firm pressure with the index fingers as you read the value. Because the medial epicondyle is lower than the lateral epicondyle, the measured distance may be somewhat oblique (figure 5.25).

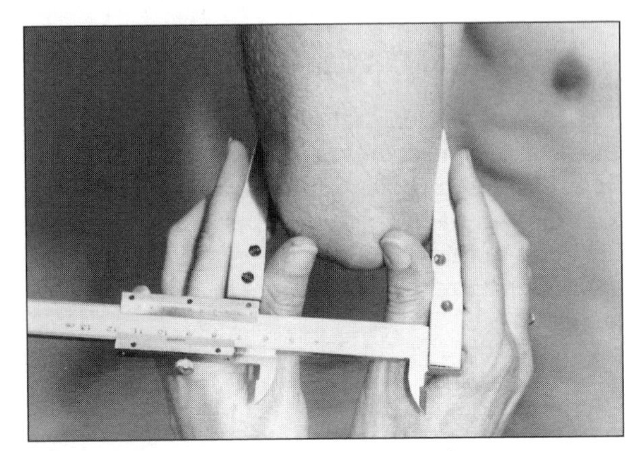

Figure 5.25 Humerus breadth (width).

Femur Breadth (Width)

Definition: The distance between the medial and lateral epicondyles of the femur when the subject is seated and the leg is flexed at the knee to form a right angle with the thigh.

Location: With the subject seated and the calipers in place, use the middle fingers to palpate the epicondyles of the femur beginning proximal to the sites. The bony points first felt are the epicondyles.

Place the caliper faces on the epicondyles so that the arms of the calipers are near horizontal. Maintain firm pressure with the index fingers until you read the value (figure 5.26).

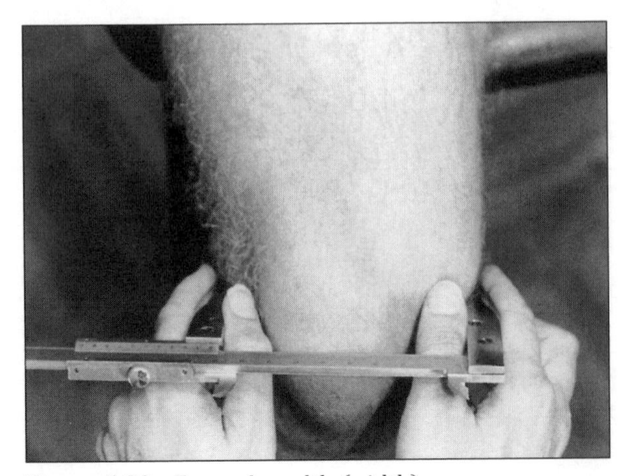

Figure 5.26 Femur breadth (width).

Technical Error of Measurement

In 1940, Dahlberg became the first to propose that for quantifying the errors associated with a particular measurement procedure, the ideal approach is to make an infinite number of measures on a single individual to obtain a normal distribution of values with a mean ($\bar{X}$) and standard deviation (σ_i, where the subscript "i" refers to the individual). By probability, there is a 95% likelihood that any future single measure for that individual will lie within the mean $\pm 2\sigma_i$. The standard deviation is a particularly useful indicator of dispersion about the mean, since σ has units that are the same as those of the original measure.

If an infinite number of *paired* measures are made on an infinite number of subjects, Dahlberg (1940) showed that the differences between the two series of measures would have a mean of zero and that the standard deviation of the *differences* (σ_d) would be equal to $\sigma_i \times \sqrt{2}$. Or, rearranging this equation, the standard deviation of an individual measure is $\sigma_i = \sigma_d / \sqrt{2}$.

The ideal approach of infinite measures or infinite subjects is impossible, so Dahlberg (1940) proposed the use of duplicate measures on a group of subjects to approximate the "standard error of a single determination," calculated as $\sigma_d / \sqrt{2}$. The term technical error of measurement (TEM) is now used to describe the same calculation, which Knapp (1992) traces to Utermole et al. (1983).

The textbook definition of the standard deviation (σ) is $\sigma = \sqrt{\Sigma di^2 / n}$, and the definition of the standard error of the mean (SEM) is $SEM = \sigma / \sqrt{n}$.

By simple rearrangement one can show that the TEM (also called the "standard error of a single determination") is described by $TEM = \sqrt{\Sigma di^2 / 2n}$.

Although use of the term TEM is somewhat restricted to the field of anthropology and anthropometry (Knapp 1992), in other disciplines, including exercise science (Baumgartner 1989), the term standard error of measurement (SEM) has been adopted. The SEM is calculated as:

$$SEM = \sigma\sqrt{1 - r}$$

where r is the intraclass correlation coefficient and should not be confused with the Pearson r that is the interclass correlation coefficient (see Baumgartner 1989). Note that TEM and SEM are identical and that they can be used interchangeably. However, in this book the term TEM is used exclusively because, compared to that for the SEM, the calculation of TEM (discussed in the next section) is relatively easy to understand for the case of two groups of subjects (on next page see "Technical Error of Measurement Calculation Example").

Irrespective of the name, the TEM (or SEM) gives an indication of the precision (or rather imprecision) associated with the measure—that is, the error of the method due to *both* biological and technical factors. The TEM is a value in the same units as the variable measured, and it indicates that the error of a single measure will be plus or minus the TEM two thirds of the time. The interpretation of TEM is described later on in this chapter; a more comprehensive description of TEM is provided by Pederson and Gore (1996). In terms of anthropometry, the TEM is usually determined by making double measures on a sample of about 20 subjects, and it is important to select a range of physiques similar to that for the subjects you intend to measure most of the time (e.g., body builders, marathoners, children, persons who are elderly). The precision of a trainee anthropometrist can also be established by comparing measures made by the trainee with those of a *criterion* anthropometrist (that is, one who does not make any systematic error).

Technical Error of Measurement Calculation

This section presents absolute and relative TEM calculations. In most cases we use the relative TEM in our analysis, although the absolute TEM is required in the first step. These calculations are best done using a spreadsheet or computer program such as LifeSize (see the appendix).

- **Absolute TEM.** The TEM is calculated as follows:

$$TEM = \sqrt{\Sigma di^2 / 2n}$$

- **Relative TEM.** The %TEM expresses the error as a percent of the means of the original paired measures. It is calculated as follows:

$$\%TEM = (TEM/[M1 + M2]/2) \times 100$$

M1 is the mean of the first series of measurements;

M2 is the mean of the second series of measurements.

When to Calculate Technical Errors of Measurement

The TEMs are calculated to determine the accuracy among team members or between trainees

Technical Error of Measurement Calculation Example

Triceps skinfold

Subject	Series X1	Series X2	X1 – X2	d^2
1	9.9	9.8	0.1	0.01
2	5.5	5.3	0.2	0.04
3	20.2	19.2	1.0	1.00
4	11.1	11.5	–0.4	0.16
5	29.2	29.2	0.0	0.00
6	20.0	20.3	–0.3	0.09
7	20.2	19.2	1.0	1.00
8	27.2	26.8	0.4	0.16
9	14.1	14.5	–0.4	0.16
10	8.1	7.9	0.2	0.04
11	11.1	11.5	–0.4	0.16
12	13.7	13.1	0.6	0.36
13	31.8	32.7	–0.9	0.81
14	16.6	16.3	0.3	0.09
15	5.2	5.3	–0.1	0.01
16	12.5	12.1	0.4	0.16
17	28.1	27.5	0.6	0.36
18	11.4	12.2	–0.8	0.64
19	8.6	8.4	0.2	0.04
20	5.2	5.2	0.0	0.0

Calculation

$n = 20$

M1 = Mean of measurement series X1 = 15.485

M2 = Mean of measurement series X2 = 15.400

Sum $d^2 = 5.29$

$$\text{TEM} = \sqrt{(5.29/40)}$$
$$= 0.36 \text{ mm}$$

$$\%\text{TEM} = (\text{TEM}/[\text{M1} + \text{M2}]/2) \times 100$$
$$= 0.36/15.4425 \times 100$$
$$= 2.3\%$$

These results show that the error of a single measure for this measurer will be plus or minus 0.36 mm two-thirds of the time. The relative TEM is less than the 5% that is acceptable for skinfold measurements. This person is a very precise measurer of the triceps skinfold.

and the criterion anthropometrist. The absolute and relative TEMs can be determined for any of the following comparisons, using one of the methods listed for obtaining paired values.

1. Comparisons
   - Single measurer for one variable within or between measurement sessions
   - Single measurer for one group of variables within or between measurement sessions
   - Between measurers for a selected variable
2. Paired values
   - Value 1 with value 2
   - Mean of values 1 + 2 with mean of values 3 + 4
   - Median of values 1, 2, 3 with median of values 4, 5, 6
   - Mean of values 1, 2, 3 with mean of values 4, 5, 6

The generally accepted levels of precision are 5% relative TEM or less for skinfolds and about 1% for other anthropometry.

Interpretation of the Technical Error of Measurement

We now need to use this TEM data to decide whether the changes in the sum of skinfolds for an athlete are real or merely within our measurement error. When two sums measured by a single anthropometrist are compared, each has the same TEM. To evaluate the significance of the changes, one must remember the following:

$$\text{Standard Error of the Difference Between Two Measurements} = \text{TEM} \times \sqrt{2}$$

~68% Confidence Interval for the Error-Free Change = Difference ± Standard Error

~95% Confidence Interval for the Error-Free Change = Difference ± 2 Standard Error

It is critical to remember that hypothesis testing is simply a matter of "accept" or "reject" and does not involve interpreting magnitude. Therefore, one cannot compare the 68% level of confidence error-free change of one example with that of another.

Example A

Consider the case for an athlete whose skinfold total was initially 48 mm and then 6 weeks later was measured at 45 mm. The difference between the two measures is –3 mm. Earlier the anthropometrist who made these measures had established that their TEM for a sum of seven skinfolds was 2.0 mm. One should remember that both the first and second measure will have a TEM of 2.0 mm.

Step 1. Calculate the error-free change:

$$\text{Standard Error of Measurement} = \text{TEM } (2.0) \times \sqrt{2} = 2.8\text{mm}$$

Error-Free Change at ~68% Level of Confidence = (–3 + 2.8) to (–3 – 2.8) = –0.2 mm to –5.8 mm

Error-Free Change at ~95% level of confidence = (–3 + [2 × 2.8]) to (–3 – [2 × 2.8]) = 2.6 mm to –8.6 mm

Step 2. Conduct a hypothesis test to accept or reject the change as real:

- The change for the sum of seven skinfolds was –3 mm.
- At a ~68% level of confidence, –3 mm lies within the error-free change of –0.2 to –5.8. Therefore, the change between trial 1 and trial 2 for this athlete is accepted as a real change.
- At a ~95% level of confidence, the error-free interval includes 0. One of the possible error-free changes in the skinfold total is zero or no change. Therefore the change between trial 1 and trial 2 for this athlete is rejected and is not likely to be a real change.

Example B

The same anthropometrist, TEM = 2.0 mm, measured a cyclist's sum of seven skinfolds twice, and the difference between the two measures was –10.0 mm.

Step 1. Calculate the error-free change:

Error-Free Change at ~68% Level of Confidence = (–10 + 2.8) to (–10 – 2.8) = –7.2 mm to –12.8 mm

Error-Free Change at ~95% Level of Confidence = (–10 to [2 × 2.8]) to (–10 – [2 × 2.8]) = –4.4 mm to –15.6 mm

Step 2. Conduct a hypothesis test to accept or reject the change as real:

- At a ~68% level of confidence, –10 mm lies within the error-free change. Therefore the change between trial 1 and trial 2 for this athlete is accepted as a real change.
- At the ~95% confidence level, the error-free interval does not include 0. Therefore the change between trial 1 and trial 2 for this athlete is accepted as a real change.

Measurement Tolerance

The accepted anthropometric TEM values are as follows:

- Body mass, 0.1 kg
- Height (stretch), 3 mm
- Skinfolds, 5%
- Bone widths, 1 mm
- Girths (limbs), 2 mm
- Girths (trunk), 3 mm

These values are rather strict and are the ideal. Measurers should aim for individual precision and should calculate and report their TEMs for research papers.

chapter 6

Blood Sampling and Handling Techniques

Graeme Maw, Simon Locke, David Cowley, and Patricia Witt

In clinical settings, the examination of blood is well established. Pendergraph (1988) dates its diagnostic history to the 19th century—shortly before the publication of the first *Manual of Clinical Diagnosis* (Todd 1908)—but laments its lack of tradition compared to the letting of blood for therapeusis; this he dates to antiquity, citing the invention of the syringe in Egypt around 280 B.C.

In comparison, blood testing in exercise and sport science is relatively new. In the 1920s, Hill wrote of the connection between blood lactate concentration and the balance between aerobic and anaerobic metabolism (Hill and Lupton 1924), but it was not until the nationalistic promotion of athletics that blood tests became common in sport. Then, in the 1960s and 1970s, scientists experimented with the blood's oxygen-carrying capacity and proclaimed anaerobic threshold as the impetus for scientific training (Ekblom et al. 1972; Mader et al. 1976). Blood testing is now intrinsic in athlete preparation—whether for fitness testing, monitoring training, or medical diagnosis (see Urhausen and Kindermann 1992 for a review)—and is currently being debated for the detection of illegal doping practices (Munster 1993; Andersen 1995).

Unfortunately, in contrast to what occurs in the rigorous clinic, methods of blood sampling and handling in the exercise setting have perhaps not kept pace with their popularity; at the very least, they have not yet been documented for methodological scrutiny. Texts abound describing accepted practice for clinical blood sampling, handling, storage, transport, and disposal (e.g., Pendergraph 1988; Garza and Becan-McBride 1989; World Health Organization [WHO] 1993); standards exist within every hospital in accord with national health-related guidelines (e.g., New South Wales Department of Health 1987; National Health and Medical Research Council [NHMRC] 1996); and accreditation/qualification is generally required to perform such invasive procedures. However, this information has been slow to reach exercise scientists.

Blood sampling and handling in exercise and sport must overcome hardships not faced in clinical practice—that is, it must be flexible enough to operate in both laboratory and field settings. However, this should not be an excuse for hygienic or ethical malpractice. In light of recent concerns about viral infection, sport itself has developed codes of practice to deal with bloody situations (see Australian National Council on AIDS [ANCA] and Australian Sports Medicine Federation 1994), and blood testing was introduced by the International Skiing Federation and the International Amateur Athletic Federation only after very careful consideration. The Norwegian Confederation of Sports, which conducted blood testing at the 1994 Lillehammer Winter Olympic Games, considered the following points, among others, as central to the issue of blood tests in sport (Andersen 1995):

- Blood sampling requires invasive techniques that may be associated with pain and anxiety.
- Blood is not a waste product, so it should not be drawn indiscriminately.

- Blood sampling carries a risk of disease transmission unless proper procedures are employed.
- Medical personnel may be required to perform the blood sampling procedure.
- Blood testing increases the probability that disease will be detected.

In addition, discussion concerned matters of consent, specifically whether consent was implied by an athlete's voluntary participation in the scrutinized event or whether additional signed agreements were necessary.

It is clear that sampling and handling blood are now ingrained in the exercise setting and will probably expand to refine athlete preparation. However, the process should not be approached nonchalantly: hygienic practices must be observed, and ethical issues such as informed consent and confidentiality must be addressed. It is equally clear that these issues have largely been addressed in the clinical literature, which (logically) should therefore be incorporated into the exercise field.

This chapter presents information that is established and accepted clinically and is pertinent to blood sampling and handling in the exercise setting. It does not address decisions regarding when or what to test and how to interpret results, except with regard to ethical concerns. The focus instead is on sampling techniques and safe handling and disposal practices. Readers will be referred to required laboratory standards and to clinical texts containing further detail. In line with the National Committee for Clinical Laboratory Standards (NCCLS 1991), safe practice will emphasize prevention of hepatitis B (HBV) and transmission of human immunodeficiency virus (HIV), as this perhaps represents the worst-case scenario: HBV, for example, has an infectious concentration approximately 106 times that of even HIV and hepatitis C (NCCLS 1991). Finally, although discussion centers on blood handling and sampling practices, the recommendations should be applied to the handling of all body fluids, as HBV has been found in semen, urine, cerebrospinal fluid, saliva, and tissue as well as in blood (NCCLS 1991).

The Practitioner

As will be established, the safe handling and sampling of blood and body fluids is dependent more on correct practice than on surroundings or equipment. The competencies of the practitioner are therefore paramount in all this chapter's procedures.

Competencies

Although the skill required for and risks associated with blood sampling depend on the method employed, one should never forget that blood letting per se is an invasive procedure. The person responsible must therefore have all the requisite skills before being allowed to practice independently. Indeed, phlebotomy—the practice of drawing blood—is considered a distinct profession in the United States, where it is supported by its own professional organization, the National Phlebotomy Association.

Although academic qualifications in phlebotomy do not exist, clinical laboratories can be accredited, for example in Australia by the National Association of Testing Authorities, to take blood. Hence, their technicians are seen to be competent and come under the jurisdiction of the laboratory's code of practice. Accredited laboratories are also eligible to train and accredit fellow phlebotomists. Such in-house training courses are generally accessible on request to outside personnel; and, although they deal largely with venous puncture and cannulation, it is advisable for anyone required to sample capillary blood to learn safe practices also. Whether blood is taken by capillary or venous puncture is irrelevant—safety and handling procedures do not vary.

Training in phlebotomy covers many of the safety procedures discussed later in this chapter. However, as a professional exercise, it also implies a concomitant level of behavior. The American Society for Medical Technology includes in its code of ethics the statement that the technician will behave "at all times in a manner appropriate to the dignity of the profession" (Pendergraph 1988, p. 9). This includes maintaining honesty and confidentiality, appropriate personal appearance (especially with regard to cleanliness), and suitable communication skills: subjects must be informed of procedures at all times.

With appropriate training, it is acceptable for others besides medically qualified personnel to conduct capillary blood sampling. Similarly, venous sampling, by single venepuncture, is generally acceptable from skilled technicians. In reviewing procedures for doping control at the 1994 Lillehammer Olympics, the Norwegian Confederation of Sports concluded that trained bioengineers were actually more suitable venepuncture practitioners than doctors, as they were in continual practice whereas the involvement of physicians might be sporadic (Andersen 1995). Similarly, following a suitable course of accreditation, the exercise scientist in regular practice might be equally capable of performing venepuncture.

Venous cannulation, on the other hand, while appropriate for many exercise test procedures, is viewed less clearly in medical circles. While some institutions allow trained technicians to insert cannulas and draw blood, others insist that it is a medical procedure to be conducted only by physicians. In many countries there is no legislated guide; hence, it is recommended that each institution seek ethical guidance from the appropriate board or coopt medical assistance. The decision may ultimately be one of establishing legal liability. For example, many nurses carry professional indemnity designed to cover them should a subject pursue a grievance; senior hospital physicians are similarly covered and pass on this insurance when delegating responsibility to subordinates. An exercise or sport scientist is unlikely to carry medical indemnity but could be similarly covered if the institution's medical supervisor has vouched for his or her competence. One must consider the actual technicalities of drawing blood and potential emergency situations, such as dealing with anaphylactic shock.

Arterial blood sampling and venous or arterial infusion are always procedures for medical personnel and should never be attempted by those unqualified for the task.

Immunization

Practitioners, in addition to dealing with the safety and comfort of the subject, must also care for their own health and safety. Again, later sections of the chapter will deal with many of these issues (such as wearing gloves), but matters of immunization are worth considering before one attempts any actual blood handling. It does not appear possible to make immunization compulsory for blood-handling personnel, as most vaccines carry some risk of side effects (Dejonghe and Parkinson 1992). However, the WHO (1993) recommends checking all new clinical laboratory staff for immunization against diphtheria, HBV, measles, mumps, poliomyelitis, tetanus, tuberculosis, and, in the case of females, rubella (for protection during pregnancy); in addition, immunization against influenza and a recent chest x-ray are desirable. The check may take the form of a written questionnaire conducted in confidence by the institution's occupational health and safety unit and kept on record for future reference. Once this has been completed and staff have been offered the necessary immunizations and advised of the associated risks and benefits, boards for workers' compensation generally assume shared liability for any ensuing infection, even if the offer of immunization was declined.

General Safe Practice

Regardless of the setting in which blood is being handled, there are general practices that must be observed to minimize the risks to both the practitioner and the subject. Many of these are requisite to attaining clinical laboratory accreditation and should be documented in the laboratory policy manual (Garza and Becan-McBride 1989). It is worth noting that even in hospital settings, approximately 5% of patients acquire some form of infection in addition to that with which they entered (Castle 1980). Much of this transmission can be linked to microorganisms transported on human hands, clothing, or instruments. Therefore it is critical not to underestimate the need for diligence to eliminate similar transmission in the varied exercise setting.

At the same time, the WHO (1993) urges a sense of proportion: "Absolute safety is unattainable. . . . Well run laboratories are not specially dangerous workplaces, and one should avoid 'overkill' from disproportionately stringent precautions or excessive reliance on physical safeguards" (p. 15). Most exercise and sport science laboratories would be considered as posing only "Basic Biosafety Risk" (level 1; WHO 1993), requiring little specialist safety equipment. Ultimately, their safety cannot be guaranteed, and it is the behavior and practice of the user that will prevail.

The following lists precautions that all blood-handling personnel in the exercise setting should understand clearly and practice consistently. This will ensure good microbiological practice and facilitate smooth operations (after New South Wales Department of Health 1987; Pendergraph 1988; ANCA 1990; WHO 1993).

- All blood and serum specimens should be treated as potentially infectious.
- Gloves should be worn when one is handling blood specimens or items that may have been contaminated with blood or other body fluids; latex gloves have been shown to reduce the transfer of blood by 46% to 86% during needle-stick injuries (Mast et al. 1993). Gloves should be changed immediately if they become contaminated or physically or chemically damaged. Broken skin is

particularly susceptible to infection and therefore should be protected with extra vigor.

• Gowns should be worn if there is a possibility that clothing might be soiled with blood or other body fluids.

• Eye protection (glasses/goggles) should be worn if there is a possibility that blood or other body fluids might splash during handling, for example during removal of lids from containers.

• Hands should be washed thoroughly, using an accepted protocol (see "Handwashing Protocol," page 90), after removal of gowns and gloves. Hands should be washed immediately after having become contaminated with blood; before one leaves the blood-handling site for any purpose; before one eats, drinks, smokes, or applies makeup; before one changes contact lenses; or before one engages in any activity that entails contact with a mucous membrane.

• Hands should be kept away from the face, nose, eyes, and mouth, as should pens, pencils, and other objects, to prevent self-infection with infectious agents. Specimens or reagents should never be pipetted by mouth.

• Eating, drinking, smoking, or applying makeup should be prohibited in blood-handling areas. Food or beverages should not be stored in refrigerators used for specimen or reagent storage.

• Accidental wounds from sharp instruments should be avoided: such instruments should not be bent or resheathed (see later discussion of needle-stick injuries) but instead placed promptly in a rigid puncture-resistant container that is marked with a prominent biohazard symbol and used solely for such disposal. All instruments used to draw blood should be thus disposed of, avoiding the risk of reusing contaminated material.

• Accident reports should be made after any accident or work-related injury, particularly any injury resulting from needle punctures.

• Work areas should be decontaminated after spills and at the end of each work session using an accepted procedure. A 1% concentration of sodium hypochlorite prepared fresh daily is effective for cleaning up blood spills (see "Disinfecting Agents," page 90). Work areas should also be kept free of materials that are not pertinent to the specific task.

• If blood-collecting trays and underpads are disposable, they should be properly disposed of after use. If they are not disposable, they should be disinfected after spills and at the end of each work session, using 1% sodium hypochlorite.

• The laboratory should be covered by a pest control program; animals not involved in the work of the laboratory should not be permitted inside.

Informed Consent

As mentioned earlier, one of the skills of the practitioner should be communication, with the athlete being informed of procedures at all times. This extends, as with all biomedical research, to describing the process, risks, discomforts, and benefits of the blood test beforehand and allowing the athlete to make an informed decision to continue participation (WHO 1976). It is a fact that blood sampling without consent can be considered assault! The rationale and procedures for informed consent are covered fully in chapter 2.

In many instances, consent might be implied by a subject's participation in an organized training or testing program. For example, one could argue that an athlete participating in elite competition or training is obliged to submit to doping control that may include blood tests, and therefore has consented simply by joining the program (Andersen 1995); hospitals frequently operate on a similar premise of assumed consent. However, this does not preclude the need for information or the subject's right to withdraw at any time (WHO 1976). Thus, in exercise and sport, informed consent is considered essential from any athlete (or guardian if the athlete is a minor) who might be subjected to blood tests; permission from the coach is not considered sufficient.

The Norwegian Confederation of Sports concluded that before blood tests could be introduced in doping control, written informed consent should be gained from athletes entering the competition (Andersen 1995). Thus, the athlete should be informed of the test requirements before entering the competition or program or joining the sport organization; otherwise the voluntary nature of the consent to blood testing is compromised. For example, without prior consent an athlete joining a competition and then refusing a blood doping-control test would be treated as guilty even if the individual had been unaware of the requirement before entering the competition.

A similar approach would be feasible with athletes entering an organized sport program that is likely to involve blood testing. Informed consent detailing the process of blood sampling and handling, again approved by the relevant board

Handwashing Protocol

Here is the acceptable procedure for washing hands or exposed skin accidentally contaminated with blood or body fluids (after Pendergraph 1988):

1. Wash with a good liquid antimicrobial detergent soap; hand-wipe towelettes and cleansing foams are not considered sufficient, as they do not provide the necessary dilution and detergent actions and their use is not generally followed by rinsing.
2. Rinse well with water.
3. Apply a solution of 50% isopropyl or ethyl alcohol; leave on the skin surface for at least 1 min.
4. Wash again with the liquid soap and rinse with water.

Disinfecting Agents

Sodium hypochlorite (household bleach) is considered an intermediate-level disinfectant whose dilution can be manipulated, largely dependent on the nature of the contaminated surface. If the surface is porous and cannot be cleansed before disinfection, then a 1% solution may be necessary; if the surface is hard and smooth and has been adequately cleaned, then a 0.05% solution may suffice (the usual commercial concentration of household bleach is 5.25%); even at 0.05%, sodium hypochlorite has been shown to inactivate HBV and HIV in 10 and 2 min, respectively (NCCLS 1991). Throughout this chapter, sodium hypochlorite is discussed as a 1% solution, as this is the concentration recommended for disinfecting biologically hazardous spills. One must recognize that the concentration soon decays, due to the instability of free chlorine, and that fresh disinfectant should therefore be prepared each day or be available as the need arises. Unfortunately, sodium hypochlorite is also corrosive to some metals (e.g., aluminum), and on such surfaces should be replaced by a more suitable substance. Alternatives include alcohols at 70% concentration, iodophors that are registered as hard-surface disinfectants, or phenolic disinfectants; aldehyde bases may be irritating and potentially toxic and should not be used on laboratory surfaces (NCCLS 1991).

of ethics, could be sought at the start of a scholarship and renewed periodically as procedures change or scholarships expire. For example, the Queensland Academy of Sport obtains informed consent from all its athletes at the start of each scholarship year, describing the sport science interaction, the physical testing procedures, and the likely nature of blood sampling and handling; athletes are encouraged to ask questions then and throughout the year. The right of athletes to withdraw their consent at any time is reflected in the consent form in chapter 2.

Procedures for handling blood test results should also take into account the maintenance of subject confidentiality as reflected in the consent form in chapter 2. There may be an argument for disclosing information to supervising bodies or persons such as national sporting organizations, coaches, or sport science coordinators, but this should be agreed upon in advance at the time of consent. Australian Swimming Incorporated, for example, has included in its draft athlete consent document the provision that physiological test results can be released to the national head coach, the national sport science coordinator, and the manager of any team to which the swimmer is selected (Australian Swimming Incorporated, personal communication, 13 December 1995); the consequence of refusing consent might be nonselection to the team.

PROTOCOLS: Blood

Procedure for Capillary Puncture

Capillary puncture can be used to collect small amounts of blood from subjects, up to approximately 150 μl at a time, with minimal discomfort or inconvenience. This is particularly attractive in the exercise setting because of the micro nature of much of the blood analysis. However, as stated earlier, regardless of how easy the procedure is, one must never forget that capillary puncture constitutes invasive blood letting and should therefore be approached with the same precautions as other blood-handling procedures.

Typically, capillary puncture during exercise uses one of two sites, a fingertip or an earlobe. Each has advantages and disadvantages; the choice perhaps depends on the type of activity (e.g., rowers may experience discomfort in having fingertips pricked). With earlobes, one should take care to stay clear of the cartilage, as any ensuing infection must be surgically removed and is likely to cause permanent damage. As both fingertips and earlobes are end-arterial sites, once suitably hyperemized they can be expected to provide blood of similar composition; the composition will not, however, be the same as that of venous blood (El-Sayed et al. 1993).

Multiple pricking of either fingertips or earlobes may cause soreness and/or scarring that is uncomfortable and disfiguring, and may later change the composition of the blood. Soreness will be accompanied by an inflammatory response, while scarring may leave tissue unperfused for later blood tests. However, procedures such as scalpel incisions instead of multiple pricks are not recommended, as the extent and healing of the wound cannot be controlled; suitable alternative puncture instruments include the Microtainer (Becton Dickinson, described in the next section). With careful handling, such as hyperemization and covering, such punctures can be kept patent for prolonged periods—perhaps for 30 min or the duration of a typical incremental exercise test. Alternatively, venous cannulation may be an option for multiple samples (in cases in which the exercise allows) to avoid the discomfort and risks of several punctures, once its methodological and ethical issues have been addressed.

Equipment:

- 70% alcohol prep pads.
- Cotton wool pads or tissues.
- Sterile lancet, with tip up to 2.5 mm long. These come in several forms: a handheld lancet, an autolancet such as an Autolet (Owen Mumford) or SoftTouch monolet (Boehringer Mannheim), or a safety-flow lancet such as a Microtainer (Becton Dickinson) with a flat blade for a wider puncture. The latter is particularly recommended when the requirement is for multiple blood samples because, with careful handling, the puncture can be kept patent for a reasonable period. Note that lancets are sterilized by manufacturers. However, once their covering has been removed they should not be allowed to touch anything until they have punctured the skin. If any contact accidentally occurs before puncture, the lancet must be discarded and replaced with a new one. Similarly, a lancet used once on a subject must not be reused on the same subject or any other subject, even after a short delay.
- Rubefacient cream, such as Finalgon (Boehringer Mannheim), to warm/hyperemize the sampling site.
- Capillary tubes and/or microcontainers.
- Surgical tape.
- Disposable gloves.
- Biohazard sharps and waste-disposal containers.

Procedure:

1. Select the puncture site.

2. Warm/hyperemize the site with the application/massage of a rubefacient cream.

3. Thoroughly clean the site using an alcohol pad. Clean with a circular motion from the center to the periphery, allow to dry, and then wipe with a cotton pad or tissue; the presence of residual alcohol will quickly hemolyze the blood.

4. Hold the site firmly and make the puncture in one continuous, deliberate, perpendicular motion. Punctures in the fingertip should be into the pulp and across, rather than parallel to, the fingerprint (figure 6.1); punctures in the earlobe should be in the flesh, avoiding any cartilage.

5. Wipe away the first drop of blood using a clean cotton pad or tissue, as it may be contaminated with other body fluids.

6. Apply moderate pressure to ensure adequate blood flow; but to avoid collection of interstitial fluid as well as blood, do not squeeze vigorously.

7. Collect the blood into the appropriate container (e.g., by capillary action into an appropriate capillary tube).

8. If further samples are required or if the subject has to immediately return to exercise, cover

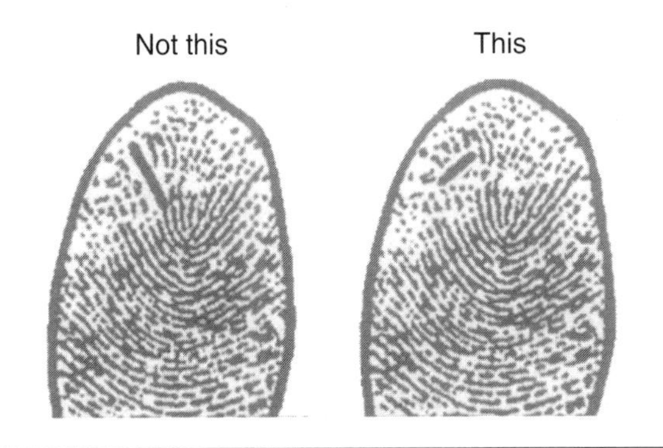

Figure 6.1 Use a transverse rather than a parallel finger prick.

the wound with surgical tape. Otherwise, apply light pressure with a cotton pad or tissue until the bleeding stops; then cover with a sticky plaster. Subjects can remove this themselves after approximately 30 min.

9. Dispose of all contaminated material into appropriately marked containers. Note that once an item has been used it is considered contaminated, whether blood is visible or not, and therefore must be disposed of or sterilized if intended for reuse. Equipment such as lancets and tissues must be used once only and then disposed of even when repeated measurements are being made on the same subject.

Procedure for Venepuncture

Venepuncture—the drawing of blood from a superficial vein—is the preferred method of blood letting when the requirement is for more volume than that in a capillary sample. For example, venepuncture is common for medical diagnostics or for study of multiple elements. However, it requires greater skill and patience (from both the practitioner and the subject) than does capillary puncture and presents greater risks to subject health. Venepuncture carries increased risks of hematoma, blood loss, and infection; and with only a few superficial veins available, one must take care to preserve their condition. For these reasons, venepuncture should be performed only by well-trained and accredited practitioners as discussed earlier. It also has limited use when serial blood samples are required during a single test session, but could then, under the right circumstances, be followed by venous cannulation (see later section on the procedure for venous cannulation).

As with capillary puncture, a number of collection devices are available for venepuncture, depending on preference and desired outcome. For example, venous blood can be collected into a sterile syringe for later distribution, or it can be collected directly into evacuated tubes inserted into a multidraw tube holder. The latter system is preferable when multiple elements are to be analyzed from separate tubes; but because of the speed of blood withdrawal, it may be unsuitable for subjects with fragile veins. Similarly, with use of a needle and syringe, the size of the equipment should be matched to the patient, such that small needles (e.g., 21 gauge) and small syringes (e.g., 5 to 10 ml) are used to puncture small veins to ensure that the pressure of blood withdrawal does not damage the vein. Indeed, 21-gauge needles should suffice for most specimen collection; larger ones (e.g., 18 gauge) are perhaps needed only for large blood donations (Garza and Becan-McBride 1989).

Equipment:

- Tourniquet.
- 70% alcohol prep pads.
- Cotton wool pads or tissues.
- Sterile needle of appropriate gauge. Needles are sterilized by manufacturers. However, note that the precautions associated with the use of lancets, outlined earlier, also apply to needles. A needle used in an unsuccessful venepuncture must be discarded before a second attempt is made.
- Appropriate evacuated tubes and multidraw holder, such as the Vacutainer system, or syringe.
- Sterile adhesive plaster.
- Disposable gloves.
- Biohazard sharps and waste-disposal containers.

Procedure:

1. Position the subject. When possible, the subject should be comfortably seated or supine for the test, as well as for the preceding 15 to 20 min; puncture and blood flow will be facilitated if the puncture site is below the level of the heart. Avoid postures such as standing that may impose the risk of fainting. In any case, the arm should be extended to form a straight line from the shoulder to the wrist. Note that changes in posture close to the time of sampling will affect the composition of the ensuing blood (Maw et al. 1995; Harrison 1985). Reliability of blood sampling is also improved with standardization of pretest diet, exercise, stress, time of day, and environmental temperature (Garza and Becan-McBride 1989).

2. Prepare all equipment, assembling the tube and needle holder and locating the alcohol pads, cotton pads, and adhesive plaster, before applying the tourniquet.

3. Select the puncture site. The three main veins used for venepuncture are the cephalic, basilic, and median cubital veins on the front of the forearm (figure 6.2). Veins in the hands, legs, and feet can be used, but they are difficult and potentially dangerous to puncture. Frequently they are not fixed and are surrounded by an abundance of nerves and bacteria; hence, medical supervision is recommended for these procedures. Do not draw blood from above the site of an intravenous infusion, as the resultant sample will not be representative of systemic composition.

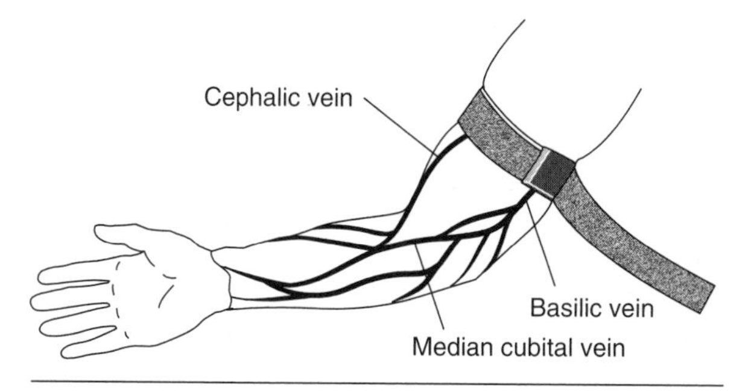

Figure 6.2 Veins of the anterior forearm.

4. Apply the tourniquet. Wrap the tourniquet firmly around the arm approximately 5 to 10 cm above the sample site and feel for the position of the vein. Having the subject make a fist and gently squeeze usually makes veins more prominent; but vigorous pumping should be avoided, as this will change the composition of the blood. Take your time to find the best vein, but never leave the tourniquet on for longer than 1 or 2 min.

5. Thoroughly clean the site using an alcohol pad. Clean with a circular motion from the center to the periphery, and then allow to dry; residual alcohol will quickly hemolyze the blood, but should be allowed to evaporate rather than being wiped away. If the puncture proves difficult and you have to repalpate the site, you must repeat the cleaning process.

6. With one hand, grasp the subject's arm approximately 5 cm below the puncture site. Pull the skin tight with pressure from the first two fingers on either side of the vein to prevent rolling.

7. Insert the needle, with the bevel upward, at an angle of approximately 15° to the subject's arm and in direct line with the vein. Puncture the skin and the vein in one motion.

8. Push the evacuated tube as far as it will go into the holder or draw slowly on the syringe plunger. The order of collecting into evacuated tubes may be important for quality assurance because there is a risk of contaminating a subsequent tube with the additive from a tube just filled. For example, if a tube containing the potassium salt ethylenediamine tetra-acetic acid (EDTA) is filled prior to a tube for electrolyte evaluation, it is possible that the electrolyte's potassium reading could be falsely increased. The recommended order of tube collection or distribution from a syringe is given by Pendergraph (1988, p. 69) (see sidebar).

9. Once blood collection is complete, release the tourniquet. Note that although the presence of a tourniquet will increase blood pressure it is unlikely to change the composition of the blood sample, as veins are impermeable to filtrate; however, the release of pressure allows filtrate to return through the capillaries, diluting the ensuing venous blood. Thus, if a tourniquet has been applied to begin venepuncture, it should remain in place until all required blood has been collected. Similarly, if a venepuncture is unsuccessful, the tourniquet should be removed and a second attempt delayed for 10 min to allow blood composition to restabilize.

Recommended Order of Tube Collection

Order of draw for the evacuated tube system	Order of draw for the syringe system
1. Plain tube(s)	1. Additive tubes
2. Additive tubes	a. Citrate (coagulation)
a. Citrate (coagulation)	b. EDTA
b. EDTA	c. Other anticoagulated
c. Other anticoagulated	2. Plain tube(s)

If one fails to observe the proper order of tube draw, two problems may occur:

- In the evacuated tube system one can contaminate a tube with the additive from a previous tube.
- When a syringe is used, one may cause microclots, resulting in false hematologic results.

Pendergraph notes, however, that if

- "electrolyte levels are determined on plasma drawn in a green stopper tube containing lithium or ammonium heparin," and
- "coagulation studies or a complete blood count (CBC) is also ordered,

the heparinized specimen should be collected before the citrated or EDTA tube to avoid possible . . . contamination. . . . It has also been suggested that a plain (red stopper) tube that contains a clot activator should be treated as an additive tube in the order of draw" (Pendergraph, 1992, 80-81).

Disposal of Contaminated Material

The disposal of laboratory waste is regulated by local and state governments; hence, each institution should determine the locally accepted procedure. However, certain generalizations apply that should be universally observed. As already discussed, gloves should be worn during handling of any potentially contaminated waste. Needles should not be bent, broken, or resheathed before being discarded. If a needle must be removed from a syringe, gloves should be worn and a "one-handed technique" employed, such as that facilitated by notches on biohazard sharps containers. All needles and other sharps should be disposed of into rigid puncture-resistant containers marked with a prominent biohazard symbol (figure 6.3) and used solely for such disposal; these should be kept close to the relevant work area. Other infectious waste should be disposed of into leak-proof bags or vessels marked specifically for biological disposal (figure 6.3). Then, incineration of the containers and bags is likely to be the preferred method of terminal disposal, although autoclaving and disposal as general waste may be acceptable (ANCA 1990, NCCLS 1991). Blood and body fluids can be discarded into a sewer, with care taken to avoid splashing (ANCA 1990).

Figure 6.3 Universal biohazard symbol.

10. Place a cotton pad over the puncture site—with the needle still inserted—and quickly remove the needle.

11. Apply pressure to the site until bleeding has stopped, and then apply the adhesive plaster.

12. Label all blood samples immediately with the subject's name and date of birth (i.e., two personal points of reference), sample date, and test code.

13. Dispose of all contaminated material into appropriately marked containers. Note that once an item has been used it is considered contaminated whether blood is visible or not and hence must be disposed of or sterilized if intended for reuse. Needles should be removed from the holders or syringes using the notch in the sharps container, or else the entire needle and syringe should be placed in the container. Needles should never be removed using fingers and should never be resheathed, bent, or broken, as so doing is the major cause of needlestick injuries.

Unfortunately, venepuncture is not always straightforward, and occasionally there is difficulty in obtaining blood—sometimes because of difficulty in locating the vein or inserting the needle and sometimes because of blockages, such as valves or walls, in the vein itself. If for any reason venepuncture proves unsuccessful, only one second attempt should be made before the process is aborted and a supervisor is informed (Pendergraph 1988); a new sterile needle must be obtained before the second attempt is made. The supervisor, or someone else who has been designated, may make another attempt; but it may be wise at this time to defer to medical assistance.

Procedure for Venous Cannulation

Occasionally it may be necessary in exercise and sport science to take serial blood samples of sufficient volume to analyze multiple elements. Although it is possible to do this using multiple venepunctures, the associated discomfort and vein damage make venous cannulation a preferable alternative. Similarly, well-executed venous cannulation might be considered less traumatic than serial capillary puncture or capillary incision. However, placement of an indwelling instrument is often considered more invasive than a simple puncture, making cannulation a matter for medical consideration. As discussed earlier, cannulation is therefore often restricted to medical personnel, although some institutions take the approach that a well-practiced technician actually performs better than an occasionally involved physician. If cannulation is desired, as it may be in research or laboratory testing, the logistics should first be approved by the appropriate board of ethics.

In many instances in the field (e.g., in swimming), venous cannulation may be impractical or may unduly increase athlete risk if the cannula were to become snagged. One should therefore carefully consider the placement of cannulas in applied settings from the viewpoint of ease of sampling as well as athlete safety. Similarly, the use of indwelling metal "butterfly" needles is not recommended at all because of the potential harm of dislodging a metal implement compared to a soft silicon cannula. Once in place, however, the cannula enables blood withdrawal without repeated application and removal of a tourniquet, so that the resultant sample is perhaps more reflective of true venous composition. After removal of a tourniquet one must wait a minimum of 10 min for blood composition to restabilize before collecting further blood (see earlier discussion of the procedure for venepuncture).

Equipment:

- Tourniquet.
- 70% alcohol prep pads.
- Cotton wool pads or tissues.
- Sterile cannula of appropriate gauge. In most exercise settings, a 22-gauge cannula not longer than 25 mm should suffice. Cannulas are sterilized by manufacturers. Note, however, that the same precautions apply to the use of cannulas as to the use of lancets, discussed earlier.
- Three-way tap and silicon tubing for attachment to the cannula.
- Sterile saline for cannula flushing.
- Appropriate evacuated tubes and multidraw holder, such as the Vacutainer system, or syringes.
- Sterile adhesive plaster.
- Disposable gloves.
- Biohazard sharps and waste-disposal containers.

Procedure: The procedure for inserting a venous cannula is essentially the same as that described earlier for performing venepuncture. What follows therefore is a description of the additional steps required for cannulation, with the reader referred to the section on venepuncture for the preliminary procedures

1. Cannula insertion follows steps 1 to 7 as described for the procedure for venepuncture.

2. Once blood "flashback" is visible in the tip of the cannula, remove the tourniquet and withdraw the introducing needle.

3. Attach the tubing and three-way tap.

4. Secure the cannula, placing a small cotton pad under the exposed head to keep it at approximately 15° to the arm; then apply adhesive tape and a covering bandage.

5. Flush with sufficient sterile saline to remove all traces of blood from the cannula, tube, and tap. This step should be repeated immediately after every blood sample to prevent blood clots, with the resultant supernatant withdrawn and discarded prior to any collection of blood. (Note that the use of heparinized saline is rarely necessary to prevent clotting, and is always considered a medical procedure as it involves infusion of a drug; it should therefore never be performed except by medically qualified personnel.)

6. Blood collection and cannula removal follow steps 8 and 10-12, respectively, as described for the procedure for venepuncture.

Transportation of Blood Samples

In connection with exercise and sport science it is often necessary to transport blood samples from their site of collection to the laboratory for analysis. This may be done by the practitioner or can be done by commercial carrier. Problems relating to the packaging and transportation of biological specimens have been reported at post offices as well as receiving laboratories (New South Wales Department of Health 1987). The exercise of due care is therefore essential in all transportation of biological specimens. The NCCLS and Australia Post have published requirements governing the packaging of biological specimens, which may be pertinent in cases in which specimens are self-transported or sent through the postal system. The following summarizes these requirements as general precautions that should govern packaging or moving of blood (after New South Wales Department of Health 1987; Garza and Becan-McBride 1989, p. 190; NCCLS 1991).

- Specimens should be properly labeled and in containers that protect individuals from contamination.
- Containers should be able to withstand leakage of contents, pressure and temperature changes, and rough handling. Ideally, the specimens should be placed in a primary screw-top container surrounded by a second, shock-absorbent package that will contain the specimen should the primary cover break.
- Specimen requisition forms or special instructions should accompany the specimen, in separate leak-proof bags.
- Containers should be clearly labeled as to their contents and the state of the contents—for example, "Frozen Blood Samples."

• A leaky or broken specimen should be handled cautiously as discussed earlier (see "General Safe Practice," page 88) and discarded as soon as possible; the sample should be recollected for analysis.

Management of Accidents and Injuries

As previously mentioned, blood sampling is not always straightforward, in which case it falls to the practitioner to deal with any situations that arise. These may include fainting, spills and splashes, and needle-stick injuries to either the patient or the practitioners themselves.

Fainting

It is not uncommon for subjects to become dizzy or faint at the thought or sight of blood. It is therefore important to keep aware of the condition of a subject throughout the procedure by maintaining good, open communication. If a subject becomes faint, remove the needle (according to the procedure for venepuncture) and move him or her to a supine position with the feet slightly elevated. Maintain communication, and stay with the person until you are assured of recovery.

Hematoma

Occasionally the area around a puncture will start to swell and discolor, indicating that blood is leaking into the surrounding tissues. This may occur because the needle has fully penetrated a vein or has slipped out slightly, or because of insufficient pressure on the puncture site after removal of the needle. Either way, remove the needle and any associated forces (e.g., tourniquation or squeezing) immediately, and apply firm downward pressure directly to the puncture site for approximately 5 min.

Blood Spills and Splashes

Blood spills should be cleaned up immediately, and the whole work area should be cleaned at the end of each session. Persons who are cleaning should wear thick gloves so that they can remove and discard broken glass or other objects with minimal risk. Hands should be washed thoroughly once the gloves are removed.

Minor blood spills should be wiped away with 1% sodium hypochlorite, prepared fresh daily. Make the site "glistening wet" and then dry or leave the disinfectant to evaporate. Major spills should first be absorbed with paper towels or a granular absorbent, as the presence of large amounts of protein may impair the disinfectant (NCCLS 1991). The site should then be covered with further towels and soaked in 1% sodium hypochlorite for 30 min in order to allow the disinfectant to act. Any remaining blood should be wiped up with clean towels and 1% sodium hypochlorite. Rinse the site with water to remove any remaining chemicals, and then dry to prevent slipping. All materials used in the decontamination process should be disposed of in biological hazard bags.

In most cases, blood splashed onto closed skin cannot result in infection. However, even under these circumstances, immediate hand washing should occur to eliminate even the slightest infection risk (see "Handwashing Protocol," page 90).

Blood splashed into the eye should also be washed out immediately with running tap water. The head should be turned to the side, and water should run across the affected eye for at least 5 min. The injured party should then report to the nearest accident and emergency center and explain the circumstances of the case.

Needle-Stick Injuries

Needle-stick injuries commonly result from attempts to resheath, bend, or break used needles (ANCA 1990). However, needle-stick-type injuries can also be caused by exposed lancets, capillary tubes, or glass of any type. Throughout this chapter, the phrase "needle-stick injury" is therefore used to refer to any puncture injury caused by a sharp object.

Needles should not be resheathed, bent, broken, or cut by hand. Nor should they be removed from disposable syringes; the syringe with the needle in place should be discarded as a whole. If a needle must be removed from a syringe, gloves should be worn and a "one-handed technique" employed, such as that facilitated by notches on biohazard sharps containers. Any technique that requires the use of two hands—that is, one holding the needle and syringe assembly and one holding the removal device—should be avoided, as the accident risk is high.

In the event of a needle-stick injury with a contaminated object, the injury should be washed immediately with water and treated with 1% povidone iodine or 70% alcohol. The injured party should then report to the nearest accident and emergency center and explain the circumstances. When appropriate the person may then be subjected to testing for hepatitis B and C and HIV antibodies, with follow-up tests scheduled for eight weeks later (New South Wales Department of Health 1987). In emergency situations in which the injured party is unable to give consent to the blood tests, testing may be performed without consent and an explanation given later. Although fewer than 0.3% of HIV-infected needle sticks lead to cross infection (Tokars et al. 1993), this possibility should be discussed with staff prior to involvement in blood handling.

To further avoid needle sticks, do not pick up contaminated broken glassware directly by hand; rather, use a mechanical device such as a vacuum or sweeping brush. The broken pieces should then be placed in a puncture-resistant sharps container for biological disposal.

chapter 7

Measuring Flexibility for Performance and Injury Prevention

Donna Harvey and Craig Mansfield

Measuring flexibility is more complex than simply conducting a sit-and-reach test. People often use the concept of flexibility loosely in describing the movement of a dancer or gymnast who performs splits or who bends over backward, but the elements of flexibility need to be defined. The relationship of flexibility to an athlete's performance and predisposition to injury can be considered and analyzed from a macro to a molecular level (Hennessy and Watson 1993; Hutton 1992).

Flexibility is a component of fitness that sport scientists and physiotherapists measure to gain an impression of a person's physical capacity. To date, research has not revealed a clear and accurate relationship between flexibility and muscular performance or an athlete's susceptibility to injury (Safran et al. 1989; Knapik et al. 1992; van Mechelen et al. 1992). This may be attributable to the lack of valid and reliable measures of flexibility, the range of injuries including acute and overuse injuries, the specificity of components of physical performance, and the large numbers of subjects needed for sufficient power to show a relationship.

Several well-controlled studies have suggested relationships between flexibility, performance, and injury or have identified a single factor correlating to an injury outcome (Reid et al. 1987; Lysens et al. 1989). The practical application of these data is limited because although acute injuries can result from a single factor, such as a collision, most overuse injuries result from a cumulative effect of several factors.

Krivickas and Feinberg (1996) used a revised protocol and scale for assessing muscular tightness and joint hypermobility in order to compare these characteristics with lower limb injury. In women athletes, lower limb injury was unrelated to muscular tightness or joint mobility; but in male athletes, injuries were associated with increasing muscular tightness and with lower joint laxity measures, that is, more stable joints. Joint hypermobility or laxity has previously been considered a positive risk factor for injury (Jones 1997). The measures of joint mobility in the Krivickas and Feinberg study included general tests of joint hypermobility, such as knee and elbow hyperextension, rather than specific joint hypermobility such as patella laxity or deficiency of the collateral or cruciate ligaments of the knee. Generalized overall joint mobility may not be specific enough to be a predictive factor for lower limb injury. Definitions and measurements of flexibility need to be accurate and valid if relationships between flexibility and optimal injury-free performance are to be clearly demonstrated.

Acknowledgments Australian Sports Injury Prevention Taskforce; Ms. Margaret Grant

Definitions of Flexibility

Definitions of flexibility in the literature vary but basically address the possible range of movement of a joint. Flexibility can be considered and described in a variety of ways, and identification of these parameters may permit a better understanding of flexibility.

Flexibility is generally expressed and measured relative to a joint. This creates some confusion in the literature; for example, hamstring flexibility may be referred to as a range of hip flexion or knee extension, depending on the test used, because the muscle group crosses two joints (see hip and knee flexibility protocol, page 112).

Referring to a joint range of movement (ROM) as a measure of flexibility does not allow easy differentiation and understanding of the specific restraints to movement. For example, when hip extension is limited, tightness of the iliopsoas or rectus femoris may limit range (see hip and knee flexibility protocol). The question is "Does it matter?"

If the range of a joint is limited, or not within normal limits for the athlete for his or her sport, the intervention focuses on stretching exercises. If the stretching exercises are general, then the flexibility measures need only be general—of the hamstrings, for example. However, if a specific muscle within a group is tight, then a test differentiating the components is necessary—for example, differential testing of the gastrocnemius and soleus muscles. Therefore, it is important to be able to use flexibility measures that can identify the limiting factor or factors, both general and specific, so that interventions are appropriate. Inappropriate use of stretching can lead to joint hypermobility and joint damage.

Terminology at present is confusing; until standardized terms and testing protocols are used to describe and assess flexibility, it will continue to be difficult to compare results across athletes and sports.

Active and Passive Flexibility Constraints

Constraints to flexibility or joint range of motion can be active or passive. Active restraints involve only the muscle-tendon unit, whereas passive constraints include the joint anatomy, capsule, and ligaments, as well as structures such as joint menisci (Stanish et al. 1990). It is important to define the limits of flexibility as accurately as possible, as this will influence the selection of intervention measures, such as stretching or stability exercises.

Hutton (1992) classifies mechanisms of resistance to movement and consequently flexibility into four headings:

1. Neurogenic constraints
2. Myogenic constraints
3. Joint constraints
4. Skin, subcutaneous connective tissue, and frictional constraints

Neurogenic Constraints

The muscular system is responsive to neural control that maintains a resting tone, facilitating the function of postural and phasic muscles. Muscle tone is influenced by supraspinal pathways and locally through reflex loops linked to the muscle spindle and Golgi tendon organs (Hutton 1992; Stanish et al. 1990). When the neural system is intact, resistance to movement is effected relative to the speed of movement or the stretching effect on connective tissue. Proprioceptive neuromuscular facilitation (PNF) stretching exercises have been developed to reduce the neural resistance to stretching exercises, thus facilitating gains in joint range (Wilkinson 1992).

In athletes with conditions such as cerebral palsy, the resting muscle tone is affected and there is significant resistance to movement. This can lead to joint contractures and limited joint range. Flexibility assessment measures need to be modified for athletes with neurological disorders. Interventions to reduce muscle tone in these athletes should be discussed with the athlete's physiotherapist or medical practitioner.

Myogenic Constraints

Within the muscle-tendon unit, there are also passive and active components. The active component is a result of cross-bridge interaction between the actin and myosin filaments within the muscle fiber, which is influenced by the position of the joint and by neural factors (Stanish et al. 1990). Other resistance to muscle extensibility has been referred to as "muscle stiffness" (Wilson et al. 1991b; Hutton 1992). Stiffness reflects the capacity of a muscle to absorb or dissipate forces such as the stress of force production from muscle activity. The effect of muscle stiffness is most relevant in the midrange of motion where there is significant overlap of the actin and myosin, rather than at the extremes of muscle-tendon range, where connective tissue limitations have more influence.

Muscle stiffness may be altered with different modes of exercise. In analyzing the effects of several methods of stretching on acute changes in flexibility, Hutton (1992) reported that muscle stiffness can be increased after concentric muscle actions and decreased after isometric or eccentric exercises. Therefore, to reduce muscle stiffness and to enhance flexibility gains from stretching, isometric and eccentric exercises pre-stretch would be most effective. This may influence the type of PNF stretch that is appropriate. For example, using an eccentric or isometric contraction preceding a stretch may lead to lower muscle stiffness and result in a greater improvement in range.

Joint Constraints

The configuration of the joint can significantly affect the movement possible. Multidimensional "ball-and-socket" joints such as the shoulder and hip are restricted by factors such as the depth of the acetabulum and the presence of a labrum, as well as by ligamentous and capsular restraints. Mortise joints such as the ankle and hinge joints or the proximal interphalangeal joints in the hand are limited in mobility by the joint architecture. As the bony anatomy of a joint cannot be altered conservatively, it is important to determine the "end feel" or limitation to joint movement so that techniques for improving joint range address modifiable factors. Joint constraints can limit movement, but these factors are not easily altered.

Surgery is an option for athletes who are hypermobile, that is, in cases such as dislocating patella or ruptured cruciate ligaments. Limited joint range due to joint changes or pathology should be assessed by a medical professional before any intervention is prescribed.

Skin, Subcutaneous Tissue, and Frictional Constraints

Skin and subcutaneous tissue may provide some mechanical and frictional resistance to joint movement. This is most evident after injury or surgery when scarring and adhesions limit the movement of the skin over underlying tissues. Stretching and manual techniques such as massage can increase mobility of the tissues. A specific joint mobility assessment, which includes the integrity of the skin and connective tissue, should be undertaken by a medical practitioner or physiotherapist.

Flexibility and Performance

People most often think of flexibility as a component of physical fitness that is important for prevention of athletic injury. In order to improve flexibility, it is necessary to elongate the active and passive constraints that limit joint range of motion through various stretching exercises. The viscoelastic properties of these constraints allow permanent deformation following force application over time, thus increasing the resting length of the tissue.

If you asked most coaches and athletes why they stretch, most commonly the answer would be "to warm up" or "to prevent injury." Previous research on flexibility has concentrated mostly upon the efficacy of stretching regimes or techniques and their associated effects on injury rates. The increase in athletic performance that can be achieved through stretching and improved flexibility, however, is much too often overlooked or unknown.

The effects of injury on participation in training and competition are obvious, and depend partially on the severity of the injury. By reducing the occurrence of injury with stretching, we provide an opportunity for enhanced performance.

Stretching During Warm-Up: Its Effects on Performance

Athletes, from recreational to elite, perform stretching exercises as part of a warm-up routine. Although long-term gains in flexibility are not generally the aim of warm-up stretching, there are useful short-term gains. Commonly thought of as an important aspect of preventing injury, stretching during warm-up also plays a role in improving athletic performance.

Many beneficial physiological mechanisms achieved during warm-up are temperature dependent (Shellock and Prentice 1985), and it is possible that stretching to improve flexibility is partially responsible for these mechanisms. During a single static stretch, a slow eccentric contraction of the muscle occurs that causes a rise in muscle temperature due to increased blood flow (Ciullo and Zarins 1983). Chemical contractility of the muscle is then enzymatically enhanced, maximizing the power of subsequent positive work. This enhanced contractility due to increased temperature may last for up to 1.5 h after a single stretch. Bergh (1980) observed that muscle contraction appeared to be faster and more forceful when the muscle temperature was elevated. The internal viscosity of a muscle decreases with increasing temperature, leading to improved mechanical efficiency (DeVries 1980). There are also claims that improved blood flow, substrate delivery, metabolite removal, oxygen availability, and neural function result from increased tissue temperature, playing a role in enhanced performance following warm-up activities and stretching.

Flexibility and Maximal Strength

Improvements in muscle strength due to improved flexibility have been confirmed in a limited number of studies. Wilson et al. (1991a) demonstrated that an eight-week flexibility program to train the pectoral and deltoid musculature increased shoulder joint flexibility as well as producing a significant improvement in one-repetition maximum concentric bench press. Similarly, Worrell et al. (1994) reported that after three weeks of hamstring stretches, isokinetic eccentric torques during knee flexion at 60°/s and 120°/s, as well as isokinetic concentric torque at 120°/s, were significantly improved. No improvement in isokinetic concentric torque was found, however, at 60°/s. One possible explanation for improved athletic performance and strength after flexibility training is increased availability of free intracellular Ca^{2+} (Yamashita et al. 1993).

Hortobagyi et al. (1985) found both direct and indirect benefits subsequent to stretching exercises for the quadriceps and hamstring muscles. While there was a significant increase in hip range of motion (flexion) and no significant change in the maximal voluntary contraction of the knee extensors, additional benefits were an improvement in knee extension speed, decrease in half relaxation time, and increase in stride frequency during a maximal stationary sprint. The authors were unable to explain these results conclusively, attributing the effects to changes in the knee extensor mechanical profile and modified recruitment profiles. One possibility is that reduced stiffness of the hamstring muscles in midrange allowed more efficient application of quadriceps torque in overcoming the external load. Therefore, the increased flexibility of the hamstrings effectively "allowed" extension to occur more rapidly.

Flexibility and Stretch-Shorten Cycles

The stretch-shorten cycle (SSC) forms the basis of many of our everyday activities and is quite important in a host of sporting endeavors. These cycles benefit human movement by improving its efficiency, allowing development of greater tension within the musculotendinous unit for the same energy cost. During an SSC, a muscle is initially stretched rapidly during an eccentric contraction before commencing the final concentric contraction to produce the desired movement. This stretch initiates the monosynaptic stretch reflex mediated by the muscle spindle, resulting in augmented contraction of the agonist muscle. During the initial stretch, energy is also stored in the passive, noncontractile components of the muscle. This stored energy is available as elastic recoil to assist the contractile components of the muscle, thus improving the overall tension within the muscle.

Worrell et al. (1994) cite two factors that determine the amount of elastic energy absorbed by muscles during the SSC: the speed of the eccentric contraction and the length of the muscle. Through increase in the length of the muscle, greater forces can be absorbed and thus more energy returned during the final concentric contraction. Ciullo and Zarins (1983) state that improving the limits of flexion and extension at a joint allows more effective storage and usage of elastic energy. Care must be taken, however, as a muscle is optimized to operate at a certain resting length and has its own specific length-tension relationship. Muscles generate their maximal concentric tension at a length 1.2 times their resting length (Norkin and Levangie 1992); past this length, active tension decreases due to insufficient sarcomere overlap. The effect of stretching and increased flexibility on the length-tension relationship of muscles is yet to be determined.

Some hypothesize that flexible muscles are able to make the most of the SSC effect, achieving greater elastic energy utilization from the initial stretch and hence generating more tension (Wilson et al. 1992; Worrell et al. 1994). A significant negative correlation has been shown between maximal stiffness and static flexibility (Wilson et al. 1991a), and thus stiffness may be reduced with flexibility exercises. This may appear to decrease the available energy stored during an SSC, but the more compliant muscle will stretch further under a given stretch force and thus store greater energy (Shorten 1987; Wilson et al. 1992). Increases in static flexibility achieved through stretch training have resulted in significantly increased performance during an SSC bench press movement (Wilson et al. 1992). The authors concluded that this increased capacity to perform an SSC movement resulted from decreased muscle stiffness and the subsequent increased storage and release of elastic energy during the initial stretch.

Flexibility and Efficiency of Movement

The effect of flexibility on movement efficiency is controversial. Godges et al. (1989) observed an increase in gait efficiency at 40%, 60%, and 80% of $\dot{V}O_2max$ with improvements in hip flexibility resulting from stretching programs. Gleim et al. (1990) concluded that tighter subjects (walkers and joggers) were more economical and faster than looser, more flexible subjects while oxygen consumption was measured on a treadmill. In contrast, DeVries (1963) observed that acute increases in flexibility had little or no effect on economy or energy expenditure during a 100 m sprint.

The efficiency of skeletal muscle contracting from its resting length has been reported to be in the vicinity of 25% (Dickinson 1929). Efficiency appears to improve with increased gait speed, reported at 40% during walking, 45% during running at 8 km/h, and 80% during running at 32 km/h (Cavagna 1977). The explanation for this observation is that much of the positive work of the muscles is produced from the storage and release of elastic energy during the previous SSC. It would appear, therefore, that if improvements in flexibility result in greater energy storage and subsequent release during an SSC, flexibility would contribute to lowering the energy costs of movement and increasing efficiency.

It appears logical that improved flexibility of an antagonist muscle (and thus decreased muscle stiffness) would lead to improved agonist efficiency, as there will be less resistance to movement. The results of Hortobagyi et al. (1985), as described earlier, may be attributed to this effect. But other factors complicate this issue, such as those constraints mentioned previously in the chapter along with other physiological mechanisms.

Flexibility and Range of Motion

So far this discussion has concerned the effects of stretching and improved flexibility on the intrinsic chemical and mechanical nature of the muscle and the resultant implications on tension generation. How then, do flexibility and the subsequent gain in range of motion affect athletic performance?

Quite simply, athletes require sufficient flexibility to achieve the optimum movement patterns and static positions necessary for sport participation. In many sports from gymnastics to swimming, flexibility plays a vital role in success. For example, not only are sprinters who have poor flexibility in their hamstring and gluteal muscles at risk of numerous musculoskeletal injuries, but, due to a lack of hip flexion or knee extension at terminal swing and early stance phase, their stride length will decrease and so will their speed. The swimmer with poor shoulder (rotation) flexibility will be unable to reach the optimum catch position, reducing the effectiveness of the pull-through phase of the stroke. Consequently, the swimmer is at risk of overuse injury and may employ compensatory mechanisms (such as excessive body roll), thus further reducing performance. Poor hip, knee, and ankle flexibility may also impede swimming performance through an inefficient kicking mechanism and loss of optimal streamlining (Blanch 1997).

Summary

Among the many variables that influence athletic performance, flexibility is an important aspect. Through intrinsic effects on the mechanical and biological properties of the muscle and its connective tissue, flexibility gains may improve performance. Improved maximal strength, greater ability to utilize the SSC effectively, augmented efficiency, and correct movement patterns through the required range of motion all result from greater flexibility. Couple these benefits with reduced injury, and your athletes are ready to train and compete at their best.

Flexibility and Injury Prevention

Knowledge of sport-specific biomechanics enables the sport scientist and coach to determine efficient movement patterns and the flexibility necessary for optimal performance. Poor flexibility may result in uncoordinated or awkward movement, leading to decreased efficiency, decreased performance, and possible injury (Shellock and Prentice 1985). In sports such as gymnastics, diving, and aerobics, extremes of flexibility are required, compared to sports such as archery, equestrian, and skiing. Research to date has not been able to clearly demonstrate a relationship between flexibility and injury; therefore the components of flexibility will be discussed relative to the mechanisms of soft tissue injury.

Muscle Strain

The most common types of injury in sport are muscle strains and ligament sprains (Garrett 1996; Finch 1995). Muscle strain is the injury most associated with flexibility issues. Muscle strains generally occur during eccentric exercise, which permits production of greater forces, or when the muscle is either stretched passively or activated during stretch (Garrett 1996).

In a rabbit model, when an isolated muscle-tendon unit is stretched to failure, the point of rupture is usually at the muscle-tendon junction

(Garrett 1996). When a muscle is stretched during contraction, the site of rupture is also at the muscle-tendon junction; however, the active muscle is able to absorb up to 100% more force than the passive muscle before rupture. This suggests that muscle activity can be a protective factor (Garrett 1996; Stanish et al. 1990). As the muscle is able to develop greater forces in midrange, this "protection" would be most beneficial in midrange and not at the end of range, where muscle-stretching exercises focus.

In a clinical study of 10 athletes with acute hamstring injury, the site of injury was localized, using computed tomography scan, to the muscle-tendon junction of the biceps femoris at the common tendon junction (Garrett et al. 1989). The architecture of the hamstring muscle is such that areas of musculoskeletal junctions occur throughout the entire muscle belly length; but the injuries were generally in the biceps femoris muscle, and mostly proximal and lateral. All injuries occurred while the athlete was sprinting or kicking a ball—activities that involve ballistic hip flexion and knee extension (Garrett et al. 1989). These clinical data from athletes experiencing hamstring strains support the research results from rabbit models.

Stretching and warm-up exercises address the myogenic constraints and are generally well-accepted measures for preventing muscle strain injury. In the description of myogenic factors related to flexibility, it was suggested that eccentric and isometric exercise reduce muscle stiffness. A warm-up usually consists of some easy exercise to increase the blood flow to the muscles before stretching. As well as increasing the temperature of the muscles, the exercise-induced changes in muscle stiffness may also augment gains in muscle extensibility.

Again in a rabbit model, the viscoelastic properties of the muscle have been shown to increase as a result of slow cyclical stretch (Garrett 1996). Ten stretches of 30 s holds were repeated with a resultant 17% reduction in tension, with the greatest effect in the first four stretches. The author repeated this experiment for innervated and denervated muscle with similar results. This suggests that the viscoelastic properties of the muscle-tendon unit at the end of range are not greatly influenced by neural control. Therefore, sustained stretches at the end of range may be more effective than PNF stretches for acute changes in the viscoelastic properties of the muscle-tendon unit.

Muscle Imbalance-Related Injury

Injuries related to flexibility can be either directly associated, as in a muscle tear, or indirectly associated, as in a bicipital tendinitis at the shoulder from tightness of the shoulder internal rotators and pectoral muscles (Pieper and Schulte 1996). As the human body is a complex and dynamic interaction of the complete musculoskeletal system, the effect of limited flexibility in a muscle or group of muscles may be manifested proximally or distally (Loudon et al. 1996; Riegger-Krugh and Keysor 1996).

Pieper and Schulte (1996) investigated muscular imbalances in elite swimmers and their relationship to injury. In swimmers with low back pain there was a positive correlation with imbalance in the low back, pelvic, and hip regions. Typically there was shortening of the iliopsoas, and a shortening of the rectus femoris was also evident in swimmers who experienced knee pain. Therefore in assessing the influence of flexibility in an athlete, it is important to consider the direct and indirect influences of changes in muscle length on injury outcomes.

Ligament Injury

Ligament injuries are common in sports that involve the likelihood of twisting forces through the joints. Athletes who are hypermobile, that is, those in whom the joint ligamentous and capsular restraints are lax, are more prone than others to injury (Jones 1997). In a joint that is normal, it would be of no benefit to consider increasing joint mobility or improving performance by stretching ligaments or the joint capsule. Ligaments are passive joint restraints, and their integrity is essential for injury prevention.

When ligaments have been stretched or ruptured, reconstructive surgery or conservative exercise measures may make the joint more stable, but these measures are not within the scope of this chapter.

Summary

Muscle strain injuries are common; and whether the muscle is injured from passive or active forces, the site of injury is usually the muscle-tendon junction. Muscle-tendon junctions can occur throughout the length of the muscle belly. Injuries can occur as a result of excessive stretch or from a stretch while the muscle is activated, such as in kicking a ball. The plasticity of the muscle-tendon unit and the stiffness of a muscle can affect the capacity of the muscle to absorb forces without injury. Athletes require optimal flexibility to be able to perform their sport, as well as to protect the body from excessive muscular forces. It appears that improved flexibility protects the athlete; therefore measures of flexibility are required to gain an overall impression of the "fitness" of the athlete.

Flexibility Measures: Musculotendinous Components

Protocols for flexibility measures in this chapter address only the musculotendinous aspects of flexibility, as this is the factor most responsive to conservative intervention used by sport scientists or physiotherapists. Orthopedic measures for joint ranges are well documented in texts, with normal ranges established for the general population. However, norms for specific athlete populations are available only in research papers (Agre and Baxter 1987; Reid et al. 1987).

Protocols in this chapter have been chosen on the basis of several criteria:

- The tests measure musculotendinous components of flexibility.
- The tests do not measure joint mobility components such as ligamentous laxity.
- The test protocols are valid measures of flexibility and are well described.
- There is some indication of normal data for athletic populations.
- The tests can identify general and specific limitations to musculotendinous flexibility.

Some common tests previously used for measuring flexibility have not been included in this chapter—the sit-and-reach test, for example. The research to date does not demonstrate a relationship between sit-and-reach flexibility results and injury or performance outcomes. The appropriateness of this test must be questioned (Sinclair and Tester 1993; Jackson and Baker 1986). Although it may give an overall impression of lower back and hamstring flexibility, the sit-and-reach test fails to differentiate between low back flexibility; hamstring length; neural tension; anthropometric measures of arm, trunk, and limb length; and calf flexibility.

Recent test results from elite junior netball players included the sit-and-reach measure and the active knee extension (AKE) test for hamstring length (see description of the latter test further on in this chapter). The results revealed that some athletes with a normal sit-and-reach score were well below the norm of the group for hamstring flexibility using the AKE test (Hoare 1997). These data also substantiate the validity of the AKE test for measuring hamstring flexibility.

The flexibility tests presented in this chapter are by no means perfect. Further research is needed to refine these tests, to investigate the validity and reliability of the tests for diverse athletic populations, and to establish the relationships between these tests and both injury and performance outcomes.

PROTOCOLS: Flexibility

Shoulder Flexibility: Passive Internal Rotation

Purpose: To assess the passive flexibility of the glenohumeral external rotators (supraspinatus, infraspinatus, teres minor) and posterior joint capsule.

Equipment:

- Modified goniometer with spirit level (figure 7.1)
- Firm bed or plinth
- Second examiner

Landmarks:

- Olecranon process of ulna, ulnar styloid process

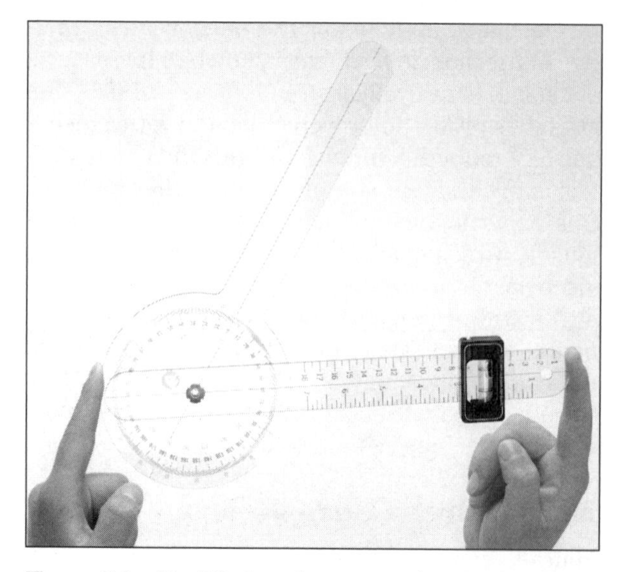

Figure 7.1 Modified goniometer with spirit level.

Protocol:

1. Athlete lies supine with arm to be tested at 90° abduction and elbow flexed 90° (figure 7.2).

2. Examiner passively internally rotates the shoulder via the forearm and wrist, ensuring that the posterior aspect of the shoulder maintains contact with the plinth.

3. Rotation is performed until either (a) the athlete complains of pain, (b) the humeral head begins to protrude excessively anteriorly (posterior shoulder not in contact with bed), or (c) no further range can be acquired.

Record: At limit of test, goniometer is aligned with landmarks to record the degree of internal rotation relative to the vertical. The reason for cessation is also recorded (a-c above).

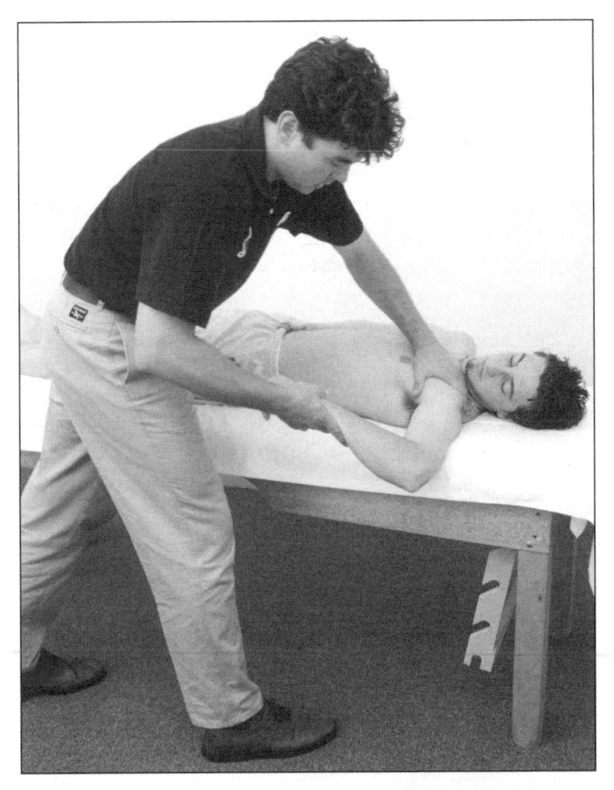

Figure 7.2 Supine glenohumeral internal rotation.

Shoulder Flexibility: Passive External Rotation

Purpose: To assess the passive flexibility of the glenohumeral internal rotators (pectoralis major, teres major, subscapularis, latissimus dorsi) and anterior capsule tightness or laxity.

Equipment:

- Modified goniometer with spirit level
- Firm bed/plinth
- Second examiner

Landmarks:

- Olecranon process of ulna, ulnar styloid process

Protocol:

1. Athlete lies supine with arm to be tested at 90° abduction and elbow flexed 90° (figure 7.3).

2. Examiner passively externally rotates the shoulder via the forearm and wrist, ensuring that the posterior aspect of the shoulder maintains contact with the plinth.

3. Rotation is performed until either (a) the athlete complains of pain or apprehension, (b) the athlete begins to extend the thoracic spine, or (c) no further range can be acquired.

Note: Care must be taken with athletes who have had previous shoulder dislocation or instability/laxity, as the test position is quite provocative and possibly unstable in this population.

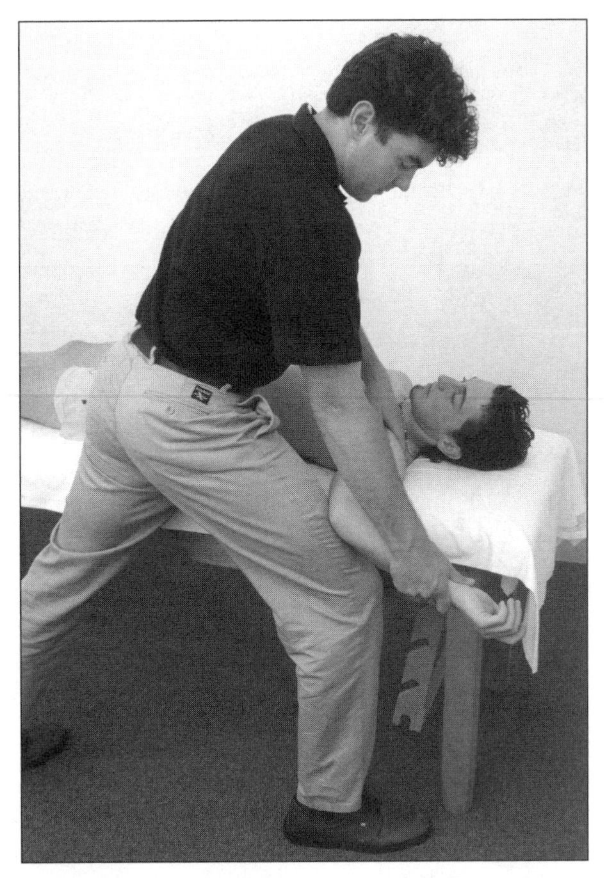

Figure 7.3 Supine glenohumeral external rotation.

Record: At limit of test, goniometer is aligned with landmarks to record the degree of external rotation relative to the vertical. The reason for cessation is also recorded (a-c, previous page).

Hip Flexibility: Hip Internal/ External Rotation in Neutral

Purpose: To assess ranges of passive hip rotation in neutral hip position.

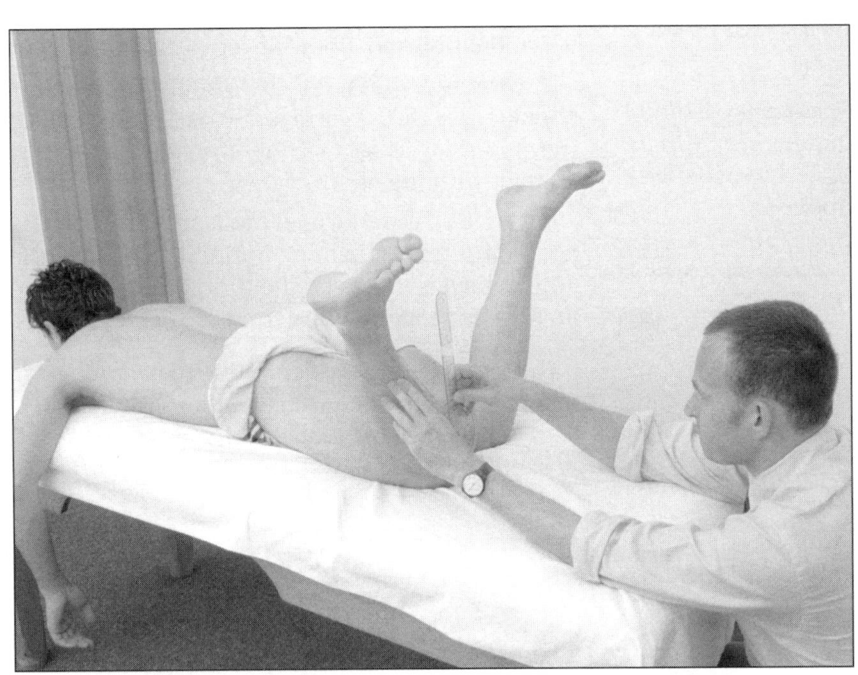

Figure 7.4 Prone hip internal rotation and measurement.

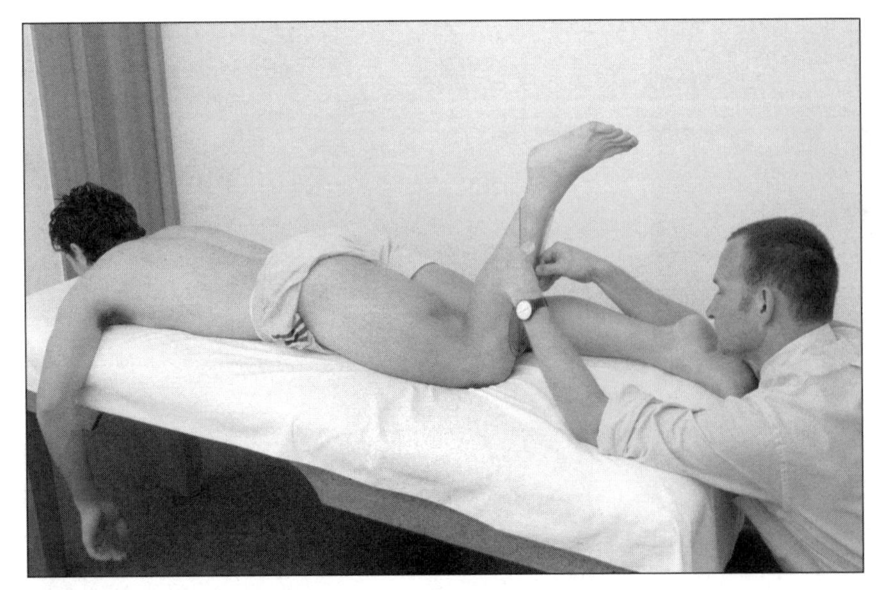

Figure 7.5 Prone hip external rotation and measurement.

Equipment:

- Goniometer with spirit level
- Firm bed/plinth

Landmarks:

- Inferior pole of patella, tibial tuberosity

Protocol:

Internal Rotation

1. Athlete lies prone with knees bent to 90°, chin resting on bed, arms by sides.
2. Let ankles move away from each other as far as possible passively (gravity assists), ensuring that hip does not flex (figure 7.4).

External Rotation

1. Athlete lies prone with one knee bent to 90°, chin resting on bed, arms by sides. Keep knees together and anterior superior iliac spine (ASIS) on bed throughout test.
2. Let ankle drop toward opposite side as far as possible (figure 7.5).

Record: Angle formed by the line of the tibia relative to vertical in each position.

Hip Flexibility: Hip Internal/External Rotation in 90° Flexion—Supine and Sitting

Purpose: To assess ranges of active and passive hip rotation in 90° hip flexion.

Equipment:

- Goniometer with spirit level
- Firm bed/plinth

Landmarks:

- Anterior border of tibia, tibial tuberosity

Protocol:

Internal Rotation

Supine:

1. Athlete lies supine with hip and knee bent to 90°, other leg extended, head resting on bed, arms by sides.

2. Tester passively internally rotates hip while maintaining 90° of flexion of hip and knee.

Sitting:

1. Athlete sits with hips and knees bent to 90° at end of bed, hands resting by sides, palms on bed.

2. Keep knees bent to 90° and together throughout test. Actively internally rotate both legs as far as possible (figure 7.6).

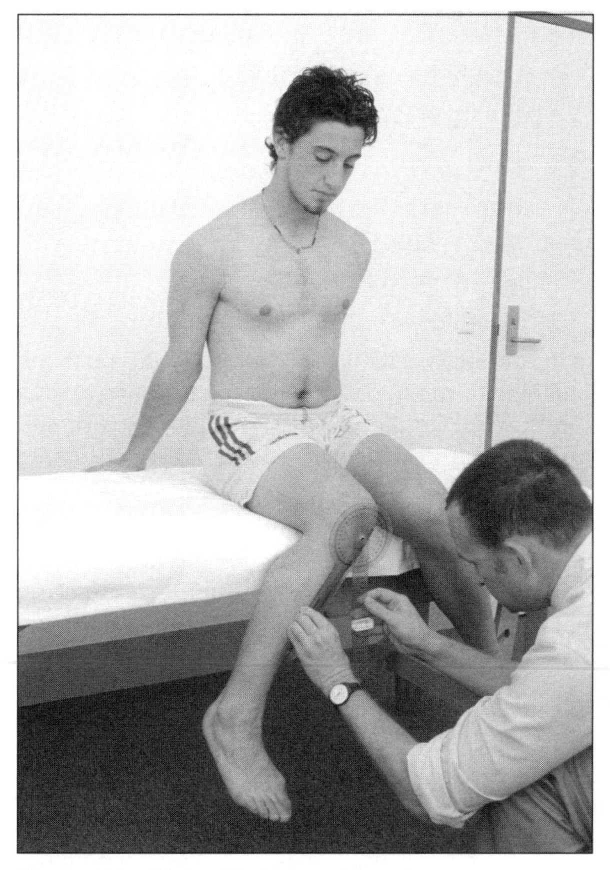

Figure 7.6 Sitting hip internal rotation.

External Rotation

Supine:

1. Athlete lies supine with hip and knee bent to 90°, other leg extended, head resting on bed, arms by sides.

2. Tester passively externally rotates hip while maintaining 90° of flexion of hip and knee.

Sitting:

1. Athlete sits with one hip and knee bent to 90° at end of bed, other hip abducted to side of bed and knee bent to 90°; ensure that ASISs are level.

2. Hands resting by sides, palms on bed.

3. Actively externally rotate hip as far as possible (figure 7.7).

Record: Supine—visual estimation of angle of excursion of the line of the tibia from cephalad-caudad alignment at starting position.

Sitting—angle formed by the line of the tibia relative to vertical.

Implications:

1. Piriformis is an external rotator, flexor, and abductor during extension and up to 60° of flexion. Past this point in flexion, because of the "inversion of muscular action" that occurs at the hip

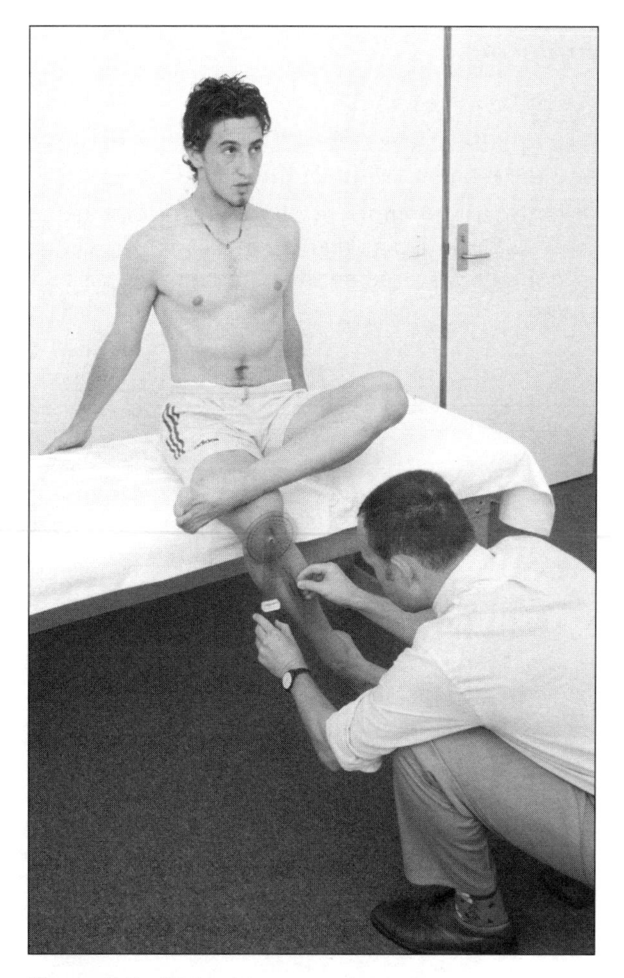

Figure 7.7 Sitting hip external rotation.

joint, piriformis becomes an internal rotator, extensor, and abductor (Kapandji 1970). Thus, piriformis may limit either internal or external rotation of the hip joint, depending upon the degree of flexion of the testing position.

2. Ligamentous tension (iliofemoral and pubofemoral) is reduced in the position of flexion, allowing greater range.

Hip Flexibility: Figure Four Test for the Hip Joint

Purpose: To identify anterior capsule tightness of the hip joint, short adductors, and internal rotators.

Equipment:

- Firm plinth/bed wide enough to support knee in test position
- In some cases a gym mat on the floor

Landmarks:

- ASIS

Protocol:

1. Athlete lies prone.

2. Athlete places sole of foot on inside of opposite knee (hip in external rotation and abduction).

3. Athlete attempts to push hip to floor using gluteals. Maintain medial aspect of ankle and knee in contact with the surface throughout the test (figure 7.8).

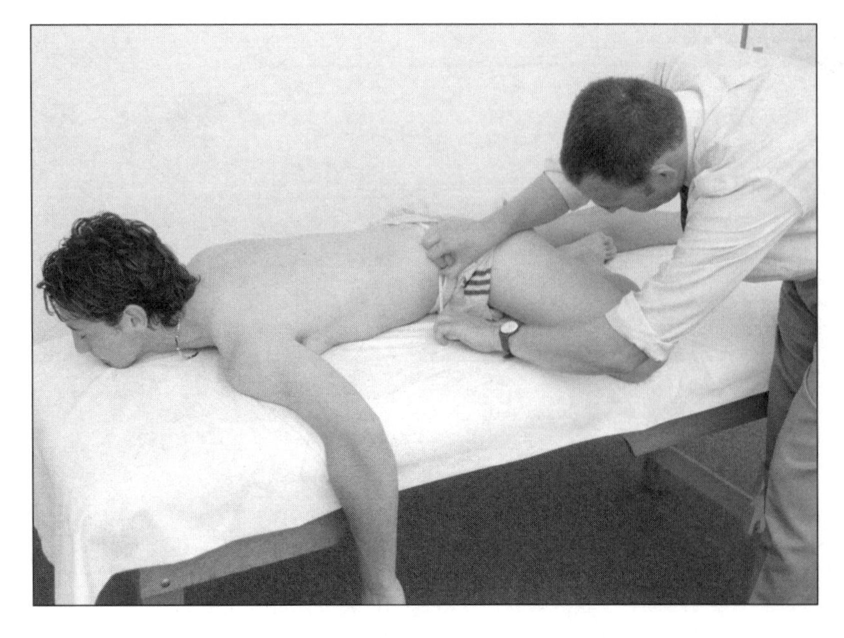

Figure 7.8 Figure four test in prone position.

Record: Distance of ASIS from floor and area of pain/discomfort reported by athlete.

Implications: A distance of more than 3 cm indicates tightness of the hip joint anterior capsule (hip musculature and contralateral sacroiliac joint). Further assessment of other structures is required to differentiate hip joint capsule restriction.

Hip Flexibility: Modified Ober's Test

Purpose: To assess flexibility of the iliotibial band (ITB) and tensor fascia lata (TFL).

Equipment:

- Firm, high bed/plinth
- Long-arm goniometer

Landmarks:

- Posterior superior iliac spines (PSISs)
- Straight line joining PSISs
- Midline of long axis of femur posteriorly

Protocol:

1. Athlete lies on side, leg to be tested uppermost (figure 7.9).

2. Flex untested hip and knee to 90°.

3. Underside arm placed posteriorly.

4. Athlete is instructed to contract lateral trunk muscles to maintain spine in neutral lateral flexion (keep PSIS line vertical). A pillow positioned under the athlete's trunk may help keep the spine in neutral.

5. Examiner stands behind athlete and supports uppermost leg in neutral adduction/abduction with one hand while stabilizing pelvis with other.

6. Examiner maintains thigh in neutral hip joint flexion/extension/rotation and the knee in 20° flexion throughout test.

7. Examiner adducts thigh as far as possible while keeping the pelvis stable to prevent lateral pelvic movement.

8. The test is repeated with the knee in 90° of flexion.

Record: Angle between PSIS line (vertical) and midline of femur at the end point of hip adduction (when lateral pelvic movement begins).

Implications: If the angle is less than 20° adduction, the ITB is tight and may predispose the athlete to patellofemoral dysfunction and ITB friction syndrome.

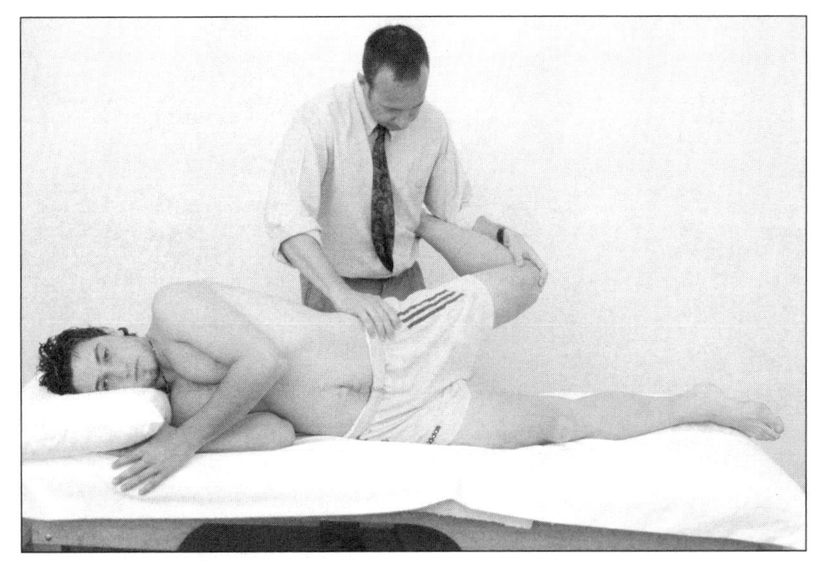

Figure 7.9 Modified Ober's test.

Hip Flexibility: Long Adductors

Purpose: To assess the length of the long adductors, namely, gracilis and adductor magnus.

Equipment:

- Long-arm goniometer

Landmarks:

- ASIS, superior pole of patella

Protocol:

1. Athlete supine on plinth, leg not to be tested hooked over edge of plinth with knee at 90° so as to prevent excessive pelvic motion.
2. Examiner passively abducts extended leg with knee straight, maintaining the hip in neutral or slight internal rotation, until athlete complains of pain (figure 7.10).

Record: Center goniometer over ASIS, stationary arm aligned with opposite ASIS and movable arm with superior border of patella. Record range of pain-free abduction.

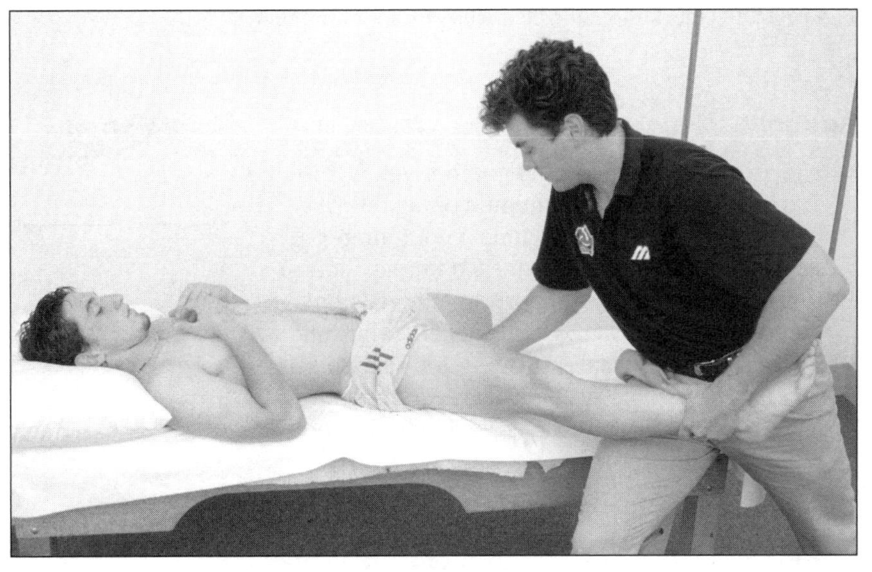

Figure 7.10 Long adductor flexibility.

Hip Flexibility: Short Adductors

Purpose: To assess the length of the short hip adductors (adductor brevis, longus, and magnus; pectineus).

Equipment:

- Tape measure
- Firm plinth

Landmarks:

- Head of fibula

Protocol:

1. Athlete lies supine.
2. Bend knees to 90° flexion with hip joints at 45° flexion, with feet touching each other.
3. Let knees drop downward so that soles of feet may be touching, hip abducted and externally rotated (figure 7.11).

Record: Use tape measure to measure distance from head of fibula to surface of plinth.

Hip and Knee Flexibility: Modified Thomas Test

Purpose: To assess the flexibility of the hip flexors (iliopsoas predominantly), quadriceps (single and double joint muscles), and TFL/ITB.

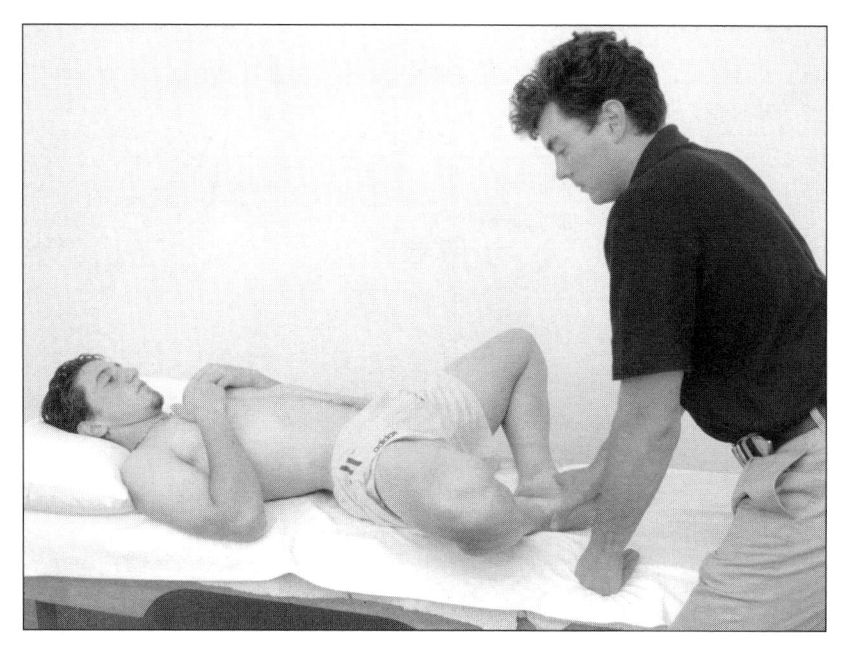

Figure 7.11 Short adductor flexibility and measurement.

Equipment:

- Firm plinth and modified goniometer. A third arm that is flexible (aluminum or flexible plastic) is attached to a standard goniometer at 90° to the stationary arm (use tape or glue). This arm, used when measuring the ITB angle, allows the line across the ASISs to be held when the contralateral hip is fully flexed. A spirit level goniometer is also recommended so that the horizontal axis can be determined accurately.

Landmarks:

- Both ASISs, greater trochanters, point bisecting the superior border of the patella, knee joint line laterally between the lateral femoral condyle and fibula head, and lateral malleolus

Protocol:

1. The athlete sits perched on the end of a plinth and rolls back to lying with both knees held to the chest (this ensures that the lumbar spine is flat on the plinth and that the pelvis is in posterior rotation).

2. The athlete holds the contralateral hip in maximal flexion with the arms while the tested limb is lowered toward the floor (figure 7.12).

3. The athlete is asked to relax the hip and thigh muscles, so a passive end-point position is obtained, due to gravity alone.

Iliopsoas:

1. The angle of hip flexion is measured with the athlete in the test position.

2. The goniometer is centered over the greater trochanter, with the fixed axis directed vertically using the spirit level. This allows measurements relative to the horizontal plane.

3. The mobile arm is pointed toward the lateral knee joint line, representing the line of the femur.

Record: The hip angle is recorded relative to the horizontal, or 0° axis, as a positive or negative angle; that is, 7° represents a hip flexed above the horizontal, and –12° represents a hip that is below the horizontal.

Implications: If a hip flexion angle of at least –7° is not obtained, then the iliopsoas is consid-

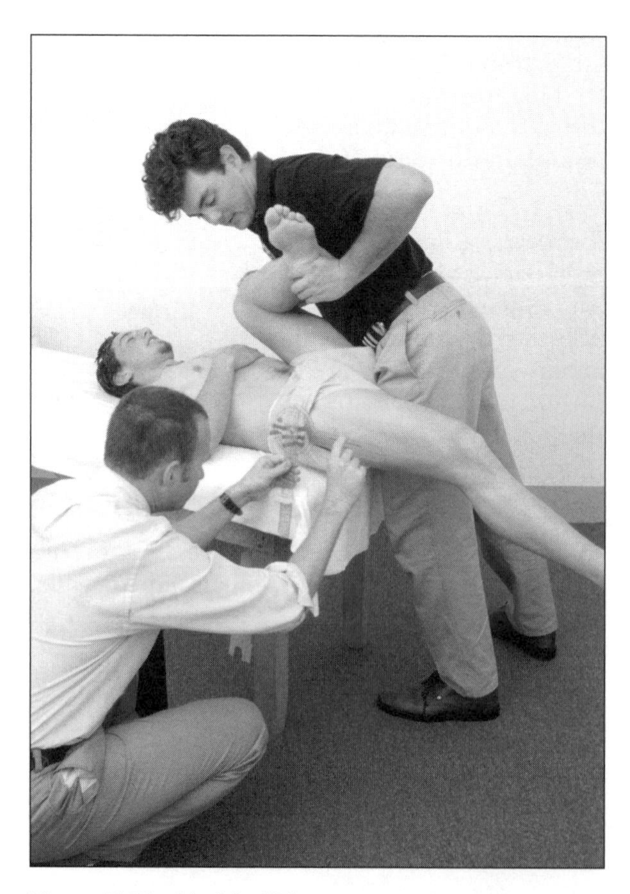

Figure 7.12 Modified Thomas test—hip extension angle.

ered tight. Norms obtained to date reveal a mean angle of −11° for rowers and tennis players, −12° for basketball players, and −14° for runners.

Quadriceps:

1. The passive length of the quadriceps is determined by measuring the knee flexion angle. Note that no overpressure is applied—purely gravity.
2. The goniometer is centered laterally at the knee joint line as described in protocols (figure 7.13).
3. The fixed arm is aligned with the length of the femur toward the greater trochanter.
4. The mobile arm points toward the lateral malleolus of the fibula.

Record: The knee flexion angle.

Implications: Norms obtained to date reveal a mean angle of 56° for rowers, 53° for basketball players, and 50° for runners and tennis players. As the knee angle does not approach 90° flexion as long as the contralateral limb is held in maximal flexion close to the chest, a significant and consistent force of gravity is still applied to all athletes.

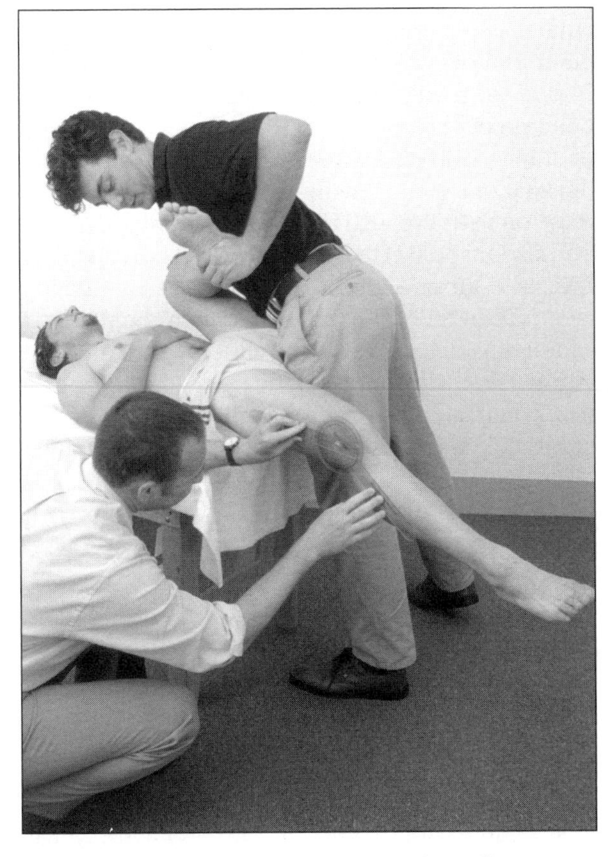

Figure 7.13 Modified Thomas test—knee flexion angle.

For differentiation between iliopsoas, rectus femoris, and vastus lateralis and medialis tightness, the following is proposed:

Hip Flexion > −7° and Knee Flexion < 45° =
Tight Rectus Femoris, Iliopsoas, or Vasti

Push thigh into extension; if this causes knee extension, rectus femoris is tight. Increase knee flexion; if this causes hip flexion, rectus femoris is tight. Increase knee flexion; if there is no hip flexion, iliopsoas is tight.

Normal Hip Angle, Knee Flexion < 45° = Tight Vasti

Iliotibial Band/Tensor Fascia Lata:

- Hip abduction angle is measured with the center of the goniometer over the ipsilateral ASIS.
- The flexible arm is directed toward the opposite ASIS.
- The remaining arm lies along the length of the femur.
- This procedure measures the "real," not apparent, angle of hip abduction.

Record: The angle of hip abduction is measured, with the line perpendicular to that of the ASISs considered as 0°. A positive angle represents hip abduction.

Implications: When one thigh is held maximally flexed to the body, the movement is comparable to trunk side flexion. This alters the line of the pelvis so that it is no longer perpendicular to the lower limbs. Therefore, when the modified Thomas test position is established, even though the leg appears to be in a line with the trunk, it is actually in a position of relative abduction.

Norms for hip abduction to date are between 15° and 16° for rowing, basketball, running, and tennis.

To discriminate between the iliopsoas and ITB, the following is recommended:

Hip Flexion Angle > −7° (e.g., −1°) and Hip Abduction > 19° = Tight TFL/ITB

Hip Flexion Angle > −7° and Hip Abduction < 19° = Tight Iliopsoas

Intratester Reliability Estimates:

	Intraclass correlation coefficient (ICC) =	
Iliopsoas	0.93	SEM = 1.48°
Quadriceps	ICC = 0.94	SEM = 1.85°
TFL/ITB	ICC = 0.91	SEM = 1.16°

Hip and Knee Flexibility: Active Knee Extension Test (90/90 or Active Knee Extension Test)—Hamstrings

Purpose: To assess hamstring length and range of AKE in a position of hip flexion as this is required in running, kicking, and striding activities.

Equipment:

- Goniometer with extended arms and spirit level
- Firm plinth/table

Landmarks:

- Inferior border of lateral malleolus and head of fibula

Protocol:

1. Athlete lies supine, head resting on table (no pillow), arms crossed on chest.
2. Passively flex hip of testing leg until thigh is vertical (use spirit level to align) (figure 7.14).
3. This position is maintained throughout the test by support behind the posterior thigh.
4. Maintain opposite leg in fully extended position throughout test by "pushing" heel away from body.
5. Keep foot relaxed and actively straighten knee until thigh begins to move from vertical position.
6. It is important to note that in cases in which full knee extension is achieved without thigh movement, the knee is flexed while the thigh is moved to 30° past the vertical position (i.e., 120° hip flexion). With a relaxed foot, the knee is again straightened until the thigh begins to move.

Record: The angle from complete knee extension at which the thigh begins to move is recorded by aligning the goniometer with the landmarks described above and the vertical plane. Flexion values are recorded as negative. In cases in which the hip is further flexed to 120° flexion, the measurement is recorded as 120 – x, where x = angle of knee extension deficit.

Other: The straight leg raise (SLR) test is widely used as an indicator of hamstring muscle length, but there is confusion regarding the validity of the test. The AKE test is thought to be a more reliable and valid test of hamstring length, having an interrater reliability coefficient of r = 0.99 in a controlled experimental setting (Gajdosik and Lusin 1983).

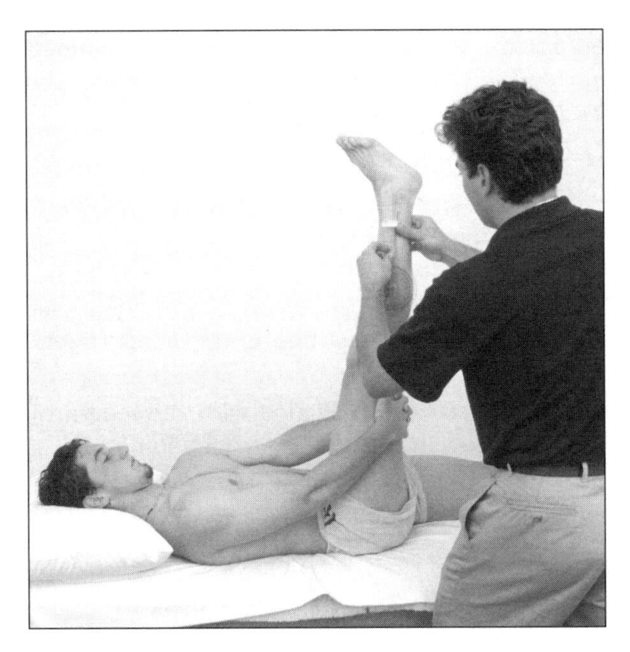

Figure 7.14 Active knee extension test.

Problems with the SLR include (1) whether the limitation of movement is due to neural or muscular tissues and (2) the degree of pelvic movement that occurs (Cameron and Bohannon 1993). The mean posterior pelvic rotation from resting position occurring with SLR has been reported to be 24.9° (Bohannon 1982, as cited in Kane and Bernasconi 1992), while mean posterior pelvic rotation with the AKE test was found by Kane and Bernasconi (1992) to be 9.6°. Thus, it appears that AKE is a more accurate measure of hamstring length than SLR. The SLR is also commonly used as a neurological test, as it causes elongation of and tension within the sciatic nerve, lumbosacral trunk, and sacral nerve roots after the first 25-30° of hip flexion.

Ankle Flexibility: Dorsiflexion (Weight-Bearing Soleus and Gastrocnemius)

Purpose: To assess range of talocrural dorsiflexion in a functional position for weight-bearing sports; indirectly measure length of soleus and gastrocnemius muscles.

Equipment:

- Semicircular (180°) goniometer with spirit level

Landmarks:

- Inferior tip of lateral malleolus
- Midline of lateral aspect of head of fibula

Protocol:

Soleus

1. Athlete in stride stand position, no shoes (figure 7.15).
2. Maintain heel contact with floor throughout test.
3. Maintain subtalar joint in neutral throughout test to prevent pronation.
4. Bend knee forward in line with the second toe until heel contact decreases or pain is experienced in or around the ankle joint.

Gastrocnemius

1. Repeat test, maintaining knee extension throughout. This will indicate length of gastrocnemius muscle (figure 7.16).

Record: Angle formed by shaft of tibia relative to vertical.

Implications:

1. Soleus flexibility should be 30-40°, and less than 30° is considered abnormal. Gastrocnemius flexibility should be 20-30°, with less than 20° considered abnormal. The difference between the two measures should be close to 10°.

2. Restriction of dorsiflexion at the talocrural joint may be due to gastrocnemius-soleus complex tightness. Athletes with restriction of dorsiflexion often compensate by pronating the foot in weight bearing. This can cause biomechanical changes in the lower extremity and predispose the athlete to overuse injuries.

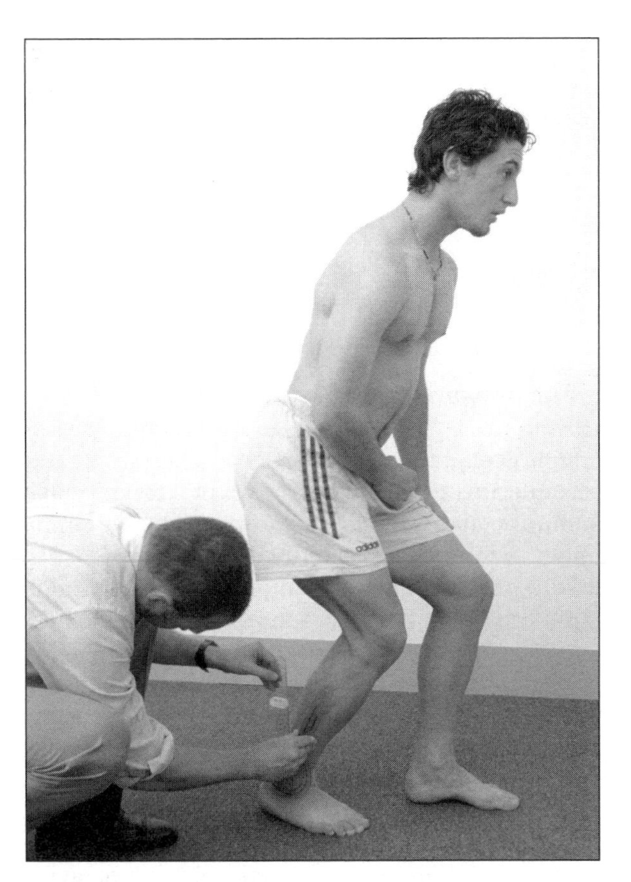

Figure 7.15 Dorsiflexion—soleus muscle flexibility.

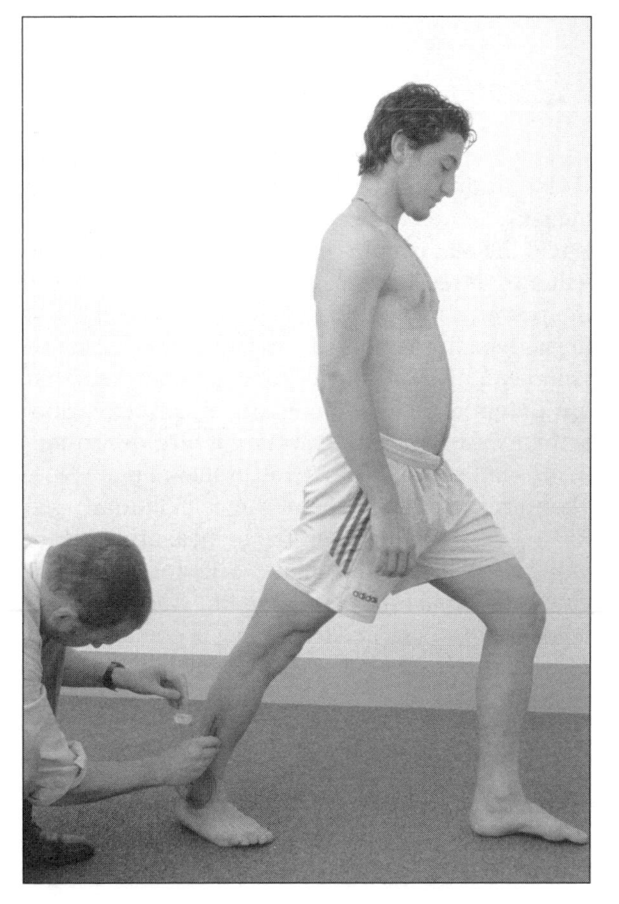

Figure 7.16 Dorsiflexion—gastrocnemius muscle flexibility.

chapter 8

Determination of Maximal Oxygen Consumption ($\dot{V}O_2max$) or Maximal Aerobic Power

Robert Withers, Christopher Gore, Greg Gass, and Allan Hahn

This chapter is a technical note for the sport scientist, addressing quite selected aspects of the measurement of maximal oxygen consumption. While the aim is not to provide a rationale for the value of assessing or interpreting maximal aerobic power, the chapter provides a description of measurement systems with particular emphasis on volume-monitoring devices and gas analyzers that are potentially the sources of largest error. Subject and laboratory preparation is also described, as are important calibration issues. Finally, the chapter provides details of the calculations used for data reduction when one is measuring either expired or inspired volume. The sport-specific chapters in this manual cover the specific rationale for measuring maximal oxygen consumption, as well as the specific test protocols.

Theoretical Rationale

Oxygen consumption ($\dot{V}O_2$) is the product of the cardiac output ($\dot{Q}$) and arteriovenous oxygen difference ($CaO_2 - C\bar{v}O_2$):

$$\dot{V}O_2 = \dot{Q} \times (CaO_2 - C\bar{v}O_2)$$

It is therefore reflected by both central and peripheral physiological variables and can be calculated via the direct Fick method:

$$\dot{V}O_2 = \dot{Q} \times (CaO_2 - C\bar{v}O_2)$$

$$\dot{Q} = \text{Heart Rate} \times \text{Stroke Volume}$$

$$\therefore \dot{V}O_2 = (\text{Heart Rate} \times \text{Stroke Volume}) \times (CaO_2 - C\bar{v}O_2)$$

The measurements of stroke volume and arteriovenous oxygen difference are highly invasive and ethically difficult to justify for the routine determination of $\dot{V}O_2max$. However, oxygen consumption and the related variables can be determined by open-circuit indirect calorimetry. This involves measuring the pulmonary ventilation and comparing inspired and expired carbon dioxide and oxygen concentrations.

During an incremental exercise test, the increase in $\dot{V}O_2$ is essentially linearly related to increasing power output (figure 8.1). Eventually a point is reached at which the oxygen consumption will not increase despite an increase in power output, thereby indicating that $\dot{V}O_2max$ has been attained. Oxygen consumption may then plateau or even

Acknowledgments The technical advice of Tom Stanef (South Australian Sports Institute) about ergometer calibration and the photography provided by Ashton Claridge (Multimedia Department, the Library, Flinders University) are gratefully acknowledged.

decline. The $\dot{V}O_2max$ is a stable and highly reproducible characteristic of an individual and is either unaffected or reduced by only 2-3% by a variety of physical stresses including blood loss (500 ml), moderate to severe dehydration, high ambient temperature, pyrogenic fever, and acute starvation (Rowell 1974). However, these stresses have known effects on heart rate and stroke volume, with the result that the validity of predicting $\dot{V}O_2max$ from the heart rate response to submaximal exercise is low and should therefore be discouraged.

$\dot{V}O_2max$ can be expressed absolutely as liters per minute (L/min) or relative to body mass per minute ($ml \cdot kg^{-1} \cdot min^{-1}$). The latter is a better indicator of average running speed during a 5 km race than a scaled predictor ($ml \cdot kg^{-2/3} \cdot min^{-1}$) that is designed to remove the confounding effect of body mass (Nevill et al. 1992). $\dot{V}O_2max$ data are presented in liters per minute when total power output is important, as in rowing, whereas $ml \cdot kg^{-1} \cdot min^{-1}$ is generally used when one is reporting $\dot{V}O_2max$ for activities in which the subject's weight is unsupported (e.g., running). A wide variation in $\dot{V}O_2max$ can be expected. It is therefore not unusual to find a range of 30-80 $ml \cdot kg^{-1} \cdot min^{-1}$ for 20- to 30-year-old males and 25-65 $ml \cdot kg^{-1} \cdot min^{-1}$ for similarly aged females.

$\dot{V}O_2max$ decreases with age and physical inactivity, and males generally have higher $\dot{V}O_2max$ values than females. The high $\dot{V}O_2max$ values of elite middle- and long-distance athletes are due to a combination of genetic endowment and training. Bouchard et al. (1986) have attributed 40% of the variation in $\dot{V}O_2max$ to genetic factors. The literature indicates that physical training can increase $\dot{V}O_2max$ by 5-30%, but greater improvements have been observed in cardiac patients, persons of very low initial fitness, and those who achieve substantial weight loss (American College of Sports Medicine Position Stand 1990).

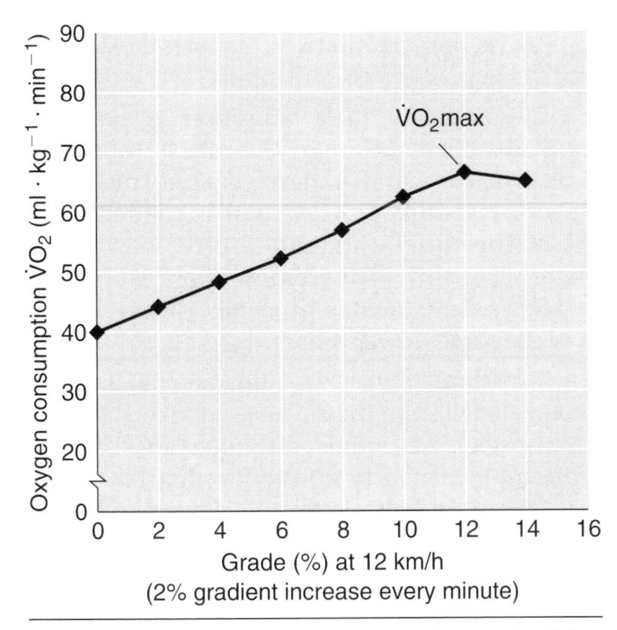

Figure 8.1 Oxygen consumption on the treadmill for a 20-year-old male middle-distance runner.

Since increases in $\dot{V}O_2max$ are associated with adaptations in both central and peripheral physiological variables, it is important that athletes simulate the movement pattern of competition by performing the test on a sport-specific ergometer (e.g., treadmill for runners, cycle ergometer for cyclists, rowing ergometer for rowers). However, with highly trained athletes, the $\dot{V}O_2max$ is a relatively weak predictor of race performance because of the importance of other factors such as mechanical efficiency, lactate threshold, anaerobic capacity, and motivation.

Test Equipment: Ergometers

This section provides information and guidelines regarding the calibration of various types of ergometers.

Treadmills

1. Zero grade on the treadmill should be checked with a spirit level and shims placed under the treadmill feet as required. The various elevations can then be calibrated:

$$\%\ \text{Grade} = \frac{\text{Vertical height}}{\text{Horizontal distance}} \times 100$$

2. The speeds can be verified by marking a line across the belt, measuring the length of the belt, and counting the number of belt revolutions per minute:

$$\text{Speed (km/h)} = \frac{\text{rpm} \times 60 \times \text{belt length (m)}}{1000}$$

3. The accuracy for elevation and speed should be within 0.1% and 0.1 km/h, respectively, of the indicated setting.

Cycle Ergometers

1. Mechanically braked ergometers such as the Monark, which use a balance, may be statically calibrated by suspending a known weight from the balance at the point of belt attachment after the zero has been checked. However, dynamic calibration of

mechanically, electromagnetically, and air-braked cycle ergometers is unequivocally the method of choice, since this calibrates the complete system including the frictional resistance of the transmission from the pedal shaft (Cumming and Alexander 1968; Russell and Dale 1986; Telford et al. 1980). A dynamic calibration rig, which is depicted in figure 8.2, can be either purchased from VacuMed (see appendix for details) or fabricated in accordance with the recommendations of Woods et al. (1994).

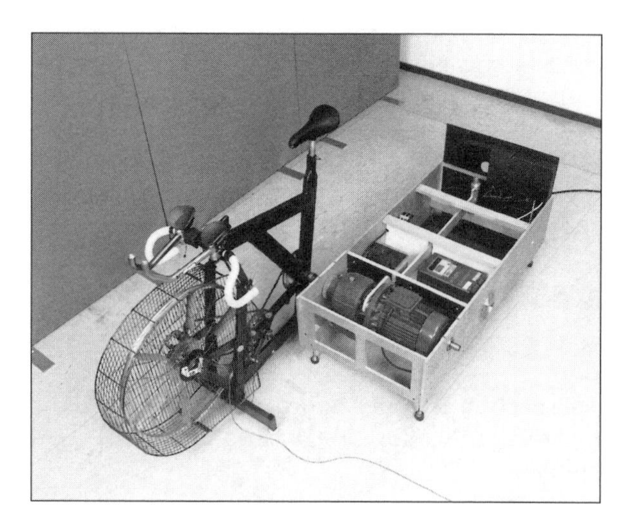

Figure 8.2 The dynamic calibration of an air-braked cycle ergometer.

This calibration rig is commercially available from the South Australian Sports Institute (see appendix for details).

2. With a mechanically braked cycle ergometer, pedal cadence must be controlled. This can be accomplished by using a revolution counter in conjunction with either a metronome or a cadence indicator.

3. The frictional load in the transmission of a mechanically braked cycle ergometer can usually be kept to a minimum by regular servicing. This involves chain lubrication, tension adjustment so that the chain moves vertically about 1 cm, checking all sets of ball bearings for damage, and greasing them.

4. The advantages of electrically braked ergometers are that the power output, which should range from 0 to 750 W, is relatively independent of pedal cadence, and small increases in power output can be made more accurately. The first advantage can also be achieved by gearing air-braked and mechanical ergometers. However, gearing changes can affect the physiological responses to a specific power output.

Kayak Ergometers

1. A first-principles torquemeter for the dynamic calibration of cycle ergometers can also be used to verify the K1 kayak ergometer manufactured by Roger Cargill (see appendix for details). While previously only the flywheel load was calibrated, the South Australian Sports Institute (SASI) has developed a modification that enables one to determine the total load presented to the paddler. The calibration therefore now includes the flywheel load plus that presented by the retractable (bungee cord) drive mechanism of the kayak ergometer. This development, which is displayed in figure 8.3, allows calibration over a range of stroke rates so that one can establish a more accurate relationship of power and flywheel velocity. As in calibration of air-braked cycle ergometers, a third-order polynomial can be derived to estimate power from the frequency of the flywheel.

2. Initial investigations have shown that the previous calibration method underestimated the true mean power when one is using the SASI software (version 1.0a). This software, which previously employed a calibration regression based on flywheel load only, was found to underestimate mean power by approximately 20% on the kayak ergometers that were fitted with the toothed belt drive. When used on the same ergometer fitted with the rope drive, the same software underestimated the true mean power by 35%.

3. The modified SASI calibration rig was also used to evaluate the original Cargill software (version 1.1). The latter underestimated true mean power by 15-25 W (10-30%) across the range 50-400 W when used with the belt-drive kayak ergometer. The rope-drive ergometer resulted in greater underestimates of mean power—20-40% over the same power range.

4. These preliminary data indicate that it is necessary to calibrate the kayak ergometers against a first-principles torquemeter to establish a calibration regression or correction that takes into consideration the total true load exerted by the paddler. After dynamic calibration, the error in measurement of mean power of kayak paddlers should be less than 3% across a typical range of 50-400 W.

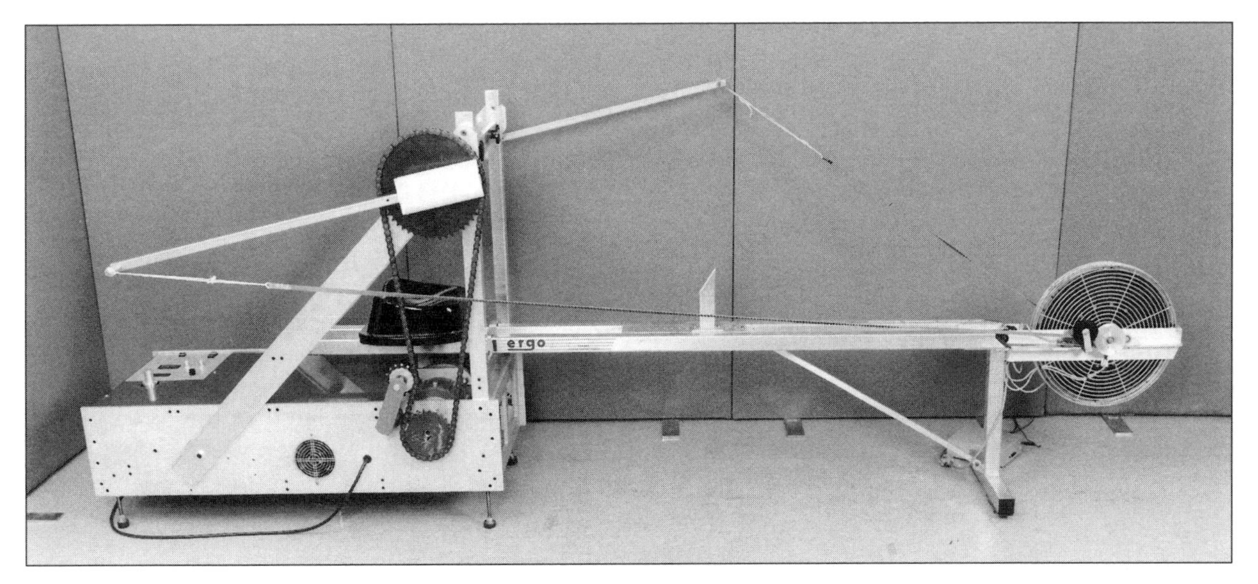

Figure 8.3 The dynamic calibration of an air-braked kayak ergometer.

This calibration rig is commercially available from the South Australian Sports Institute (see appendix for details).

Rowing Ergometers

1. The protocols for testing rowers (see chapter 21) recommend the Concept IIc rowing ergometer. The manufacturer claims that, provided appropriate maintenance procedures are undertaken regularly, the measured power output is within ±2% of the true value.

2. While it is not possible at present to dynamically calibrate rowing ergometers, a method to approximately verify power output has been developed by Ms. Peggy McBride of the Australian Institute of Sport (AIS) in Canberra. Power is calculated from measurements of the force applied to the chain that connects the "oar" handle to the ergometer flywheel, the duration of the force, and the horizontal linear displacement of the handle:

$$\text{Power} = \frac{\text{Force} \times \text{Distance}}{\text{Time}}$$

The force applied to the chain is measured via a precalibrated quartz force link mounted such that there is no interference with normal stroke length.

3. A custom-designed device, which consists of a cog in contact with the chain and a 360° potentiometer, is used to determine the chain travel and consequently the horizontal displacement of the handle throughout the stroke cycle. Force and displacement data are continuously sampled by a computer at 50 Hz per channel.

4. This system allows precise determination of the external mechanical work performed over a given time and hence power. Readers should note that this power output is always greater than that indicated by the Concept IIc display, even though the latter purportedly gives an average power output.

Test Systems

The traditional approach is the Douglas bag method (Douglas 1911). The only tubing is that from the respiratory valve to the Douglas bags or meteorological balloons in which the expired gas is collected. The bags should be flushed with expired gas beforehand and then completely emptied with an exhaust pump. They should also be periodically checked for leaks by filling with expired gas and serially monitoring the $\%CO_2$ and $\%O_2$. Continuous collection of expired gas can be facilitated by connecting three Douglas bags to a four-way stopcock. Adequate time should be allowed for the tubing to be washed out with expirate before any gas is collected.

While an excellent semiautomated systems approach has been described by Wilmore and Costill (1974), completely automated systems can be either purchased commercially or fabricated (Sainsbury et al. 1988).

If the expirate passes through a mixing chamber and the oxygen consumption is changing as during a $\dot{V}O_2$max test, then the temporal misalignment between the CO_2/O_2 determinations and the inspired volumes should be examined at different flow rates. This delay will be a function of the time constant of the mixing chamber (Time Constant [min] = Mixing Chamber Volume [L]/$\dot{V}_E$[L/min]) and the transit time from the mixing chamber to the gas analyzers. While the latter is a constant, the former may be negligible during the high pulmonary ventilations of maximal exercise because it is inversely related to flow rate, and only three time constant intervals are necessary for the change in gaseous composition to attain approximately 95% of its final value. Nevertheless, for the lower workloads of a $\dot{V}O_2$max test, it may be necessary to align the expired gas concentrations with the inspired volumes they represent. Readers may wish to consult the excellent treatment of mixing chambers in Wasserman et al. (1994, pp. 441-442).

Volumetric and Gas Analysis Equipment

The following sections deal with equipment and measurement issues related to the airways system, gas volume, and gas analysis.

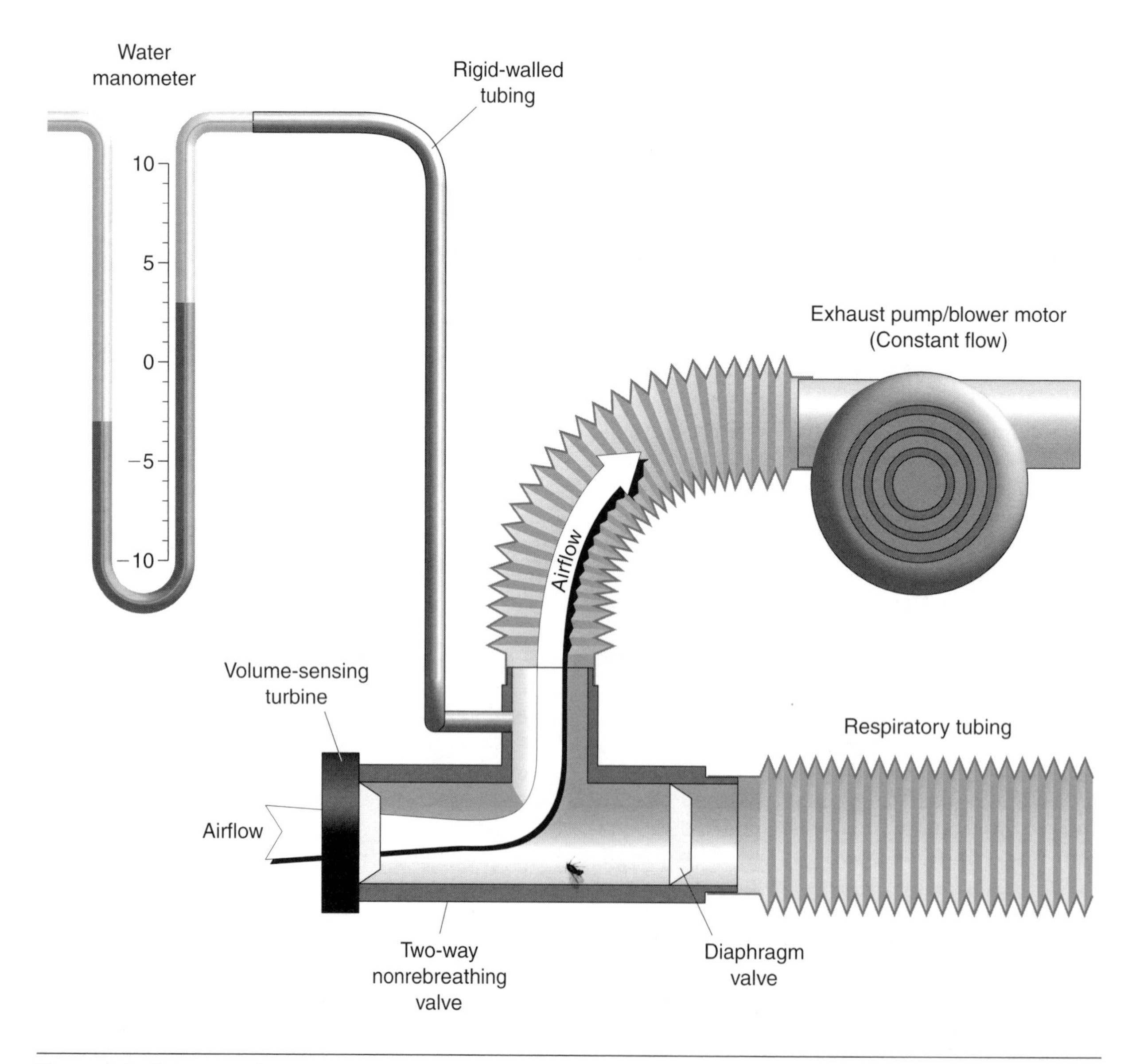

Figure 8.4 Schematic for measurement of pressure due to inspiratory resistance of a respiratory circuit. The measurement procedure requires that the rigid-walled tubing pierce the valve body so that both sides of the manometer are open—the left side to ambient pressure and the right side to the airflow pressure inside the nonrebreathing valve.

Airways System

1. The overall pressure due to inspiratory and expiratory resistance of the system must be checked with a flow meter, manometer, and exhaust pump (figure 8.4); in each case the resistance should be less than 6 cm H_2O at flows up to 300 L/min (Jones 1997).

2. Respiratory valves should have a low resistance, should have dead space less than 100 ml excluding the mouthpiece, and should not leak (e.g., Hans Rudolph 2700 T valve and 2730 Y valve). The dead space, which is the volume that is common to both inspiration and expiration, is inversely proportional to the resistance. A balance must therefore be struck between these two variables, and this will largely depend on the flow rate.

3. Resistance is proportional to length and inversely proportional to the radius raised to the fourth power. Hence, tubing should be >30 mm in internal diameter and not longer than 1.5 m on either the inspiratory or expiratory side. Sudden angulation of the tubing should be avoided.

4. VacuMed manufactures corrugated tubing that minimizes kinking; the smooth inside surface also minimizes turbulence. Tubing of this type is strongly recommended.

5. Regular checks for leaks should be made with a manometer by placing gas in the circuit under pressure. Leaky junctions may be located by using a liquid leak detector such as Snoop, which is available from Swagelok (see appendix for details).

Gas Volume

1. The instrument that measures gas volume should have accuracies of ±2.0% for pulsatile flows and ±1% for continuous flows (Gardner 1979).

2. A water-sealed spirometer (e.g., Tissot or Stead-Wells) is the primary calibration standard against which other secondary calibration standards, such as syringes and sinusoidal artificial lungs, should be initially checked. However, the constancy of the cross-sectional volume of the spirometer's bell should still be verified throughout its elevation by withdrawal of saturated gas with a calibration syringe. It is also advisable to use a manometer to confirm that the gas pressure within the bell corresponds with that of the atmosphere throughout its elevation.

3. The calibration procedure must be specific to the method of measuring the volume (Hart and Withers 1996). Thus if the measuring device is on the inspiratory side of the circuit, calibrations should be conducted with a Tissot spirometer and syringe or sinusoidal artificial lung using pulsatile flows of 50, 100, 150, 200, and 250 L/min. If the Douglas bag method is used in conjunction with an exhaust pump drawing the expirate through a gas meter, calibrations should be conducted at a constant flow rate.

4. Dry gas meters such as the Parkinson-Cowan CD4 and American Meter Company DTM325 are popular because they are much cheaper than other volume-measuring devices such as pneumotachographs and turbine volume transducers. The American Meter Company dry gas meter has been shown to measure continuous flows of water vapor-saturated air with an error of <1.0%, whereas sinusoidal flows of 8-100 L/min were misread by <1.0% and the error was still within 2.0% at 140 L/min (Hart et al. 1992). However, this instrument requires at least 25 L to be passed per measurement to negate the inaccuracy caused by the alinearity of its bellows within one revolution (10 L) of the dial pointer (Hart et al. 1992). Also, the addition of a shaft encoder to the meter's pointer to facilitate a digital readout increases rotational inertia, which causes a 4-5% underreading (Hart et al. 1992). The internal mechanism therefore needs to be recalibrated in accordance with the manufacturer's instructions (Beck 1970).

5. The PK Morgan Mark II Ventilometer comprises a turbine volume transducer and a control/readout that contains either a VENTX 5A or VENTX 6A program. It is calibrated via four strokes of a 1 L syringe. While accuracy can be affected by the syringe rate during calibration and the investigator must beware of entrainment (Hart and Withers 1996), tests (Hart et al. 1994) have indicated that

- readings were highly reproducible;
- the VENTX 5A read continuous flows within an error of ±0.5%—the corresponding error for the VENTX 6A was –0.7% to 2.1%; and
- the sinusoidal flow errors for the VENTX 5A and VENTX 6A programs ranged from –3.9% to 1.3% and –4.1% to –1.1%, respectively, for minute ventilations spanning 15-200 L/min. These errors decreased to –1.9% to 1.3% for the VENTX 5A at minute ventilations >80 L/min. Also, omission of the two lowest tidal volumes (1.0 and 1.5 L) for the VENTX 6A narrowed the error range to –3.1% to –1.5%.

6. Metabolic carts of the latest generation use pneumotachographs to determine volume. They measure the decrease in pressure of airflow across a screen or through a tube. Bernouilli's law (flow is proportional to the square root of the pressure difference) then enables one to convert a pressure change to a volume. These instruments have been

demonstrated to have good precision and accuracy up to flow rates of 120 L/min BTPS (body temperature, ambient pressure, and saturated with water vapor) (Porszasz et al. 1994).

7. Gas collection should be carried out for a minimum of 30 s at each workload.

8. Gas temperature and barometric pressure should be measured to ±0.2° C and ±0.5 mm Hg, respectively. If a mercury barometer is used, a correction factor for metal expansion at room temperature should be applied (table 8.1). The National Association of Testing Authorities (1988) has produced an excellent technical note on the calibration of barometers.

9. The timing devices should be accurate to ±0.1 s.

Gas Analysis

1. Electronic oxygen and carbon dioxide analyzers should have an absolute accuracy of at least 0.05%. It is important that they are warmed up for at least an hour to eliminate electrical drift.

2. The analyzers must be calibrated prior to any testing and at regular intervals over the expected physiological range of measurement (O_2: 18-15%; CO_2: 3-5%) with either gases of alpha grade that have been prepared gravimetrically or gases of beta grade that have been verified chemically on a Lloyd-Haldane or micro-Scholander analyzer. For example:

	CO_2 (%)	O_2 (%)
Atmospheric air	0.03	20.93
Tank 1	3.25	18.05
Tank 2	5.08	14.85

Wasserman et al. (1994, p. 448) approve of just a two-point calibration check; they recommend atmospheric air and a previously verified calibration gas whose CO_2 and O_2 concentrations are in the middle of the anticipated range of the mixed expirate. Nevertheless, if this method is adopted, the accuracy of the analyzers should still be checked regularly using multiple gases.

3. Since most analyzers are pressure dependent and therefore flow dependent, it is essential that the resistance to flow be identical for calibration and measurement.

4. The gas sample is usually passed through a desiccant column of anhydrous calcium sulfate or calcium chloride. The calculations in the final section of this chapter therefore assume that the CO_2

Table 8.1 Temperature Correction Factors for Mercury Barometers (Glass Scale)*

	mm Hg														
°C	640	650	660	670	680	690	700	710	720	730	740	750	760	770	780
12	1.33	1.35	1.37	1.39	1.41	1.43	1.45	1.47	1.49	1.51	1.54	1.56	1.58	1.60	1.62
13	1.44	1.46	1.48	1.51	1.53	1.55	1.57	1.60	1.62	1.64	1.66	1.69	1.71	1.73	1.75
14	1.55	1.57	1.60	1.62	1.65	1.67	1.69	1.72	1.74	1.77	1.79	1.82	1.84	1.86	1.89
15	1.66	1.68	1.71	1.74	1.76	1.79	1.81	1.84	1.87	1.89	1.92	1.94	1.97	2.00	2.02
16	1.77	1.80	1.82	1.85	1.88	1.91	1.94	1.96	1.99	2.02	2.05	2.07	2.10	2.13	2.16
17	1.88	1.91	1.94	1.97	2.00	2.03	2.06	2.09	2.11	2.14	2.17	2.20	2.23	2.26	2.29
18	1.99	2.02	2.05	2.08	2.11	2.15	2.18	2.21	2.24	2.27	2.30	2.33	2.36	2.39	2.43
19	2.10	2.13	2.17	2.20	2.23	2.26	2.30	2.33	2.36	2.40	2.43	2.46	2.49	2.53	2.56
20	2.21	2.24	2.28	2.31	2.35	2.38	2.42	2.45	2.49	2.52	2.56	2.59	2.62	2.66	2.69
21	2.32	2.36	2.39	2.43	2.47	2.50	2.54	2.57	2.61	2.65	2.68	2.72	2.76	2.79	2.83
22	2.43	2.47	2.51	2.54	2.58	2.62	2.66	2.70	2.73	2.77	2.81	2.85	2.89	2.92	2.96
23	2.54	2.58	2.62	2.66	2.70	2.74	2.78	2.82	2.86	2.90	2.94	2.98	3.02	3.06	3.10
24	2.65	2.69	2.73	2.77	2.82	2.86	2.90	2.94	2.98	3.02	3.06	3.11	3.15	3.19	3.23
25	2.76	2.80	2.85	2.89	2.93	2.98	3.02	3.06	3.11	3.15	3.19	3.23	3.28	3.32	3.36
26	2.87	2.92	2.96	3.00	3.05	3.09	3.14	3.18	3.23	3.27	3.32	3.36	3.41	3.45	3.50
27	2.98	3.03	3.07	3.12	3.17	3.21	3.26	3.31	3.35	3.40	3.45	3.49	3.54	3.59	3.63
28	3.09	3.14	3.19	3.24	3.28	3.33	3.38	3.43	3.48	3.53	3.57	3.62	3.67	3.72	3.77
29	3.20	3.25	3.30	3.35	3.40	3.45	3.50	3.55	3.60	3.65	3.70	3.75	3.80	3.85	3.90
30	3.31	3.36	3.41	3.46	3.52	3.57	3.62	3.67	3.72	3.77	3.83	3.88	3.93	3.98	4.03

*These values must be subtracted from the observed height in mm Hg.

Reprinted from K. Diem and C. Lentner, 1974, *Scientific tables* (Basel: Ciba-Geigy Ltd.), 255.

and O_2 percentages are for dry expirate. If the water vapor is not removed, this dilution of the CO_2 and O_2 reduces their concentrations as follows (Wasserman et al. 1994, pp. 458-460):

$$\% \text{ of True Value } = \frac{P - pH_2O}{P} \times 100$$

Hence, if after "drying," the humidity of the mixed expirate is 30% (pH_2O = 5.26 mm Hg at 20° C; assume P = 750.0 mm Hg), then true CO_2 and O_2 values of 4.0% and 17.0% will be reduced to 3.97% and 16.88%, respectively. The O_2 in this example will therefore be spuriously elevated by ~3% if no correction is made. It should also be noted that Nafion tubing, which only removes water vapor such that an equilibrium is achieved with ambient humidity, does not fully dry the gas unless a countercurrent of dry air is drawn continuously past its exterior.

5. Verifying room air: while accurate gas analyzers are critical for determining the percentages of O_2 and CO_2 in expired gas, accurate measurement of inspired gas percentages is also important, because in simple terms $\dot{V}O_2$ is calculated (see the final section of this chapter) as:

$$(\% \ O_2 \text{ in dry inspirate} \div 100 \times \dot{V}_I) - (\% \ O_2 \text{ in dry expirate} \div 100 \times \dot{V}_E)$$

An error in either the inspired or the expired gas fraction will therefore cause errors in the calculated $\dot{V}O_2$. In an attempt to circumvent this problem, the software of some $\dot{V}O_2$ systems allows the operator to alter the inspired values according to those measured by the gas analyzers during the calibration procedure. The measured values are then used to calculate $\dot{V}O_2$ instead of the standard room air values of 20.93% and 0.03% for O_2 and CO_2, respectively. However, this approach should be reviewed carefully, because a 0.035% error in O_2 will affect the calculated $\dot{V}O_2$ by ~1.0%. The following data illustrate this point. Samples of inside and outside air for Laboratory 1 were analyzed at Laboratories 1 and 2. At Laboratory 1 there appeared to be a 0.07% reduction in the ambient O_2 (table 8.2 on next page). When the same foil bags of inside and outside air were analyzed at Laboratory 2 (table 8.2), the results indicated that the air in Laboratory 1 was not contaminated. Rather, the results from Laboratory 2 indicated a potential alinearity problem with the O_2 analyzer at Laboratory 1 that caused this analyzer to read room O_2 as 20.85%/20.86% instead of 20.93%. Unless chemical gas analysis can be used to establish unequivocal contamination of room air, it is recommended that any $\dot{V}O_2$ software default values for inspired O_2 and CO_2 be set at 20.93% and 0.03%, respectively.

General Procedures

In addition to calibration of individual items of equipment, which has been covered in the previous six pages, accurate $\dot{V}O_2$max data depend on controlled preparation of the laboratory and subject. This section also covers the criteria for attaining $\dot{V}O_2$max and its biological variability, measurement error, and the calibration of indirect calorimetry systems.

Environmental and Laboratory Conditions

The temperature of the laboratory must fall between 18 and 23° C with a relative humidity <70%. Both these variables should be recorded on the data sheet.

The Subject

Be sure that you have observed all the items on the Pretest Preparation Checklist (page 13 in chapter 2). Note that the determination of $\dot{V}O_2$max should preferably be made at least 2 h after the ingestion of food.

1. Each subject must have become familiar with the breathing apparatus, the ergometer, and if a treadmill is used, the emergency stop procedure. This familiarization should preferably take place on a day prior to the test.

2. A trimmed piece of Elastoplast or similar material placed on the nose will ensure that sweating does not cause the noseclip to slip off during the test. Check the patency of the noseclip by asking the subject to try to expire through the nose.

3. If electrodes rather than a heart rate monitor are worn, a stress-testing vest will secure these and the lead cables to the body, thereby improving the quality of the electrocardiogram.

4. The subject and the investigator must have a communication system in place so that the subject can indicate the termination of the test.

5. The subject should have a cool-down period after the $\dot{V}O_2$max test. This is to minimize the venous pooling of blood, assist with the removal of lactate, and provide a transition from vigorous exercise to rest. The subject could therefore be asked to exercise at a low workload until the heart rate recovers to below 120 beats/min.

Table 8.2 Interlaboratory Comparisons of Percentages for Inside and Outside Air

	Nominal values		Lab 1 Measured values		Lab 2 Measured values	
	O_2	CO_2	O_2	CO_2	O_2	CO_2
†Lab 1 cal. Gas 1—BOC α gas	15.05*	2.51	15.05	2.49	n/a	n/a
†Lab 1 cal. Gas 2—BOC α gas	16.47	3.75	16.49	3.69	n/a	n/a
†Lab 1 cal. Gas 3—BOC α gas	17.94	5.00*	17.94	5.00	n/a	n/a
†Lab 1 cal. Gas 4—special BOC α gas	20.93	0.03	20.89	0.01	20.92	0.02
Lab 1 air from sample drying tube of $CaCl_2$	20.93	0.03*	20.86	0.03	n/a	n/a
Foil bag of Lab 1 air	20.93	0.03	20.85	0.02	20.92	0.04
Foil bag of Lab 1 outside air	20.93	0.03	20.86	0.01	20.93	0.03
‡Lab 2 calibration gas	15.10*	5.08*	n/a	n/a	15.10	5.08
Lab 2 air from sample drying tube of $CaSO_4$	20.93	0.03	n/a	n/a	20.93	0.03

*Set or zeroed/spanned analyzers on this gas.

Laboratory 1 calibrated analyzers with four α standard gases† (BOC Gases Australia Limited) and used software to fit a linear regression before the foil bags were measured. The analyzers at Laboratory 2 were calibrated on one chemically analyzed gas mixture‡ and room air that had been previously chemically analyzed at 20.93% v/v O_2 and 0.03%v/v CO_2.

Criteria for $\dot{V}O_2$max

1. A plateau in $\dot{V}O_2$ despite increases in workload.
2. An R value >1.10.
3. A 5 min postexercise blood lactate of >8.0 mmol/L.
4. Points 2 and 3 are merely supplementary to the main criterion (point 1) and do not in themselves indicate that $\dot{V}O_2$max has or has not been attained.

The strictness of the first criterion is currently under debate (Howley et al. 1995). The most frequently quoted leveling criterion is that of Taylor et al. (1955), who advocated an increment of <2.1 ml · kg^{-1} · min^{-1} for an increase in treadmill elevation of 2.5% at 11.3 km/h. However, it is not often appreciated that this criterion was based on the $\bar{X} \pm$ SD of 4.2 ± 1.1 ml · kg^{-1} · min^{-1} for the increase in $\dot{V}O_2$max that was associated with a step increment in their protocol. Taylor et al. stated that there was only a small chance of making an error in deciding that $\dot{V}O_2$max had been attained if the increase was less than two standard deviations for the expected mean rise in $\dot{V}O_2$. The absolute leveling criterion therefore depends on the magnitude of the workload increment.

5. Increases in work rate should be selected so that the incremental part of the protocol is completed within 8-12 min (Buchfuhrer et al. 1983). Such a procedure has been demonstrated to yield the highest $\dot{V}O_2$max, but differences with durations outside this range are small (Buchfuhrer et al. 1983). Unpublished data from the AIS and the SASI have demonstrated that both an incremental protocol of seven 4 min rowing stages and a 5 min all-out cycling performance test produce $\dot{V}O_2$max values equal to those attained during a continuous 1 min incremental protocol of approximately 12 min duration.

6. If the main criterion for $\dot{V}O_2$max is not reached, the subject should be asked to return two to three days later for an additional incremental exercise test. After a standard warm-up, commence this test three levels below the terminating point on the preceding one, using increments in workloads that are 50% of those in the original test.

7. All volumes expired/inspired on which $\dot{V}O_2$max is calculated should be over 60 s. If data are collected for shorter durations, adjacent sampling periods (e.g., 2 × 30 s) should be averaged.

Biological Variability of $\dot{V}O_2$max

1. The measurement of $\dot{V}O_2$max includes both technical error related to equipment calibration and biological variability, since the subject may not reproduce exactly the same effort when assessed on two or more occasions. Although the literature contains little information about the biological variation in $\dot{V}O_2$max, Katch et al. (1982) reported that the combined technical error and bio-

logical variability for $\dot{V}O_2$max was 5.6%. They furthermore concluded that biological variability accounted for 90% of the total error.

2. The Laboratory Standards Assistance Scheme of the Australian Sports Commission now requires laboratories that seek certification to submit data for duplicate measures of $\dot{V}O_2$max on a group of athletes. The intraindividual variability of these data is a combination of technical or equipment error and biological variability. Calculation of the technical error of measurement (TEM), which is the standard error of a single score (Dahlberg 1940), enables quantification of the precision or reliability with which a laboratory can measure $\dot{V}O_2$max.

3. Technical error of measurement data for $\dot{V}O_2$max from five certificated Australian laboratories yielded a mean of 2.2%, which is less than half the value reported by Katch et al. (1982). This suggests that the biological variation in $\dot{V}O_2$max is likely to be approximately 2% with well-calibrated equipment and well-habituated subjects. Technical error of measurement data can also be used to determine the probability that a change in $\dot{V}O_2$max is a true difference as a consequence of training or detraining rather than a result associated with measurement error and/or biological variation:

- Standard error of the difference between two successive measurements = TEM $\times \sqrt{2}$.
- 95% confidence interval for a time change = 1.96 $\times$ standard error. Hence, differences between two $\dot{V}O_2$max measurements falling at or outside this confidence interval are real changes at p 0.05.

Measurement Error Analysis

Even if one could eliminate the biological variability in the $\dot{V}O_2$max of a human subject, the need for precision of measurement is illustrated clearly by a mathematical analysis of the effects on the calculated $\dot{V}O_2$max of changes in ventilation, gas fractions, temperature, barometric pressure, and relative humidity. Such an analysis permits all but one variable to be held constant so that it is possible to determine this variable's effect on the calculated $\dot{V}O_2$max. This analysis (table 8.3) indicates that small changes in either of two variables—ventilation ($\dot{V}_I$) and the fraction of oxygen in the dry expirate (F_EO_2)—will most greatly impact the calculated $\dot{V}O_2$. A 5% error in the measured ventilation translates directly to a 5% error in $\dot{V}O_2$, while a 1% error in F_EO_2 (e.g., 0.18% O_2) translates to a 6.5% error in the $\dot{V}O_2$. On the other hand, a 1% error in barometric pressure (7.55 mm Hg), which is extremely large compared with errors obtained from a suitably calibrated mercury barometer, translates to only a 1% error in $\dot{V}O_2$. Furthermore, the errors in ventilation and F_EO_2 can be cumulative; for example, if ventilation is 5% high and F_EO_2 is 1% low, then the calculated $\dot{V}O_2$ will be 11.7% in error. This analysis illustrates that careful calibration of both the ventilation device and the O_2 analyzer are critically important if one is to minimize errors.

Calibration of Indirect Calorimetry Systems

The individual components of an indirect calorimetry system should be calibrated before and after each test, and it is also important to check the function of the overall system on a regular basis.

Component Calibration

The calibration of indirect calorimetry systems often involves the technique of component calibration. Calibration of the ventilation device is conducted with either pulsatile or constant flows to span the physiological flow rate of 10-250 L/min BTPS. Separate calibration of the gas analyzers at three points establishes their linearity over the range for dry expirate of 15-18% for O_2 and 3.0-5.0% for CO_2. The gold standard for ventilation calibration is a water-sealed spirometer such as a Tissot gasometer or Stead-Wells spirometer. Both these instruments have a precision cylinder with a known and constant cross-sectional area; hence, the volume and bell factor can be calculated from first principles:

$$\text{Volume} = \pi r^2 \times h$$

where:

r = cylinder radius

h = vertical displacement of cylinder

Bell factor (ml/mm) = volume (ml)/h (mm)

The gold standard for gas analysis is the manometric chemical method, as performed with the Lloyd Haldane or micro-Scholander analyzer; but commercial companies with precision scales provide high-quality mixtures that are determined gravimetrically (from the molecular masses of the component gases) with a claimed accuracy of ±0.02%.

System Calibration

Biological approaches to calibration can provide an idea of the function of an indirect calorimetry

Table 8.3 The Effect of Likely Variation in Ventilation, Gas Percentages in Dry Expirate, Temperature, Barometric Pressure, and Relative Humidity (RH) Upon Calculated Indirect Calorimetry Values

Likely error	F_EO_2 (% v/v)	F_ECO_2 (% v/v)	$\dot{V}_IATP$ (L/min)	$\dot{V}_TSTPD$ (L/min)	$\dot{V}_EBTPS$ (L/min)	$\dot{V}_ESTPD$ (L/min)	$\dot{V}O_2$ (L/min)	Error in $\dot{V}O_2$ versus reference values (%)	$\dot{V}CO_2$ (L/min)	R	Temp (°C)	P bar (mm Hg)	RH (%)
Reference values	**17.51**	**3.79**	**150.00**	**136.10**	**166.63**	**136.70**	**4.542**	**0.00**	**5.146**	**1.133**	**22.0**	**755.0**	**50.0**
+5% $\dot{V}_I$	17.51	3.79	***157.50***	142.90	174.96	143.53	4.770	+5.00	5.403	1.133	22.0	755.0	50.0
+1% F_EO_2	***17.69***	3.79	150.00	136.10	167.00	137.00	4.249	–6.46	5.158	1.214	22.0	755.0	50.0
+1% F_ECO_2	17.51	***3.83***	150.00	136.10	166.71	136.76	4.532	–0.23	5.195	1.146	22.0	755.0	50.0
+1% P bar	17.51	3.79	150.00	137.48	166.54	138.08	4.589	+1.01	5.198	1.133	22.0	***762.55***	50.0
+1% temp	17.51	3.79	150.00	135.99	166.51	136.60	4.539	–0.07	5.142	1.133	***22.22***	755.0	50.0
+1% RH/pH_2O	17.51	3.79	150.00	136.08	166.61	136.68	4.542	–0.02	5.145	1.133	22.0	755.0	***50.5***
Cumulative error													
$\dot{V}_I$ +5% and F_EO_2 – 1%	***17.34***	3.79	***157.50***	142.88	174.55	143.19	5.076	+11.73	5.390	1.062	22.0	755.0	50.0

*Reference values are shown in **bold** and the variable that has been altered in subsequent analyses is shown in ***bold italics***.

system in general. The calibration machines commercially available at present (see appendix) are not appropriate for use with elite athletes, but discussions are underway with manufacturers.

Biological Calibration. Calibration of the critical components before and after each test is a crucial step in the calibration of an indirect calorimetry system, but it is also important to attempt an integrated system calibration on at least an annual basis or when a new system is purchased. Biological approaches to calibration are a useful and expedient first step toward integrated calibration. Tables such as those published by Åstrand and Rodahl (1986) indicate "ballpark" figures for steady-state oxygen consumption at a variety of cycling work rates; for example, the $\dot{V}O_2$ = 1.5 L/min at 100 W, 2.8 L/min at 200 W, and 4.2 L/min at 300 W. While these figures ignore interindividual variation in the mechanical efficiency of cycling and assume calibrated cycle ergometers, they can provide a basic guide to the overall function of an indirect calorimetry system.

Calibration Machines. There have been a number of attempts to manufacture a mechanical device that can deliver to an indirect calorimetry system precise gas fractions and ventilations that mimic those of an athlete. Huszczuk et al. (1990) describe a calibrator that simulates $\dot{V}O_2$ up to 5.0 L/min by mixing room air and 21% CO_2 (balance N_2). However, the commercially available version from Medical Graphics Corporation (see appendix for details) simulates $\dot{V}O_2$ and ventilation of up to only 2.7 L/min and 120 L/min, respectively. This suggests that the instrument is targeted at hospital laboratories and not those that specialize in the physiological monitoring of elite athletes. A further limitation is that the commercial calibrator can be used only on systems that measure expiratory volume.

The AIS has recently developed a "$\dot{V}O_2$max" calibrator (Gore et al. 1997) that has the simulated capacity of an athlete and can calibrate both $\dot{V}_E$ and $\dot{V}_I$ systems. Compared with values measured by criterion indirect calorimetry systems, those from the calibrator demonstrated an accuracy of ~±2% and a precision of ~±1% for $\dot{V}O_2$ ranging from 2.9 to 7.9 L/min and ventilation ranging from 89 to 246 L/min. The "$\dot{V}O_2$max" calibrator provides a method to interrogate any $\dot{V}O_2$ system in terms of the accuracy of the component O_2 and CO_2 analyzers as well as the ventilation device and software used for data reduction.

Calculations

The remainder of this chapter shows the $\dot{V}O_2$ calculations when either the expired volume or inspired volume are measured.

When Expired Volume Is Measured

1. The volume of expired gas is measured under ATPS (ambient temperature and pressure, saturated with water vapor) conditions.

2. The volume of expired gas measured over one minute is known as either the pulmonary ventilation or minute ventilation ($\dot{V}_E$) and is usually expressed in L/min BTPS. It may be computed by either (a) multiplying the $\dot{V}_E$ L/min at ATPS by the appropriate conversion factor (Diem and Lentner 1974, p. 259) or (b) using the combined gas laws formula. Note that in this latter method, the temperature in degrees Celsius (°C) is converted to absolute temperature in Kelvin (K) by adding 273.16:

$$\frac{P_1 \times V_1}{T_1} = \frac{P_2 \times V_2}{T_2}$$

$$\therefore \dot{V}_E \text{ L/min BTPS} = V_1 \times \frac{P_1 \times T_2}{P_2 \times T_1}$$

where:

$V_1 = \dot{V}_E$ L/min ATPS

P_1 = barometric pressure of the ambient air minus the partial pressure of water vapor at the temperature of the expired gas (see table 8.4)

T_2 = 310.16 K (273.16 K + 37° C for body temperature)

P_2 = barometric pressure of the ambient air in mm Hg minus 47.1 mm Hg for the partial pressure of water vapor at body temperature

T_1 = 273.16 K plus the temperature of the expired gas in °C

3. The volume of the expired air needs to be adjusted to STPD (standard temperature of 0° C, standard pressure of one atmosphere or 760 mm Hg, and dry, indicating the absence of water vapor) for determining the $\dot{V}O_2$. To do this, multiply the $\dot{V}_E$ ATPS L/min by the appropriate conversion factor (Diem and Lentner 1974, pp. 260-269). The combined gas laws formula can also be used:

$$\frac{P_1 \times V_1}{T_1} = \frac{P_2 \times V_2}{T_2}$$

$$\therefore \dot{V}_E \text{ L/min STPD} = V_1 \times \frac{P_1 \times T_2}{P_2 \times T_1}$$

where:

$V_1 = \dot{V}_E$ L/min ATPS

P_1 = barometric pressure of the ambient air minus the partial pressure of water vapor at the temperature of the expired gas (see table 8.4), which is assumed to be completely saturated with water vapor

T_2 = 273.16 K

P_2 = 760 mm Hg

T_1 = 273.16 K plus the temperature of the expired gas in °C

4. The concentrations of O_2 and CO_2 in dry atmospheric air are constant at 20.93% and 0.03%, respectively, but these may be altered if the testing environment is not properly ventilated. It is also known that the remaining gas (79.04%), which comprises primarily N_2, does not participate in physiological reactions. The volume of inspired gas ($\dot{V}_I$) can therefore be calculated using what has frequently been called the Haldane transformation (Haldane 1912) but should, according to Poole and Whipp (1988), be more correctly attributed to Geppert and Zuntz (1888):

$$(\dot{V}_I \text{ L/min STPD}) \times (79.04) = (\dot{V}_E \text{ L/min STPD}) \times (\%\ N_2 \text{ in dry expired gas})$$

$$\therefore \dot{V}_I \text{ L/min STPD} = \frac{(\dot{V}_E \text{ L/min STPD}) \times (\%\ N_2 \text{ in dry expired gas})}{79.04}$$

where:

$\%N_2$ in dry expired gas = 100 – ($\%O_2$ in dry expired gas + $\%CO_2$ in dry expired gas)

5. $$\text{Volume of } O_2 \text{ inspired} = \dot{V}_I \text{ L/min STPD} \times \frac{20.93}{100}$$

$$\text{Volume of } O_2 \text{ expired} = \dot{V}_E \text{ L/min STPD} \times \frac{(\%\ O_2 \text{ in dry expired gas})}{100}$$

$$\therefore \dot{V}O_2 \text{ L/min STPD} = \text{volume } O_2 \text{ inspired L/min STPD} - \text{volume } O_2 \text{ expired L/min STPD}$$

6. $$\dot{V}O_2 \text{ ml} \cdot \text{kg}^{-1} \cdot \text{min}^{-1} = \frac{\dot{V}O_2 \text{ L/min STPD}}{\text{Body mass in kg}} \times 1000$$

7. The respiratory exchange ratio (R) may then be calculated:

$$R = \frac{CO_2 \text{ produced}}{O_2 \text{ uptake}}$$

$$R = \frac{\dot{V}CO_2 \text{ L/min STPD}}{\dot{V}O_2 \text{ L/min STPD}}$$

where:

$$CO_2 \text{ produced } (\dot{V}CO_2 \text{ L/min}) = \left(\dot{V}_E \text{ L/min STPD} \times \frac{(\%\ CO_2 \text{ in dry expired gas})}{100}\right) - \left(\dot{V}_I \text{ L/min STPD} \times \frac{0.03}{100}\right)$$

8. The ventilatory equivalents for oxygen and carbon dioxide can also be calculated. By convention these are expressed as:

$$\frac{\dot{V}_E}{\dot{V}O_2} = \frac{\dot{V}_E \text{ L/min BTPS}}{\dot{V}O_2 \text{ L/min STPD}}$$

$$\frac{\dot{V}_E}{\dot{V}CO_2} = \frac{\dot{V}_E \text{ L/min BTPS}}{\dot{V}CO_2 \text{ L/min STPD}}$$

When Inspired Volume Is Measured

1. If the volume-measuring device is on the inspiratory side of the circuit, it is necessary to measure the relative humidity with a hygrometer, and the computations are as follows:

$$\dot{V}_I \text{ L/min STPD} = V_1 \times \frac{P_1 \times T_2}{P_2 \times T_1}$$

where:

$V_1 = \dot{V}_I$ L/min ATPH (ambient temperature, pressure, and humidity)

P_1 = ambient pressure in mm Hg minus pH_2O, which depends on the temperature and relative humidity (thus the pH_2O at 24° C and 40% relative humidity is 8.95 mm Hg; that is, 40% of 22.38 mm Hg, which is the pH_2O when gas is completely saturated with water vapor at 24° C; see table 8.4 on next page)

T_2 = 273.16 K

P_2 = 760 mm Hg

T_1 = 273.16 K plus the temperature of the inspired gas in °C

2. $$\dot{V}_E \text{ L/min STPD} = \frac{\dot{V}_I \text{ L/min STPD} \times 79.04}{\%\ N_2 \text{ in dry expired gas}}$$

where:

$\%N_2$ in dry expired gas = 100 – ($\%O_2$ in dry expired gas + $\%CO_2$ in dry expired gas)

3. $$\dot{V}_E \text{ L/min BTPS} = V_1 \times \frac{P_1 \times T_2}{P_2 \times T_1}$$

where:

$V_1 = \dot{V}_E$ L/min STPD

P_1 = 760 mm Hg

T_2 = 310.16 K (273.16 K + 37° C for body temperature)

P_2 = ambient pressure in mm Hg minus 47.1 mm Hg for the partial pressure of water vapor at body temperature

T_1 = 273.16 K

4. All the subsequent calculations are performed as outlined in steps 5 to 7 for measurement of expired volume.

Many laboratories now have facilities for online acquisition, reduction, and display of the variables just discussed. This involves interfacing the electrical outputs from the physiological recorders with a microcomputer via an analog-to-digital converter (12 bit).

Table 8.4 Partial Pressure (mm Hg) of Water Vapor in Saturated Gas

°C	0	0.1	0.2	0.3	0.4	0.5	0.6	0.7	0.8	0.9
12	10.51	10.58	10.65	10.72	10.79	10.87	10.94	11.01	11.08	11.15
13	11.23	11.30	11.38	11.45	11.52	11.60	11.68	11.75	11.83	11.91
14	11.98	12.06	12.14	12.22	12.30	12.38	12.46	12.54	12.62	12.70
15	12.78	12.87	12.95	13.03	13.12	13.20	13.29	13.37	13.46	13.54
16	13.63	13.72	13.81	13.89	13.98	14.07	14.16	14.25	14.34	14.43
17	14.53	14.62	14.71	14.81	14.90	14.99	15.09	15.18	15.28	15.38
18	15.47	15.57	15.67	15.77	15.87	15.97	16.07	16.17	16.27	16.37
19	16.47	16.58	16.68	16.79	16.89	17.00	17.10	17.21	17.32	17.42
20	17.53	17.64	17.75	17.86	17.97	18.08	18.19	18.31	18.42	18.53
21	18.65	18.76	18.88	18.99	19.11	19.23	19.35	19.46	19.58	19.70
22	19.82	19.95	20.07	20.19	20.31	20.44	20.56	20.69	20.81	20.94
23	21.07	21.19	21.32	21.45	21.58	21.71	21.84	21.98	22.11	22.24
24	22.38	22.51	22.65	22.78	22.92	23.06	23.19	23.33	23.47	23.61
25	23.76	23.90	24.04	24.18	24.33	24.47	24.62	24.76	24.91	25.06
26	25.21	25.36	25.51	25.66	25.81	25.96	26.12	26.27	26.43	26.58
27	26.74	26.90	27.05	27.21	27.37	27.53	27.70	27.86	28.02	28.18
28	28.35	28.52	28.68	28.85	29.02	29.19	29.36	29.53	29.70	29.87
29	30.04	30.22	30.39	30.57	30.75	30.92	31.10	31.28	31.46	31.64
30	31.83	32.01	32.19	32.38	32.56	32.75	32.94	33.13	33.32	33.51
31	33.70	33.89	34.08	34.28	34.47	34.67	34.87	35.07	35.26	35.47
32	35.67	35.87	36.07	36.28	36.48	36.69	36.89	37.10	37.31	37.52
33	37.73	37.95	38.16	38.37	38.59	38.81	39.02	39.24	39.46	39.68
34	39.90	40.13	40.35	40.58	40.80	41.03	41.26	41.49	41.72	41.95
35	42.18	42.41	42.65	42.89	43.12	43.36	43.60	43.84	44.08	44.33
36	44.57	44.82	45.06	45.31	45.56	45.81	46.06	46.31	46.56	46.82
37	47.08	47.33	47.59	47.85	48.11	48.37	48.64	48.90	49.17	49.43

Reprinted from K. Diem and C. Lentner, 1974, *Scientific tables* (Basel: Ciba-Geigy Ltd.), 257.

chapter 9

Protocols for the Physiological Assessment of Team Sport Players

Lindsay Ellis, Paul Gastin, Steve Lawrence, Bernard Savage, Andrea Buckeridge, Andrea Stapff, Douglas Tumilty, Ann Quinn, Sarah Woolford, and Warren Young

This chapter provides generic descriptions of the test procedures and equipment common to a number of the field and court sports covered in later chapters. Most of the tests in this chapter are "performance" oriented, useful for obtaining an indication of athlete status. The physiological assessment of team sport athletes generally involves a series of tests that are easy to administer in the field environment, or sometimes in the laboratory, and that typically require little specialized equipment. Results give the coach an indication of an individual's strengths and weaknesses in relation to the various components of fitness being assessed, as well as providing a measure of progress the athlete has made in response to prescribed training programs. However, athletes should be brought into the laboratory for a more precise assessment of physiological fitness.

The following tests of speed, aerobic power, agility, anaerobic power/capacity, flexibility, leg power, and abdominal strength represent a general field testing battery. Other chapters describe sport-specific field tests. In all cases, the specific test procedures outlined in each sport-specific chapter should be the ultimate reference point for testing athletes from a particular sport.

The technical error of measurement (TEM) is essential information for the correct interpretation of results from one test to the next. For more detail on TEM calculation and interpretation, refer to page 83.

Environment and Subject Preparation

The test environment and subject preparation appropriate to specific sports are described in each sport chapter; however, the following generic guidelines may be relevant.

1. **Test environment.** Although the tests under discussion are often referred to as field tests, there are advantages in performing as many of them as possible in a controlled environment, such as in a laboratory or hall. Climate and floor surface will then remain reasonably constant, and distances can perhaps be marked permanently. Thus, unless stipulated in the sport-specific chapter, it is recommended that tests be conducted indoors. If tests such as sprints and the multistage fitness test are performed outside, it is essential that test conditions be as controlled as possible. If tests are done at the same time of day, there is a greater chance that temperature, humidity, wind, and radiation will be less variable. Details of these conditions should, in any case, be recorded. Footwear should be specific to the sport played and the testing surface used.

2. **Subject preparation.** Changes in the fitness of the subject should be the only variable that produces a change in the score on any test. Therefore one should encourage consistency in the condition in which a subject presents for testing.

Chapter 2 contains both appropriate checklists for the tester's use and an excellent pretest preparation questionnaire for the athlete to fill out.

3. **Equipment and tester.** The same equipment and tester should be used each time a test is conducted. Where possible, calibration of testing equipment should be completed immediately before each testing session or at regular intervals throughout the testing year. Aspects such as the encouragement given to subjects and the rest period allowed between repetitions of a single test or between different tests should remain constant from one test occasion to the next.

4. **Scheduling tests.** For any structured yearly training program, the coach and sport scientist should decide on and plan all testing dates in advance. Generally appropriate testing times are at the beginning and end of each training phase, allowing assessment of the effectiveness of each phase. The coach can then use this information to plan the next training phase at both the team and the individual level.

Equipment Checklist

This section lists the appropriate equipment for each of the tests. In cases in which the equipment is critical to test administration, readers are referred to the appendix for the brand and supplier contact details of at least one manufacturer.

1. Acceleration and speed test
   - ☐ Light gates accurate to 0.01 s and dual beam.
   - ☐ Measuring tape.
   - ☐ An indoor nonslip surface is recommended, as this is more conducive to better test precision.
   - ☐ Suitable light gates are available from Swift Performance Equipment; TAG Heuer systems can be purchased from AST Stopwatches and Alge Timing systems from Cambrian Timing. For details on all of these, see the appendix.
2. Aerobic power—multistage fitness test
   - ☐ An indoor nonslip surface at least 25 m in length.
   - ☐ CD or audio cassette player.
   - ☐ CD supplied with *20m Shuttle Run Test* (Australian Coaching Council 1998), or audio cassette supplied with *Multistage Fitness Test* (Brewer et al. 1988). See the appendix for information about these products.
   - ☐ Measuring tape (at least 22 m).
   - ☐ Marker cones.
   - ☐ Approximately 1-1.5 m width of floor space per person.
   - ☐ Stopwatch.
3. Agility—505 test
   - ☐ Light gates, dual beam, with 0.01 s resolution.
   - ☐ Measuring tape.
   - ☐ Masking tape and/or cones.
   - ☐ An indoor nonslip surface.

See the appendix for information about light gates (Swift Performance Equipment), TAG Heuer systems, and Alge Timing systems.

4. Anaerobic power/capacity—sprint and fatigue index tests
   - ☐ Repco Exertech-10 cycle ergometer (or a highly geared cycle ergometer—46:14/46:14), calibrated in the range of 500-1500 W.
   - ☐ Exertech work monitor unit (or software to acquire data at ~40 Hz).
   - ☐ Toe clips and heel straps to ensure that subject's foot cannot pull off pedal.
   - ☐ Two stopwatches.

Note that the Repco Cycle Company no longer manufactures this ergometer or work unit. Alternative cycles that may be suitable are Kingcycle and the SRM Performance Ergometer (see appendix for details).

5. Flexibility—sit-and-reach test
   - ☐ A box 30 cm in height, at least 30 cm in breadth, and having a vertical footrest surface on its front side. A horizontal rod is affixed to the top of the box, centered to the footrest. It is marked at 1 cm intervals, with the zero mark level with the front surface of the box. There should be 20 cm of the rod scale overhanging the front of the box while 30 cm runs along the breadth of the box.
6. Leg power—countermovement jump
   - ☐ Contact mat (78 × 52 cm).
   - ☐ Timing module.
   - ☐ Computer with appropriate software.

A suitable system is available from Swift Performance Equipment (see appendix).

7. Leg power—vertical jump
   - ☐ Yardstick jumping device.

This system is available from Swift Performance Equipment (see appendix).

Or:

- ☐ Vertec jumping device.

This device is available from Sports Imports (see appendix).

Or:

- ☐ Wall-mounted vertical jump board marked in centimeters.
- ☐ Magnesium carbonate chalk.
- ☐ Cloth to wipe board clean.

8. Muscular strength—abdominal stage test
   - ☐ 2.5 and 5 kg weights.

PROTOCOLS: Team Sports

Acceleration and Speed Test

See table 9.1 for normative data.

Rationale: Athlete acceleration and speed are assessed with timing gates placed at different distances. (Refer to the chapters on specific sports for the relevant distances and data analyses.) Historically, a large number of sports have assessed acceleration and speed over 20 m, with 5 and 10 m split times. In many team sports, players rarely run more than 20 m in a straight line during a game or match. Thus, testing speed over distances greater than 20 m is often irrelevant as players do not have the opportunity to achieve such stride patterns during match play. Additionally, the standing start used in the sprint tests is specific to many sports that require players to run relatively short distances from a standing start.

Test Procedure:

1. Set timing gates at 0, 5, 10, and 20 m intervals.
2. Mark a starting line (0 m) and a finishing line (20 m) with masking tape and/or cones.
3. The starting position is with front foot up to the starting line.
4. The subject may start when ready, thus eliminating reaction time.
5. The subject sprints as fast as possible through to the finish line, making sure not to slow down before the finish gate.
6. Split times (at 5 m and 10 m) and final time (20 m) for three trials are recorded to the nearest 0.01 s.
7. The best time for 5, 10, and 20 m is used as the final result even if these times come from different trials.

Aerobic Power—Multistage Fitness Test

See table 9.2 (page 133) for normative data.

Rationale: The multistage fitness test has been selected as the test of aerobic power for a number of reasons. First and most importantly, the test has been found to be a sufficiently accurate estimate of aerobic power (Brewer et al. 1988; Leger and Lambert 1982). Secondly, the activity is similar to that of many team sports with respect to the stop, start, and change-of-direction movement patterns. Finally, it is a very time-efficient test with which a whole team or squad can be assessed simultaneously.

Test Procedure:

1. Calibrate the speed of the cassette tape drive according to the instructions at the beginning of the tape. (A 60 s standard time period is provided on the *Multistage Fitness Test* [Brewer et al. 1988] cassette tape. Run the cassette on your cassette player, and with a stopwatch [accurate to 0.10 s] check whether the duration of the standard time period is actually 60 s long. If it is shorter or longer than 60 s, correct the 20 m running distance as:

$$s = 20 \times \frac{t}{60}$$

where s is the corrected distance [m] and t is the time [s] measured by stopwatch.)

Note that a CD version of the *20m Shuttle Run Test* (Australian Coaching Council 1998) is also available, and CD player technology is far superior to that of cassette players so that tape stretch is not an issue. Most CD players, when lacking sufficient power, will simply stop playing rather

Table 9.1 Normative Data: Acceleration and Speed Scores for Australian Athletes

Sport	Squad	n	5 m sprint time (s)			10 m sprint time (s)			20 m sprint time (s)		
			Mean	SD	Range	Mean	SD	Range	Mean	SD	Range
Basketball	Female										
	State and national	110	1.20	0.08	0.96–1.40	–	–	–	3.48	0.21	3.04–4.34
	ACT	7	1.21	0.07	1.12–1.28	2.05	0.11	1.92–2.21	3.53	0.21	3.32–3.87
	Male										
	State league	28	1.06	0.06	1.00–1.18	1.79	0.07	1.68–1.95	3.05	0.11	2.86–3.26
	ACT	13	1.02	0.04	0.97–1.10	1.74	0.05	1.66–1.83	3.00	0.09	2.88–3.16
Cricket	Female										
	Test	40	–	–	–	2.07	0.1	1.9–2.3	3.52	0.2	3.2–3.9
	Male										
	ACT	11	–	–	–	1.9	0.1	1.8–2.1	3.2	0.1	3.0–3.4
	AIS pace bowlers	5	–	–	–	1.76	0.0	1.75–1.79	2.99	0.1	2.88–3.07
	AIS batsmen	6	–	–	–	1.78	0.1	1.70–1.86	3.04	0.1	2.82–3.19
Hockey	Female										
	SASI	20	–	–	–	2.00	0.08	1.87–2.21	–	–	–
	ACT	10	–	–	–	2.09	0.10	1.91–2.23	–	–	–
	Senior	409	–	–	–	1.98	0.07	1.80–2.24	–	–	–
	Male										
	SASI	15	–	–	–	1.86	0.05	1.77–1.93	3.18	0.10	3.02–3.34
	Senior	131	–	–	–	1.81	0.07	1.61–2.00	–	–	–
Netball	Female										
	Australian Open	16	1.16	0.06	1.09–1.27	1.98	0.08	1.87–2.13	3.40	0.12	3.19–3.60
	ACT	15	1.17	0.05	1.17–1.31	2.03	0.08	1.91–2.25	3.48	0.16	3.25–3.90
	AIS U/21	15	1.15	0.07	1.06–1.30	1.96	0.10	1.85–2.16	3.37	0.15	3.17–3.61
	QAS U/19	22	1.16	0.05	1.00–1.30	1.99	0.05	1.88–2.18	3.41	0.09	3.25–3.62
	SASI U/19	9	1.20	0.02	1.17–1.24	2.03	0.04	1.96–2.08	3.49	0.08	3.30–3.58
	NSWIS U/18	17	1.19	0.05	1.12–1.32	2.01	0.08	1.9–2.2	3.47	0.15	3.27–3.74
	U/17	76	1.23	0.08	1.08–1.52	2.05	0.09	1.89–2.37	3.49	0.15	3.04–3.93
	SASI U/17	11	1.18	0.04	1.12–1.26	2.09	0.33	1.92–3.06	3.41	0.12	3.26–3.56
Rugby union	Male										
	ACT	14	–	–	–	1.80	0.08	1.65–1.96	3.10	0.13	2.95–3.48
Soccer	Female										
	NSWIS	23	1.11	0.03	1.05–1.16	1.91	0.06	1.80–2.00	3.29	0.06	3.1–3.47
	SASI	47							3.47	0.19	2.47–3.87
	Male										
	NSWIS	37	1.03	0.05	0.93–1.98	1.74	0.04	1.66–1.81	3.04	0.09	2.83–3.24
	SASI	45	1.12	0.05	1.03–1.26	1.88	0.06	1.75–2.02	3.21	0.09	3.04–3.41
	SASI U/16	12	1.14	0.05	1.04–1.25	1.88	0.05	1.77–1.99	3.19	0.08	3.06–3.31
	SASI U/15	18	1.12	0.04	1.05–1.18	1.88	0.05	1.77–1.95	3.18	0.09	3.04–3.31

(continued)

Table 9.1 *(continued)*

Sport	Squad	n	5 m sprint time (s) Mean	SD	Range	10 m sprint time (s) Mean	SD	Range	20 m sprint time (s) Mean	SD	Range
Tennis	Female										
	NSWIS	3	1.14	0.06	1.08–1.20	1.95	0.07	1.88–2.01	3.43	0.09	3.36–3.54
	AIS	6	1.15	0.05	1.11–1.24	1.94	0.05	1.89–2.02	–	–	–
	Male										
	NSWIS	4	1.05	0.03	1.00–1.08	1.80	0.04	1.76–1.85	3.11	0.11	3.03–3.26
	AIS	5	1.05	0.03	1.02–1.10	1.78	0.03	1.75–1.81	–	–	–

ACT, AIS, NSWIS, QAS, and SASI are all state-level athletes, one level below national-level athletes.

ACT = Australian Capital Territory, AIS = Australian Institute of Sport, NSWIS = New South Wales Institute of Sport, QAS = Queensland Academy of Sport, SASI = South Australian Sports Institute.

than run slow or fast. Nevertheless, users may also check timing pips on the CD for the 20 m shuttle run test using a digital stopwatch.

2. Measure the "20 m" distance and mark it clearly with cones.

3. Allow the players to warm up by running and stretching.

4. Start the tape and ensure that the players listen carefully to the instructions. Begin the test at Level 1.

5. The tape emits a single beep at various intervals. A player must try to be at the opposite end of the 20 m track by the time the next beep sounds. After approximately each minute, the time interval between beeps decreases and running speed has to increase correspondingly.

6. The player needs to always place one foot on or behind the 20 m mark at the sound of each beep. Players who fail to reach the line at the sound of the beep must receive a warning that they will be eliminated if they are not at the opposite end of the 20 m track at the sound of the next beep.

7. When near exhaustion, players falling short of the 20 m line twice in succession have their test terminated and their score recorded. Their score is the level and number of shuttles immediately previous to the beep on which they were eliminated.

8. After completing the test, subjects should cool down by walking followed by stretching.

Agility—505 Test

See table 9.3 (page 135) for normative data.

Rationale: The basic movement patterns of many team sports require the player to perform sudden changes in body direction in combination with rapid movement of limbs. The whole-body movement can be in the horizontal plane, as when the player is dodging, or in the vertical plane, as when a player is jumping or leaping (Draper and Lancaster 1985). The ability of the player to use these maneuvers successfully in the actual game will depend on other factors such as visual processing, timing, reaction time, perception, and anticipation. Although all these factors combined are reflected in the player's on-field "agility," the purpose of most agility tests is simply to measure the ability to rapidly change body direction and position in the horizontal plane.

The 505 test is a relatively simple test that measures the time for a single, rapid change of direction over a short "up-and-back" course with a running start. The test is designed to minimize the influence of velocity while accentuating the effect of acceleration immediately before, during, and after the change of direction. Therefore, unlike what occurs with many other agility tests, the results are not contaminated by the influence of individual differences in running velocities before and after the directional changes (Draper and Lancaster 1985). Furthermore, the 505 test is one of the few tests that has been validated in the context of a team game; in this case, the sport was cricket (Draper and Pyke 1988).

Agility Test Preparation:

1. Using the measuring tape and masking tape, mark out the points of the course according to the diagram in figure 9.1.

2. Set up the timing lights at the 5 m mark to form a "gate" approximately 2 m wide.

Table 9.2 Normative Data: Multistage Fitness Test Scores for Australian Athletes

		Level + shuttle (number)			Predicted $\dot{V}O_2max$ ($ml \cdot kg^{-1} \cdot min^{-1}$)		
Sport	**Squad**	**n**	**Mean**	**Range**	**Mean**	**SD**	**Range**
Basketball	Female						
	ACT	11	9+10	8+2 to 11+5	46.3	4.0	40.2 to 51.4
	NSWIS U/17	15	9+9	7+7 to 12+13	46.3	–	38.8 to 53.7
	SASI	20	10+7	8+10 to 12+3	48.7	–	42.8 to 54.4
	WAIS and national	99	11+1	8+2 to 14+1	50.3	–	40.3 to 60.9
	Male						
	ACT	13	13+7	12+2 to 15+2	59.1	3.3	54.0 to 64.6
	SASI	28	13+4	10+3 to 14+13	58.1	–	47.5 to 64.2
Cricket	Female						
	Test	62	10+2	5+9 to 12+5	47.2	–	32.9 to 55.0
	Male						
	Test	27	11+3	9+6 to 13+4	50.9	–	45.0 to 58.1
	State	59	12+6	8+9 to 15+3	55.3	–	42.5 to 65.0
	ACT	11	11+9	9+1 to 14+5	52.7	6.3	43.6 to 62.0
Hockey	Female						
	ACT	10	11+5	9+7 to 15+3	51.5	–	45.5 to 65.1
	NSWIS	17	10+9	8+8 to 13+1	49.6	–	42.4 to 57.6
	SASI	29	9+11	6+10 to 12+5	46.6	–	36.3 to 55.0
	WAIS	98	11+1	9+1 to 13+5	50.3	–	43.5 to 58.4
	Senior	232	11+3	8+6 to 13+6	50.9	–	41.5 to 58.7
	Male						
	ACT	13	12+8	10+7 to 14+6	55.8	4.0	48.7 to 62.2
	NSWIS	18	12+7	9+11 to 14+3	55.8	–	46.8 to 61.4
	WAIS	48	13+1	8+9 to 15+1	57.3	–	42.5 to 64.5
	Senior	195	13+1	9+10 to 15+2	57.3	–	46.3 to 64.7
Netball	Female						
	National	16	11+2	9+1 to 13+5	50.6	3.9	43.6 to 58.6
	ACT	16	8+9	6+1 to 11+3	45.0	5.1	33.3 to 51.1
	AIS	15	11+3	7+10 to 14+6	50.8	5.6	39.5 to 62.4
	QAS U/19	18	10+4	9+7 to 11+5	47.8	3.1	43.6 to 54.6
	SASI U/19	9	10+1	8+2 to 11+3	46.9	–	40.3 to 50.9
	NSWIS U/18	17	9+5	7+1 to 12+11	45.2	5.7	37.4 to 55.4
	SASI U/17	10	10+11	8+10 to 13+1	50.0	–	42.8 to 57.3
	U/17	76	9+11	7+1 to 13+1	46.5	–	36.6 to 57.3
Rugby union	Male						
	ACT	15	12+7	8+9 to 15+5	55.7	4.9	42.4 to 65.6

(continued)

Table 9.2 *(continued)*

Sport	Squad	Level + shuttle (number) n	Mean	Range	Predicted $\dot{V}O_2max$ ($ml \cdot kg^{-1} \cdot min^{-1}$) Mean	SD	Range
Soccer	Female						
	National	19	10+9	9+1 to 13+12	49.4	4.1	43.6 to 60.0
	Junior	31	10+1	8+2 to 12+2	47.3	3.8	40.5 to 54.2
	Youth	15	9+7	6+7 to 11+7	45.2	4.8	35.4 to 52.2
	NSWIS	23	11+5	7+1 to 14+11	50.0	6.0	39.0 to 61.0
	SASI	47	9+8	6+1 to 12+5	45.2	4.8	35.4 to 52.2
	Male						
	Olympic	22	13+8	11+4 to 15+5	59.3	–	51.1 to 65.4
	AIS youth	16	13+5	10+8 to 15+1	58.6	4.0	49.0 to 64.3
	National U/17	37	13+0	10+6 to 15+0	56.9	3.6	48.3 to 64.0
	NSWIS	37	12+6	9+1 to 15+13	56.0	4.7	46.4 to 65.3
	SASI	59	12+12	10+3 to 14+7	57.0	–	47.5 to 62.6
Tennis	Female						
	AIS	6	11+10	10+8 to 13+1	53.2	3.2	49.3 to 57.4
	NSWIS	3	9+5	8+2 to 10+8	46.0	4.1	41.4 to 49.3
	Male						
	AIS	5	13+7	12+2 to 15+6	59.1	5.1	53.7 to 62.2
	NSWIS	4	11+4	11+1 to 13+7	53.1	3.5	50.4 to 58.2

ACT = Australian Capital Territory, AIS = Australian Institute of Sport, NSWIS = New South Wales Institute of Sport, QAS = Queensland Academy of Sport, SASI = South Australian Sports Institute, WAIS = Western Australian Institute of Sport.

1. ACT, AIS, NSWIS, QAS, SASI, and WAIS are all state-level athletes, one level below national-level athletes.

2. Australian sports use only level and shuttle number since this is the relevant performance measure. A direct measure of $\dot{V}O_2max$ should be made in the laboratory to derive a true measure of aerobic power.

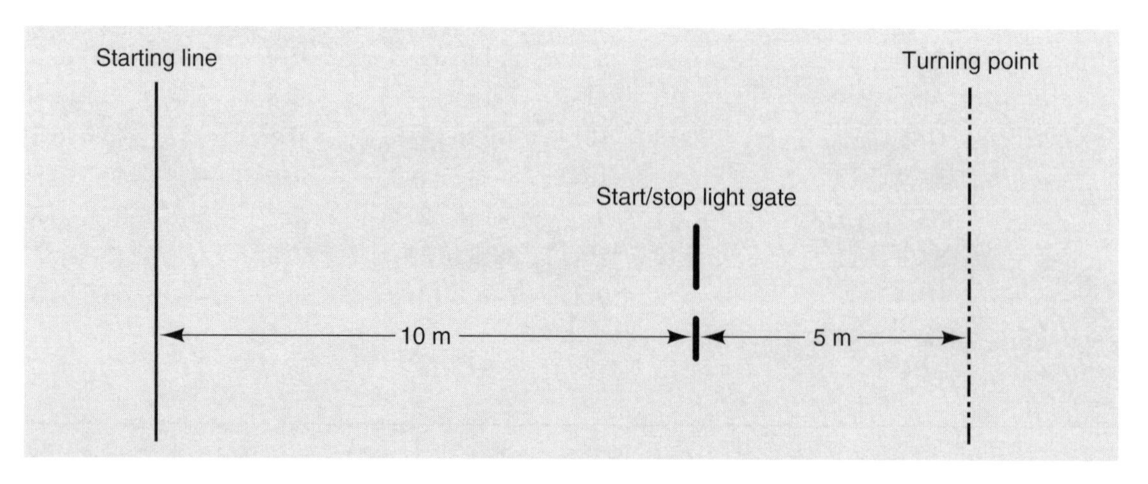

Figure 9.1 Equipment setup for the 505 agility test.

Agility Test Procedure:

1. Instruct the subject to assume the starting position at the start line.
2. After a signal that the light gates are set, the subject should start when ready.
3. Subjects should sprint from the starting line, through the light gates to the zero line, where they are required to turn on either the left or right foot and then accelerate off the line back through the light gates.
4. Subjects may slow down only after passing through the light gate for the second time.
5. Record to the nearest 0.01 s the time taken to cover the 10 m distance from the light gate to the turning point and back.
6. Subjects complete three trials turning on their preferred foot. Alternatively, they may perform three trials turning on both the left and right feet, for a total of six trials.
7. The fastest time is recorded as the best score.

Table 9.3 Normative Data: 505 Agility Test Data for Australian Athletes

			Right foot (s)			Left foot (s)		
Sport	**Squad**	**n**	**Mean**	**SD**	**Range**	**Mean**	**SD**	**Range**
Basketball	Female							
	ACT	7	2.59	0.16	2.39–2.84	2.56	0.18	2.38–2.88
	Male							
	ACT state league	13	2.20	0.09	2.08–2.34	2.21	0.11	2.03–2.37
	State league	26	2.18	0.08	2.03–2.38	2.23	0.08	2.03–2.40
Hockey	Female							
	ACT	10	2.51	0.12	2.30–2.75	2.48	0.10	2.28–2.64
	SASI*	20	2.57	0.10	2.36–2.78	–	–	–
	WA	124	2.49	0.12	2.22–2.99	–	–	–
	Senior*	284	2.47	0.13	2.25–2.96	–	–	–
	Male							
	ACT	15	2.28	0.14	2.10–2.60	2.27	0.06	2.18–2.40
	WA*	52	2.26	0.12	2.03–2.54	–	–	–
	Senior*	74	2.30	0.11	2.10–2.78	–	–	–
Netball	Female							
	National senior	16	2.47	0.13	2.22–2.65	2.48	0.18	2.18–2.74
	ACT	15	2.49	0.10	2.35–2.75	2.49	0.08	2.40–2.63
	AIS U/21	12	2.40	0.10	2.33–2.54	2.44	0.10	2.36–2.60
	QAS U/19	22	2.47	0.11	2.32–3.65	2.48	0.09	2.30–3.71
	SASI U/19	9	2.57	0.05	2.48–2.67	2.57	0.08	2.43–2.72
	SASI U/17	11	2.50	0.07	2.40–2.67	2.51	0.08	2.39–2.66
Tennis	Female							
	AIS/VIS	12	2.38	0.08	2.21–2.46	2.43	0.09	2.31–2.58
	NSWIS	3	2.56	0.04	2.52–2.59	2.53	0.05	2.49–2.58
	Male							
	AIS/VIS	11	2.25	0.06	2.17–2.36	2.24	0.07	2.14–2.37
	NSWIS	4	2.31	0.06	2.27–2.40	2.23	0.03	2.20–2.26

*Tested on preferred foot only.

ACT = Australian Capital Territory, AIS = Australian Institute of Sport, SASI = South Australian Sports Institute, NSWIS = New South Wales Institute of Sport, QAS = Queensland Academy of Sport, VIS = Victorian Institute of Sport, WA = Western Australia.

ACT, AIS, NSWIS, QAS, SASI, VIS, and WA are all state-level athletes, one level below national-level athletes.

Anaerobic Power/Capacity: Front-Access Ergometer Sprint and Fatigue Index Tests—10-Second Maximal Ergometer Sprint

See table 9.4 for normative data.

Rationale: The 10 s cycle ergometer test was selected as a laboratory test to measure explosive/alactic power (Telford et al. 1989). Although it is a nonspecific test for running, Telford et al. (1989) showed the ability of this test to identify anaerobic power across a series of sports. This makes it a useful measure to assess the requirement in team sports to perform maximal or near-maximal sprints of short duration.

Test Procedure: With use of a Repco work monitor, the following procedure is appropriate.

1. Set the work monitor to the "High" range and press "Reset" to remove any previous readings.
2. At the completion of 10 s of work (accomplished as described immediately below; points 1-3,) press the "Hold" button as the subject continues pedaling to "wind-down."
3. Record the kilojoules on the display of the work monitor while the "Hold" button is still depressed. Then release the "Hold" button and depress the "Watts" key to ascertain the peak watts attained during the test.
4. Note that the work monitor gives a value in kilojoules that should be converted to joules by multiplying by 1000. The joule and watt values are then divided by the subject's body mass to give work in J/kg and power in W/kg.

With use of an integrated computer data acquisition and analysis system to generate peak power and work values, the following are the appropriate procedures:

1. After sufficient warm-up, including a practice start requiring the athlete to accelerate from a standing position to top speed (2-3 s) and then rest for 60 s, the subject should assume a standing, stationary position, with pedals at a 45° angle to the horizontal. The preferred foot should be up and forward to ensure maximum power on starting.
2. Instruct the subject to maximally accelerate to maximum power on the command "Three, two, one, go" and to maintain this power for 10 s. At the end of the sprint, give the command "Stop."
3. Verbal encouragement is allowed, but to avoid distraction of the subject this must have been part of the test familiarization process. Testers also need to ensure that their "Go" and "Stop" commands are clear and are distinct from the verbal encouragement used during the test.
4. Upon completion of the 10 s, the subject should continue pedaling to "wind-down."
5. Work in joules and peak power in watts are recorded at the end of the test. These figures are then divided by the subject's body mass to give work in J/kg and power in W/kg.

Note: If using the Repco work monitor, measure the work completed in the first 6 s by briefly depressing the "Hold" button and reading the accumulated work score. This is then recorded and used as a control measure for the repeat-effort test (see next section).

Limitations: One must acknowledge athlete motivation and the nonspecific nature of the exercise when interpreting the data.

Table 9.4 Normative Data: 10 s Work and Peak Power Scores for Australian Athletes

Sport	Group	n	Mean	SD	Range
			10 s work (J/kg)		
Basketball	Female	79	111.5	13.5	76–135
Hockey—field players	Female	465	129.7	10.4	106.0–168.9
	Male	359	150.5	13.5	106.1–187.0
			Peak power (W/kg)		
Basketball	Female	89	13.8	1.8	9.3–16.7
Hockey—field players	Female	474	16.08	1.42	12.11–19.10
	Male	355	18.93	1.75	12.21–23.13

Anaerobic Power/Capacity: Front-Access Ergometer Sprint and Fatigue Index Tests—5 × 6-Second Repeat-Effort Test (Cycle Ergometer)

See table 9.5 for normative data.

Rationale: This test is designed to identify the ability of the athlete to repeat short maximal efforts with minimal rest (Ward 1991; Fitzsimons et al. 1993). The test score is not highly correlated to either aerobic power, peak power, or 30 s anaerobic capacity, which therefore suggests that it measures a specific capacity. The test involves five repetitions of 6 s maximal sprints on the cycle ergometer, with 24 s recovery between each effort.

Test Procedure: The test procedures have been detailed for data collection using a Repco work monitor. If using computerized data acquisition, the reader should disregard references to the work monitor.

1. Subjects must start each sprint from a stationary position, with their pedals at a 45° angle.
2. The preferred leg should be uppermost and forward to enable a more powerful push-off.
3. Give the same starting commands before each sprint, including "Ready" and "Go" commands.
4. If using handheld timing, start two stopwatches on the first "Go" command. One is to indicate continuous time; the other is to measure each individual sprint.
5. The subject must pedal maximally for the full 6 s in each individual sprint.
6. It is important to remember that the first sprint must be within 95% of the score obtained for the first 6 s of the 10 s test. If it is not, the test is terminated and started again after a minimum 3 min rest period.
7. At the end of each 6 s sprint, press the "Hold" button on the work monitor unit (if using the Repco work monitor) and give the subject the "Stop" command.
8. Measure and record work and peak power for each sprint.
9. Reset watch measuring work time at the end of each sprint.
10. The subject may sit down and pedal slowly between each sprint.
11. Give a command to the subject to assume the starting position 6-7 s prior to end of the rest period.
12. Sprint start time is every 30 s on the running clock.
13. Reset the work monitor before each sprint once the subject has assumed the stationary start position.
14. Start the watch used to measure each individual sprint time on the "Go" command or when the subject begins pedaling, whichever occurs first.

Data Reduction:

1. Total work done by the subject (i.e., sum of five sprints) is expressed in joules per kilogram of body mass.
2. Work decrement is calculated as the sum of the five measured efforts divided by the potential maximal amount of work done (expressed as the highest work effort multiplied by 5). This value is then expressed as a percentage and subtracted from 100% to give a value for decrement. For example:

$$\begin{aligned}&\text{Work Decrement}\\&= 100\% - ([5000\ \text{J} + 4900\ \text{J} + 4800\ \text{J} + 4200\ \text{J} + 3900\ \text{J}] \div [5000\ \text{J} \times 5] \times 100)\\&= 100\% - (22{,}800\ \text{J} \div 25{,}000\ \text{J}) \times 100\\&= 100\% - 91.2\%\\&= 8.8\%\end{aligned}$$

3. Power decrement is calculated in the same manner as work decrement.

Limitations: The test is invalidated by pacing; therefore strict control must be maintained over the effort the athlete is exerting. It is important that the subject be encouraged to give a maximal effort for each sprint.

Interpretation: As this test was designed to identify ability to repeat maximal efforts with limited recovery time, the work decrement score is the more important variable. However, work decrement should be evaluated with respect to total work. That is, low work decrement and high total work constitute the best possible result, while low decrement associated with a low total work score would indicate that an athlete requires more short-term anaerobic power training. High decrement associated with high total work indicates a possibility that the athlete has poor ability to repeat efforts.

Flexibility—Sit-and-Reach Test

See table 9.6 for normative data.

Rationale: Flexibility refers to the range of motion that can be performed at a specific joint and reflects the ability of the muscle-tendon units to elongate without physical restrictions of the joint

Table 9.5 5 × 6 s Work, Work Decrement, and Power Decrement Scores for Australian Athletes

Sport	Group	n	Mean	SD	Range
			5 × 6 s total work (J/kg)		
Basketball	Female	87	298.6	44.2	202.0–372.0
Hockey—field players	Female	428	349.0	27.0	282.2–427.3
	Male	337	394.9	32.4	297.2–471.0
			Work decrement (%)		
Basketball	Female	80	7.6	4.2	1.5–32.0
Hockey—field players	Female	413	9.4	3.6	3.0–23.8
	Male	333	12.1	4.6	3.4–28.3
			Power decrement (%)		
Basketball	Female	80	6.4	2.9	1.1–21.3
Hockey—field players	Female	398	8.0	2.9	3.0–20.8
	Male	330	9.7	3.9	3.0–24.1

(Hubley-Kozey 1991). The accurate assessment of flexibility is joint specific and requires specialized equipment such as goniometers referenced precisely to anthropometric landmarks as outlined in chapter 7. The sit-and-reach test, although requiring a combined joint action movement, gives a crude approximation of flexibility around the hip joint. Despite the major limitations of this test (see page 104), a number of sports continue to use it because large amounts of normative data are available and test administration is expedited.

Test Procedure:

1. Prior to this test, subjects are asked to stretch their hamstrings and lower back.

2. Subjects sit on the floor in front of the sit-and-reach box/apparatus with the legs fully extended, placing their bare feet against the vertical surface of the apparatus.

3. One hand should be placed over the top of the other with the palms facing down, fingertips overlapping, and fingers outstretched; the elbows should be straight.

4. The subject leans forward as far as possible, sliding his or her hands along the ruler of the sit-and-reach box. It may be necessary for the tester to apply gentle pressure above the knees to ensure that leg extension is maintained. Full stretch is held for at least 2 s to avoid the effect of bouncing.

5. The distance the fingers pass beyond the toes is recorded as a positive score. If the fingers fall short of the toes or zero line, the score is recorded as a negative score. The best of three trials is recorded to the nearest 0.5 cm.

Leg Power—Countermovement Jump and Vertical Jump

See table 9.7 for countermovement jump normative data and table 9.8 for vertical jump normative data. Figure 9.2 shows the stages of the countermovement jump.

Rationale: The countermovement jump (CMJ) is a vertical jump without the arm swing. The arm swing in the traditional vertical jump-and-reach test has been shown to contribute approximately 10% to the jump height (Luhtanen and Komi 1978). Strength training of the shoulder muscles that swing the arms has also been shown to improve vertical jump performance (Narita and Anderson 1992). These findings indicate that improvements in the vertical jump score may not be entirely due to the function of the leg extensor muscles, which is what the vertical jump test is thought to measure (Young 1994). With elimination of the arm swing, the intent is that the CMJ reduces the skill/coordination requirement of the test and focuses the effort on the leg extensor muscles. Therefore the CMJ is preferable to the vertical jump as a test of leg power and for monitoring the effects of strength-training programs that target the leg muscles (Young 1994).

Although the CMJ has a very sound rationale for testing leg power, relatively few data have been collected so far (table 9.7). One reason may be the relative fragility of contact mats, but more robust versions are now available (see section on equipment earlier in this chapter).

Overall jumping ability incorporating arm swing, trunk extension, and leg extension can

Table 9.6 Sit-and-Reach Flexibility Scores for Australian Athletes

Sport	Squad	n	Mean	SD	Range
Basketball	Male				
	State	9	4.4	11.2	–15.5 to +19
Cricket	Female				
	Test	62	13.4	5.7	+3.5 to +29.5
	Male				
	Test	27	11.6	7.6	–12 to +22
	State	59	12.1	7.0	–5.5 to +28
Netball	Female				
	Australian Open	13	15.5	5.8	+4 to +25
	AIS U/21	12	17.3	6.2	+4.5 to +24
	SASI U/19	9	16.3	5.3	+6 to +21.5
	SASI U/17	12	11.6	5.9	–2 to +20
	U/17	76	11.5	6.7	–15 to +28
Sailing	Female				
	National	6	15		+8.0 to +20
	Male				
	National	39	14		–10 to +25.5
Soccer	Female				
	State	46	15.0	7.0	–6 to +28.5
	Male				
	SASI	12	7.4	12.4	–4 to +24
Softball	Female				
	State	59	12.1	7.0	–5.5 to +28

AIS = Australian Institute of Sport, SASI = South Australian Sports Institute.
AIS and SASI are state-level athletes, one level below national-level athletes.

be assessed from the vertical jump test. This may be relevant for sports in which height jumped is an important aspect of the game (e.g., volleyball, netball, basketball). Sport-specific modifications to the vertical jump test are described in the relevant chapters. Note that data from the Western Australian Institute of Sport indicate that jumps approximately 5 cm higher are attained when the same athletes are tested on the Yardstick as compared to a wall-mounted board. Therefore, it is imperative that data sheets specify the type of equipment used to conduct the test.

Test Procedure—Countermovement Jump:

1. The subject stands on the contact mat with hands on hips. Instruct the subject to keep the hands on the hips for the duration of the jump.

2. Instruct the subject to jump for maximum height and to execute a dip or countermovement immediately before the upward propulsion.

3. Instruct the subject to land on the balls of the feet in an upright extended position (i.e., full extension at the hips, knees, and ankles).

4. Once contact with the ground has been made, the knees are allowed to bend to soften the impact of landing. (If the knees bend before landing, a falsely elevated "flight time" is recorded that falsely increases the calculated jump height.)

5. No specific instructions should be given regarding the depth or speed of the countermovement; this could influence the results.

6. However, subjects should be encouraged to "experiment" during warm-up or practice trials to find their optimal countermovement conditions.

7. Generally, a very shallow or deep or slow countermovement does not produce the best jump height.

8. As soon as the subject returns to the mat from a jump, the computer screen immediately displays the height attained, which should be conveyed to the subject as feedback.

9. An unlimited number of trials should be allowed in order to reveal the true best performance of the subject. This usually requires four to eight jumps with a complete recovery (15 s or more) between trials.

10. To enhance the precision of the test (test-retest consistency), it is recommended that the subject display a reasonable level of consistency (e.g., variations of less than 1.5 cm between the top three jumps).

11. The score used as the subject's performance is the mean of the best three jumps.

12. An alternative is to retain the single best jump as the score.

Figure 9.2 The countermovement jump: (a) preparation, (b) countermovement, (c) jump, and (d) landing.

Table 9.7 Countermovement Jump Scores for Australian Athletes (cm)

Sport	Squad	n	Mean	SD	Range
Athletics	Female				
	NSWIS	25	39.6	4.2	29–46
	Male				
	NSWIS	33	51.3	6.3	33–63
	Club sprinters	16	45.0	5.1	37–56
Handball	Male				
	National	11	41.5	4.0	34–52
Netball	Female				
	AIS	11	33.3	2.1	29–37
Rugby	Male				
	AIS	9	40.5	4.9	35–50
Volleyball	Female				
	VIS	9	35.4	3.4	31–41

AIS = Australian Institute of Sport, NSWIS = New South Whales Institute of Sport, VIS = Victorian Institute of Sport. AIS, NSWIS, and VIS are state-level athletes, one level below national-level athletes.

Test Procedure—Vertical Jump:

Yardstick

1. The subject should stand side-on to the Yardstick jumping device.

2. Keeping the heels on the floor, the subject reaches upward as high as possible, fully elevating the shoulder to displace the zero reference vane.

3. An arm swing and countermovement are used to jump as high as possible with the subject displacing the vane at the height of the jump. The takeoff must be from two feet, with no preliminary steps or shuffling.

4. Record the distance between the reach and jump vane to the nearest 1 cm.

5. The subject performs a minimum of three trials but may continue as long as he or she is making improvements. The best of the trials is recorded.

Wall-Mounted Board

With use of a wall-mounted vertical jump board, the procedure is virtually identical to that with the Yardstick device.

1. After sufficient warm-up, the subject stands against a wall (either left or right side is permitted) with feet flat on the floor and the body close to the wall.

2. The subject chalks the fingertips, elevates the shoulder, and stretches out the arm and hand closest to the board, leaving a mark at the height of full stretch. The position of the initial mark is recorded.

3. From a stationary starting position and crouching to whatever depth is preferred, the subject must take off from two feet with no preliminary steps or shuffling. The subject leaps up as high as possible, leaving a chalk mark on the measuring board with the inner hand.

4. As this test involves some skill, three attempts are allowed with a minimum of 30 s between attempts. The distance between the initial mark and the highest jump is recorded.

Table 9.8 Vertical Jump Scores for Australian Athletes (cm)

Sport	Squad	n	Mean	SD	Range
Basketball	Female				
	State and national[W]	132	46.6	5.6	35–60
	ACT[Y]	9	46.3	4.0	40–51
	Male				
	ACT[Y]	28	69.5	4.8	60–82
Cricket	Female				
	Test[Y]	62	41.3	5.1	31–53
	Male				
	Test[Y]	27	52.6	9.5	22–78
	SA[Y]	59	54.8	8.3	38–69
	ACT[Y]	11	53.1	8.9	35–66
Hockey	Female				
	NSWIS[W]	14	44.6	5.1	35–53
	WA[W]	68	42	4	34–50
	Senior[W]	491	46	4	33–56
	Male				
	ACT[Y]	19	56	6	46–66
	WA[W]	21	56	8	40–68
	Senior[W]	363	56	5	40–77

(continued)

Table 9.8 *(continued)*

Sport	Squad	n	Mean	SD	Range
Netball	Female				
	Australian Open[Y]	16	53.4	4.6	45–61
	ACT[Y]	14	42	5.7	31–56
	AIS U/21[Y]	15	49.0	7.0	39–63
	QAS U/19[Y]	22	46.0	4.0	41–55
	SASI U/19[W]	9	47.6	4.6	40–53
	NSWIS U/18[Y]	17	47.6	5.4	63–40
	SASI U/17[W]	10	48.7	3.4	41–54
	U17[Y]	76	44.5	5.5	35–56
Soccer	Female				
	National[W]	19	42	3	36–47
	Junior[W]	31	39	6	22–50
	Youth[W]	18	38	4	30–44
	NSWIS[Y]	23	46	4.7	39–57
	Male				
	Olympic[W]	22	57	–	50–68
	National U/17[W]	37	52	5	42–61
	AIS youth[W]	16	56	5	44–64
	NSWIS[Y]	37	47	5.8	36–62
	SASI state[Y]	44	52.1	5.2	43–66
	SASI U/16[Y]	12	52.6	3.9	46–59
	SASI U/15[Y]	20	51.7	4.3	47–60
Softball	Female				
	ACT[Y]	12	47	6	39–56
Tennis	Female				
	AIS[Y]	6	40.3	5.2	32–48
	NSWIS[Y]	3	40	3	37–43
	Male				
	AIS[Y]	6	53.2	4.8	47–59
	NSWIS[Y]	4	54	3	50–58

Y = Yardstick protocol, W = wall-mounted board protocol.

ACT = Australian Capital Territory, AIS = Australian Institute of Sport, SASI = South Australian Sports Institute, NSWIS = New South Wales Institute of Sport, QAS = Queensland Academy of Sport, WA = Western Australia.

ACT, AIS, NSWIS, SASI, QAS, and WA are state-level athletes, one level below national-level athletes.

Muscular Strength—Abdominal Stage Test

See table 9.9 for normative data. Figures 9.3 through 9.8 illustrate six of the seven stages of the abdominal strength test.

You should be aware that strength testing is not a simple matter. Chapter 10 discusses the subject at some length, including the mechanisms of acute and chronic adaptation to strength training and ensuring accuracy and reliability of measurements (see page 147).

Rationale: The abdominal stage test is a graded test for abdominal strength. Each of the seven stages becomes progressively more difficult as the positions of the hands and arms are modified. These modifications plus the use of 2.5 and 5 kg disks places increasing stress on the abdominal musculature. Strict control over the technique being utilized provides a reliable indicator of abdominal strength. The aim in this test is to accomplish as many of the stages as possible.

Test Procedure:

1. The starting position for all stages is lying supine on the floor with a 90° bend at the knee. The feet—without shoes—should be comfortably apart, in contact with the floor, and not held.
2. The subject is allowed up to three attempts to pass each stage.
3. Each of the stages is completed successively.
4. All movements are to be conducted in a smooth, controlled manner.
5. The subject's score is the last stage completed successfully.
6. Any attempt is unsuccessful if the subject

- lifts either foot partially or totally off the floor,
- throws the arms and or head forward in a jerky manner,
- moves the arms from the nominated position,
- lifts the hips off the floor,
- fails to maintain a 90° angle at the knee, or
- is unable to complete the nominated sit-up.

The Seven Stages of Abdominal Strength

- **Stage one: Palms over knees.** Arms straight with hands resting on thighs. Move forward until the fingers are touching the patella.

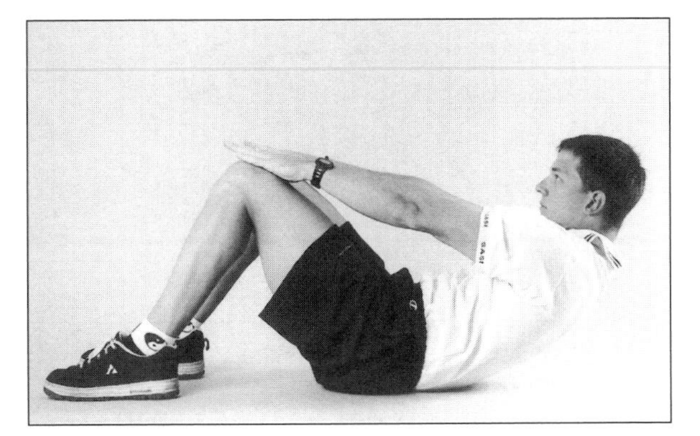

Figure 9.3 Stage one of abdominal strength.

- **Stage two: Elbows over knees.** Arms straight with hands resting on thighs. Move forward until elbows are touching the patella.

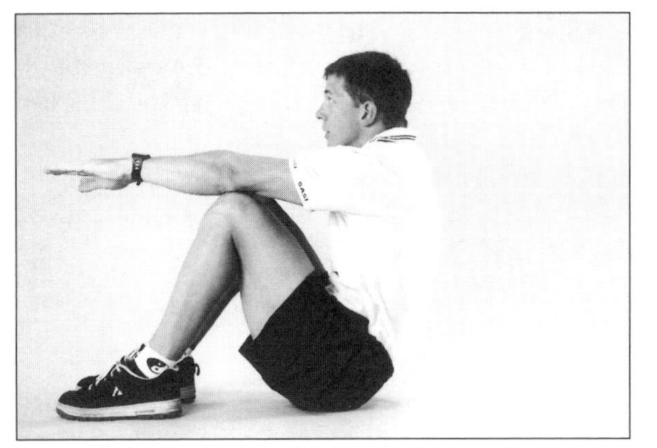

Figure 9.4 Stage two of abdominal strength.

- **Stage three: Forearms to thighs.** Arms across and in contact with the abdomen with hands gripping opposite elbows. Move forward until the forearms touch midthighs.

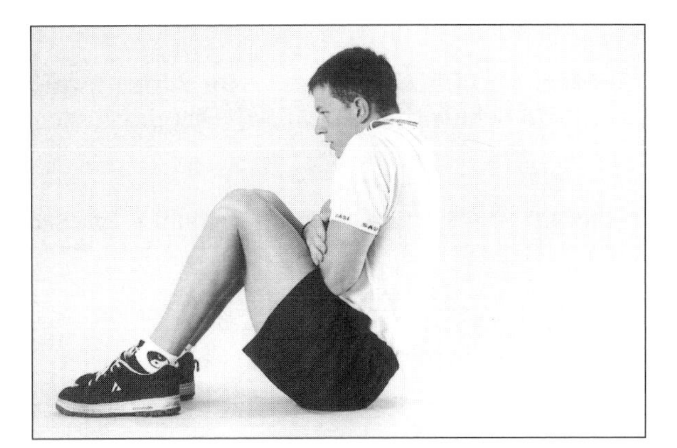

Figure 9.5 Stage three of abdominal strength.

- **Stage four: Elbows to midthighs.** Arms across and in contact with the chest with hands gripping the opposite shoulders. Move forward until the elbows touch the midthighs.

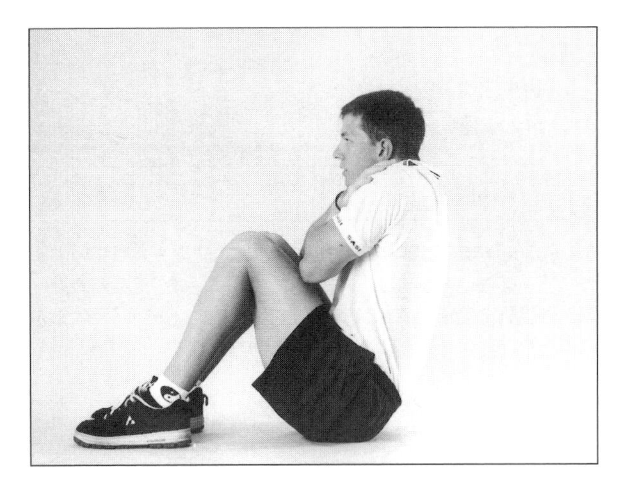

Figure 9.6 Stage four of abdominal strength.

• **Stage five: Chest to thighs.** Arms bent behind the head with the hands gripping opposite shoulders. Move forward until the chest touches the thighs.

Figure 9.7 Stage five of abdominal strength.

• **Stage six: Chest to thighs with 2.5 kg mass.** Arms bent behind the head with the hands crossed and holding a 2.5 kg mass. Move forward until the chest touches the thighs.

Figure 9.8 Stage six of abdominal strength.

• **Stage seven: Chest to thighs with 5 kg mass.** Arms bent behind the head with the hands crossed and holding a 5 kg mass. Move forward until the chest touches the thighs.

Table 9.9 Normative Data: Abdominal Stage Scores for Australian Athletes

Sport	Squad	n	Mean	SD	Range
Cricket	Female				
	Test	20	5	1.0	2–7
	Male				
	AIS Academy—batsmen	11	3	1.3	2–6
	AIS Academy—pace bowlers	9	4	0.9	4–6
	Sheffield Shield	59	4	1.1	1–5
	Test	27	4	0.4	3–5
Netball	Female				
	Australian Open	9	5	1.2	4–7
	QAS U/19	18	3	1	2–7
	NSWIS U/18	17	5	1	3–6
Soccer	Female				
	NSWIS	23	4	1.1	1–5
	Male				
	NSWIS	37	3	1.3	1–6

AIS = Australian Institute of Sport, NSWIS = New South Wales Institute of Sport, QAS = Queensland Academy of Sport.

AIS, NSWIS, and QAS as well as Sheffield Shield are state-level athletes, one level below national-level athletes. Test cricket players are international level.

part III

Strength and Power Assessment

chapter **10**

Introduction to the Assessment of Strength and Power

Peter Abernethy and Greg Wilson

Strength and power are critical to the performance of many athletic tasks. However, their assessment is currently problematic because of the enormity of the area and the limited work that has been completed within the field to date. (For a more complete discussion of this topic see Abernethy et al. 1995.) Among the key issues are a better understanding of the mechanisms of acute and chronic adaptation to strength training and ensuring that the measurement techniques are accurate and reliable. This book presents three different abdominal strength tests: the abdominal stage test described in chapter 9 and used by cricket (chapter 15), netball (chapter 20), and softball (chapter 26); the abdominal stability test used by rugby and tennis (chapters 22 and 28); and the abdominal endurance test used by sailing (chapter 24). This variety reflects the need for further validation studies. One should also note that each sport has its own sport science advisory committee that selects tests considered relevant to that sport. Given the controversy surrounding strength assessment, it is encouraging to see that various sports are evaluating the efficacy of a variety of test procedures. The reader should keep in mind that although the material presented in this volume represents the current thinking within the field, these ideas and the related procedures often still require deliberation and in some cases validation.

The development of better strength and power assessment procedures is contingent upon

- gaining improved insights into the mechanisms underpinning the acute responses and chronic adaptations to strength and power activity,
- accurately describing the effect of various conditioning regimens upon strength and power, and
- ensuring that measurement techniques are both reliable and valid (Abernethy et al. 1995).

Advancements will occur simultaneously across these three fields, and thus sport scientists are urged to undertake research, or at least be aware of developments, in these three areas.

This chapter introduces the field of strength and power assessment and describes one possible process that may aid in selecting and delimiting the use of strength and power protocols.

Definitions, Types, and Purposes of Assessment

Strength and power are sport-specific indexes. An athlete's strength (maximum force or torque) and/or power (rate of mechanical work, force × velocity or strength × speed) during competition are applied in particular postures, during particular movements, and at particular movement speeds. This raises the fundamental issue of the ecological validity of various forms of dynamometry, protocols, and procedures. For example, how valid is it to infer performance outcomes from tests involving movement patterns dissimilar to those in particular sporting contexts? This does not mean that one cannot obtain valuable information from various methodologies; rather it means that the relationship between sporting performance and the test should be scientifically established and not assumed.

Sport science has typically used three modes of dynamometry (i.e., isometric, isoinertial [isotonic], and isokinetic dynamometry). Subsequent chapters contain detailed descriptions of these modalities; consequently their discussion here is brief.

Two isometric indexes commonly assessed by sport scientists are the maximum voluntary contraction (Abe et al. 1992; Fry et al. 1991; Murphy et al. 1994) and the rate of force development (Aura and Viitasalo 1989; Jaric et al. 1989; Young and Bilby 1993). Maximum voluntary contraction is the maximum force developed during a maximal contraction that does not result in a change of joint angle, while isometric rate of force development represents the rate of force development over the initial 60-100 ms of a maximal isometric effort. Isoinertial dynamometry involves weight lifting during one or more repetition maximum tasks; the number of repetitions completed and the speed of body weight resistance activities; heights attained during vertical jumping (standing, countermovement, and drop jumps) with and without additional mass; and the distances that objects of various mass are thrown. Traditionally, these tasks have been described as isotonic (i.e., involving constant tension). However, such a description is inaccurate, as changes in joint angle along with changes in speed of movement mean that the level of tension is constantly in flux. "Isoinertial" (i.e., involving constant mass) may better describe the loading of these activities as it is characterized by a constant mass of the body, bar, or implement. Theoretically, during isokinetic assessment the limb moves at a constant speed. Data collected during isokinetic dynamometry can be used to plot torque power velocity and angle curves; to develop peak, whole-curve, and time-based indexes; and to make comparisons between agonists and antagonists and between contralateral limbs (Marshall and Taylor 1990; Marshall et al. 1990; Taylor et al. 1991).

Finally, two common considerations regarding the three forms of dynamometry are (a) any and all factors that may compromise reliability and internal validity (Thomas and Nelson 1990) and (b) participant safety. These issues must be addressed prior to assessment of athletes.

Data collected during strength and power assessment should be used in at least one of four ways:

1. To profile a sport (i.e., to determine the significance of strength and power in the optimal performance of a particular athletic pursuit)
2. To identify talent (i.e., to identify athletes who are better suited to particular sports than to others)
3. To diagnose strength (i.e., to identify an individual athlete's deficiencies and develop specific ameliorating training interventions)
4. To determine the effectiveness of training and rehabilitation interventions (Sale 1991; Schmidtbleicher 1992)

The purpose(s) fulfilled by a particular protocol should be scientifically established for a given sporting group prior to its extensive use on athletes. A good example of the preliminary work required in relation to protocol application is the work of Pryor (1995). Finally, it is important to remember that no single protocol or form of dynamometry provides information in relation to all these purposes for one athletic pursuit, let alone for all athletic pursuits (Abernethy et al. 1995).

A Process

Given that no single protocol or form of dynamometry provides all the information desired, there is a need to develop systematic approaches to strength and power measurement and the interpretation of resultant data. One such process is described as follows.

What Are We Measuring?

Do isometric, isoinertial, and isokinetic dynamometry approaches measure the same qualities within strength and power performance? Are these assessment modalities similarly sensitive to the effects of training and detraining? Do we need to differentiate between strength and power measures within a modality? These fundamental questions have received only a limited amount of consideration (Abernethy et al. 1995; Abernethy and Jürimäe 1996; Baker et al. 1994; Hortobagyi et al. 1989; Komi et al. 1982; Murphy et al. 1994; Pearson and Costill 1988).

Changes in strength and power in one form of dynamometry are not necessarily linked to changes of a similar magnitude for another modality. For instance, we recently found that changes in isoinertial (triceps extension and triceps push-down), isometric, and isokinetic (1.04-5.20 rad/s) strength following 12 weeks of isoinertial training of the triceps brachii were dissimilar in magnitude when transformed to the common effect size (ES) metric (ESs of 0.2, 0.5, and 0.7 are considered to be small, moderate, and large, respectively) (Abernethy and Jürimäe 1996). The triceps extension (ES = 1.50), triceps push-down (ES = 1.95), and isometric (ES = 1.30) indexes were all associated with large ESs, while the changes in isokinetic scores (ESs: 0.3-0.45) were far more modest. In addition, changes in the strength and power scores in one modality as a consequence of resistance training or detraining may be poorly correlated with changes in another form of dynamometry. For example, the Pearson product moment correlations between triceps extension and isometric (r = –0.31) and isokinetic strength (1.04 rad/s, r = –0.06; 3.13 rad/s, r = 0.10; 5.20 rad/s, r = 0.17) scores were poor. Baker et al. (1994) have reported similar findings with use of isometric and isoinertial modalities. Furthermore, despite the close relationship between strength and power indexes, the inference that they are the same or that they parallel one another within a given assessment modality (Hortobagyi et al. 1989) appears, on occasion, to be incorrect (Abernethy et al. 1996). Clearly, these questions have profound implications for the assessment of strength and power in athletes, and require further investigation.

Rationale for and Considerations With Strength and Power Assessment

We contend that good strength and power assessment is predicated upon a sound understanding of

- the mechanisms underpinning strength and power development (Abernethy et al. 1994; Huijing 1992; Sale 1992),
- the various forms of dynamometry,
- the athletes being assessed (the nature of their athletic pursuit, documented strengths and weaknesses of athletes from the particular sport, and the heterogeneity within the sample), and
- the procedures used to enhance strength and power.

Careful consideration is in order if the assessment of strength and power is to be of value to the athlete and coach. Key questions we need to answer about various protocols and forms of dynamometry are the following:

- How reliable is a particular measurement procedure?
- What is the correlation between the test score and either the whole or a part of the athletic performance under consideration?
- Does the test item discriminate between the performance of members of heterogeneous and/or homogeneous groups?
- Is the measurement procedure sensitive to the effects of training, rehabilitation, and/or acute bouts of exercise?
- Does the technique provide insights into the mechanisms underpinning strength and power performance and/or adaptations to training (Abernethy et al. 1995)?

It is both surprising and disturbing that many "sport scientists" do not state a priori what they want to determine through strength and power assessment. The reader needs to realize that we are not yet able to prescribe assessment procedures that allow us to address these five questions for every sport. Thus sport scientists must clearly indicate whether their role in the strength and power assessment of various athletes is as a researcher (see research phase of figure 10.1) or as a provider of sport science to athletes (see athlete assessment phase of figure 10.1). As illustrated in figure 10.1, the roles of the five questions listed, and the ways in which they are addressed, will

Phase I: Research

1. What do you want from your test?
2. Develop the rationale for the test (preferably based upon mechanistic research).
3. Ascertain the safety of the proposed test. (If risk of injury is too great, then terminate development.)
4. Determine how challenges to internal validity affect the test.
5. Ascertain whether the test is reliable. (If test is unreliable, terminate its development, as an unreliable test is an invalid test.)
6. Ascertain whether the test is valid. (If the test is invalid, determine whether through serendipity it addresses any other questions. If not, reject the test and move to step 8.)
7. Develop an appropriate database (e.g., normative data).
8. Communicate findings to the research and sport science fraternities.

(continued)

Figure 10.1 Sport scientists should distinguish between research and athlete assessment roles when collecting strength and power data.

Phase II: Athlete Assessment

1. What athletes do you want to assess?
2. What types of strength and power information do you want?
   Has the research phase been completed?
   If not, there is a need to re-enter the research phase.
   If it has, then there is a need to revisit this base work.
3. Preparation for athlete assessment prior to the test day:
   (a) Select protocol(s) and form(s) of dynamometry. Selection based on research.
   (b) Verify reported level of reliability and consider factors that may modify this.
   (c) Re-evaluate the reported levels of safety.
   (d) If necessary, complete assessor training.
4. On the day(s) of subject's familiarization: ensure that sufficient learning has occurred so that this threat to internal validity does not distort data.
5. At the time of subject assessment:
   (a) Ensure that equipment is calibrated and operational.
   (b) Ensure that subject completes an appropriate warm-up.
   (c) Ensure that subject receives appropriate instructions.
6. At the conclusion of the assessment:
   (a) Confirm data have been stored.
   (b) Recalibrate equipment and note whether there has been any drift.
7. Retrieve and analyze data:
   (a) Communicate results and their implications to the coach and athlete.
   (b) Where data are extraordinary, report this to the scientific community.
   (c) Where data appear to be unreliable or invalid, re-enter the research phase.

Figure 10.1 *(continued)*

vary with these contexts. The quality of research and service may be compromised where practitioners do not distinguish between the research and athlete assessment phases. Figure 10.1 presents the sequence of steps that an individual in either the research or athlete assessment phase should complete. Clearly, the research phase should always precede the athlete assessment phase.

Conclusion

In summary, strength and power assessment should be based upon a sound knowledge of what and who is being measured, how the measurement is conducted, and how various training regimens may affect the measures and the athlete. The question of what is being measured is fundamental to good sport science, as protocols and dynamometry may vary with the purpose of assessment. It is critical for practitioners to determine whether they will be approaching the assessment from a research or service-provider perspective (figure 10.1). There are limitations with, and should be delimitations in, the use of isometric, isoinertial, and isokinetic dynamometry. Furthermore, the different sensitivities of testing modalities and protocols, along with the fact that they appear to be measuring different qualities of strength and power, will probably necessitate the use of a number of tests. However, no assessment should be undertaken without full consideration of factors that may compromise reliability, internal validity, and athlete safety.

chapter 11

Limitations to the Use of Isometric Testing in Athletic Assessment

Greg Wilson

Isometric tests require subjects to produce a maximal force or torque against an immovable resistance that is in series with a strain gauge, cable tensiometer, force platform, or similar device whose transducer measures the applied force. These are very popular tests of muscular function and have been among the most widely used methods of strength assessment over the last 50 years. Indeed, in a comprehensive review of the strength literature, Atha stated: "Strength can be defined simply as the ability to develop force against an unyielding resistance in a single contraction of unrestricted duration" (1981, p. 7). Researchers have utilized isometric testing of muscular function and have extrapolated dynamic strength from isometric strength (Costill et al. 1968; Rasch 1957; Sale and Norman 1982; Young and Bilby 1993).

Isometric tests are generally performed to quantify the maximal force (or torque) and/or the maximal rate of force (or torque) developed (RFD). The maximal RFD is typically quantified as the greatest slope of the force-time curve over a time interval of 5 ms (Wilson et al. 1993) through to 60 ms (Christ et al. 1994). Other methods include determining the time to reach a certain level of absolute force, such as 500 N, or the time to achieve a relative force level, such as 30% of maximum (Hakkinen et al. 1985). Alternatively, the force or impulse (force × time) value in a specified time, such as 30 or 100 ms, has been used to quantify the RFD during an isometric test (Baker et al. 1994; Tidow 1990).

Isometric assessment procedures have been popular for several reasons:

- They are easily standardized and hence reproducible. Indeed, a number of studies have reported high levels of reliability with the use of isometric testing procedures (Bemben et al. 1992; Hortobagyi and Lambert 1992; Viitasalo et al. 1980).
- They are simple tests that require very little technique or skill and hence can be used with untrained and trained subjects.
- They are straightforward to administer and safe for subjects to perform.
- They use relatively inexpensive equipment.

Relationship to Performance

Despite the popularity of isometric tests of muscular function, a number of researchers have reported that these tests have a relatively poor relationship to dynamic performance measures that presumably would require a relatively high level of strength. A summary of this research (table 11.1) indicates that the relationship between isometric measures of muscular function and dynamic performance could be best described as poor. Many of the relationships were nonsignificant; and of those that were significant, the correlation coefficients were typically in the order of r = ~0.5, indicating that only approximately 25% of the variance was common.

Table 11.1 Summary of Research Reporting Relationship Between Isometric Tests and Dynamic Performance

Isometric test	Performance measure	Subjects	Correlation coefficient (r)	Reference
Variety of measures of upper body Fmax	Throwing velocity	Elite male water polo players (n = 21)	0.11–0.55	Bloomfield et al. (1990)
Max torque in a knee extension	VJ	Male university students (n = 38)	0.35	Considine and Sullivan (1973)
Fmax and RFD in a knee extension	VJ Maximum squat	Strength-trained males (n = 22)	VJ 0.1–0.52 Squat 0.57	Baker et al. (1994)
Fmax and RFD in a bench press action	SSPT 1RM bench press	Strength-trained males (n = 13)	SSPT 0.22–0.38 1RM 0.47–0.78	Murphy et al. (1994)
RFD in a bench press action	SSPT	Strength-trained males (n = 13)	0.42	Pryor et al. (1994)
RFD in a squat action	Peak power 6 s cycle	Strength-trained males (n = 30)	0.38	Wilson et al. (1995)
Fmax and RFD in a squat action	30 m sprint time	Athletic males (n = 15)	0.08–0.46	Wilson and Murphy (1995)
RFD in a squat action	VJ	Novice males (n = 18)	0.07	Young and Bilby (1993)
Fmax and RFD in a leg press action	Maximal running velocity	Male sprinters of varying ability (n = 25)	Fmax 0.62 RFD n/s	Mero et al. (1981)
Arm/Shoulder actions	Sprint swimming	Male and female swimmers of mixed levels (n = 58)	0.37–0.50	Strauss (1991)
Fmax and RFD in a leg press action	VJ	Males of varying athletic ability (n = 23)	Fmax n/s RFD 0.51–0.67	Viitasalo et al. (1981)
RFD in a leg press action	VJ	Elite male weight lifters (n = 14)	0.5–0.58	Hakkinen et al. (1986)
Fmax and RFD in a variety of lower-body actions	VJ	Active male physical education students (n = 39)	Fmax 0.22–0.42 RFD 0.35–0.46	Jaric et al. (1989)
Fmax and RFD in a bench press action	Seated medicine ball throw	Untrained males (n = 24)	Fmax 0.47–0.55 RFD 0.08–0.31	Murphy and Wilson (1996)

Pearson correlation coefficients were used to assess strength of association.

1RM = one-repetition maximum, Fmax = maximum force, n/s = not significant, RFD = rate of force development, SSPT = seated shot put throw, VJ = vertical jump.

Comparison to Other Testing Modalities

Several recent studies have directly compared the relationship of various testing modalities to performance. Murphy et al. (1994) compared isometric versus isoinertial testing in a bench press action. Thirteen trained male subjects performed a series of concentric and eccentric isoinertial tests using loads of 30%, 60%, and 100% (concentric) and 100%, 130%, and 150% (eccentric) of a one-repetition maximum (1RM) lift. The peak force exerted against each of the loads was recorded. In addition, the subjects performed an isometric test in which both peak force and RFD were quantified. These tests were compared against a variety of dynamic upper-body performances, including a seated shot put throw and a maximal bench press lift. For all of the performance measures assessed, the isoinertial and not the isometric testing modality showed the highest correlation with performance measures. The disparity in relationship between the testing modalities was greatest for the seated shot put throw. While the correlation between the shot throw and the isometric RFD was 0.38, the correlation between the shot throw and the concentric 30% test was 0.86.

In a similar study, Pryor et al. (1994) used the same six isoinertial tests. However, they recorded RFD rather than peak force. Again, in all of the performance tests, the isoinertial and not the isometric tests had the highest relationship to performance. Similar to the results of Murphy et al. (1994), the disparity between the isometric and isoinertial testing modalities was greatest for the seated shot put ($r = 0.42$ vs. 0.80, respectively).

Wilson et al. (1995) examined the relationship of isometric, isokinetic, and vertical jump tests to the peak power output produced in a 6 s cycle test. Thirty male subjects with previous weight-training experience participated in the study. The relationships between the various tests of muscular function and cycling performance were determined prior to and at the completion of 10 weeks of resistance training. Both before and after the training period, the isokinetic and vertical jump tests showed the highest relationship to cycling performance ($r = 0.51\text{-}0.73$), while the pre- and posttraining values for the isometric test were 0.38 and 0.03, respectively. The isokinetic and vertical jump tests were capable of discriminating between subjects with different cycling performance levels. However, the isometric tests were not able to effectively discriminate between the different groups. Other researchers have presented similar results. Fry et al. (1991) reported that although maximal weight-lifting strength effectively discriminated between female volleyball players of varying ability, isometric strength was not different between the groups. Similarly, Abe et al. (1992) reported no difference in isometric strength levels between elite and trained female alpine skiers, although they did report differences in eccentric strength.

Wilson and Murphy (1995) compared the relationship of isometric, concentric, and stretch-shorten cycle (eccentric-concentric) tests of muscular function to 30 m sprint time in 15 athletic males. The isometric tests were not significantly related to sprint performance and could not discriminate between sprinters of differing performance levels. However, the concentric tests were significantly related to performance and were capable of discriminating between sprinters of differing ability.

A number of other studies have also presented findings that tend to demonstrate the poor sensitivity of isometric assessment methods. Baker et al. (1994) reported that strength training-induced changes in weight-lifting strength were unrelated to the changes induced in isometric strength. Similarly, Wilson et al. (1993) reported that power training enhanced cycling and jumping performance, but had no effect on the maximum isometric RFD. Fry et al. (1994) reported that overtrained subjects experienced significant decreases in isoinertial (1RM) and isokinetic force. Maximal voluntary isometric force, however, was unchanged.

Underlying Mechanisms for Poor Isometric Validity

The rationale for the superiority of dynamic—rather than isometric—tests of muscular function in their relationship to dynamic performance is based on the neural and mechanical differences between isometric and dynamic contractions. Differences in motor unit recruitment have been shown to occur within isometric tasks with changes in the direction of force application (Ter Haar Romeny et al. 1982) or the performance of different tasks by the same muscle (Ter Haar Romeny et al. 1984). There are also distinct differences in activation patterns between isometric and dynamic contractions at the same joint angle (Nakazawa et al. 1993; Tax et al. 1990). Murphy and Wilson (1996) reported significant differences in both the activity and firing chracteristics of the musculature between isometric tests and dynamic performance. Consequently, the dynamic muscular function tests may be superior to the isometric tests because they invoke a neural response that has greater similarity to the dynamic performance of interest.

There are also large mechanical differences between isometric and dynamic muscular actions. Dynamic performances that involve stretch-shorten cycle actions also involve the use of substantial quantities of elastic strain energy (Asmussen and Bonde-Petersen 1974; Komi and Bosco 1978; Wilson et al. 1991). However, isometric actions do not benefit from the use of elastic strain energy. Further, the maximum force—particularly the RFD produced in isometric tests—is heavily dependent on the stiffness of the musculotendinous system. Wilson et al. (1994) reported a correlation of 0.72 between musculotendinous stiffness and isometric RFD and 0.63 between stiffness and maximum isometric force. However, musculotendinous stiffness was not related to eccentric force production or maximum concentric force.

During an isometric contraction, the contractile component will contract as the musculotendinous unit extends (figure 11.1). The extent and the rate of contractile component shortening will be proportional to the magnitude of the muscular contraction and the stiffness of the musculotendinous unit. Conceptually, a stiffer musculotendinous unit would enhance isometric force production, as compared to a more compliant system, not only due to reduced contractile component shortening velocity, but also due to a relatively longer length of the contractile component throughout the contraction. Further, the musculotendinous unit representes the link between the skeletal system and the contractile component of the muscles; and as such its stiffness will, to some extent, determine how effectively and rapidly internal forces generated by the contractile component are transmitted to the skeletal system.

These mechanisms underlying the relationship between musculotendinous stiffness and performance should operate in both isometric and concentric conditions; however, they will have a greater influence on the isometric force production. During the concentric activity, the contractile component is shortening dominantly due to the actual movement. In contrast, during an isometric activity, the shortening of the contractile component occurs dominantly due to the extension of the musculotendinous unit. Therefore, although a stiffer musculotendinous unit may serve to reduce the overall length and rate of contractile component shortening, its effect should be less pronounced in concentric as compared to isometric activities.

Consequently, analogous to the neural differences outlined earlier, substantial mechanical differences exist between isometric and dynamic muscular actions. Hence dynamic tests of muscular function are superior to isometric tests in their relationship to performance, as performance is dynamic in nature and thus involves muscular mechanics that are similar to those imposed by the dynamic muscular function tests used in their assessment.

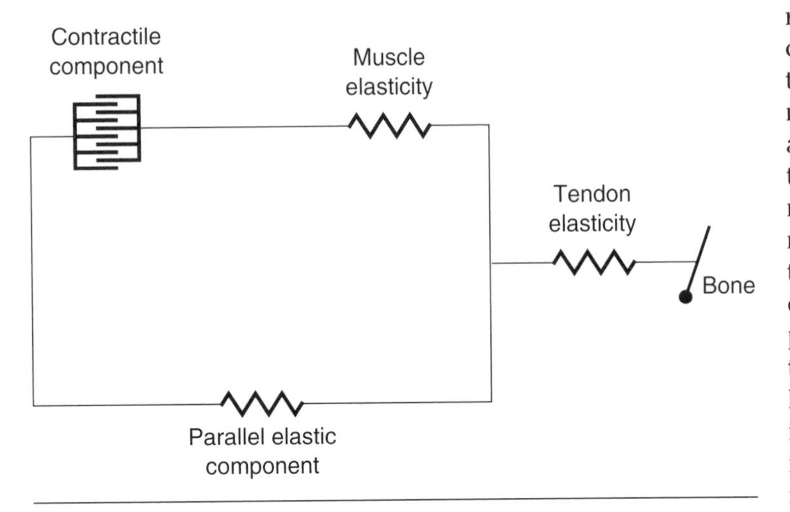

Figure 11.1 Model of muscle.

Conclusion

Isometric tests are popular tests of muscular function, often used as a basis for inferring the functional capacity of the musculature under dynamic conditions (Hakkinen and Komi 1986; Hakkinen et al. 1985; Schmidtbleicher and Buehrle 1987; Viitasalo et al. 1981). Such tests are highly reliable, are easy to administer, require little skill involvement, and are relatively inexpensive. However, a large body of research suggests that isometric muscular function tests are poorly related to dynamic performance and that dynamic forms of muscular assessment are superior. These findings appear to be attributable to the neural and mechanical differences between dynamic and isometric muscular actions. Therefore, it would appear that dynamic tests of muscular function provide a more valid assessment of the functional capacity of the musculature in dynamic movements than do isometric tests. Hence it is strongly recommended that isometric tests not be used for the purposes of athletic assessment.

chapter 12

Strength Assessment by Isokinetic Dynamometry

Tim Wrigley and Geoffrey Strauss

The aim of this chapter is to examine important aspects of isokinetic dynamometry as a tool for strength assessment. Isokinetic dynamometry is not a single, generic method: its validity, reliability, and utility vary with the assessment context. This chapter reviews equipment and reliability issues, pretest procedures and test protocols, and data collection and reporting principles. A final section suggests important directions for future research in this emerging area.

The Purposes of Strength Assessment

Sale (1991) identified four main purposes of testing strength and power of elite athletes:

1. Determining the relevance and relative importance of strength and power to performance
2. Developing an athlete profile
3. Monitoring training progress
4. Monitoring the rehabilitation of injuries

The first of these purposes underpins the application of strength testing in general—and isokinetic dynamometry in particular—to athlete assessment. Thus, this issue must be addressed first.

For many sports, the relative importance of strength and power in relation to other factors is uncertain. A given sport will lie somewhere along a continuum of strength/power importance: at one end are those sports in which these factors are critical, and at the other end are those sports in which these factors have no importance. Accordingly, sports along the continuum may be termed strength/power-*limited* or -*dependent* sports; strength/power-*related* sports; and strength/power-*independent* sports. Strength/power limits to performance may be in terms of absolute strength, or strength relative to body size, or both.

One should determine the relevance of strength/power for a given sport on the basis of the degree of association between strength measures and performance and/or the ability of such measures to discriminate between elite and subelite performers. A lack of association between strength and performance for a particular sport may indicate

1. a lack of relevance of strength/power to that performance,
2. a requirement for only a "threshold" level of strength (above which increased strength does not correlate with increased performance), or
3. the insensitivity of the assessment methods to performance-specific strength/power.

This chapter is abbreviated to fit the space requirements of the book; the reader should appreciate that many of the issues could have been discussed in much greater depth. The extensive reference list provides readers a reasonable starting point from which to further explore the complexities of isokinetic dynamometry.

Assessment methods may therefore be adopted, refined, or discarded on this basis. Different assessment methods may be chosen for the different categories of sports. Obviously one will devote more comprehensive efforts to assessment for strength/power-limited sports.

For some strength/power-limited sports with specific, clearly defined strength tasks (e.g., rowing, sprint swimming, cycling, and kayaking), there have been efforts to design very specific isokinetic testing procedures that attempt to mimic athletic tasks as closely as possible. Modifications to isokinetic dynamometers can be made for this purpose (Katch et al. 1974; Vandervoort et al. 1984; Pyke et al. 1979; Fry and Morton 1991; Klentrou and Montpetit 1991). However, in many sports, the strength requirements are less clearly defined and are often associated with the performance of a number of different tasks. It is not possible to design single, specific, strength assessment tests for such sports; these sports require generic methods of strength assessment. Many of these sports can be categorized as strength-related rather than strength-limited; that is, while performance is influenced by strength, strength is not a critical factor that directly limits performance. Examples may include many team, field, or court sports (e.g., basketball, soccer, hockey, football, netball, baseball, softball, tennis).

Independent of the overall relevance of strength/power to a given sport, an individual athlete's performance may be prejudiced by particular weaknesses. One may employ *intra*-individual assessments such as bilateral comparisons (same muscle group in opposite limbs) to identify weaknesses in particular muscle groups.

Note that the discussion in this chapter does not make a velocity-based distinction between "strength" and "power." While operational definitions of strength generally refer to isometric or low-velocity strength (measured as torque or force), power can be measured at any point on the velocity continuum. The notion that power is a high-velocity phenomenon is erroneous: peak muscle power in fact occurs at less than half peak muscle contraction velocity. Strength coaches have adopted the term "speed strength" to describe strength at high velocity and thus to avoid potential confusion. This chapter uses "strength" as the generic term to describe maximum muscle contractions at any velocity, qualified if necessary by a description of the specific conditions (e.g., velocity) under which it is recorded.

Strength-Testing Methods for the Elite Athlete

This chapter is relevant mainly to sports that can be considered strength-limited or strength-related. The discussion assumes that strength testing can be conducted in a sufficiently specific fashion for a given sport. Exercise scientists use the term specificity in several different ways. In physiological testing, specificity usually refers to the use of testing methods that bear a close physical resemblance to the movement patterns and intensity of the sporting activity. It is assumed—and sometimes known—that such methods also bear a close relationship to the physiological requirements of the sport. The ultimate proof of such specificity is a substantial correlation between test performance and athletic performance. This is the criterion for specificity that is used here.

One would expect such correlations to be higher for strength-limited sports than for strength-related sports. Also, inappropriate methods (i.e., nonspecific or error-ridden), or inappropriate data treatment with otherwise appropriate methods, would yield a lower correlation with performance. A review of the isokinetic performance correlation literature appears in Wrigley (2000); the strength assessment example presented later in this chapter refers to the portion of that literature relating specifically to soccer.

In the elite athlete context, it is likely that differences in actual competition performance levels among athletes may be relatively small. This can make it extremely difficult to demonstrate correlations between any physiological or biomechanical parameter and athletic performance (van Ingen Schenau et al. 1996). Discrimination among athletes thus requires a particularly sensitive strength measure. Without sufficient discriminating ability, a test may well indicate that a group of athletes as a whole are "strong," but this is of limited value to the individual athlete. Such a test may, however, still detect athletes who are clearly stronger or weaker than their athletic peers. A test incapable of detecting relatively small differences between muscle groups of an individual's left and right limbs, or of detecting changes over time, is likewise of questionable value.

Using Isokinetic Dynamometry

Isokinetic dynamometry has been used to assess elite athletes since its inception in the late 1960s, for each of the four purposes outlined by Sale (1991) (page 155). Furthermore, for two decades it has been the standard research tool used by physiologists to investigate dynamic muscle function of single muscle groups.

Isokinetic dynamometry can be performed under a range of conditions—of angular velocity, positioning, range of motion, contraction mode, movement sequence, and so on—from which a wide range of measurement parameters can be derived. Therefore, it is not accurate to regard isokinetic dynamometry as a single, generic method. The fact that a given aspect of isokinetic methodology has been shown to be valid and reliable, or that a given measure has been shown to be useful, does not imply that isokinetic dynamometry as a whole is reliable, valid, and useful in all contexts. Likewise, one should not regard evidence of limitations of particular dynamometers, or particular applications, as evidence against the overall utility of isokinetic dynamometry.

When one considers the currently available isokinetic dynamometers, the adjective "isokinetic" is perhaps somewhat limiting. All current dynamometer models also support isometric and isotonic modes, and some systems support more novel modes such as "isoacceleration" (Westing et al. 1991).

All of these current dynamometer testing modes may have application in the assessment and training of elite athletes. However, we have as yet no clear indication of where isometric, isotonic (Mahler et al. 1992), and isoacceleration (Westing et al. 1991) modes may yield unique information (that is, not available by isokinetic means).

It is useful to consider the particular characteristics of isokinetic and other forms of strength assessment loading in terms of the independent variables, which are fixed or controlled, and dependent variables, which are uncontrolled and therefore free to vary (table 12.1). All dependent variables should be measured if potential sources of test variability are to be assessed. However, the mere presence of multiple dependent variables increases the scope for variability, as it increases

Table 12.1 Independent and Dependent Variables in Isokinetic Dynamometry and Other Forms of Strength Assessment

	Isometric	**Isokinetic (isovelocity)**	**Isoacceleration***	**Isoinertial (e.g., free weights, Isotechnologies)**	**"True" isotonic** (i.e., isotorque or isoforce)**
Displacement	Independent [constant (zero)]	Dependent (measured)	Dependent (measured)	Dependent (usually unmeasured)	Dependent (can be measured)
Velocity	Independent [constant (zero)]	Independent [constant]	Dependent (measured)	Dependent (usually unmeasured)	Dependent (can be measured)
Acceleration	Independent [constant (zero)]	Independent [constant (zero)]	Independent [constant]	Dependent (usually unmeasured)	Dependent (can be measured)
Mass and moment of inertia	Independent [constant]	Independent [constant]	Independent [constant]	Independent [constant]	Independent [constant (except gravitational torque)]
Torque (or force)	Dependent (measured)	Dependent (measured)	Dependent (measured)	Dependent (usually unmeasured)	Independent [constant]
Number of dependent variables	1	2	3	4	3

* Available only on (some) isokinetic dynamometers.

** For example, constant-torque mode on isokinetic dynamometer.

Modified from Kroemer et al. 1990, Figure 3-3.

the number of movement strategies that the athlete can choose to employ.

The sections that follow describe the general principles of isokinetic testing and conclude with a discussion of strength assessment in soccer as an example.

Equipment

A complete list of comparative features of the various dynamometers does not appear here, as such lists tend to be out of date as soon as they are printed (e.g., Sale 1991). It is better to understand the benefits and limitations conferred by various dynamometer capabilities so that one can make an appropriate choice of testing technology. The reader may refer to Wrigley and Grant (1995) for discussion of all the factors one should consider in choosing an isokinetic dynamometer. Readers should note that a number of dynamometer manufacturers have left the market in recent years, so that only a few machines are currently commercially available. However, the machines no longer commercially available will remain in service for the foreseeable future, so the discussion here encompasses those also.

All isokinetic dynamometers are capable of testing concentric muscle contractions, with many of the later models used for athlete testing also capable of testing eccentric contractions. However, in some cases the dynamometer imposes restrictions on the way in which contraction modes can be combined (e.g., velocity or torque limitations when concentric and eccentric contraction modes are combined).

Comparison of Test Results Between Dynamometers

Tests performed on identical models of isokinetic dynamometers at different sites, using identical test protocols, yield essentially similar results from duplicate tests of the same subjects; for example, this has been shown for concentric knee extension-flexion on the Cybex II (Molczyk et al. 1991; Thigpen et al. 1990) and Cybex 340 (Byl et al. 1991) and for Cybex trunk strength dynamometers (Byl and Sadowski 1993).

Unfortunately, the literature is not particularly helpful regarding whether the results obtained on different types of isokinetic dynamometers are generally interchangeable. Although there are a number of studies on this issue (see table 12.2), most suffer from one or more limitations that make them of little practical use. Some of these limitations are outlined in table 12.2 (under "Limitations"). These studies have shown some differences between the results obtained with the different dynamometers investigated, under the particular conditions of the studies. However, the conditions were such that the results are often not generalizable. In some cases, the actual magnitude of the differences was not reported. Thus, it is not established whether data recorded on different brands of dynamometer, under commonly employed conditions, can be used interchangeably.

Despite the equivocal research, however, one would have a number of reasons to expect differences in the operating principles of some dynamometers to lead to results that may not be directly comparable. One can draw information from published comparisons of the differing operating conditions available on a single dynamometer, which also exist between dynamometers. In this way, it is possible to identify what may be the minimum range of differences to expect between the brands of dynamometer that have different operating characteristics. For example, such studies provide information on what sort of differences one might expect when comparing the results of a typical Kin-Com test (discrete, single movements, with preloading, over a moderate range of motion) to a typical "Cybex-style" test (such as might be performed on a Cybex or Biodex dynamometer with continuous, reciprocal movements) with no preload, low-inertia attachments, and a somewhat greater range of motion.

The dynamometers essentially all measure known torque loads accurately (see chapter addendum); therefore discrepancies between results from different dynamometers are likely to be due to intrinsic differences in operating characteristics, the way the test protocols are implemented, or the way the torque signals are processed. While the exact reasons for this lack of agreement are speculative at this stage, possible factors include:

1. Characteristics that users may be able to control:
   - Different test positions for a given joint or muscle group (Smith 1993; Soderberg and Blaschak 1987; Walmsley and Szybbo 1987)
   - Body stabilization differences
2. Differences in operating characteristics (intrinsic or user selected):
   - The use of continuous, reciprocal movements versus single, discrete movements in test protocols (Vyse and Kramer 1990; Grabiner and Hawthorne 1990; Keskula and

Perrin 1994; Gleeson et al. 1994; Strauss et al. 1996)

- The use of different contraction sequences (Vyse and Kramer 1990; Grabiner 1994; Kroll et al. 1996)
- Extent of "torque overshoot" (see addendum to this chapter)
- Differing pre-constant-velocity loading; that is, the use of isometric torque preloads and/or acceleration "ramping" versus free acceleration (Gransberg and Knutsson 1983; Jensen et al. 1991; Harridge and White 1993; Bobbert and Harlaar 1992; Caiozzo et al. 1981, 1982; Kramer et al. 1991b; Narici et al. 1991; Rathfon et al. 1991; James et al. 1994; Gravel et al. 1988, 1990; Stam et al. 1993; Tis et al. 1993)
- Differences in the duration of the constant-velocity phase (which are usually a consequence of different pre-constant-velocity loading, just described) (Taylor et al. 1991; Brown et al. 1995)
- Constrained range of motion limits versus unconstrained range of motion (Leslie et al. 1990; see discussion of range of motion later in this chapter)
- Direct versus indirect measurement of torque (via force measurement), requiring lever arm length measurement (and the assumption of constant length) (see "Dynamometer Lever Length," page 169)

3. Processing of data:
   - Different criteria for determining the boundaries at which a torque curve is deemed to begin and end (Wilk et al. 1992; Hoens and Strauss 1994; Rothstein et al. 1983; Brown et al. 1995)
   - Different data-filtering strategies
   - Gravitational torque compensation (differences in, or absence of, methods) (Appen and Duncan 1986; Perrin et al. 1992) (see later section on "Gravitational and Other Torques Not Due to Muscle Contraction")
   - Parameter calculation differences (for example, see later section on "Measurement Parameters" and discussion of use of "raw" torque data versus mean torque-angle data)
   - Accuracy and linearity of calibration (see next section, "Calibration and Reliability")

Until we have more comprehensive research on measurement differences between dynamometers, and a better understanding of the reasons for any lack of agreement (and hopefully resolution), different brands of isokinetic dynamometers should be regarded—like different types of bicycle ergometers for anaerobic power testing, for example—as distinct devices that may not yield similar results.

Software

The software for isokinetic dynamometers often limits how the whole system can be used. In particular, the software typically imposes the choice of test movements (continuous, reciprocal movements vs. discrete, single repetitions). This may create problems for comparison of results between dynamometers and for reproducing the contraction patterns of sporting movements (such as eccentric-concentric, stretch-shortening cycles). Relatively few dynamometers support the necessary degree of flexibility for setting up and deriving data from such test protocols.

Calibration and Reliability

Calibration of the various transducers within isokinetic dynamometers deserves much more attention than it receives from manufacturers and users alike. It is a prerequisite for test-retest reliability and for comparisons between athletes.

Calibration

Regular calibration ensures the accuracy of isokinetic measurements. Experience and some research evidence suggest that isokinetic dynamometers do not go out of torque calibration readily (Levene et al. 1991; Johansson et al. 1987, 1988). Therefore, most manufacturers suggest a calibration frequency of once a month to verify that the dynamometer is still measuring within the specified accuracy. However, the various dynamometer systems provide grossly different support for calibration of their various transducers (torque, angle, velocity).

Torque

The facilities for calibration of torque (or force) range from provision of a range of standard calibration weights with some systems—allowing calibration of a dynamometer across its measurement range—to factory calibrations that are difficult to check or change. Some dynamometers use static loading for torque calibration, while others use dynamic loading through the horizontal plane at a slow velocity (taking only the peak torque as the criterion measure).

Table 12.2 Comparison of Test Results Between Dynamometers

Study	Dynamometers	Subjects	Age	n	Movement	Position	Mode
Bandy and McLaughlin (1993)	Cybex 6000, Cybex II	Nonathletic	23 mean	10 M 10 F	Knee extensors, flexors	Seated	Concentric Continuous reciprocal
Gross et al. (1991)	Biodex, Cybex II with Isoscan II software	Nonathletic	29 mean (21–40)	5 M 5 F	Knee extensors, flexors	Seated	Concentric Continuous reciprocal
Thompson et al. (1989)	Biodex, Cybex II+ with CDRC	Nonathletic	21–46	20 M 28 F	Knee extensors, flexors	Seated	Concentric Continuous reciprocal
Wilk et al. (1987)	Biodex, Cybex, Kin-Com	Athletic	? High school	23 M	Knee extensors, flexors	Seated ?	Concentric Continuous reciprocal (incl. Kin-Com ?)
Wilk et al. (1988)	Biodex, Cybex, Lido	Nonathletic ?	26 mean	31 (M & F)	Knee extensors, flexors	Seated ?	Concentric Continuous reciprocal (incl. Kin-Com ?)
Francis and Hoobler (1987)	Lido 2 Analog. Cybex II with CDRC	Nonathletic	29 mean (22–53)	10 M 11 F	Knee extensors, flexors	Seated	Concentric Continuous reciprocal
Warner et al. (1985)	Ariel, Cybex II	Nonathletic ?	26 mean	20 F	Knee extensors, flexors	Seated ?	Concentric Continuous reciprocal
Timm (1989)	Cybex 340, Merac	Nonathletic ?	25 mean (15–41)	24 M 16 F	Knee extensors, flexors	Seated	Concentric Continuous reciprocal
Pressly et al. (1991)	Cybex II, Kin-Com II	Nonathletic ?	24 mean	10 M	Knee extensors, flexors	Seated ?	Concentric Continuous for Cybex Discrete for Kin-Com

Velocity	Parameter(s)	Gravitational torque compensation	Time between tests	Results	Magnitude and direction of differences	Limitations
60, 180, 300	Peak torque	No	1 week	ICC = 0.72–0.89	Cybex II means 10–20% < Cybex 6000 means.	6000 would not be used without gravitational torque compensation.
60, 180	Peak torque, work	Not mentioned	2 to 5 days	ICC = 0.79–0.95	Cybex scores ~15% < Biodex scores. Peak torque diffs. < work diffs.	Nonstandard software (Isoscan II). Subjective ROM data "window" used. Increased work variability attributed to ROM variability.
60, 180, 240	Peak torque, peak torque angle, flexion/extension ratio	No	10 min	Small, sig. diffs. for KE at 60, 180, 240; r = 0.92–0.97 for peak torque; r = 0.26–0.69 for flexion/extension ratio.	Cybex ~5% < Biodex	Pearson r insensitive to systematic diffs.
60, 180—all 300—Biodex, Cybex only	Peak torque	Not mentioned	Not stated	Sig. diff. for KE at 180. Sig. diff. for KE at 300. Sig. diff. for KF at 300.	Cybex mean (130 N · m) < Biodex (150) < Kin-Com (175) for KE at 180. Cybex (88) < Biodex (123) for KE at 300. Cybex (47) < Biodex (77) for KF at 300.	
60, 180, 300	Peak torque, total work, peak torque angle	Not mentioned	Not stated	Sig. diffs. between almost all machines and velocities.	Magnitudes not reported.	
60, 120, 240	Peak torque	Cybex but not Lido (!)	24 to 48 h	Sig. diffs. at all velocities except KF at 60 and KE at 240.	Lido mean (141) < Cybex (167) for KE at 60. Cybex (61) < Lido (83) for KF at 240.	Differences consistent with gravitational torque inconsistency. No upper-body stabilization.
60, 90, 180, 240, 300	Peak torque	No ?	1 week	Sig. diffs. at 90, 180, 240.	Magnitudes not reported.	
60, 120, 180, 240, 300	Peak torque, average power, total work	Yes ?	Not stated	r=0.65, 0.47, 0.53 for KE PT, AP, TW; r=0.64, 0.29, 0.51 for KF PT, AP, TW. Sig. ANOVA diffs. too.	Magnitudes not reported.	Merac not widely used.
60, 180	Peak torque	Not mentioned	Not stated	Sig. diff. for KF at 60.	Kin-Com < Cybex (mean difference 26 N · m).	

(continued)

Table 12.2 *(continued)*

Study	Dynamometers	Subjects	Age	n	Movement	Position	Mode
Greenberger et al. (1994)	Cybex II, Biodex B2000, Kin-Com 500H	Nonathletic	20 mean	2 M 10 F	Knee extensors	Seated ?	Concentric Discrete reps ?
Brett (1992)	Cybex II, Kin-Com	University PE students	20 mean	5 M 8 F	Knee extensors, flexors	Seated	Concentric Continuous reciprocal (incl. Kin-Com ?)
Cress et al. (1991)	Lido Digital, Lido Active	Nonathletic	72 mean (65–86)	25 F	Knee extensors, flexors	Seated	Concentric Continuous reciprocal
Fleshman and Keppler (1992)	Cybex II, Orthotron KT-II, both with HUMAC software	39 college athletes, 5 nonathletes	18–22	37 M 7 F	Knee extensors, flexors	Seated	Concentric Continuous reciprocal
Walmsley and Dias (1995)	Kin-Com, Cybex II, Lido	Nonathletic	22 mean (18–24)	9 M 6 F	Shoulder internal, external rotators	Supine, 90 deg abd	Concentric Continuous reciprocal (incl. Kin-Com ?)
Kimura et al. (1996)	Biodex B2000, Kin-Com, Lido	Nonathletic ?	22 mean	28 M	Shoulder internal, external rotators	Seated, 90 deg abd	Concentric Continuous reciprocal (incl. Kin-Com ?)
Timm (1994)	Cybex TEF, Cybex 6000-TMC	Nonathletic	25 mean (17–49)	2880 M 2880 F	Trunk extensors, flexors	Standing	Concentric Continuous reciprocal

Velocity	Parameter(s)	Gravitational torque compensation	Time between tests	Results	Magnitude and direction of differences	Limitations
60, 240	Peak torque, mean peak torque	Yes ?	At least 72 h	Sig. diff. at 60 for all dynos for peak torque. Sig. diff. at 240 only for Cybex vs. Biodex peak torque. r = 0.84–0.97.	Cybex (136 N · m) < Kin-Com (151) < Biodex (164). Cybex (80) < Kin-Com (88) < Biodex (94) KE at 240.	Pearson r insensitive to systematic diffs. Cybex and Biodex report max peak torque. Kin-Com reports mean peak torque.
30, 180	Peak torque, mean peak torque	No	2 days	No sig. diffs.	n/s	No upper-body stabilization.
60, 180, 240, 300	Peak torque, work	Yes		Sig. diffs. for KE and KF PT at 60, 240, 300. Sig. diffs. for KF work at 180, 240, 300.	Magnitudes only shown graphically (appear to be generally less than 10%). Velocity or movement not reported. Magnitudes not reported.	Two dynamometers from same company (not widely used). No upper-body stabilization.
60, 240	Peak torque, work, total work	Not mentioned	At least 48 h	Sig. diffs. for work.	Velocity or movement not reported. Magnitudes not reported.	Orthotron KT-II very rarely used for measurement.
60, 120, 160	Peak torque	Yes	7 days (3 tests)	Sig. diffs. for IR at 120, 160.	Kin-Com < Cybex = Lido (34 N · m/42/44 at 120; 33/40/40 at 160). Lido < Kin-Com = Cybex (28/37/39 at 60; (29/36/37 at 120; 27/33/35 at 160).	
120	Peak torque	Yes	6 days	Sig. diffs. for concentric IR. Sig. diffs. for eccentric IR. Sig. diffs. for eccentric ER.	Biodex (53 N · m) = Lido (53) < Kin-Com (88). Kin-Com (38) < Lido (58) < Biodex (68). Biodex (57) < Lido (62) < Kin-Com (103).	Seated position not standard for Lido.
30, 60, 90, 120	Peak torque	Not mentioned	2 days	No sig. diffs.	TEF means > TMC by ~20 N · m.	Dynamometers not widely used.

Torque calibration should involve the use of verified calibration loads at a slow angular velocity (i.e., regular direct calibration rather than reliance on a pre-delivery, factory calibration).

Calibration is used to verify the criterion-related validity of an isokinetic dynamometer for measuring torque. One must be certain that the criterion is itself a legitimate standard for the quantity of interest. While the use of inert weights to assess dynamometer validity may represent a valid criterion measure for torque accuracy at slow velocities, such loading may not be a valid criterion for fast velocities. Known torque loads based on inert weights at the end of a lever arm must accelerate (usually just due to gravity) up to the preset isokinetic velocity in order for a calibration measurement to be made. If either the torque or the velocity is high, the dynamometer can be forced into "overspeeding" when the load first meets the dynamometer resistance mechanism at the preset angular velocity. Examination of torque recordings of such gravitational weight loading at faster velocities (over the limited range of motion possible) reveals that they are excessively contaminated by "torque overshoot" (see chapter addendum, page 198), and thus patently dissimilar to those generally produced by active muscle contraction in compliant human muscle-joint systems.

Therefore, calibration of isokinetic dynamometers at fast velocities is not possible with conventional calibration procedures (i.e., weight loads at the end of a lever arm). Special calibration rigs based on inert weight loading via pulleys (Ericsson et al. 1982) or torque motors (Handel et al. 1996) may allow application of known torque loading over much larger ranges of motion, thus potentially allowing valid calibration after initial overshoot artifacts have subsided. However, many dynamometers now restrict axis rotation to less than 360°, which would preclude some such approaches. For all these reasons, addition of valid high-velocity calibration procedures to the standard low-velocity calibration may be impossible for most dynamometers.

On systems in which the primary measurement is torque (rather than force), the length of the calibration lever arm must be accurately known, and the torque due to the lever arm mass (acting at its center of mass, whose position must also be known) must be included in the total calibration torque. Most systems take care of this by providing standard lever arms of known torque (i.e., known mass and center-of-mass position) and precision weights that one can regard as point masses.

The addendum to this chapter covers further issues related to the validity of torque measurements in isokinetic dynamometry.

Joint Angle, Angular Displacement, and Angular Velocity

The measurement of lever arm angle is a relatively trivial electromechanical operation. In fact, measurement accuracy is not the major challenge to the validity of this measure: the major challenge is the assumption that lever arm angle is a faithful representation of joint angle. It is important to remember that the measured lever arm angle is not the real joint angle, because the dynamometer lever arm is rarely if ever perfectly aligned with the limb. However, accuracy of joint angle measurements per se is usually of minor significance to most isokinetic tests.

Accurate angular displacement measures (i.e., change in joint angle) are important in the derivation of some measurement parameters, such as work, that will be discussed further on. Accuracy of angular displacement measurement is also important because velocity is derived by differentiation of this signal in some isokinetic systems.

In general, there is little or no provision for user calibration of the lever arm angle transducer. Users must devise their own procedures for checking the accuracy of angle parameters.

Velocity calibration, although probably the most difficult on most dynamometers, is potentially as important as torque calibration, since variation in the test angular velocity between sessions will result in torque data that come from different portions of the muscle torque-velocity relationship. Where this relationship is steep, such variation could be particularly important.

Velocity calibration is straightforward only on the Cybex II dynamometer, as the axis of this dynamometer can rotate continuously through 360°; therefore it is easy to perform and time 10 revolutions at a particular velocity setting to verify accuracy. As already noted, axes of other dynamometers cannot rotate through 360°, so velocity calibration is more problematic (Taylor et al. 1991). However, ranges of movement of around 280° may allow sufficiently long periods of constant angular velocity movement and enable one to perform an accurate calibration.

Reliability of Dynamometer and Athlete

The ability of isokinetic dynamometers to measure known torques accurately and reliably has been demonstrated (see chapter addendum, page 198). The ability of human subjects to perform reliably is somewhat lower than that of the dynamometers. Most human reliability studies have concerned healthy, young subjects (see Nitschke 1992 for a review). Less is known about the ability of high-level athletes to perform reliably on such systems. In fact, researchers have paid relatively little attention to the issue of the reliability of any physiological and biomechanical testing of elite athletes.

There are two perspectives on the isokinetic reliability literature:

- According to one perspective (Nitschke 1992), relatively few studies have conducted and reported a rigorous standardized reliability trial such that others could replicate the exact protocol. That is, although a particular isokinetic test may have proven reliable, the description of the protocol may not allow replication. However, as already noted, there are restrictions on the way in which one can conduct test protocols on a given dynamometer such that there is no single isokinetic test protocol for any joint that could be replicated exactly on all modern dynamometers.
- According to the other perspective, the literature consistently demonstrates the reliability of isokinetic dynamometry, under a range of operating conditions, indicating a robust strength assessment methodology that is relatively resistant to unwanted variance. This suggests that exact replication of a published protocol is not imperative. The circumstances under which increased variability may be present in isokinetic dynamometry are also consistent throughout the literature. For example, eccentric contractions appear to be associated with slightly lower reliability than concentric contractions (see the review in Kellis and Baltzopoulos 1995). Also, tests at faster velocities appear to be somewhat less reliable than those at slower velocities (e.g., Montgomery et al. 1989).

Variability and Confidence Intervals in Isokinetic Dynamometry

For valid interpretation of isokinetic data, exercise scientists need to have a feel for the confidence intervals for the various isokinetic tests and measurements. The 95% confidence interval is estimated by twice the standard error of measurement, SEM (Verducci 1980), which is also known as the technical error of measurement, TEM (Knapp 1992). In general, more reliable tests will have a smaller TEM and a smaller 95% confidence interval. The TEM or SEM is usually calculated from the intraclass correlation coefficient (Verducci 1980), which is a preferred initial step for expressing test-retest reliability (Baumgartner 1989).

The results of a study by McCleary and Andersen (1992), one of the few studies of isokinetic test reliability for high-level athletes, provide an example of the use of the confidence interval. The authors investigated the reliability of concentric knee extension-flexion testing of male intercollegiate athletes at 60°/s on the Biodex dynamometer. They found 95% confidence intervals of around 5% to 8% of the group mean peak torque scores. It is possible to quote TEM or 95% confidence interval figures with each individual score when reporting team data, so that the potential variability in such data is apparent to athletes and coaches.

Reliability of Isokinetic Ratios

Ratios are widely used for expressing isokinetic data (e.g., torque-to-body mass ratio, agonist-antagonist torque ratio, left-right torque ratio); see "Reporting Results," page 180. Many users have simply assumed that the reliability of these ratios is similar to that of the absolute scores investigated in almost all reliability studies. On the contrary, the mathematical nature of commonly used isokinetic ratios would suggest that the quasi-independent variability of both numerator and denominator, and their expression as a quotient, would lead to lower reliability (and thus wider confidence intervals). Recent studies have indicated that such ratios do indeed have lower reliability, especially at faster velocities (Gleeson and Mercer 1991, 1996; Holm et al. 1994; Kramer et al. 1994; Wrigley et al. 1995). For example, Gleeson and Mercer (1996), in reviewing their studies of isokinetic reliability among power and endurance athletes, reported 95% confidence intervals (2 × TEM) for concentric knee flexion-extension peak torque ratios (at 60 and 180°/s) of ±13.5% of the mean ratio for the subject group, in comparison to confidence intervals of ±5% to 10% for the associated absolute peak torque scores.

Pretest Procedures

Important components of any testing procedure that need to be identified and reported are factors such as frequency of testing, athlete pretest status, ergogenic agents, time of day, pretest warm-up, positioning and stabilization, axis alignment, dynamometer lever length, gravitational and other torques, range of motion, and other dynamometer settings.

Frequency of Testing

As with all physiological testing of elite athletes, there must be a good rationale for performing a given test at any time. Isokinetic strength testing may take place on several occasions throughout the year. Often, the first testing is conducted at the beginning of preseason training to identify any weaknesses and institute remedial work. In sports with a high injury rate, it may be desirable to get an indication of an athlete's "normal" strength levels for comparison in the event of injury during the competitive season. Isokinetic strength varies throughout a season, presumably reflecting the athlete's overall training/competition intensity and the attention paid to in-season strength training, and possibly also due to overtraining (Sale 1991, figure 3.35; Hagerman and Staron 1983; Posch et al. 1989; Johansson et al. 1988; Koutedakis et al. 1994, 1995; Gibala and McDougall 1993; Martin et al. 1994; Hoffman et al. 1991; Ready 1982; Eckerson et al. 1994; Roemmich and Sinning 1996; Callister et al. 1990; Fry et al. 1994).

Athlete Pretest Status

As with all physiological tests involving intense physical effort, athletes should arrive for testing in a rested state. Chapter 2 addresses this critical element of test quality control. When it is necessary to test athletes from a squad on different days, it is critical to make every effort to avoid prejudicing some athletes by testing on days following training sessions of different intensities. Indeed, one should stress to athletes the importance of avoiding heavy training during at least the 24 h prior to isokinetic testing. It is important to make clear to athletes and coaches that failure to adhere to this guideline may adversely affect the results.

It has not been uncommon in the past for athletes to perform strength tests and other physiological tests in a single visit to the lab. Unless one can be sure that there will be no interaction between tests, the wisdom of this practice must be questioned. Athlete testing often occurs under time and cost constraints; but is it worthwhile to do more than one type of test in a session if data from all tests subsequent to the first may be compromised?

If, however, the only alternative is to perform strength testing along with other tests in one visit, it is advisable to do the strength testing first. Anecdotal experience indicates that there is no obvious decrement in treadmill $\dot{V}O_2max$, for example, after an isokinetic test protocol involving knee and hip extension and flexion (not including an isokinetic endurance test). Athletes do not generally report perceived fatigue prior to the treadmill test. This is consistent with the difference in the energy systems used, as well as the quick recovery of ATP-PC stores—the primary energy source for the strength tests. However, a rest interval of at least 15 to 30 min is probably prudent.

For female athletes, the evidence suggests that menstrual cycle phase has no significant effect on concentric and eccentric knee extensor-flexor isokinetic performance (DiBrezzo et al. 1988, 1991; Gur 1997).

Ergogenic Agents

Some results indicate ergogenic effects of pretest ingestion of caffeine on isokinetic performance in athletes (Jacobson et al. 1992), although other studies have failed to show an effect (Jacobson and Edwards 1991; Bond et al. 1986). The difference may be dose related. However, it would seem prudent to restrict caffeine intake for several hours prior to isokinetic testing. Enhanced isokinetic performance has also been observed with bicarbonate (Coombes and McNaughton 1993) and creatine ingestion (Greenhaff et al. 1993).

Time of Day

The potential influence of diurnal rhythms on athletic performance and physiological assessments has generally been ignored. However, some evidence indicates that isokinetic and other assessments may be affected by such daily rhythms. Wyse et al. (1994) found that concentric knee extension-flexion performance among collegiate athletes was generally lowest in the early morning and highest in the late afternoon. Although most measures were only slightly affected, some differed by around 10%. While further evidence is necessary to support a definitive recommendation regarding the optimum testing time, it is suggested that isokinetic testing be performed in the same part of the day for all members of a given athletic squad, and probably not in the early morning.

Pretest Warm-Up

Prior to testing, athletes should stretch the muscles surrounding the joint to be tested. They should also perform several minutes of light exercise on an appropriate ergometer. Opinion is divided on whether stretching should follow or precede the exercise warm-up; this can probably be left to personal preference. Although comprehensive studies are lacking, evidence suggests that bicycle ergometer warm-up and stretching neither inhibit nor enhance concentric knee extensor and flexor strength (Wiktorsson-Moller et al. 1983).

Athletes who will perform eccentric testing may need to devote greater attention to the stretching and warm-up. There is a further period of warm-up and familiarization on the dynamometer itself, as outlined later.

Positioning and Stabilization

All modern dynamometers have good provision for body stabilization, to ensure that movement occurs only around the joint of interest. While all dynamometer manufacturers recommend ostensibly similar positioning for the most common test, knee extension-flexion, there is considerable variation in the standard positions suggested for other joints, such as the shoulder. Even for knee extension-flexion in the common seated position, differences in hip angle, for example, will change the length of the biarticular rectus femoris and hamstring muscles; this will affect the length-tension relationship of these muscles at the knee, and therefore the torque-angle curve produced. Consequently it is important to standardize and report positioning. The following sections identify suggested standard positions for the most common joint movements tested. Figure 12.1 shows typical torque-time recordings for concentric tests of knee extension-flexion, ankle dorsiflexion-plantarflexion, and shoulder internal-external rotation.

Knee Extension-Flexion

The recommended test position is seated, with hip flexion angle of 80°. The knee extension-flexion axis is approximated by aligning the dynamometer rotational axis with the lateral femoral condyle. Restraint straps are applied for the distal thigh, pelvis, and upper torso. The shin pad is attached just above the lateral malleolus.

Isolated testing of the knee extensors may also sometimes be performed in supine, with the hip in neutral (0°). This position may be used for eccentric and concentric testing of this muscle group. Axis alignment and stabilization are often easier than in reciprocal concentric-concentric testing of agonist and antagonist muscle groups; the direction for torque application is the same, and it tends to force the thigh against the testing bench, thus providing efficient stabilization.

Conversely, isolated testing of knee flexors may be conducted in the prone position, also with the hip in neutral. The consistent direction of torque application for eccentric and concentric testing of this muscle group again aids in stabilization.

Ankle Plantarflexion-Dorsiflexion

The recommended test position is prone, with knee and hip angles of 0°. The ankle rotational axis for plantarflexion-dorsiflexion is approximated by aligning the dynamometer axis with the lateral malleolus. Body stabilization can be somewhat difficult, as the tendency for the body to slide on the testing couch must be minimized. A seated position may also be acceptable. The alternative flexed-knee test position limits the contribution of the two-joint gastrocnemius (Fugl-Meyer 1981), while soleus is unaffected; this is undesirable from a functional perspective.

Shoulder Internal-External Rotation

As a test of the anterior and posterior rotator cuff musculature, shoulder internal-external rotation is arguably the most important shoulder test for most upper limb athletes.

The recommended test position is seated, with 45° humeral abduction and 90° elbow flexion. The dynamometer axis is aligned with the long axis of the humerus. This test has traditionally been performed with a hand-grip attachment; thus wrist movements and wrist strength may affect the test. To avoid this potential confounding factor, it is likely that more dynamometers will eventually employ a wrist "clamp" or resistance pad at the distal forearm instead. There has been some discussion of the relative advantages of testing in the coronal plane versus the "scapula plane;" most testing has been performed in the coronal plane (Hellwig and Perrin 1991). The choice of 45° humeral abduction is a compromise between a position of zero (or slight) abduction and one of 90° abduction. The former is not a functional position for upper limb sports; the latter, although a functional position, is often uncomfortable (especially in external rotation) for many upper limb athletes with a history of shoulder problems.

Other shoulder test movements include flexion-extension and abduction-adduction (either in the frontal plane, or horizontal abduction-adduction in the transverse plane).

Hip Flexion-Extension

Although the hip flexors and extensors are important in many sports (kicking, jumping, sprinting), they have not often been tested by isokinetic dynamometry. With the earlier generation of dynamometers (Cybex II), part of the reason was the conservative

a

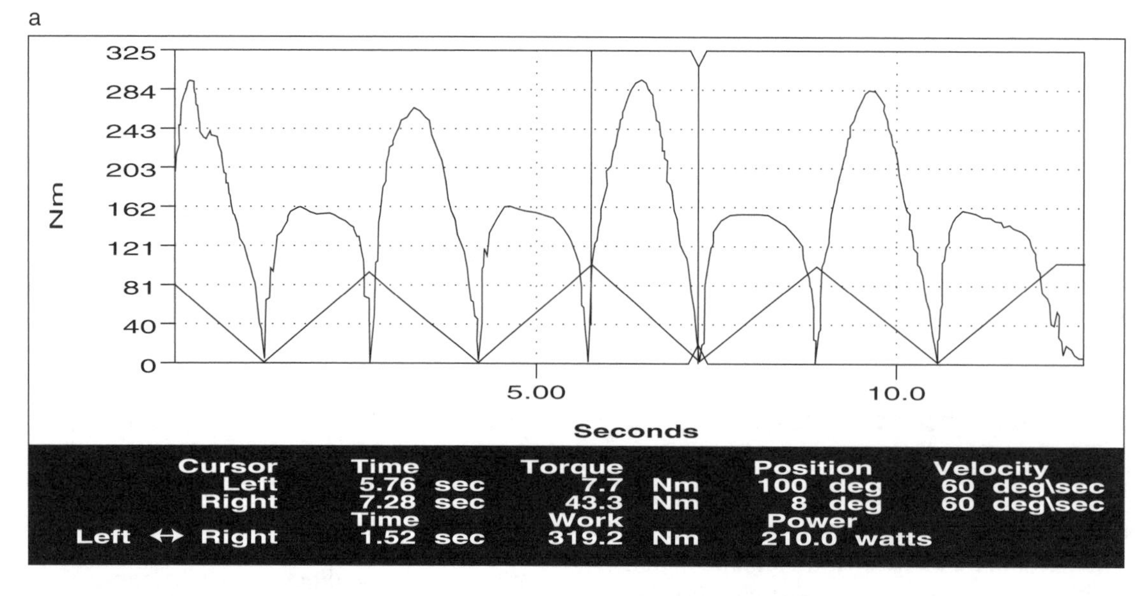

b

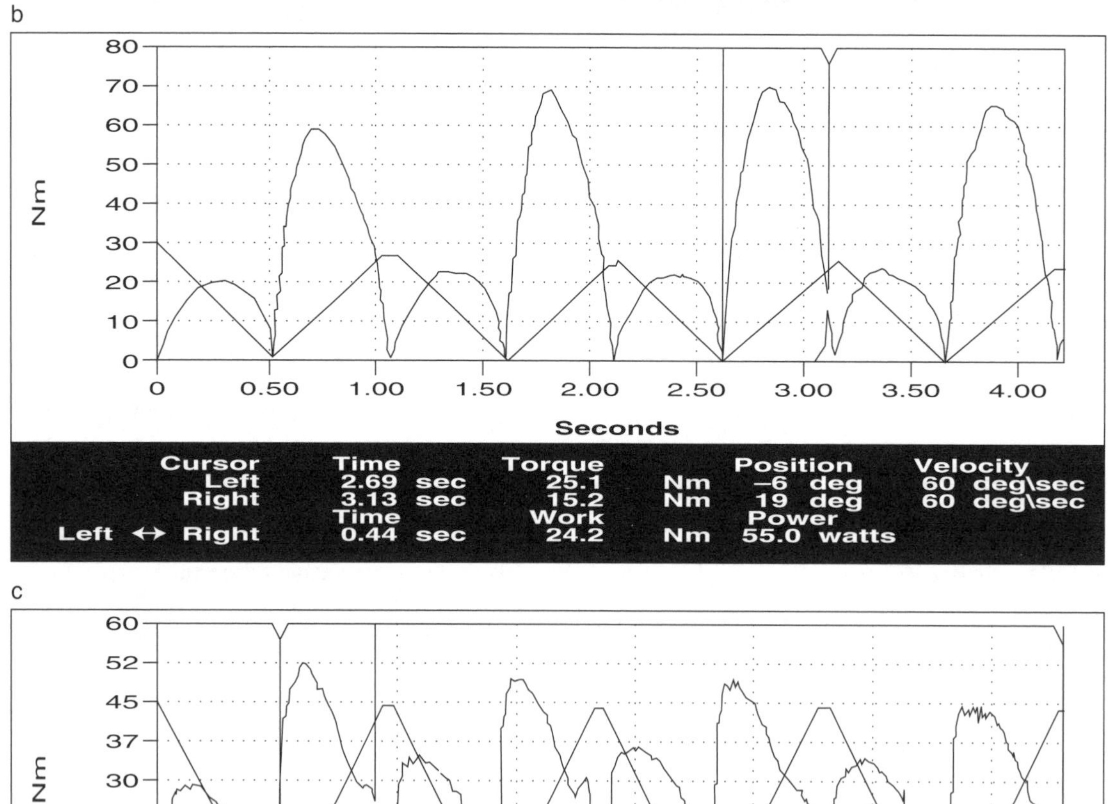

c

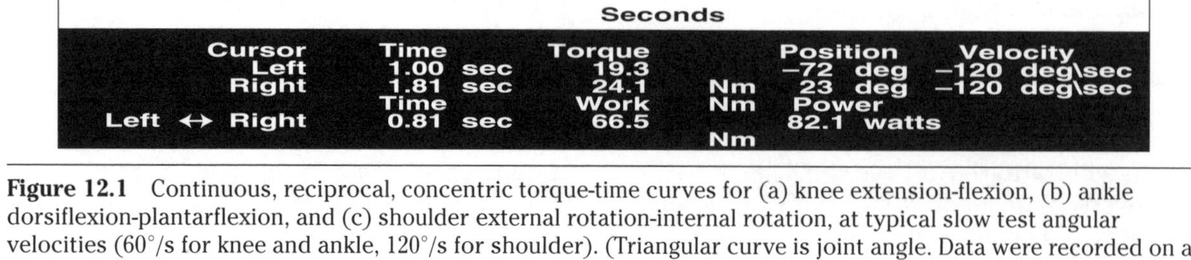

Figure 12.1 Continuous, reciprocal, concentric torque-time curves for (a) knee extension-flexion, (b) ankle dorsiflexion-plantarflexion, and (c) shoulder external rotation-internal rotation, at typical slow test angular velocities (60°/s for knee and ankle, 120°/s for shoulder). (Triangular curve is joint angle. Data were recorded on a Biodex dynamometer with no range of motion limits or torque threshold [isometric preload].)

maximum torque limit of these machines, which precluded testing of hip extension in strong athletes. This may still be a problem in some modern dynamometers, particularly for eccentric testing.

The traditional position for hip flexion-extension testing is in supine, with the dynamometer axis aligned with the greater trochanter. The means of limb attachment and stabilization supplied with the dynamometers have traditionally been less than optimal for hip extension-flexion testing. For example, the lever arm is usually attached to the distal thigh such that the shank is not fixed, thus allowing changes in knee angle. Consequently, rectus femoris and hamstring length is allowed to change across both the hip joint (desired) and the knee joint (not desired). Also, the gravitational torque (discussed later) due to the shank varies unpredictably throughout the range of motion.

Other Test Movements

Trunk strength testing (e.g., flexion-extension) would be more common if it did not require special dynamometer attachments or separate dynamometers. To a greater extent than for peripheral joint testing, each manufacturer's trunk dynamometry system is distinctly different. Beimborn and Morrissey (1988) and Dvir (1995) have provided reviews of trunk strength testing. In addition to the equipment issue, trunk testing is associated with problems related to axis location/alignment, stabilization, resistance pad placement, and gravitational torque compensation. Each manufacturer has a different approach to these issues.

Tests of the wrist and elbow are not commonly performed, but will be relevant for sports such as tennis and hockey.

Axis Alignment

It is safe to assume that the measurement of torque using isokinetic dynamometers truly represents joint torque only if the resistance moment arms of the dynamometer and the athlete's limb coincide (Dempster 1987). This is achieved by aligning the dynamometer's rotational axis with that of the joint to be tested. For many joints, this is not difficult. However, for others, such as the shoulder and trunk, locating a suitable axis and maintaining alignment throughout a test movement can prove difficult (Walmsley 1993; Stokes et al. 1990). Furthermore, if one is using reciprocal test movements, with the same contraction mode in both directions, the joint axis will tend to diverge from the machine axis, with the movement occurring in opposite directions for each movement direction. This can make adequate stabilization more difficult. This problem is less evident in testing of only one direction of movement, or in cases in which the contraction modes in two opposite test directions are dissimilar, because torque generation is then unidirectional.

Dynamometer Lever Length

One of the theoretical advantages of measuring torque in isokinetic dynamometry (as opposed to force) is its independence from dynamometer lever length. While most dynamometers measure torque directly, the Kin-Com actually measures force on the lever arm (at the point of limb attachment); and the user must enter the measured lever length (accurately!) so that the software can accurately calculate torque. To facilitate accurate torque calculation, the attachment of the Kin-Com force transducer should be placed at an exact centimeter gradation mark on the lever arm. In the dynamometers that measure torque directly (via transducers linked to the dynamometer's rotational axis), the transducers "see" not only torque generated by the subject, but also the gravitational torque induced about the dynamometer axis by the mass of the limb, lever arm, and attachments. Therefore, independence of torque measurements from lever-length influences will hold only if a gravitational torque compensation is performed (see next section on gravitational and other torques). Otherwise, variation in lever length between individuals and tests will vary the center-of-mass position of the lever arm, resulting in an uncontrolled and variable gravitational torque for the lever arm (although the torque variation would be relatively small). Dynamometers that measure force have the transducer mounted on the lever arm, where it does not "see" the gravitational torque of the lever arm (although it will measure the gravitational force of any weight on the measurement side of the transducer, including the subject's limb).

Even when data are compensated for gravitational torque (see next section), Otis and Gould (1986) have pointed out instances in which independence of torque measurements from lever-length variation may not hold. However, it is unlikely that the magnitude of the departure from this assumption is large enough to be of practical significance in most cases (Taylor and Casey 1986; Kramer et al. 1989).

Gravitational and Other Torques Not Due to Muscle Contraction

There are a number of sources of resistive and assistive torque about the dynamometer axis during isokinetic test movements (Herzog 1988):

1. Net torque applied by the test subject: this is the algebraic sum of torque due to muscle contraction (agonist minus antagonist, the latter generally silent) and any passive torque present (assistive or resistive, due to passive tissue forces acting at the joint).
2. Resistance torque applied by the dynamometer.
3. Gravitational torque due to the limb-lever arm system.
4. Inertial torque due to acceleration/deceleration of the limb-lever arm system.

The dynamometer generates (2) the resistance torque, which its transducer also measures; this torque is the algebraic sum of (1), (3), and (4). The aim of most isokinetic assessments is to report (1), the net torque applied by the test subject. To ensure that the final measured value is a true reflection of this net torque, it is therefore necessary to account for or minimize the influence of (3) and (4). The passive torque referred to in (1) has generally been ignored.

Winter et al. (1981) initially drew attention to the need to account for (3), the gravitational torque due to the weight of the subject's limb and the dynamometer lever arm. Without appropriate compensation of the data, this torque will decrease the measured torque for movements resisted by gravity and will increase the torque for movements assisted by gravity. Current models of all dynamometers incorporate procedures for gravitational torque compensation for most or all movement patterns where it is a factor (i.e., larger limb segments and longer lever lengths). However, some systems have omitted the procedure from movement patterns that should in fact have it (e.g., shoulder internal-external rotation; Perrin et al. 1992).

Most gravitational torque-compensation procedures are based on a single measurement of gravitational torque at a known angle, either statically or during slow isokinetic movement. The subject must be completely relaxed during this procedure. All test data are then automatically corrected in software by subtraction or addition of the predicted gravitational torque at each joint angle (which is calculated by multiplying the maximal gravitational torque by the cosine of the angle in relation to the horizontal)—depending on whether the movement is assisted or resisted by gravity. Hellwig and Perrin (1995) found this procedure to be reliable for the knee extension-flexion axis on the dynamometer studied.

Several concerns relate to implementation of the currently available gravitational torque-compensation procedures. By basing gravitational torque assessment on a single measurement, current systems make two assumptions:

1. That when the subject relaxes for the assessment, relaxation is indeed complete (that is, no muscle-generated torque is involved).
2. That there is no torque across the joint from passive tissue resistance/assistance at the angle of gravitational torque measurement.

For knee extension-flexion testing, the maximum gravitational torque (shank horizontal) corresponds to a position near full knee extension. In the common seated position for this test, subjects typically generate passive tension at small knee flexion angles (hamstrings and posterior knee joint capsule), in addition to the gravitational torque (van der Leeuw et al. 1989). Therefore, if gravitational torque compensation is derived from a single torque measurement at or near to the extended-knee horizontal position, the trigonometric estimation of the amount of gravitational torque compensation to be applied to recorded torque data from more flexed-knee angles will be excessive. In general, it would be more correct to refer to the procedures termed gravitational torque compensation as gravitational and passive tissue torque compensation, or "noncontractile" torque compensation.

It is advisable to make the single torque measurement for gravitational torque compensation at a joint angle where noncontractile torque is due to gravitational torque only (i.e., no passive tissue torque), but as near to the horizontal as possible (on dynamometers that allow this). The findings of Keating and Matyas (1996b) are consistent with this approach. The suggestion by Finucane et al. (1994) and Hellwig and Perrin (1995)—that gravitational torque compensation for seated knee extension-flexion testing be made near the horizontal—is not recommended, because of the potential for increased passive tissue torque in this position.

Although the significance of "reciprocal" ratios between agonist and antagonist strength has yet to be demonstrated (see later discussion of reciprocal ratios in "Reporting Results," page 180), gravitational compensation can have a profound effect on concentric ratios in the vertical plane. For example, early reports of the concentric knee flexor-knee extensor torque ratio ("ham-to-quad" ratio) showed these ratios to increase with test angular velocity; the advent of widespread gravitational compensation has subsequently shown that angular velocity has a limited influence on these ratios (Prietto and Caiozzo 1989; Fillyaw et al. 1986; Appen and Duncan 1986; Aagaard et al. 1995).

The early data collected with isokinetic dynamometers (mainly with the Cybex II) were not compensated for the effect of gravitational torque. However, all testing should now be performed with gravitational torque compensation.

Range of Motion

Standardizing range of motion for all isokinetic testing is problematic, as dynamometers use range limits differently. In some dynamometers, range limits are placed essentially as safety limits—that is, limits beyond which it might not be safe for an athlete to move. This is especially so for eccentric contractions, where the dynamometer lever arm moves of its own accord (although some dynamometers require a minimum applied force to initiate and maintain movement).

Some dynamometers use range limits to actually enforce the compulsory movement range; the dynamometer software will not accept that the movement has been completed until the range limit is reached. The Kin-Com generally operates in this way. Other dynamometers allow the subject to stop and initiate movement in the opposite direction (in bi-directional testing) even if the range limit has not been reached. With the dynamometers that use range limits to enforce the range of motion, these must be set within a range that subjects can easily reach. With the other type, the range limits for concentric tests may be set outside the subject's achievable range—subjects simply stop or change direction upon reaching the end of their movement range. On both types, the range limits for eccentric contraction must be set within the subject's movement limits, as the lever arm will move the limb to these limits in the eccentric mode.

Given these differences, a single recommendation about standardizing range of motion is not possible. However, the settings used should be clearly indicated in the reported test specifications (see "Reporting Test and Measurement Specifications," page 189). Comparison of results must take this into account, as the calculation of some isokinetic measurement parameters such as work is affected by the range of motion (see later discussion of measurement parameters, page 176).

Other Dynamometer Settings

The various dynamometers offer varying levels of user control over other aspects of isokinetic test movements. These include premovement and in-movement torque thresholds, and acceleration/deceleration "ramping" at the beginning and end of the movement range. Studies on the effect of some of these settings were listed in "Comparison of Test Results Between Dynamometers," page 158.

Because different dynamometers provide different facilities for premovement (isometric preload) and in-movement torque thresholds, as well as acceleration/deceleration control at the beginning and end of the movement range, one can make no single recommendation regarding standardization of these settings.

Isometric preloads—where available—require the subject to achieve a certain torque level before the lever arm will commence movement. These may be set at an arbitrarily chosen torque, or at a percentage of the subject's isometric maximum torque at the start angle, for example (Keating and Matyas 1996a). In cases in which no isometric preload is employed, there may be submaximal muscle activation in the initial movement range; this will be most evident at faster velocities (Harridge and White 1993). However, given that the ability to reach maximal muscle activation quickly is an important performance factor in many sports, it is not clear that it is necessary to use an isometric preload to keep this phenomenon from influencing isokinetic test results of athletes. The inclusion of an isometric preload assumes greater importance for persons with slower rates of muscle activation and for those who are generally weaker.

Instructions, Familiarization, and Practice

In addition to the warm-up, athletes should have sufficient practice repetitions on the isokinetic dynamometer prior to each test velocity. It is essential that athletes fully appreciate the requirements of the test. Some take longer than others to grasp these.

As with all physiological testing, athletes should receive brief, simple instructions that emphasize exactly—and only—what the athlete must do; explanations of the isokinetic concept, torque, and so on are not necessary. Explanation of the results may not be possible or appropriate at the time of testing. Therefore, one should ensure that provisions are adequate for giving the athletes and coach appropriate feedback at a later date (see "Reporting Results," page 180).

Athletes should perform several graded, submaximal repetitions at each test velocity whereby they build up to a near-maximal effort on the last practice repetition. A simple instruction ("a bit harder") as the subject begins each repetition should ensure an adequate increase in effort with each repetition. In testing at fast velocities, it is important to emphasize the "explosive" nature of the effort required. Two practice repetitions at each fast test velocity are usually adequate. The lack of the required explosive effort is usually immediately evident. On-screen graphic display, if available, will aid in judging the readiness of an athlete to perform true maximal exertions repeatably when the actual test begins.

For all test repetitions, the athlete should be exhorted to push/pull "as hard and fast as possible" and to complete the full range of motion. Strong verbal encouragement should be provided during the actual test. Experience suggests that maximum efforts can generally be performed very consistently. Therefore, erratic torque curve shape may suggest a less-than-maximal effort.

Because athletes will generally find eccentric contractions more difficult to perform maximally and reliably than concentric contractions, they may need extra practice. In eccentric testing, variability may be more likely to reflect a failure to grasp the requirements of the test than a less-than-maximal effort. Only strict attention to "quality control" can ensure the reliability of eccentric test data.

Test Protocols

Isokinetic testing protocols have been proposed and adopted on a largely arbitrary basis, often according to manufacturers' recommendations. Currently, the protocol(s) possible on a given dynamometer are dictated to a large extent by the software, and not all systems have equivalent protocols. This is unfortunate, as these software-enforced protocols are unlikely to be ideal in all circumstances. As noted earlier, no single protocol is exactly replicable on all dynamometers. The major distinction between isokinetic dynamometry protocols is discussed further on in "Continuous, Reciprocal Movements Versus Discrete, Single Movements."

Because of the operational differences within and between dynamometers, it is vital to clearly record and report the exact details of all test conditions. "Reporting Test and Measurement Specifications" later in this chapter (page 189) outlines reporting requirements.

Contraction Modes

Most isokinetic testing has involved, and will continue to involve, concentric contractions. The assessment of eccentric contractions did not become possible until more than a decade after the advent of isokinetic dynamometry. Eccentric assessment has been widely available for over a decade now, and its use continues to evolve. The indications for adding eccentric assessment to a standard, concentric assessment are strength-limited sports with an important eccentric component. These are often sports involving eccentric-concentric "stretch-shortening cycles." Examples include sprinting (especially hamstrings) and many throwing sports, such as baseball (especially shoulder internal rotators).

When both concentric and eccentric strength are to be assessed, concentric testing usually takes place before eccentric testing.

Continuous, Reciprocal Movements Versus Discrete, Single Movements

Most dynamometers (e.g., Cybex, Biodex, Lido, Merac) compel continuous performance of test movements (e.g., knee extension followed immediately by knee flexion). On these dynamometers, tests typically involve three to five repetitions. Most testers have used such continuous movements to measure concentric contractions in both directions of movement in order to assess reciprocal muscle groups (agonist and antagonist). Examples of this sequence for knee extension-flexion, ankle plantarflexion-dorsiflexion, and shoulder internal-external rotation were shown in figure 12.1. Since dynamometers capable of measuring eccentric contractions have become available, continuous movements have also been used to measure concentric-eccentric and eccentric-concentric contraction sequences for the same muscle group.

In contrast, assessments on the Kin-Com dynamometer are typically performed as discrete movements in a single direction, followed by a pause. The software requires the tester to decide whether or not to keep the current repetition for the final torque (or average torque) versus angle curve that is subsequently calculated. In such tests, subjects again usually perform three to five of these "interrupted," discrete, single repetitions.

At present, therefore, two distinct styles of protocols are required—one for each of the two classes of dynamometer.

One must be somewhat pessimistic about the possibility of comparing data recorded under such different contraction sequences (Grabiner and Hawthorne 1990; Vyse and Kramer 1990), although some studies have indicated insignificant differences (Keskula and Perrin 1994; Mawdsley 1985; Duncan 1987). Some contraction modes may be more subject to differences than others (Strauss et al. 1996; Gleeson et al. 1994). The few studies that have investigated the reliability of the two approaches have shown no clear difference (Keskula and Perrin 1994; Gleeson et al. 1994; Heinrichs et al. 1995). However, further work in this area is required.

These two approaches are most likely to induce differences in the data early and late in the movement, where the presence or lack of a pause, and/or subsequent premovement isometric preload and acceleration ramping, will likely induce different contractile responses (Jensen et al. 1991; Harridge and White 1993; Bobbert and Harlaar 1992; Gransberg and Knutsson 1983; Kramer et al. 1991b; Narici et al. 1991; James et al. 1994; Grabiner and Hawthorne 1990; Vyse and Kramer 1990; Tis et al. 1993; Gravel et al. 1988). Measures taken from the entire torque curve (e.g., average torque, work, average power—see "Measurement Parameters," page 176) are probably more likely to show disagreement between the two forms of dynamometer operation than a single measure taken from around the middle of the curve (e.g., peak torque). One may reduce the influence of such factors by limiting the torque data to those recorded during the true constant-velocity period (Wilk et al. 1992; Rothstein et al. 1983; Hoens and Strauss 1994).

Angular Velocity

The most common test protocol involves three to five concentric contractions, first at a "slow" velocity and then at a moderately "fast" velocity. On the dynamometers that utilize continuous, reciprocal contractions, for example, concentric knee extension-flexion tests are most commonly performed at 60°/s and 240°/s. If the dynamometer supports faster velocities, a third velocity of 360°/s may also be used. The aim of a general protocol such as this is to characterize low- and high-velocity strength with a relatively short protocol. Common concentric test velocities for different joints and movement patterns are shown in table 12.3.

The Kin-Com dynamometers that employ discrete, single-direction movements have lower maximum velocities than the other dynamometers. Depending on the model of the Kin-Com, the maximum velocity may not reach 240°/s. Therefore, the second, moderately fast test velocity will often be 180°/s.

Joints with smaller ranges of motion require slower velocities. This is so because at faster velocities, a greater portion of the range of movement is necessary for the limb to reach the isokinetic velocity at the start of the movement and to slow the limb at the end of range. For any given velocity, if the total range of motion is small, too great a proportion of the movement will be taken up in this way, and little of the range of motion will be spent at the isokinetic velocity. A rule of thumb for choosing the slowest test velocity is that the velocity should not be lower than the range of motion divided by 2.

Body segments with large mass (e.g., trunk; the thigh and shank in hip flexion-extension and abduction-adduction testing) will also require slower test velocities. Such large masses are associated with significant inertial torques at higher velocities, which may result in large artifacts throughout the torque recording.

Eccentric testing is not generally performed at higher velocities (i.e., greater than 180°/s). While eccentric contractions may be required in high-speed sports, on an isokinetic dynamometer these contractions are very difficult to coordinate. This suggests that the skill required to generate high-velocity eccentric contractions on a dynamometer is a "novel skill," and not the same as that associated with high-speed eccentric contractions in sport. Furthermore, the eccentric torque-velocity relationship (unlike the concentric relationship) is such that eccentric torque is usually relatively constant across the velocity spectrum (figure 12.2).

Test Order

Isokinetic testing has traditionally proceeded from slow-velocity tests to fast-velocity tests. Wilhite et al. (1992) showed this to be the most reliable sequence, while Timm and Fyke (1993) found no effect of test-velocity order. It would seem most prudent to adhere to the traditional practice of test-velocity ordering from slow to fast.

Rest Intervals

Since the energy supply for isokinetic strength tests involving a small number of repetitions comes primarily via the ATP-PC system, a rest interval of 40 to 60 s should be allowed between bouts. This is especially important at slow concentric velocities, for which torque generation, contraction duration, and perceived exertion are greatest. Test bouts involving discrete, single repetitions (rather than continuous, reciprocal contractions) may use shorter rest intervals. Tests at faster velocities can probably be separated by shorter time intervals, as the (concentric) contractile force is lower and the contraction duration shorter.

Table 12.3 Common Concentric Test Velocities

Joint	Movement	Slow velocity (deg/s)	Fast velocity (deg/s)
Shoulder	Extension-flexion	60, 120	180, 240, 300, 360
	Abduction-adduction	60, 120	180, 240, 300, 360
	Internal-external rotation	60, 120	180, 240, 300, 360
Elbow	Extension-flexion	60, 120	180, 240
Wrist/Forearm	Pronation-supination	30, 60	90, 120, 180
	Extension-flexion	30, 60	90, 120, 180
Hip	Abduction-adduction	30, 60	90, 120
	Extension-flexion	60	90, 120
	Internal-external rotation	30, 60	
Knee	Extension-flexion	60	180, 240, 300, 360
	Internal-external rotation	30, 60	120, 180
Ankle	Plantarflexion-dorsiflexion	30, 60	120, 180
	Inversion-eversion	30, 60	120, 180
Trunk	Extension-flexion	15, 30, 60	90, 120

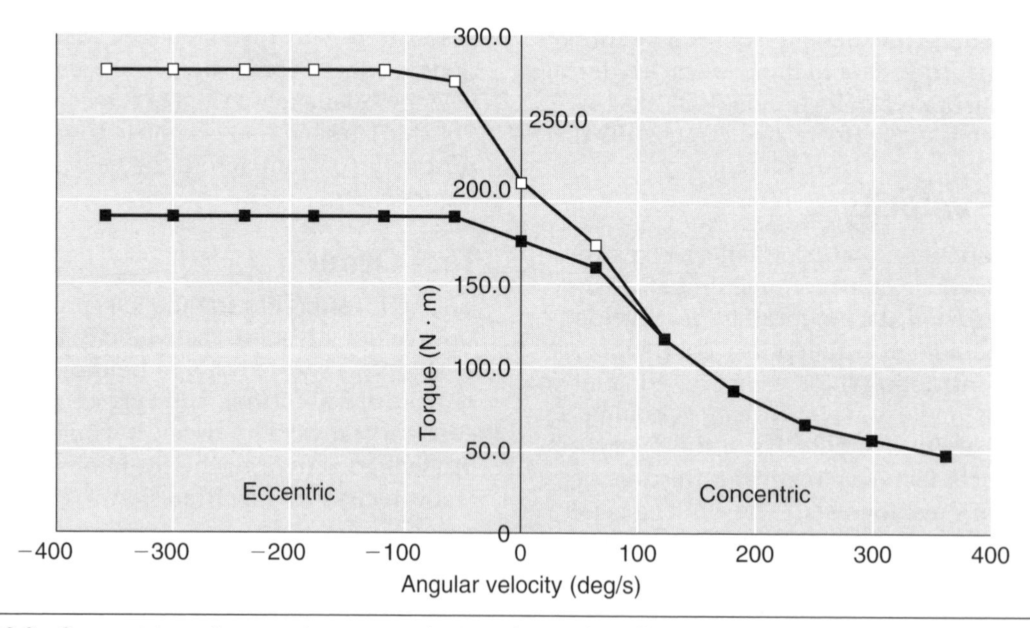

Figure 12.2 Concentric and eccentric torque-velocity relationship. (Solid squares show the shape of the relationship as measured on human subjects under volitional maximum contractions. Open squares show the general nature of the relationship recorded in vitro with electrical stimulation of animal muscle, and also with electrical stimulation of human muscle.)

Torque-Velocity and Power-Velocity Relationships

While isokinetic tests are generally performed at more than one discrete angular velocity on the velocity continuum, the results are typically interpreted as if they were discrete and independent. That is, the type of analysis usually employed effectively ignores the fact that tests at multiple velocities are part of a continuous torque-velocity relationship. However, there are several ways in which one can assess the torque-velocity relationship from data recorded at three or more angular velocities. Most methods relate only to the concentric torque-velocity relationship. See the discussion later under "Reporting Results," page 180.

Total Limb Strength

Nicholas et al. (1976) introduced the notion of "total leg strength" in clinical contexts in which injury at one joint may be reflected in weakness at joints some distance away. The concept has been extended to "total arm strength" (Ellenbecker 1991).

Because of the time it takes to perform tests at multiple joints, this notion has not been widely explored. However, its application to elite sport performance should be encouraged. It is unlikely that the single muscle groups routinely tested isokinetically can fully explain elite performance. Figure 12.3 presents an example of isokinetic test results for the hip, knee, and ankle joints from athletes in different sports (Bartonietz 1994).

Visual Feedback

The research on the possible influence of real-time visual feedback on isokinetic performance is somewhat equivocal. However, it seems that such feedback may enhance slow-velocity isokinetic test performance to some extent—and to a lesser extent for high-velocity tests (Baltzopoulos et al. 1991; Hald and Bottjen 1987; Figoni and Morris 1984; Hobbel and Rose 1993). Both eccentric and concentric contractions may benefit (Carlson et al. 1992; Kellis and Baltzopoulos 1996). A global recommendation to employ real-time visual feedback is further complicated by the fact that not all dynamometers provide it, that those providing it do so in different forms, and that not all testing positions enable the athlete to view the computer screen.

Anaerobic Endurance Testing—Concentric Endurance

In addition to the typical test protocol based on three to five repetitions as already described, endurance/fatigue testing, usually at a relatively fast concentric velocity (e.g., 180 or 240°/s) over 20-40 repetitions, may also be performed if indicated. Fatigue rates differ with different test velocities (Clarke and Manning 1985; Barnes 1981; Mathiassen 1989). Despite the fact that from a physiological standpoint these are clearly tests of anaerobic endurance, they have been used with athletes for whom anaerobic endurance of this (maximal) intensity and duration is of questionable relevance (although more commonly this occurs in clinical environments).

A number of issues must be considered in relation to the valid conduct of such testing. Experience suggests that some subjects may "pace" themselves to some extent during the test (although athletes may be more highly motivated to maintain maximal effort). Consequently, the torque recording may not show a consistent decline with time but rather periods of decreased and increased effort. Also, the choice of an appropriate index of test performance can be problematic. This issue is addressed in the "Data Collection" section.

For athletes in sports requiring repeated bouts of high-intensity effort, interspersed with periods of relative inactivity, a "recovery test" can assess this capability (Tesch and Wright 1983; Tesch et al. 1985; Schwender et al. 1995). For example, Tesch and colleagues described a maximal, five-repetition, concentric knee extension test at 180°/s, performed 40 s after the 50-repetition "Thorstensson test" at 180°/s.

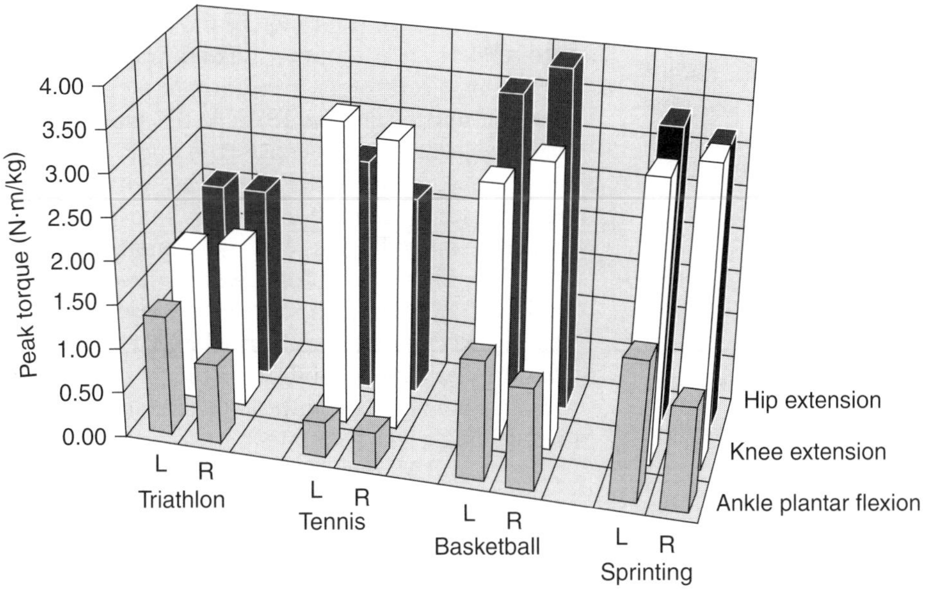

Figure 12.3 Concentric isokinetic strength profiles for hip, knee, and ankle muscle groups of athletes from four different sports.

Modified from Bartonietz 1994.

Physiological testing of elite athletes often includes other tests of peak anaerobic power and endurance, such as "all-out" bicycle ergometer tests. Many studies have demonstrated relatively high correlations between isokinetic endurance test measures and bicycle ergometer anaerobic power tests (Wrigley, 2000). Therefore, performing anaerobic endurance tests with both an ergometer and an isokinetic dynamometer may be redundant, unless the aim is to look more closely at isolated single-joint (isokinetic) endurance in comparison to multijoint (ergometer) endurance.

Anaerobic Endurance Testing—Eccentric Endurance

Assessment of eccentric endurance has been rare, and its significance is uncertain. Eccentric fatigue occurs at a much slower rate than concentric fatigue for the same muscle group (Gray and Chandler 1989; Mathiassen 1989; DeNuccio et al. 1991; Emery et al. 1994; Tesch et al. 1990). There is some evidence that its assessment may reveal information not apparent from concentric testing (Johansson 1992; Westblad et al. 1996); however, further work is necessary.

Evidence suggests that eccentric endurance tests can be conducted reliably (Westblad and Johansson 1993). However, the potential for delayed onset muscle soreness—associated with high-volume eccentric work—suggests that one should contemplate such protocols with some caution.

Data Collection

Important data collection issues include the choice of measurement parameters, use of units and conversion factors, the appropriateness of maximum versus mean values, the problem of redundancy, and parameterization for endurance testing.

Measurement Parameters

The measurements one can make from an isokinetic test at a given velocity fall into three categories:

1. "Peak" measures—for example, peak torque, angle-specific torque, peak power
2. "Average," or "whole (torque)-curve" measures—for example, average torque, average power, work
3. "Time-based" measures—for example, time to peak torque, "torque acceleration energy" *(sic)*

It is advisable to take at least one measure from each of the first two categories, as the basis on which an interpretation of the test is made (see section on reporting results, page 180). The time-based measures in the third category are the least reliable (Madsen 1996; Barbee and Landis 1984) and the least commonly available in commercial dynamometer software. Rate of torque production measurements are sometimes included as components of isometric tests.

Figure 12.4 shows the most common isokinetic measurement parameters for a hypothetical knee extension torque versus angle curve generated at 60°/s, over a range of motion of 90°.

Torque

Torque measurements can be derived for both the 'peak' and 'average' parameter categories.

- Peak torque (Newton meter, N · m) is the highest torque achieved in the test movement. This has been the most commonly used measure in isokinetic dynamometry. Its use is most appropriate in instances in which the relationship between torque and joint angle for a particular muscle-joint system has a clear peak within the range of motion, with lower torque at either extreme of the range of motion. In cases in which the torque-angle relationship is ascending or descending over the range of motion (Kulig et al. 1984), peak torque may not be the most appropriate measure; a whole-curve measure may be more suitable.

- Angle-specific torque (N · m) is the torque at a specified angle in the range of motion. This measure reflects an attempt to make torque measurements at a relatively constant muscle length (Perrine and Edgerton 1978; Yates and Kamon 1983), regardless of velocity (the angle at which peak torque occurs tends to increase with increasing concentric velocities). There is some evidence that angle-specific torque measurements may not be as reliable as peak torque measures (Kannus 1994), although not all studies are consistent in this regard (Arnold and Perrin 1993). However, it does appear that reliability problems may be associated particularly with angle-specific torques taken at angles early or late in the range of motion (Gleeson and Mercer 1996).

- Average torque (N · m) is the mean level of torque for the whole torque curve. This is the fundamental whole-curve measure just identified. Unfortunately, only a few dynamometer software systems currently provide it. On systems that do not report average torque but do calculate average

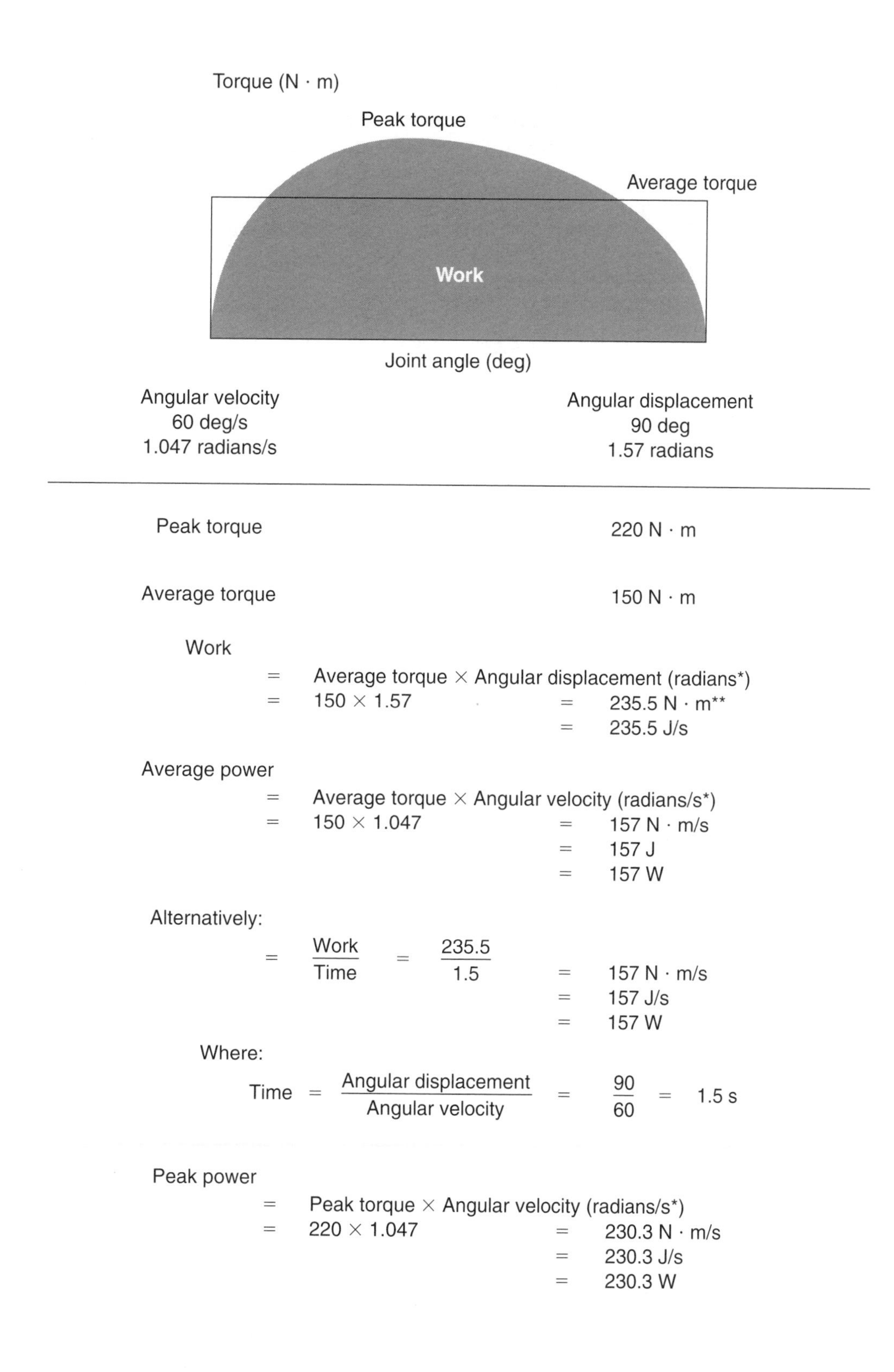

* Angular displacement and angular velocity, in these calculations, must be in radians and radians/second, respectively. The radian, being the ratio of two distances, is a "unit-less" unit, and cancels out when part of a compound unit.

** N · m of work is not the same as N · m torque.

Figure 12.4 Metric/SI units for common measurement parameters in isokinetic dynamometry.

power, average torque can be derived by dividing the average power (watts = N · m/s) by the angular velocity (in radians/s); dimensionless radians are dropped, and the result is in Newton meters. The average torque measurement may be calculated from data in the constant-velocity "window," with the nonisokinetic phases of movement truncated from the range of movement (Hoens and Strauss 1994; Wilk et al. 1992).

Work

The derivation of work (joules, J) in isokinetic testing has been a major source of misunderstanding (Wrigley 1989). Work done in a rotational movement is analogous to its linear form: force multiplied by displacement. Work in a rotational movement is derived from average torque (N · m) multiplied by the angular displacement (which must be in radians rather than degrees); radians are dimensionless units, so the result is Newton meters of work, which are equivalent to joules. Work is equivalent to the area under the torque versus angle curve—not the area under a torque versus time curve, as has commonly been described (the latter is in fact angular impulse). Because the calculation of work includes angular displacement, one should not compare work values between tests unless the measured ranges of motion are numerically equal (see previous section on range of motion). For example, without this control, work scores will often show left-right ratios different from the corresponding ratios for peak torque.

Power

Power is the rate at which work is done—which is equivalent to the product of torque and angular velocity (or force and linear velocity). Average or instantaneous power can be calculated.

- Average power (watts, W) is derived by dividing work (J) by the time (duration) of the contraction (watts = joules/s). This is equivalent to taking the mean of the product of instantaneous torque (N · m) and instantaneous velocity (in radians/s) at every time interval throughout the movement (dimensionless radians are dropped, and the resultant N · m/s = watts). In cases in which velocity can reasonably be assumed to be constant, average torque may be multiplied by this velocity to yield average power. Average power is less dependent on range of motion than work; therefore, average power and average torque are the preferred whole-curve measures of isokinetic performance. Unfortunately, the use of average power is complicated by the fact that some dynamometer software calculates it as the highest average power obtained for any single torque curve, while other software calculates it as the average power over all torque curves (repetitions). The latter is not as useful a measure as the former, except perhaps in an endurance test.

- Instantaneous power (watts) is the product of torque and the angular velocity. Therefore, peak power (peak torque multiplied by the test angular velocity) or angle-specific power may be calculated. Although these measures will follow the same general trend as average power (in relation to velocity), there will not be exact correspondence between the trends, and of course the actual values for peak power will be higher (Wrigley and Grant 1995, figure 17.4).

Units and Conversion Factors

Much of the literature on isokinetic dynamometry emanates from the United States, where Imperial units are still in common use. Consequently it is often necessary to convert such units to metric/SI units; for these and other commonly required conversion factors, see table 12.4.

Maximum Versus Mean Values

At various times in the history of physical performance testing, discussion has focused on whether one should report a mean performance score or the maximum performance score. If the aim is to measure maximum performance, then the maximum score is indeed the best criterion, although it may not always be as reliable as the mean score (Baumgartner and Jackson 1991; Johnson and Meeter 1977). If assessment of average, or "typical," performance is the aim of the testing, then the mean score is the appropriate choice. However, use of the mean score can be complicated by the fact that some isokinetic measures may increase over the first several repetitions, before stabilizing (Mawdsley and Knapik 1982; Arrigo et al. 1994).

Use of "Raw" Torque Data Versus Mean Torque-Angle Data

Standardization of isokinetic test procedures is complicated by incompatibilities among procedures imposed by different dynamometer manufacturers. One example of this is the derivation of measurements from either (1) each individual recorded "raw" torque curve or (2) a mean torque-angle curve (based on the mean torque generated at each angle in the range of motion, for all repetitions). Most dynamometers calculate all param-

Table 12.4 Unit Conversions

From	To	Multiply by
Torque	Preferred unit:	Newton meter (N · m)
N · m	ft · lb	0.737562
ft · lb	N · m	1.355818
Force	Preferred unit:	Newton (N)
N	kg · f *	0.1020
kg · f	N	9.80665
kg · f	lb · f *	2.20462
Work	Preferred unit:	Joule (J)
J	N · m **	1.0
N · m	J	1.0
J	ft · lb **	0.737562
ft · lb	J	1.355818
Power	Preferred unit:	Watt (W)
J/s	W	1.0
J/s	N · m/s	1.0
W	N · m/s	1.0
Angular displacement	Preferred unit:	Degree (deg) or radian (rad)
deg	rad	0.017453
rad	deg	57.29578
Angular velocity	Preferred unit:	Degrees per second (deg/s) or radians per second (rad/s)
deg/s	rad/s	0.017453
rad/s	deg/s	57.29578

*The kilogram and the pound, while strictly units of mass, are sometimes used as units of force; in such instances, they can be given a qualifier (i.e., kilogram · force [kg · f]), although this is not mandatory.

**N · m and ft · lb of work are not the same as N · m and ft · lb of torque.

eters from individual raw curves or provide data from both methods. In the Kin-Com dynamometer, however, parameters are derived from a calculated mean torque curve (unless the tester "accepts" only one recorded torque curve). Although some work on knee extension testing has indicated similar reliability for peak and angle-specific torques derived by the two methods (Arnold and Perrin 1993), further studies are necessary on this issue.

Information Redundancy in Multiple Measurement Parameters

There are many measurements one can take from an isokinetic test. Do all these measures represent unique information content, or is some of the information redundant? Are some measures more sensitive than others to important characteristics of performance-related strength or training adaptations? Does the way in which the data is expressed (e.g., various ratio expressions, or absolute scores vs. scores relative to body mass) add to or obscure information content? Are some measures more reliable than others? Some of these issues have already been discussed; others are covered here. The issue is also considered further on within the context of reporting results.

A number of studies have demonstrated relatively high correlations between some of the isokinetic measures, generally obtained in healthy, young, active subjects (Kannus 1992; Kannus and Jarvinen 1989; Woodson et al. 1995). Although the authors have concluded that some parameters contain redundant information that is available in other parameters, such conclusions may be premature. For example, peak torque and average torque for concentric knee extension at 60°/s will generally show a significant relationship, with most subjects demonstrating about 50% higher peak torque than average torque. However, there is considerable variation in this relationship even though the group correlation between the parameters may be high. The ratio of these two measures is a simple index of torque curve shape, which may be related to performance. Some research (Rajala et al. 1994; Froese and Houston 1985) suggests the potential importance of differences in torque curve shape. Thus, at present one should view with caution conclusions regarding isokinetic measurement parameter redundancy based merely on correlation studies conducted with nonelite subjects.

All calculated parameters should be examined for useful information content. More often than not, the dominant trends evident in the data will be present in all measured parameters. This strengthens the veracity of the test results. Where such consistency exists, only one parameter (e.g., peak torque) may need to be reported. Where different parameters reveal different trends, one should attempt to explain the differences in terms of muscle-joint function or an aberrant test. For example, as noted earlier, work measures may vary simply due to range of motion differences. Alternatively, if different findings for different parameters appear to be legitimate, they may be the basis of specific training recommendations focused on altering one or more parameters.

Some have suggested that testing at multiple velocities is somewhat redundant. However, as pointed out earlier, a full assessment and analysis across the torque-velocity and/or power-velocity relationship are potentially useful. Furthermore, testing at multiple velocities allows one to confirm the presence of interlimb differences for a muscle group. At faster velocities, such differences are usually relatively smaller than at slow velocities; that is, tests at faster velocities may be less sensitive to differences than those at slow velocities. Thus these smaller interlimb differences may be more difficult to distinguish from the inherent human variability in isokinetic performance. However, testing across a range of velocities may allow verification of the consistent presence of any imbalances across several increasing velocities (albeit usually of decreasing magnitude), thus confirming the assessment. The discussion of bilateral comparisons under "Reporting Results" covers interpretation of interlimb differences in more detail.

Endurance Test Measurement Parameters

As already noted, the choice of measurement parameters for endurance tests entails certain problems. Most suggested measurements are fatigue indexes of some description (Thorstensson 1976; Tesch 1980). The percent decrease in peak torque and/or the number of repetitions completed prior to 50% fatigue (i.e., 50% decline in peak torque from initial value) were the indexes commonly used in the past. Several studies have shown these measures to be reliable (Thorstensson and Karlsson 1976; Burdett and Van Swearingen 1987).

It is now common to use indexes based on work done: for example, the ratio of work done in the last third of the bout to work done in the first third. While each of these cumulative work values may possess good reliability, their expression as a ratio may not be so reliable (Burdett and Van Swearingen 1987; Pincivero et al. 1997; Grana and Frontera 1993). Kannus et al. (1992) found this ratio to be insensitive to endurance training, because initial and final work increased proportionally. Total work and average power may also be measured. However, these measures may not discriminate between subjects who start at a high work/power level and finish low and those who start at a moderate level and maintain this level.

As noted earlier, a potential complication with work measurements is that they depend not only on the average torque, but also on the range of motion (angular displacement). Therefore, differences between two tests in the amount of work done can simply reflect differences in the range of motion, which may or may not be physiologically significant.

When Winter et al. (1981) drew attention to the error resulting from failure to compensate torque recordings for the effect of gravity, the error was most apparent for work recordings in endurance tests, since the effect of the uncompensated torque error on work accumulates with each repetition.

Reporting Results

Interpretation of isokinetic results is typically based on *intra*individual assessments ("bilateral" and "reciprocal" comparisons) and sometimes also on *inter*individual assessments (absolute scores and scores relative to body mass). The measurement parameters that form the basis of these comparisons were discussed earlier. As already noted, at least one parameter is usually taken from each of the two categories, peak measures and average or whole-curve measures.

Intraindividual Assessment—Bilateral and Reciprocal Comparisons

Intraindividual comparison refers primarily to bilateral comparison of results between left and right limbs. It can also refer to "reciprocal" comparisons between agonist and antagonist muscle groups in the same limb, and to comparisons between concentric and eccentric contractions in the same muscle group.

Bilateral Comparisons

Bilateral comparisons between the same muscle group in opposite limbs have been interpreted under the assumption that "weaknesses" or "imbalances" identified by this index may be associated with decreased performance or an increased risk of injury. There is little evidence to support these notions. It may, however, be reasonable to infer that athletic performance may be affected by significant interlimb imbalances, which are usually due to previous injury.

When a bilateral comparison reveals an apparent weakness in a particular muscle group in one limb for an athlete with an injury history, its magnitude is usually greatest at slower velocities. As test velocity increases, the difference typically diminishes. When an athlete shows only a small to moderate left-right imbalance at a slow velocity, there may be no apparent imbalance at faster velocities. However, the converse is rarely true: that is, experience suggests that athletes do not show "true" interlimb imbalances at fast velocities when none is present at a slow velocity (although this occurs occasionally in clinical situations, usually in nonathletes). This observation should prompt consideration of the possibility of an aberrant result. As already noted, an advantage of testing at three or more velocities is that one can more readily verify the consistency of findings across velocities.

In persons with a history of injury in both limbs, the bilateral comparison may be confounded, and it may be difficult to judge the extent of the remaining weakness(es) in one or both limbs. As will be discussed further on in "Reciprocal Comparisons," the significance of such comparisons between agonist and antagonist muscle groups has probably been overemphasized. However, in the case of a bilateral injury history, an unusual reciprocal muscle group strength ratio for both limbs may provide evidence of weaknesses in both those limbs (not apparent in the bilateral ratios). For this reason, it is important to have some feel for the usual range of such reciprocal ratios in subjects who do not have particular muscle group weaknesses.

Experience suggests that very few athletes exhibit interlimb differences for which there is no obvious explanation in terms of a current or previous injury or of the unilateral demands of their sport. This rarity argues against the notion that such imbalances might be present without explanation and therefore possibly be predictive of a future injury.

Limb Strength Dominance. Evidence suggests that interlimb differences in isokinetic lower limb strength for a given muscle group of up to 10% may be relatively common in uninjured individuals (Daniel et al. 1982; Vagenas and Hoshizaki 1991; Agre and Baxter 1987; Grace et al. 1984; Knapik et al. 1991, 1992; Chin et al. 1994; Sapega 1990). Differences of this magnitude do not appear to be associated with an increased risk of injury. Differences within this range may simply reflect the inherent variability of the score for each limb, as well as the expression of these variable scores in the form of a ratio (see "Reliability of Isokinetic Ratios," page 165).

Bilateral comparison of isokinetic results between limbs is usually based on a premise of strength equality between the limbs as the normal state. Most often this has been assumed, but limb dominance in strength (i.e., "natural" interlimb differences greater than 10%), where it exists, confounds the notion. There is more evidence of limb dominance for upper limb strength than for lower limb strength; for example, there is no evidence of lower limb strength dominance in soccer, as discussed later in the chapter. In contrast, dominance is often evident in the shoulder musculature in unilateral sports (Brown et al. 1988; Chandler et al. 1992; Cook et al. 1987; Wilk et al. 1993; Alderink and Kuck 1986; Pedegana et al. 1982; Hinton 1988; Perrin et al. 1987). However, not all upper limb muscle groups appear to be equally subject to limb dominance (Alderink and Kuck 1986; Ellenbecker 1991, 1992; Wilk et al. 1995; Koziris et al. 1991). For example, in nationally ranked male tennis players, Ellenbecker (1991) reported that concentric strength dominance of the playing arm was particularly evident for shoulder extension and flexion, internal rotation, wrist flexion and extension, and wrist supination (figure 12.5); however, external rotation strength (tested in 90° abduction, coronal plane) was not greater on the playing side.

Dominance can certainly complicate the interpretation of bilateral comparisons for upper limb sports. For example, if a tennis player with a history of shoulder problems records playing-side strength similar to or less than that of the nonplaying side, it would be reasonable to suggest that the playing-side strength is deficient, probably due to the injury history. However, difficulties arise if the playing side is stronger: how

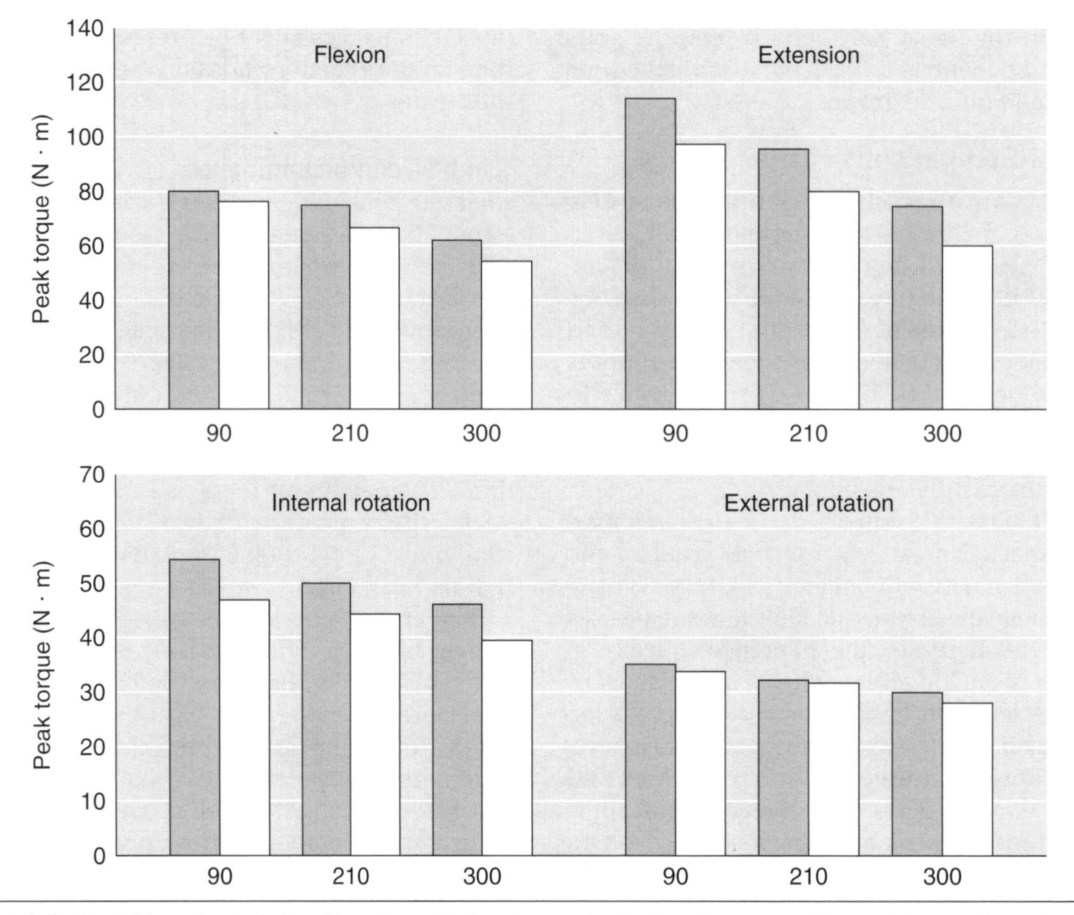

Figure 12.5 Variable extent of shoulder strength dominance for isokinetic concentric peak torque for extension, flexion, and internal and external rotation at 90, 210, and 300°/s.

Subjects were elite male tennis players (data from Ellenbecker 1991). Solid bars, playing arm; open bars, nonplaying arm.

much stronger is "enough?" Furthermore, experience in testing groups of such athletes suggests that players without a history of shoulder problems may not always demonstrate a stronger playing arm. How should their results be interpreted in comparison to those of athletes who do demonstrate playing-side dominance? A possible explanation for such findings is the likelihood that, in strength-related sports (as opposed to strength-limited sports), some athletes may rely on strength to a greater extent than others. This may lead to greater limb dominance. Alternatively, a lack of dominance may be due to bilateral strength training.

In cases such as these, one must resort to all available evidence to confirm a test interpretation. Thus agonist-antagonist reciprocal muscle group strength ratios in the upper limb may be given more weight than they would be in the lower limb, where their significance has not been established (as noted later in the discussion of reciprocal comparisons). For example, when interpretation of the bilateral comparison is problematic, one may look for the expected trend for reciprocal ratios on the playing side. In the example from Ellenbecker (1991), internal rotator strength was greater on the playing side, while external rotation strength was similar in the two limbs. This pattern should thus result in an increased internal-external rotation strength ratio on the playing side.

Sports that one might assume to be associated with greater strength of one limb may not in fact show the association. For example, as previously mentioned, evidence is lacking for what might be an expected strength dominance in the kicking leg in soccer. On the other hand, sports in which the notion of dominance is less obvious may exhibit dominance. There are isolated reports of dominance in knee extensor strength in volleyball (Puhl et al. 1982), on the oar side in rowing (Kramer and Leger 1991; Kramer et al. 1991a), and between the stance and kick leg in baseball pitchers (Tippett 1986). Such findings further suggest caution in the interpretation of all bilateral comparisons, especially when the interlimb difference is relatively small.

Ratio Expressions for Bilateral Comparisons. The most straightforward form of bilateral, interlimb ratio simply expresses the score for the weaker limb as a percentage of the score for the stronger limb—that is, weaker limb/stronger limb (%). Some isokinetic software systems use a more complicated ratio for the bilateral comparison, which is usually labeled a "deficit." This ratio is calculated as follows (or in an equivalent form):

$$\text{Deficit} = \frac{\text{Uninvolved limb score} - \text{involved limb score}}{\text{Uninvolved limb score}} \times 100\ (\%)$$

As these software systems are often designed primarily for clinical use, this terminology is used to signify an injured limb ("involved") and an uninjured limb ("uninvolved"). Most such systems assume that the uninvolved limb will be tested first. In healthy athletes, there is no strong rationale for testing one limb before the other. However, the first limb tested will usually be designated "uninvolved" by default. Users must be aware of some scope for mathematical variation in calculated "deficit" results resulting from an arbitrary choice of limb testing order. The "deficit" formulation has complex numerical properties that are not linearly related to the more straightforward weaker/stronger ratio. Designation of the "involved" and "uninvolved" limb can have an unintended effect on the magnitude of the "deficit" computed between limbs, because this choice sets the denominator.

Therefore, it is probably best to base interlimb comparisons only on the simple "weaker limb/stronger limb" formulation. If there is a clear dominant limb that is relevant to the athlete's sport (e.g., playing arm in tennis), then all athletes' ratios should be computed based on dominance (i.e., dominant/nondominant or vice versa). If "deficits" are to be reported, they should always be based on (stronger – weaker)/stronger; this can be reported in terms of which is the "weaker limb" and how much it is "weaker by."

Some regard a 10% difference between limbs as the threshold for a "significant" interlimb weakness on bilateral comparison. In light of the apparent prevalence of normal ratios in this range (as noted above), as well as the increased error variance inherent in ratio expressions (see earlier section, "Reliability of Isokinetic Ratios," page 165), it is advisable to use this threshold with caution. A more conservative threshold of 15% may be more appropriate for identifying "true" weaknesses between limbs. The authors hope that information in this chapter will aid in error reduction and therefore allow the use of lower thresholds with better discriminating ability.

Reciprocal Comparisons

Reciprocal comparisons are based on the expression of an agonist muscle group score as a percentage of the antagonist group score for the same limb. Traditionally, these ratios express the expected weaker group as a percentage of the expected stronger group. When examining reciprocal ratios, one must consider the interdependence of bilateral and reciprocal ratio scores. Obviously, an apparently low or high reciprocal ratio between agonist and antagonist strength in one limb usually reflects a left-right imbalance in one of the muscle groups, in order for the reciprocal ratio to be biased in that way. Therefore, the imbalance identified by the bilateral comparison is in effect the primary imbalance. The reciprocal ratio simply reflects this; in that sense, it is secondary.

In the past, some thought that particular concentric peak torque ratios (e.g., knee flexor-extensor or "ham-to-quad" ratio) might be associated with increased risk of injury, such as hamstring or knee injury. There is little good evidence to support such notions. Furthermore, there is little evidence that performance in sport is related to particular agonist-antagonist ratios. However, more research work in these areas is required. Figure 12.6 shows concentric knee flexor-knee extensor peak torque ratio data at 60°/s, collected by Rankin and Thompson (1983) on a wide range of U.S. college athletes; the data show little apparent difference between vastly different sports.

Note that many considerations important in the interpretation of bilateral and reciprocal comparisons, as just discussed, strengthen the case for testing both limbs in athlete assessment. Single-limb isokinetic tests should be avoided.

Eccentric-Concentric Ratios

The eccentric-concentric torque ratio for a particular muscle group should exceed 100%, as eccentric strength is greater than concentric strength (see Dvir 1995 for a review of this ratio). Because concentric strength declines with velocity whereas eccentric strength usually remains relatively stable, this ratio usually increases with velocity (Kramer and Balsor 1990; Mikesky et al. 1995).

The investigation of eccentric-concentric ratios among athletes is in its infancy. Differing relative

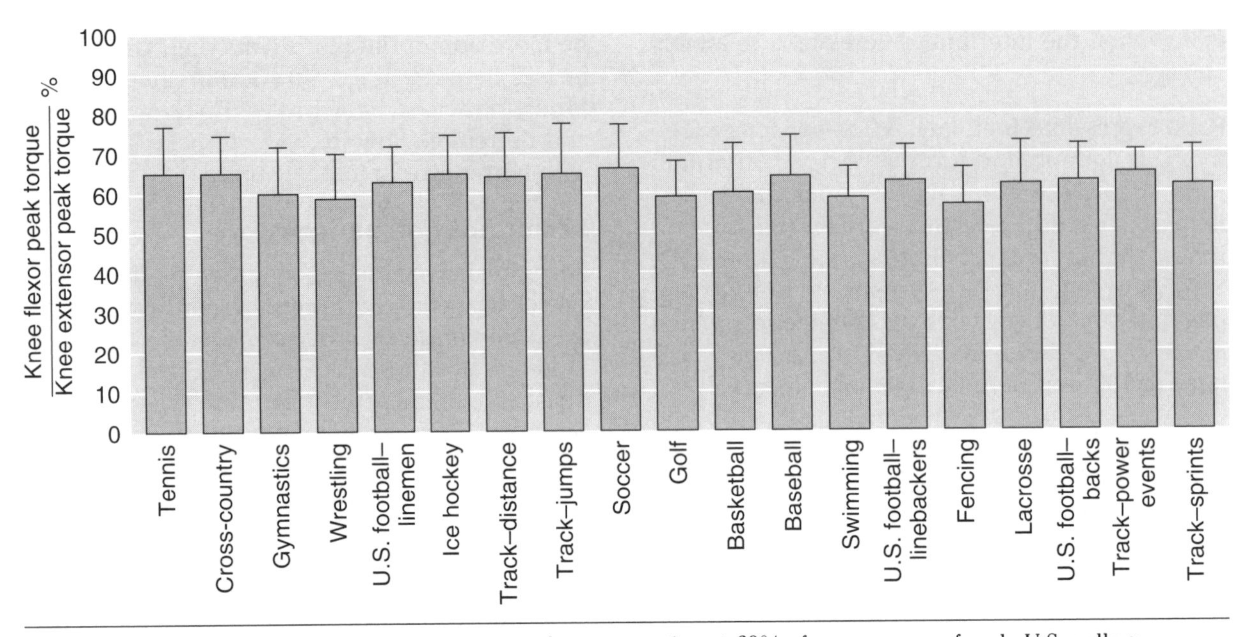

Figure 12.6 Concentric knee flexor-extensor peak torque ratios at 60°/s, from a range of male U.S. college athletes.

Data from Rankin and Thompson 1983.

eccentric-concentric torque-velocity performance among different athletes (Tesch 1995) is among the findings reported to date. Abe et al. (1992) found that slow (30°/s) eccentric knee extensor strength discriminated between elite and nonelite female alpine skiers, but concentric strength did not; this difference would be reflected in a higher eccentric-concentric ratio for the elite skiers. Koutedakis et al. (1995) have shown that the eccentric-concentric ratio can be increased in overtrained athletes as a result of a relative decline in concentric strength but maintained eccentric strength.

While eccentric-concentric ratios are usually constructed from torque scores from the same muscle group, ratios have also been described that are based on the eccentric strength of one muscle group expressed as a percentage of concentric strength of a different muscle group (Dvir et al. 1989; Dvir 1991; Brady et al. 1993; Aagaard et al. 1995). The two muscle groups are not always the traditionally assumed agonist-antagonist groups examined by reciprocal testing. Although still largely speculative at this stage, this approach may be more specific to the functional roles and interactions of particular agonist-antagonist pairs—in relation to injury mechanisms or athletic performance—than the traditional concentric-concentric ratios.

The way in which eccentric and concentric contractions are recorded is likely to significantly influence the calculated ratio. In particular, measurement of the two contraction modes in a continuous sequence, with no pause, is likely to produce a different result than measurement of two discrete, isolated movements with a pause (see the earlier discussion of continuous, reciprocal movements versus discrete, single movements, page 172). The measurement parameter used to construct the ratio is also likely to influence the ratio's magnitude.

Summary

The bilateral comparisons for each muscle group should be inspected first (peak and whole-curve measures); these will be the major basis of the intraindividual test assessment. The patterns apparent in the reciprocal comparisons will generally confirm the findings based on the bilateral comparisons.

Normal findings based on intraindividual comparisons can be summarized as follows:

- Bilateral comparisons for each tested muscle group not greater than approximately 10% to 15% (except for cases in which strength dominance in one limb is expected for a particular sport). In order to be considered significant, imbalances in bilateral ratios should be present at more than one velocity (although possibly of smaller magnitude at faster velocities).

- Reciprocal comparisons within the usual range for each agonist-antagonist pair tested. This provides secondary confirmation of the bilateral comparisons, but such ratios are probably rarely significant in their own right.
- If there are results for both concentric and eccentric testing, eccentric-concentric torque ratios should be greater than 100% for each muscle group tested, increasing with increased velocity.

In addition to any findings inconsistent with those listed, any other atypical responses should be reported (e.g., joint or muscle pain experienced during the test).

Interindividual Assessment—Absolute and Relative Scores

Isokinetic data from an athlete may be compared to those from teammates, or occasionally normative data may be available from previous testing or in the literature. Comparison within a team or to previous test results is preferable to the use of published data, as it does not involve difficult judgments regarding the validity of comparisons with a "normative" data set for the athlete(s) in question. Although the literature contains a considerable amount of data, not all of this information is useful. When using published data, one must be particularly careful of the following potential barriers to valid comparisons:

- Type of athlete
- Age, height, weight
- Time during season
- Pretest status (e.g., heavy training, injury excluded?)
- Dynamometer used
- Exact test movement performed
- Body position and stabilization
- Test velocity used
- Which limb(s) tested
- Range of motion
- Gravity correction performed
- Use of preload and/or ramping
- Continuous versus discrete repetitions
- Contraction type(s) and sequence
- Torque measurement parameter used
- Use of maximum or mean scores

All these factors should of course be controlled, and reported, in all standardized isokinetic testing (see "Reporting Test and Measurement Specifications," page 189).

Success in strength-based performances in some strength-limited and strength-related sports is based on absolute strength: the greater the torque or force developed, the greater the likelihood of performance success. In such sports the athlete applies force to an implement (e.g., throwing events in track and field) or an opponent (e.g., body contact sports). In other sports, particularly those in which the athlete must propel their own body mass faster, further, or higher, strength relative to that body mass is most important (e.g., gymnastics). Strength-limited sports requiring absolute strength tend to be dominated by large athletes. Strength-limited sports requiring relative strength tend to be dominated by smaller athletes, as relative strength is greater in smaller individuals—because increased body size is not associated with a proportional increase in strength.

Some sports have mixed requirements (that is, they require both absolute and relative strength) or present other special constraints on performance (Sale and Norman 1982). For example, a tennis player requires absolute strength for shot production and relative strength for acceleration and general speed around the court. A high jumper requires high relative strength for jumping ability, but must also be tall. A basketball player requires these characteristics, but also absolute strength for maintaining position against opponents around the basket.

Absolute Strength Scores

Absolute strength, then, is of interest when a sport requires the application of force to an implement or opponent. This usually occurs at the end of an athlete's limb (i.e., at the hands or feet). Thus isokinetic strength can be reported as force at the end of the relevant limb. A measurement of dynamometer lever arm length—or, better still, limb length—should be taken in order to convert the recorded isokinetic torque to force at the relevant point on the limb. Although the Kin-Com dynamometer measures force at the point of limb attachment directly, torque should initially be calculated; data can then be expressed either way, as necessary.

One sees an example of the use of force rather than torque in several studies showing a substantial relationship between isokinetic force and soccer kicking performance (Cabri et al. 1988; DeProft et al. 1988). For knee extension-flexion, the dynamometer lever arm was attached to the players' limbs just above the lateral malleolus. The force measured at this point would have represented a proportional approximation of the force that might be applied to a soccer ball (without consideration of the limb inertial properties).

It might seem that for sports in which the athlete's body mass is supported (e.g., cycling, rowing), absolute scores should be used. However, in some of these sports, body size may still increase resistance to movement (e.g., rolling resistance, air resistance); so expressions of relative strength, in relation to body mass, will also be important.

Relative Strength Scores—Data Interpretation for Athletes of Differing Body Size

Although many assume that torque-to-body mass ratios are helpful for making comparisons between athletes, it is important to recognize the problems associated with inappropriate approaches to body mass scaling.

Torque-to-Body Mass Ratios. The most commonly used relative isokinetic strength expression is the torque-to-body mass ratio. Figure 12.7 shows such ratios for concentric knee extension peak torque at 60°/s, recorded by a wide range of male American college athletes. While there are some interesting patterns, it would appear that the means and standard deviations for many sports are similar. Thus the torque-to-body mass ratio may not clearly discriminate between sports that might all be considered as strength-limited, for example. This suggests that a similar "type" of strength is required for success in many sports. However, within those sports (especially relative strength-limited sports), performance level may still be closely related to relative strength.

Scaling of Strength Data for Body Size Differences. Users generally assume that use of torque-to-body mass ratios facilitates interindividual comparisons between athletes by appropriately accounting for differences in body mass. In exercise science, this practice is a common means of "scaling" metabolic data (e.g., $\dot{V}O_2$max in $ml \cdot kg^{-1} \cdot min^{-1}$) and mechanical data (e.g., peak power in W/kg) to account for such body mass differences. However, one should not use this practice for scaling data—including strength data—without an understanding of its implications and of possible preferable alternatives.

Practitioners are probably making one or both of two implicit assumptions when they choose to use the simple torque-to-body mass ratio (Wrigley and Grant 1995). First, they may assume that functional sporting performance is limited by body mass and that the torque-to-body mass ratio expresses strength relative to this limitation. To thus assume that this limitation relates to body mass to the power of 1.0 for all athletic performances may be simplistic. Second, users of the torque-to-body mass ratio may assume that body mass is a reflection of muscle mass and therefore an appropriate means of controlling for differences in muscle mass (Delitto et al. 1989); this is clearly the assumption when torque is expressed relative to lean body mass rather than total body mass (Housh et al. 1984; Nutter and Thorland 1987). Since the ability of muscle to generate force is related to its cross-sectional area, rather than its mass, such assumptions may be similarly questionable.

Recognition of the bias that may be introduced by inappropriate approaches to body mass scaling of human exercise performance data has been inconsistent over recent decades. The work of Tanner (1949) was apparently the stimulus for a number of important papers in the early 1970s that addressed this issue in relation to oxygen consumption (Katch 1972, 1973; Katch and Katch 1974). However, this work was effectively ignored in subsequent years, despite a whole chapter on the subject of scaling physiological data in the major exercise physiology text of the era (Åstrand and Rodahl 1977). Recently renewed interest in this important issue has come principally from British exercise physiologists (Nevill et al. 1992; Nevill and Holder 1995; Armstrong and Welsman 1994; Winter 1992). Discussion of body size scaling issues has begun to reappear in exercise science textbooks (Winter and Nevill 1996). Some of the errors of interpretation that had resulted from inappropriate scaling in exercise physiology were reviewed by Jakeman et al. (1994).

The validity of various dimensional scaling approaches is the province of "allometry," full discussion of which beyond the scope of the chapter. The application of allometry to scaling of strength measures is still developing. Readers may consult a review by Wrigley (2000) of some of the issues.

Assessment of Torque-Velocity and Power-Velocity Relationships

A number of techniques have been available for analyzing the entire torque-velocity and/or power-velocity relationship resulting from isokinetic testing at multiple velocities. Most simply, one can construct a torque-velocity and power-velocity

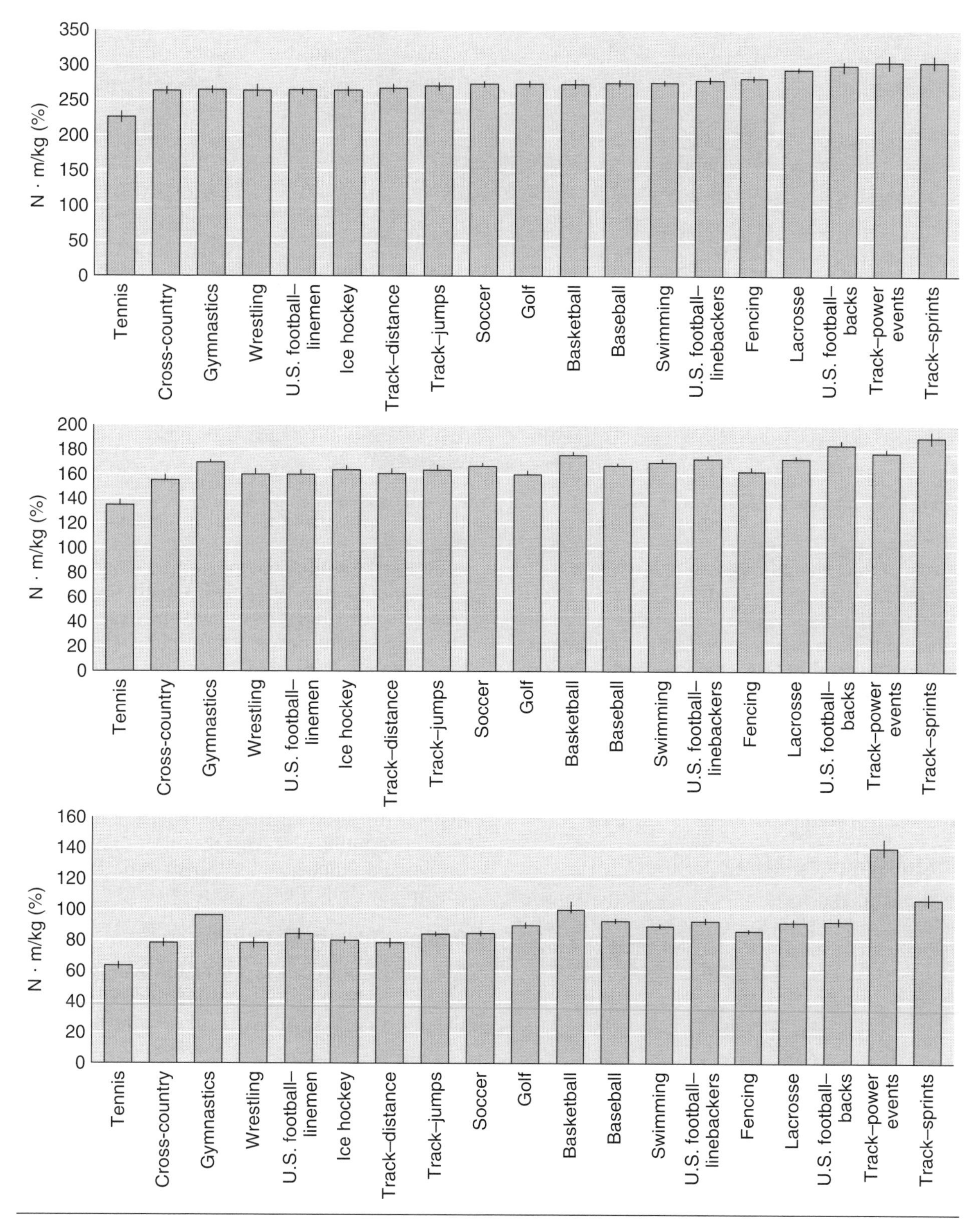

Figure 12.7 Peak torque-to-body mass ratios (N · m/kg%) for concentric knee extension at 60 (top), 180 (middle), and 300 (bottom) °/s, from a range of male American college athletes.

Data from Rankin and Thompson 1983.

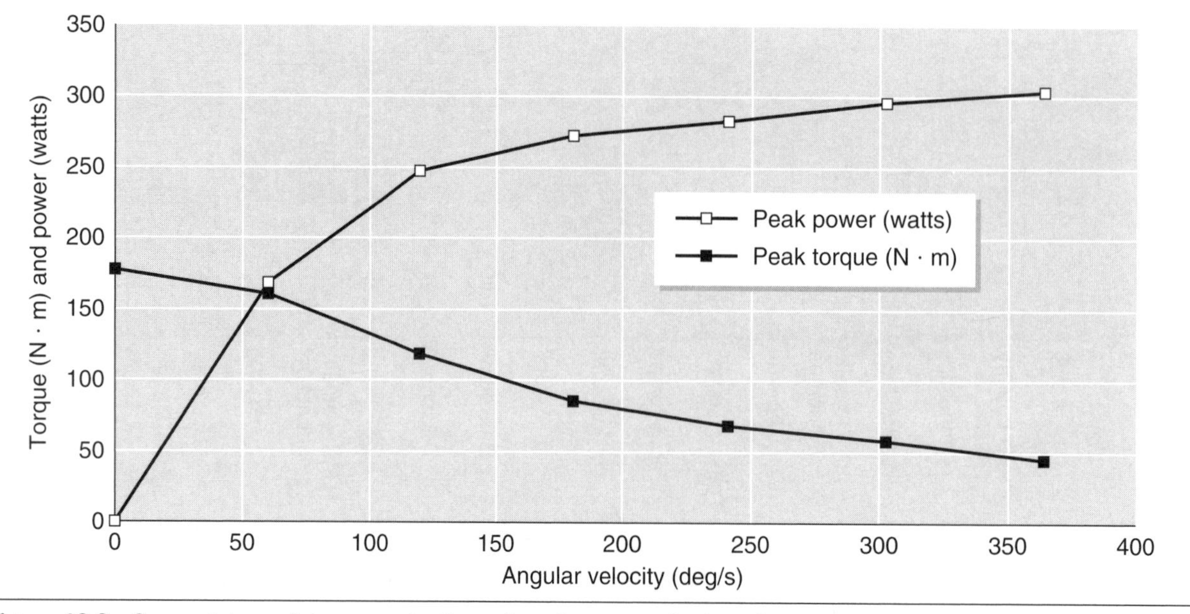

Figure 12.8 Concentric peak torque-velocity and peak power-velocity relationships.

graph; among the parameters of interest are the maximum power output and the velocity at which it occurs (figure 12.8). A more sophisticated approach involves fitting some form of mathematical function to the torque-velocity and/or power-velocity relationships, usually based on the Hill curve (MacIntosh et al. 1993; DeKoning et al. 1984; Clarke and Wysochanski 1986; Tihanyi et al. 1982; Gulch 1994). Such analyses may yield the parameters mentioned, as well as estimates of the rate of decline in torque as velocity increases.

Torque-Velocity Relationship

In the simplest form of analysis of the torque-velocity relationship, the torque recorded at faster velocities is expressed as a percentage of maximum torque, which will occur at a slow velocity (Yates and Kamon 1983). This effectively normalizes torque for differences in slow-velocity strength, whose primary determinant is muscle cross-sectional area and the muscle moment arm. This allows assessment of the relative high-velocity muscle performance of different athletes (figure 12.9). Use of such indexes to discriminate athletic performance has been reported (Oberg 1993; Oberg et al. 1986; Harman et al. 1990; Sale 1991, figure 3.16), although unsuccessful attempts have also been published (Farrar and Thorland 1987; Brown and Wilkinson 1983; Berg et al. 1986; Wilson and Murphy 1996; Piastra et al. 1990).

Power-Velocity Relationship

The calculation of peak power and average power was outlined earlier in the discussion of measurement parameters. It is not always clear which calculation method a dynamometer software system is using (particularly whether calculation of average power is from a single repetition or from all repetitions). Since power is the product of torque and angular velocity, and since angular velocity is nominally constant, the computation of power effectively occurs

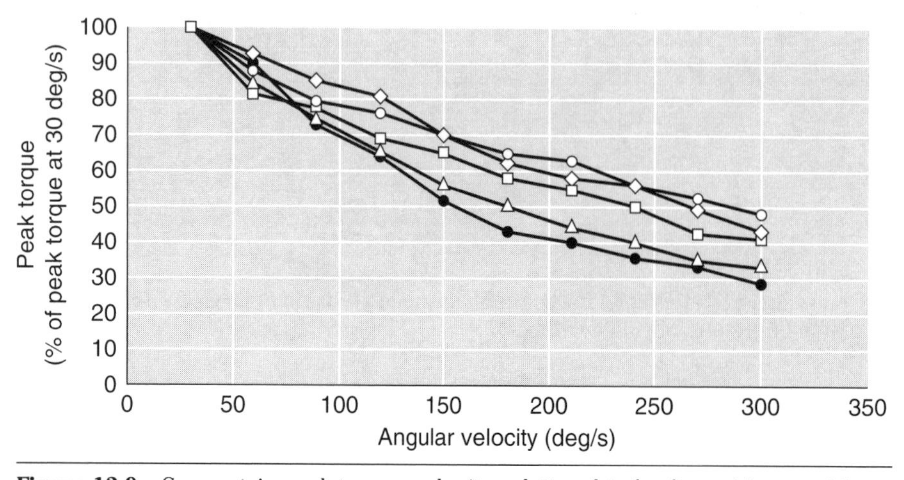

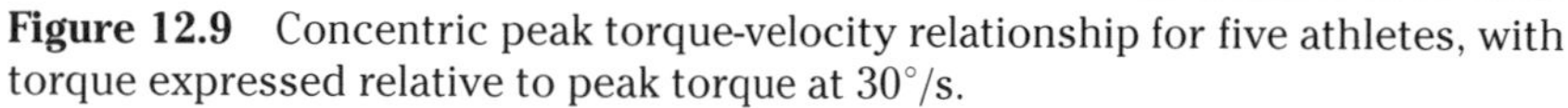
Figure 12.9 Concentric peak torque-velocity relationship for five athletes, with torque expressed relative to peak torque at 30°/s.

via multiplication by a constant. Therefore, as far as muscle function at high velocity is concerned, it would seem to make little difference whether peak torque or peak power (alternatively average torque or average power) is reported for high-velocity isokinetic tests. The one exception is construction of a power-velocity relationship. It is easier to discern relative high-velocity muscle function by examining the magnitude of the peak power and the velocity at which it occurs, than to distinguish relative performance based on examination of the exponential decline in the torque-velocity relationship. However, note that multiplication by a constant to compute power from torque will tend to amplify any error or variability in the data. Therefore, if high-velocity power output is the focus of an isokinetic assessment, one should test at several fast velocities (or repeat a test at a single fast velocity) in order to obtain an accurate indication of high-velocity power.

Feedback to Athletes and Coaches—Report Formats

Figure 12.10 is a graphic comparison sheet for isokinetic test results recorded for an Australian Football League team. This sport lies somewhere between strength-related and strength-limited sports on the continuum discussed at the beginning of the chapter. It involves a large range of body sizes and requires both absolute and relative strength. This report format provides a clear summary of relative performance for the whole team, both for absolute peak torque and peak torque relative to body mass; the single sheet is informative but not overwhelming for the coach and athletes. It includes bilateral-comparison ("weaker limb"), graphical, and numerical data for each athlete in relation to the team mean and standard deviations, as well as rankings. A sheet such as this can be provided for each measurement parameter of interest to the coach and athletes. For example, for the common protocol for isokinetic testing of agonist and antagonist muscle groups at a single joint, at two or more velocities, four such results sheets might be provided (e.g., concentric agonist peak torque at slow velocity, concentric antagonist peak torque at slow velocity, concentric agonist peak torque at fast velocity, and concentric antagonist peak torque at fast velocity).

These reports convey an overall picture of team isokinetic performance and clearly highlight athletes whose results are high or low in comparison to those of their teammates. In addition to these team sheets, one may provide results profiles for individual athletes as illustrated in figure 12.11. A written report for each athlete is still another option, although it is generally only used when an athlete has shown a particular weakness that warrants special attention or is being tested as an individual rather than part of a team.

Results typically do not include reciprocal ratios, as there is little evidence that these have any significance in relation to athletic performance (see earlier discussion of intraindividual assessment, page 180) or injury. However, when interpreted in conjunction with the bilateral comparison, a reciprocal ratio may indicate a weakness in a particular muscle group—in which case the findings would be the subject of an individual report.

Reporting Test and Measurement Specifications

To a greater extent than in most other areas of physiological testing, isokinetic dynamometry has a plethora of potentially confusing or ambiguous terminology. Clear guidelines on reporting requirements are necessary if isokinetic data are to be stored or exchanged in a manner that allows rigorous assessment of the data.

This section outlines the test specifications that should be recorded for such purposes. Because of the number of isokinetic parameters that can be measured (and the different ways in which they can be measured), it is particularly important that reporting of measurement parameters be unambiguous.

Subject Specifications

The following are the necessary subject specifications:

- Age
- Height
- Body mass
- Limb dominance (must specify how defined; e.g., kicks with left foot)
- Complete injury history (for the test joint, as well as proximal/distal joints and limbs)
- Recent illness history
- Pretest exercise (24 h)

The following subject specifications are optional:

- Other sports played (especially if unilateral)
- Time during season or training phase
- Recent athletic performances

Name	Weaker leg	Weaker by (%)	Quadriceps strength (Raw score) Best concentric knee extension peak torque (N · m) 60 deg/s	Rank	Quadriceps strength (Relative to body weight) Best concentric knee extension peak torque (N · m/kg %) 60 deg/s	Rank
Athlete 1	R	1		26		27
Athlete 2	L	15		14		10
Athlete 3	L	25		21		17
Athlete 4	L	32		8		19
Athlete 5	L	2		5		12
Athlete 6	L	5		9		14
Athlete 7	R	6		1		2
Athlete 8	L	18		6		4
Athlete 9	L	14		20		9
Athlete 10	L	2		2		5
Athlete 11	L = R	0		27		26
Athlete 12	L	5		18		20
Athlete 13	L	3		4		3
Athlete 14	-	-		15		11
Athlete 15	R	12		11		13
Athlete 16	R	4		19		18
Athlete 17	L = R	0		23		24
Athlete 18	L	15		24		8
Athlete 19	L	1		12		21
Athlete 20	L	6		15		15
Athlete 21	L	19		7		7
Athlete 22	L	15		17		23
Athlete 23	L	14		13		25
Athlete 24	L = R	0		3		1
Athlete 25	R	28		24		22
Athlete 26	L	1		10		6
Athlete 27	R	5		22		16
-	-	-		-		-
-	-	-		-		-
-	-	-		-		-

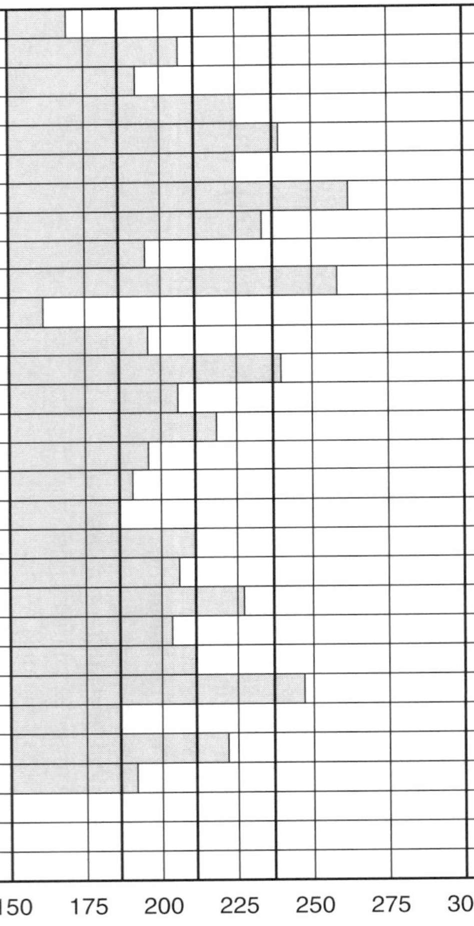

Raw score axis: 150 175 200 225 250 275 300

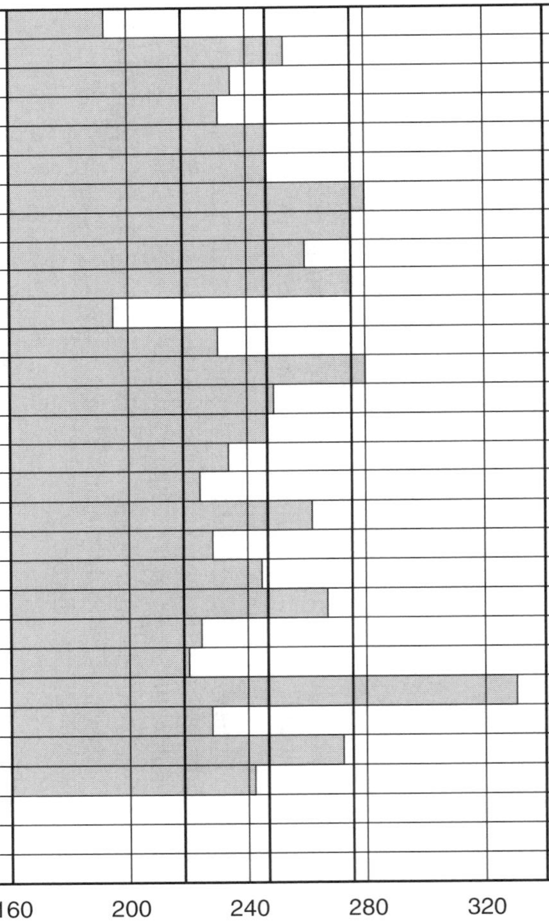

Relative to body weight axis: 160 200 240 280 320

Solid vertical bars
Left: −1 SD Middle: Average Right: +1 SD

Figure 12.10 Team isokinetic test results sheet.

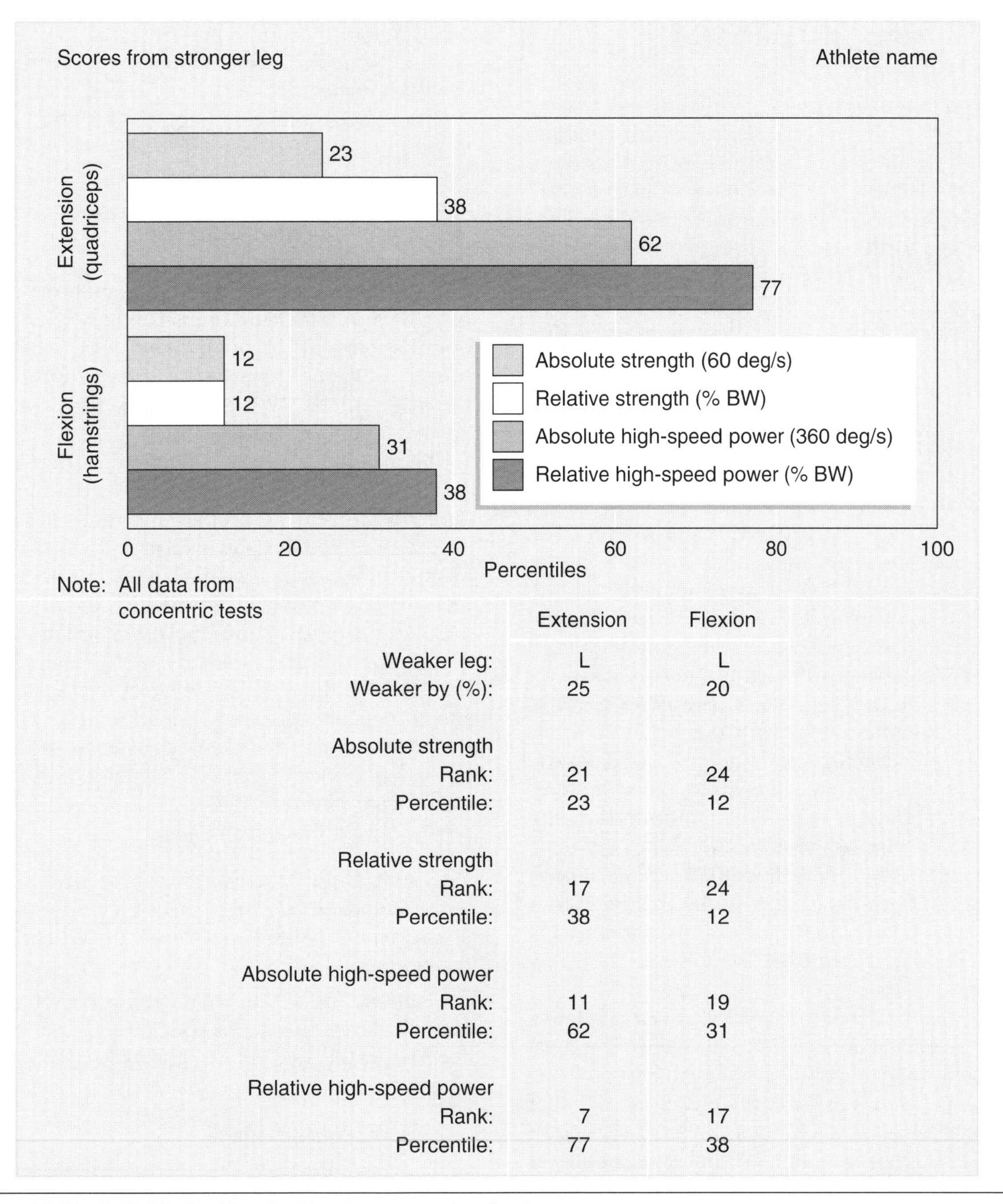

	Extension	Flexion
Weaker leg:	L	L
Weaker by (%):	25	20
Absolute strength		
Rank:	21	24
Percentile:	23	12
Relative strength		
Rank:	17	24
Percentile:	38	12
Absolute high-speed power		
Rank:	11	19
Percentile:	62	31
Relative high-speed power		
Rank:	7	17
Percentile:	77	38

Figure 12.11 Individual athlete isokinetic test results sheet.

Test Specifications

The following test specifications are necessary:

- Dynamometer
- Software version
- Test movement
- Body position
- Stabilization
- Dynamometer lever arm length (mandatory if only force is recorded or reported)
- Any variations on standard procedures

The following test specifications are optional:

- Tester
- Calibration date (procedural details are also useful to report)
- Warm-up details

Test Movement and Parameter Specifications

To avoid confusion, one must distinguish clearly between potentially ambiguous measurement parameter terminology. One area of potential confusion was discussed in relation to the use of "raw" torque data versus mean torque-angle data. That is, average torque is a whole-curve measure that is the mean level of torque generated throughout a single repetition of a particular movement. However, the default parameter calculations of the Kin-Com dynamometer software are based on a calculated torque-angle curve, which is the average torque at each joint angle from all repetitions (if more than a single repetition are "accepted") of a particular movement at a given velocity.

Describing all possible differences in operating and computational procedures is beyond the scope of routine reporting. Therefore, interpretation of data from different dynamometers will often rely on a knowledge of these differences between machines. As noted earlier, comparison of test results between different brands of dynamometers may be problematic. This applies particularly to dynamometers with markedly different operating characteristics, such as the discrete, single repetitions typically employed on the Kin-Com dynamometer versus the continuous, reciprocal repetitions assessed on the Cybex-style dynamometers (including Biodex). It is important to report test specifications in sufficient detail that users can judge when cautious comparisons between tests may be possible and when they may not be advisable.

Note that the documentation for a given dynamometer system may not make it clear exactly how measurement parameters are calculated. Furthermore, such documentation never highlights the fact that the calculation for a given measurement parameter may be different from that for the same parameter in other systems. Therefore, it is the user's responsibility to be aware of possible discrepancies.

Cybex dynamometer software (from the 300 series onward) displays Max, Average, and Best Work Rep (BWR) torque-angle curves. The Max torque curve represents the highest torque recorded at each angle (for a given angular velocity); the Average torque curve is the mean torque recorded at each angle; the BWR curve is the torque-angle curve in which the highest work value was recorded. Only the BWR curve is an actual torque curve; the Max and Average curves are calculated curves. Although each curve is displayed, calculated data come from only the Max curve or the BWR curve. The peak torque and angle-specific torque come from the Max curve, while all work and average power data come from the BWR curve.

Other dynamometers (e.g., Biodex) report the highest value for each parameter that was achieved in any single curve. Therefore, only these and Cybex dynamometers provide the highest peak torque recorded in any repetition in relatively unambiguous form.

Retrieving other equivalent parameters is more problematic. The Kin-Com is the only dynamometer that currently provides a measure of average torque. Both the Cybex and Biodex provide work recorded in the BWR (i.e., the highest work recorded in any repetition).

The average power in a single repetition is provided by the Cybex dynamometers from the BWR curve; if the velocity profiles and range of motion are essentially equivalent for all repetitions, then this average power should be the highest recorded. The measure of average power provided by the Biodex is in fact the mean average power recorded over all repetitions at a given velocity, not a single repetition. On the Kin-Com, the highest average power should occur for the same repetition that records the highest work (this likewise assumes essentially equivalent velocity profiles for all repetitions), since range of motion is constrained to be the same for all repetitions on this dynamometer.

- **Units.** Units must be clearly presented. Preferred units and conversion factors for the various measurement parameters were outlined earlier in the chapter.
- **Ratios.** The way in which any reported ratios are calculated should be specified.
- **Bilateral ratios.** The nature of the ratio calculation should be specified:

1. Weaker/stronger
2. Deficits (see discussion in "Reporting Results")
3. Injured/uninjured
4. Dominant/nondominant

In addition, the criteria used to define dominance should be specified (e.g., right foot kick). With use of the injured/uninjured form, injury history for both limbs must be provided so that this ratio can be interpreted accordingly.

- **Reciprocal (agonist/antagonist) ratios.** Reciprocal ratios are rarely a focus of test reports because of their lower significance in comparison to other measures, and indeed their dependence on other measures (i.e., bilateral ratios). Also, they can be calculated if necessary at a later date from individual data (if provided). The order of the numerator and denominator for such ratios is usually self-evident.

- **Gravitational and other noncontractile torques.** The position at which gravitational torque was recorded, and its value, should be reported.

Movement Constraints

It is also necessary to take into account several movement constraints:

- **Range of motion.** The nature of range of motion restriction must be specified. The actual range associated with a particular torque curve should appear with the data; but it must also be clear whether the movement was constrained to occur only within a fixed range or whether the athlete was free to cease movement in a given direction at his or her own volition.
- **Preload (torque or force).** Any premovement threshold torque or force that the athlete had to exceed prior to movement commencement must be specified. It must be clear whether this is an arbitrary level or a percentage of the athlete's maximum isometric contraction, for example.
- **High/Low thresholds (torque or force).** Any maximum or minimum torque or force limits set for movement should be specified.
- **Ramping.** Any acceleration control ("ramping") employed at the beginning or end of each movement direction must be specified.

Group Data Statistics

The following statistics should be clearly reported:

- Mean
- Standard deviation
- Number of data values

Reporting of percentiles is optional.

Any information on the test-retest reliability of the reported parameters should be included. This information should ideally take the form of an absolute SEM (in cases in which the reliability data come from the same subject group as the reported test data) or the standard error expressed as a percentage of the mean score recorded for the reliability group. See earlier discussion of variability and confidence intervals.

Final Note

It should go without saying those reporting results should do so carefully so that the end users of the data—coaches and athletes—do not draw inappropriate conclusions. In extreme cases, tests have been known to affect team selection; this may occur, for example, with the use of isokinetic testing to assess recovery from injury. In professional sports, this amounts to affecting the athlete's livelihood. In some countries where physical tests—including strength tests—play a role in job selection, the conduct and interpretation of these tests are subject to legislation; inappropriate use of such tests is regarded as a human rights violation (Norman 1992).

An Example of Isokinetic Dynamometry Strength Assessment

To illustrate the general principles of isokinetic assessment that have been discussed, this section focuses on isokinetic assessment for soccer. The example highlights important elements of the process of selecting and performing isokinetic assessments. Although soccer is a strength-related sport, the isokinetic testing methods outlined are essentially similar to those for strength-limited sports. This highlights the generic nature of the muscle properties assessed by isokinetic dynamometry, which are strongly related to success in many athletic performances (Wrigley 2000).

The example reflects the following aspects of isokinetic strength assessment:

- Correlation with athletic performance
- Ability to distinguish between performers at different levels
- Sensitivity to training adaptations
- Reliability of test protocols
- Availability of normative data for comparison
- Biomechanical analyses on which to base an isokinetic testing strategy

Rationale

Soccer, a strength-related sport, requires both absolute strength (e.g., for kicking and body contact with opponents) and relative strength (e.g., for running and jumping). Research evidence on soccer satisfies important criteria upon which a strength assessment strategy should be justified:

1. Isokinetic strength distinguishes between elite and subelite performers in soccer (Rochcongar et al. 1988; Oberg et al. 1986; Kirkendall 1979, 1985; Togari et al. 1988); see figure 12.12. There is also

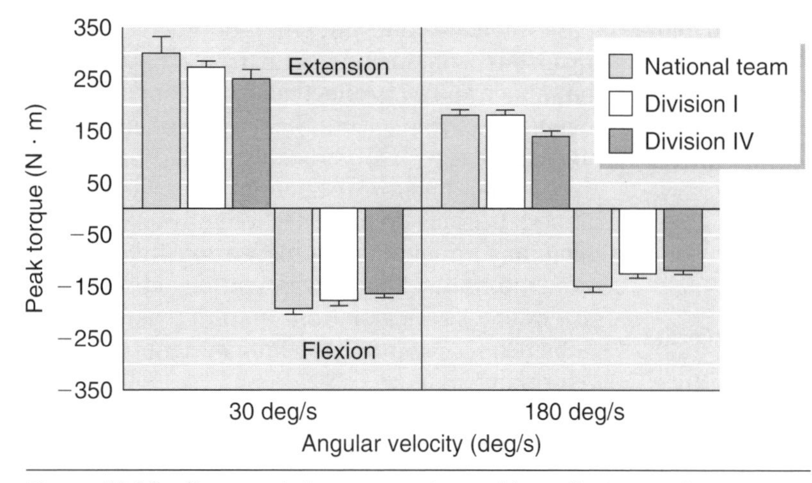

Figure 12.12 Concentric knee extension and knee flexion peak torque in Swedish soccer players.

Modified from Oberg et al. 1986.

evidence that isokinetic assessment is sensitive to team positional differences in strength (Oberg et al. 1984; Davis et al. 1992; Togari et al. 1988); however, Agre and Baxter (1987) failed to find such differences. Positional strength differences will be difficult to discern for many sports because of the lack of real differences in the strength requirements of different positional roles, particularly if the sports are not strength-limited.

2. Isokinetic knee and hip strength is correlated with the most obvious soccer-specific skill, kicking (Narici et al. 1988; Cabri et al. 1988; DeProft et al. 1988; Mognoni et al. 1994; Poulmedis et al. 1988), with only one known negative finding by McLean and Tumilty (1993). In the studies showing the correlation, the association was evident despite the fact that kicking is a ballistic, multijoint activity, whereas the isokinetic assessments in the studies were of single-joint movements and of course were performed at constant angular velocity. This evidence therefore suggests that the knee extensor strength and hip flexor strength assessed, in particular, are limiting factors in the generation of kicking velocity. Maximum kick velocity/distance is important in many instances in soccer (e.g., shots on goal, certain passes).

3. Many generic tasks in soccer are also common to other sports (e.g., sprinting, jumping). Isokinetic strength is related to the performance of these tasks (Wrigley 2000). (For example, studies supporting the association between isokinetic strength and jumping include Oddsson and Westing 1991; Appling and Weiss 1993; Zefang 1993; Piastra et al. 1990; Wiklander and Lysholm 1987; Genuario and Dolgener 1980; Podolsky et al. 1990; Ashley and Weiss 1994; Oberg et al. 1985b.)

4. Isokinetic assessment has been shown to be sensitive to changes in strength over extended training and competition phases in soccer (Sale 1991, figure 3.35). There are published and anecdotal reports of a common decline in strength over a competitive season in other sports (Johansson et al. 1988; Posch et al. 1989). Whether this is attributable to greater emphasis on strength training in the preseason than during the season, or perhaps reflects a maladaptation (i.e., overtraining), is not clear. However, isokinetic assessment may provide a basis for adjusting the relative training emphases. Decreased concentric knee extensor strength at an intermediate velocity (alone, and in comparison to eccentric strength) has recently been found to be an indicator of overtraining in elite athletes (Koutedakis et al. 1995).

5. At least one study (DeProft et al. 1988) has shown that isokinetic assessment is sensitive to changes in strength due to (nonisokinetic) strength training by soccer players. Furthermore, the strength changes detected isokinetically in that study were associated with changes in sport-specific (kicking) performance, thus indicating sensitivity to function-related strength adaptation. However, such associations may depend on the kicking ability of the players prior to training—that is, the potential for improvement (Trolle et al. 1993).

6. Normative data are available on male elite and lower-level soccer players, including juniors (Rochcongar et al. 1988; Oberg et al. 1984, 1985a, 1986; Rhodes et al. 1986; Svetlize 1996; Poulmedis 1985; Leatt et al. 1987; Togari et al. 1988; Mangine et al. 1990; Kirkendall 1979, 1985; Kramer and Balsor 1990; Tumilty et al. 1988; Chin et al. 1992, 1994; Sale 1991; Agre and Baxter 1987; So et al. 1994; Chook et al. 1986; Davis et al. 1992; Ekstrand and Gillquist 1983; Gauffin et al. 1988; Smith et al. 1994; Aagaard et al. 1994, 1995). Some female data are also available (Costain and Williams 1984; Fillyaw et al. 1986; McKay et al. 1987; Nyland et al. 1997).

7. Isokinetic assessment of the knee extensors and flexors has been shown to be reliable in representative samples of soccer players (Oberg et

al. 1986; Rochcongar et al. 1988), albeit with inadequate description of methods.

8. Biomechanical analyses support what in this case is intuitively obvious—the importance of knee extension and hip flexion, in particular, to kicking performance (Huang et al. 1982; Robertson and Mosher 1983; Roberts and Metcalfe 1968; Roberts et al. 1974).

Generic Test Protocols

For soccer, as a strength-related sport, a generic protocol of the type described in test protocols is most suitable. A standard protocol involving either continuous, reciprocal movements or discrete, single movements (usually determined by the type of dynamometer available) may be employed.

Knee Extensors-Flexors

As indicated earlier, the standard reciprocal, concentric knee extensor-flexor protocol involves tests at two or three angular velocities. The velocities are most often 60°/s and 240°/s.

Hip Flexors-Extensors

Evidence for the importance of these muscles to soccer comes from a number of studies showing them to be as important to kicking performance as the knee musculature (Narici et al. 1988; Poulmedis et al. 1988; Cabri et al. 1988; DeProft et al. 1988). Major considerations and limitations in hip testing were discussed in connection with positioning and stabilization.

Ankle Plantarflexors-Dorsiflexors

The isokinetic strength of the plantarflexor and dorsiflexor muscle groups has been studied in soccer players (Fugl-Meyer 1981; So et al. 1994; Poulmedis 1985; Poulmedis et al. 1988). Cox (1995) reviewed studies of this muscle group in other athletes, also discussing the protocols employed.

Limb Dominance

Evidence suggests that limb dominance among lower limb muscle groups of soccer players is not marked: there is usually less than a 10% difference between dominant and nondominant limbs (Brady et al. 1993; Kramer and Balsor 1990; Lai et al. 1986; Capranica et al. 1992; Leatt et al. 1987; Agre and Baxter 1987; Mangine et al. 1990; Costain and Williams 1984; McLean and Tumilty 1993; Narici et al. 1988; Ekstrand and Gillquist 1983; Svetlize 1996; So et al. 1994; Chin et al. 1994).

Torque-Velocity and Power-Velocity Relationships

A more comprehensive assessment of strength and power for soccer can be derived from testing at a larger range of angular velocities (Aagaard et al. 1994, 1995; Kirkendall 1979, 1985). For example, concentric knee extension-flexion may be tested at 60, 120, 180, 240, 300 and 360°/s (figure 12.13). The resulting data may be analyzed in a number of ways, as described in the discussion of reporting of results (page 180).

Anaerobic Endurance

Many aspects of soccer-specific physiological requirements can be usefully assessed with appropriately designed field tests. However, isokinetic testing can address targeted assessment of particular muscle groups and limbs, under standardized conditions. Anaerobic endurance testing by isokinetic dynamometry is an example of such targeted testing, which offers options not available via field testing or other forms of anaerobic ergometry. The ability to focus on particular muscle groups in isolated movements may be especially useful when an athlete is recovering from injury.

The intermittent, short-term, high-intensity efforts required in soccer suggest the importance of assessment of short-duration anaerobic endurance, as well as repeat efforts after short recovery periods. The options for these types of protocols were presented in the earlier section on anaerobic endurance testing (page 175).

Data Presentation

Data should be presented in both absolute and relative form, as both are necessary to cover the varied strength requirements of soccer. Figure 12.14 shows an example of a single player profile from a physiological test battery, including isokinetic knee extensor peak power (absolute and relative).

Important Issues for the Future

Many practices in isokinetic dynamometry continue to evolve, becoming more sophisticated and, one hopes, more rigorous. This evolution is vital if isokinetic dynamometry is to reach its full potential as a tool for athlete assessment. Most of

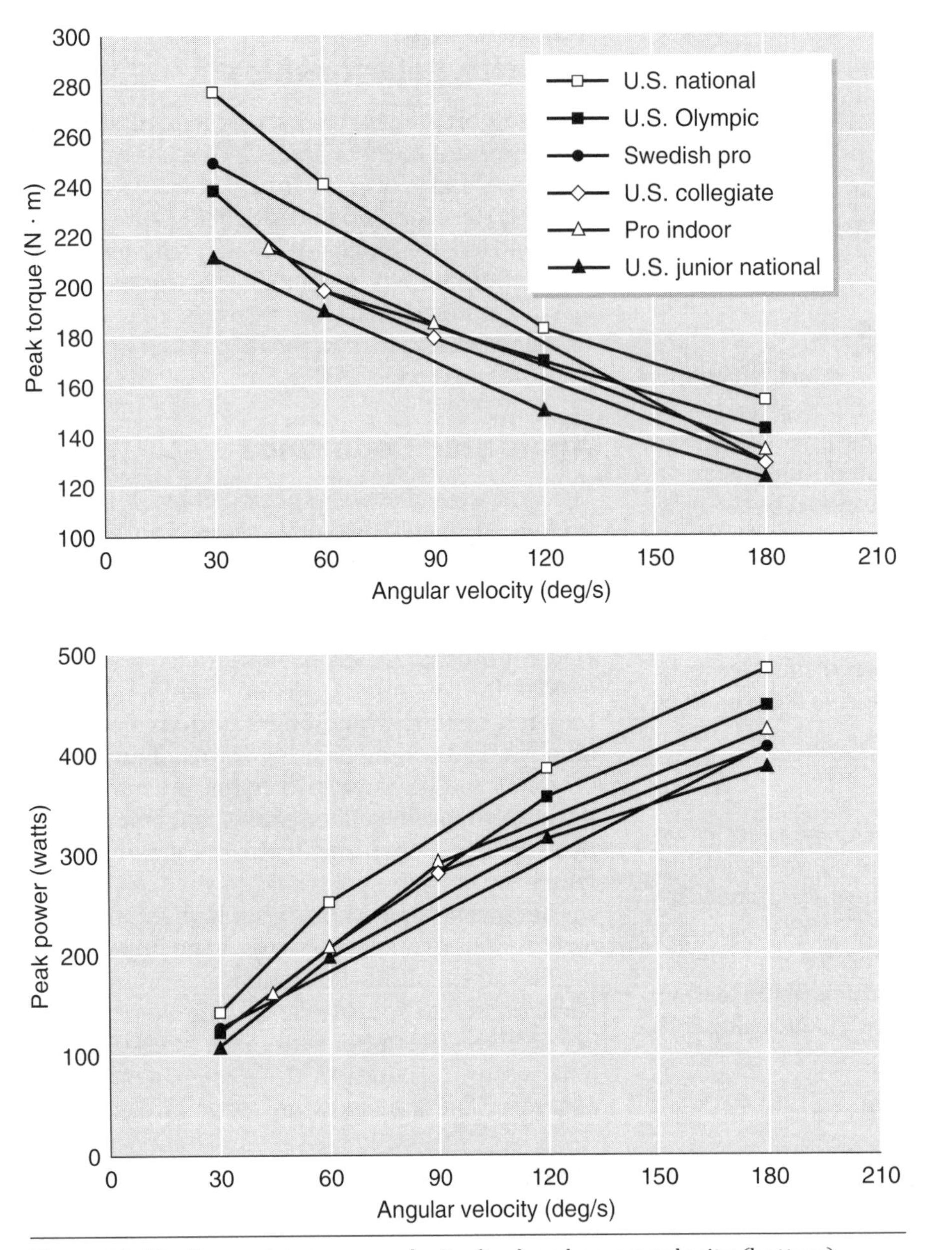

Figure 12.13 Concentric torque-velocity (top) and power-velocity (bottom) relationships for elite soccer players at various levels.

Modified from Kirkendall 1985.

the following issues have already been discussed, but their importance suggests that they are worthy of special attention. Some issues concern limitations in dynamometer capabilities that are ultimately the responsibility of the manufacturers, although users must be aware of them. Other issues concern refinements in testing and reporting practices that the users of isokinetic dynamometry must pursue.

1. **Machine comparison.** Many more studies are required to determine the validity of comparisons between different brands of isokinetic dynamometers. Agreement between dynamometers will never be complete unless manufacturers reconcile the differences in operating principles. Users should ensure that all testing adheres to valid principles and that reporting of test specifications and data is accurate and comprehensive. At this time, it is appropriate to take a conservative and necessarily pessimistic position regarding data comparisons among isokinetic dynamometers.

2. **Software.** Discrepancies between the operating modes and calculations of the various dynamometer systems appear to be unnecessarily restrictive. Therefore, in addition to the uncertainties regarding result comparisons between dynamometers that *do* report similar measurement parameters, comparison between particular types of data from some dynamometers is impossible because they do *not* calculate similar measurement parameters. Database support in all dynamometer systems is extremely limited. No systems offer easy aggregated group data calculations and reporting, which are vital for athlete assessments.

3. **Eccentric testing.** Isokinetic dynamometers have been capable of eccentric testing for over a decade. However, the possible distinctive merits of eccentric testing have yet to be widely investigated, demonstrated, or exploited in routine isokinetic testing. This chapter has highlighted initial findings that hint at the potential importance of eccentric-concentric strength ratios; but unless eccentric testing reveals unique information, there will be resistance to extending the duration of the assessment to include both concentric and eccentric muscle function.

4. **Advanced protocols.** Isokinetic testing has been dominated by concentric contractions. These

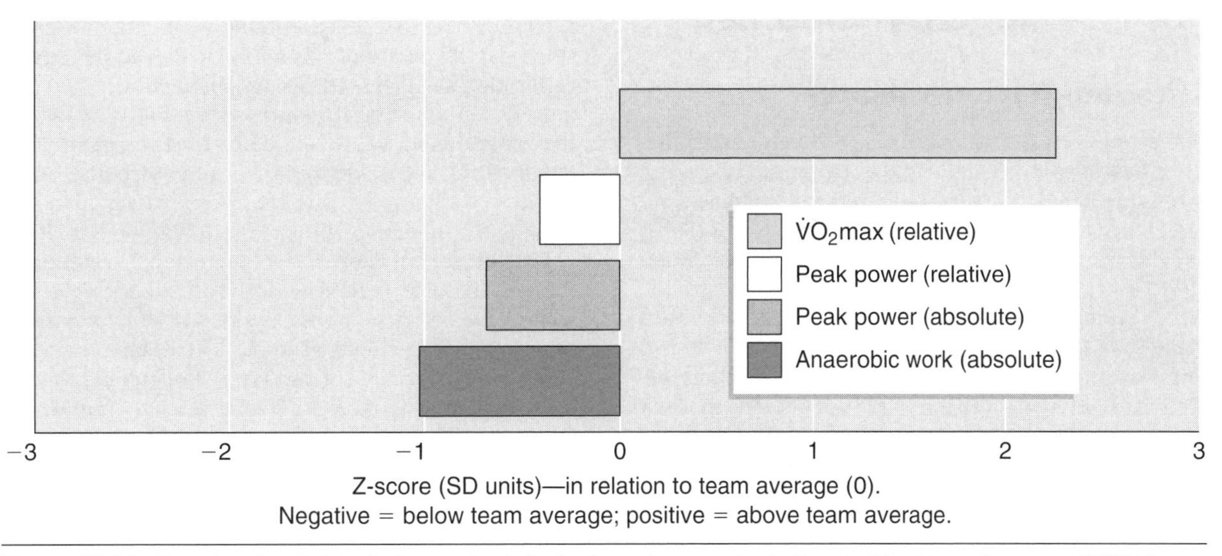

Figure 12.14 Results of individual player physiological test battery, including isokinetic peak power (360°/s).

tests can now be conducted with good validity and reliability in healthy subjects. Although there is a strong case for the relevance of such tests to athletic performance, isokinetic dynamometry has much potential to provide more sophisticated testing approaches to the particular muscle function requirements of sports; eccentric testing is an example. Isokinetic protocols to assess the so-called stretch-shortening cycle are in their infancy, and further work is necessary to investigate these protocols—the measurement issues as well as their relevance to athletic performance.

5. **Athletic performance correlation.** The case for the association between isokinetic measures and athletic performance was investigated in detail for this project, both on theoretical grounds (see chapter addendum) and via review of the literature (see Wrigley 2000). More studies are needed to further refine earlier work, investigating the relative sensitivity of different measurement parameters, with elite and subelite groups, using modern dynamometers.

6. **Data scaling.** Assessment of isokinetic test results is currently based primarily on *intra*individual comparisons; *inter*individual comparisons between individuals are often difficult to make with certainty, especially if absolute scores are not justified and relative scores must be used. The use of simple relative strength scores may not provide an unbiased estimate that fairly assesses athletes of different sizes. One hopes that a resurgence of interest in these scaling issues for other physiological assessments (e.g., oxygen consumption, ergometer power output) will also lead to further isokinetic studies of scaling.

7. **Injury relationship.** Throughout its history, isokinetic dynamometry has been used to make judgments about future injury potential or the extent of recovery in injured athletes (i.e., readiness for return to sport). Experience suggests that recorded weaknesses not explained by injury history—that might be predictive of future injury—are rare. This, along with the scarcity of research evidence, suggests that such prognostications regarding future injury are dubious; but much more quality work on this topic is necessary.

Summary

In this chapter we have outlined the important issues in the application of isokinetic dynamometry for athlete assessment in strength/power-related and strength/power-limited sports. We have identified factors important in achieving reliability and validity and offered recommendations for their incorporation in standardized assessments. The chapter has highlighted some limitations in the literature, as well as the opportunities for further research and practical applications. A peripheral but important issue that the chapter does not address is the validity of isokinetic dynamometry, which is not always understood. Thus, we include here an addendum that covers the issue in some depth.

Addendum: Validity of Isokinetic Dynamometry

Criterion-Related Validity

The strongest form of validity, criterion-related validity, is assessed by examining the association between scores on the test whose validity is being examined and direct measures of the characteristic of interest (Safrit 1973; Rothstein 1985; Mayhew and Rothstein 1985; Wood 1989). The aim of criterion-related validity is to provide direct evidence justifying the inferences one wishes to make from the test methodology in question. The two forms of criterion-related validity—concurrent and predictive validity—refer to the validity of inferences about the present time and about some time in the future, respectively. In the case of strength assessment, there are two classes of inferences that one generally wishes to make:

1. That the test is an accurate and reliable measure of "strength."
2. That the measured strength bears a close relationship to the "functional" strength required in a given sport.

These inferences come under the heading of concurrent validity. In some cases, for example in talent identification, the predictive validity of isokinetic strength assessment will also be of interest.

With respect to the first inference mentioned, there is no universally accepted definition of "strength," although most definitions refer to isometric or low-velocity strength. The advent of isokinetic dynamometry several decades ago provided a large range of potential means of "parameterizing" strength under various conditions of angular velocity, joint angle, body position, and measurement parameters (e.g., peak and average torque, work, peak and average power—see earlier discussion of measurement parameters, page 176). Thus the notion of a single definition of strength is neither likely nor particularly useful. Some of the many isokinetic measurement parameters may provide unique information, while others may provide redundant information as detailed in the discussion of measurement parameters.

Torque

The range of measurement parameters that have been mentioned are all derived from the basic measurement parameter in isokinetic dynamometry, torque (or force on some dynamometers, from which torque can be calculated). Thus one verifies concurrent validity for the first category of inference by demonstrating that an isokinetic dynamometer measures known torques accurately and reliably. The dynamometer systems have indeed been shown to be valid and reliable for measurement of known torque loads applied by inert weights under gravitational loading (Farrell and Richards 1986; Mayhew et al. 1994; Timm et al. 1992; Moffroid et al. 1969; Taylor et al. 1991). Routine verification of criterion-related validity of this sort is achieved by regular calibration as discussed earlier. Among issues that one must consider with criterion-related validity is whether the criterion is a valid "standard" for the quantity being used to assess validity; the issue of whether inert weights always meet this requirement was also discussed in connection with calibration and reliability.

Both classes of inference are important in relation to the overall validity of isokinetic dynamometry. However, some have argued that discussions of the concurrent criterion-related validity of strength assessments should most emphasize the second type of inference, that is, association of isokinetic strength with "functional strength" (Mayhew and Rothstein 1985) in sport performance. This is usually assessed by correlating isokinetic results with functional, sporting performance. As reviewed by Wrigley (2000), the literature is overwhelmingly supportive of a relationship between isokinetic strength and various athletic performances (especially jumping, throwing, sprinting, and kicking).

Furthermore, many studies have demonstrated good correlations between isokinetic strength (especially at faster velocities) and muscle fiber type (see Wrigley 2000), which is an important determinant of athletic performance.

Torque "Overshoot"

Discussion of the validity of isokinetic dynamometry has also focused on several perceived challenges to criterion-related validity. This has concerned the related issues of

1. the constancy of angular velocity (i.e., the isokinetic assumption),
2. the so-called torque-overshoot phenomenon that occurs when the test angular velocity is first reached by the athlete, and
3. the damping (filtering) of the torque curve that may be required to deal with torque overshoot.

These issues relate primarily to the first widely used dynamometer, the Cybex II with its Dual Channel Chart Recorder, whose rather simple design reflected 1960s technology. As technological approaches to these issues have improved in the newer dynamometers, the issues have faded from prominence in the literature.

Later versions of Cybex dynamometers, and dynamometers from other manufacturers, have tended to employ more sophisticated velocity-control servomechanisms (or less "responsive," mechanically damped systems) so that less overshoot occurs. Therefore, less torque signal damping/filtering is required, and the torque-angle relationship is not subject to the phase shift distortion (i.e., time shifting of the torque curve) associated with the original Cybex damping circuitry. However, the same conditions that led to marked overshoot (long limbs of large mass and moment of inertia, moving at high velocities) can still produce some overshoot artifact in the torque recordings of modern dynamometers.

Some dynamometer manufacturers have further addressed the potential torque overshoot problem by imposing an initial period of controlled, or "ramped," acceleration (Gransberg and Knutsson 1983; Rathfon et al. 1991) to prevent the period of free acceleration allowed in the Cybex II and similar dynamometers that is particularly associated with overshoot. Acceleration is reduced and gradual, so that the limb does not develop significant momentum and thus does not cause the dynamometer to overspeed when the preset test angular velocity is reached; thus torque overshoot is reduced.

Note that none of the latest models of the most common dynamometer brands (Kin-Com AP and 500H, Cybex 770/Norm and 6000, Biodex System 2 and 3) allow totally free acceleration. However, each uses a different method of velocity control prior to the attainment of the preset test velocity. The research to date (albeit with earlier dynamometers employing similar strategies) would suggest that these approaches are compatible with the production of accurate data (Taylor et al. 1991; Farrell and Richards 1986; Rathfon et al. 1991).

Some have suggested that all data recorded during these nonisokinetic phases be excluded from analysis (Wilk et al. 1992; Hoens and Strauss 1994; Rothstein et al. 1983). However, this may mean that each athlete is assessed over a different range of motion and over smaller ranges at faster velocities. Alternatively, if all measurements are taken over the minimum, "true" isokinetic range of motion recorded at any test velocity, then many potentially important recorded data (especially at slower velocities) will be discarded. Thus there is no definitive solution to the overshoot problem; compromises are inherent in every approach.

Angular Velocity

A number of studies have focused on the extent of the dynamometer lever arm's departure from constant velocity during various stages of its movement. This occurs mainly in association with torque overshoot. The studies have employed independent means of deriving velocity, such as cinematography (Sapega et al. 1982; Taylor et al. 1991; Iossifidou and Baltzopoulos 1996). The accuracy of the dynamometers' own velocity measurements has rarely been studied (Taylor et al. 1991). It is worth noting that there has been a tendency to accept alternative velocity measurement methodologies (such as those used in these studies) as a "gold standard" when studying dynamometer lever arm velocity—which they generally are not.

Challenges to the Validity of Isokinetic Dynamometry—Face Validity

The validity of isokinetic dynamometry as an assessment, rehabilitation, and training modality has occasionally been challenged, principally on two grounds: (1) that isokinetic (i.e., constant velocity) movements rarely occur during athletic performance and (2) that the range of angular velocities that isokinetic dynamometers are able to operate at is well below the maximum velocities seen in many high-speed sports. These assertions are essentially challenges to the "face" validity of isokinetic dynamometry. Rothstein has noted, "Face validity is the lowest form of validity because it reflects only whether a test *appears* to do what it is supposed to do" (1985, p. 17, emphasis added). Challenges to the face validity of isokinetic dynamometry contend that—for the two reasons outlined above—such dynamometry does not appear to measure strength in a way relevant to sport performance, especially high-speed performance. Rothstein went on to describe face validity as the "last resort" to justify use of a particular measurement or, by implication, to reject the use of a particular measurement.

Given the superiority of criterion-related validity for justifying the use of any measurement methodology, the correlation between isokinetic dynamometry and athletic performance is the best argument against challenges based only on face validity. As previously noted, the literature on this subject is compelling in its support of the association between isokinetic measures and many sporting activities (Wrigley 2000).

The "inadequate-velocity" assertion—that the range of angular velocities of isokinetic dynamometers is well below the maximum velocities in many high-speed sports—reflects a misunderstanding of the generation of high velocities during sporting movements. This issue is addressed in Wrigley (2000).

Face validity is important in one sense only. An athlete's attitude toward a test will depend on whether he or she perceives the test to be useful for the stated purpose. Therefore, if the athlete doubts the face validity of a test, then test performance may be affected. Ironically, some athletes may perceive greater face validity for tests that come close to mimicking the exact movement patterns of their sport although good reliability and criterion-related validity for such tests may not have been verified. Therefore, to ensure an athlete's confidence in an isokinetic test, it is a good idea to explain to athletes what sort of information one is seeking from the test, how the test should provide that information, and how that information may be helpful to enhancing performance in their sport.

chapter 13

Protocols for the Assessment of Isoinertial Strength

Peter Logan, Danielle Fornasiero, Peter Abernethy, and Katrina Lynch

Most movement in athletic settings involves the acceleration and deceleration of a constant mass (i.e., a limb or external load) about the associated joints or articulations. Thus the majority of sporting movements entail the development of isoinertial force (Murphy et al. 1994). The conditioning tenet of specificity (Bompa 1983) suggests, therefore, that the assessment of isoinertial strength should be an important tool for diagnosis and for the design of appropriate strength-training programs for athletes.

Isometric tests of muscle function relate to the muscle's capacity for maximal "static" force development; they relate poorly to dynamic sporting actions requiring strength (Abernethy and Jurimae 1996; Baker et al. 1994; Young and Bilby 1993). Although isokinetic strength assessments do involve dynamic muscle actions, the information they yield may help in gauging the rehabilitation of muscle and joint function during submaximal tasks but bear little relationship to sport performance (Pearson and Costill 1988; Abernethy et al. 1994). On the other hand, increasingly data are emerging that support "context specificity" (Abernethy et al. 1995) in the area of strength and power assessment. This suggests a real potential for isoinertial strength to correlate highly with dynamic sport performance (Abernethy et al. 1996; Wilson et al. 1993; Murphy et al. 1994; Viitasalo and Auro 1984).

Weight-lifting tasks have traditionally formed the basis of resistance-training prescriptions for developing dynamic muscle strength and power. These tasks have also been commonly used in protocols designed to determine changes to various indexes of strength and power. Although athletic weight lifting was once considered isotonic, it is now understood that the external loading in weight-lifting tasks is not constant throughout a given effort—again, because of acceleration and deceleration of the weight and changes in joint angle. The term isoinertial more accurately reflects the underlying muscular effort throughout a weight lifting-type task. Isoinertial loading implies a constant resistance to motion rather than merely a constant resistance or load throughout the lift (Abernethy et al. 1996).

However, comparisons between the mechanical profiles of dynamic sporting actions and traditional weight-lifting tasks prompt the conclusion that performance on the latter may provide only a limited diagnosis of an individual's capacity for developing functional isoinertial strength. This relates to the concept that weight-lifting tasks afford unidimensional feedback based only on the muscle's capacity to either overcome or yield to a given load with no temporal constraints. There is no indication of the quality of the muscle activity involved. Indeed, the high-force isoinertial muscle contractions in most athletic movements are diverse. While some actions require the development of high forces over a matter of milliseconds, others—such as a sprint running stride or a maximal bench press in competitive power lifting—are almost isometric as near-constant force is developed over a comparatively long period. A number of actions, such as kicking and jumping, involve the development of high force in a stretch-shorten cycle (SSC) and invoke intense eccentric muscle activity; other muscle actions—such as a "jab-punch" in boxing and the crouch start in sprinting—are principally concentric. It is logical, therefore, that one should carefully match a chosen mode of isoinertial strength assessment to the specific impositions of individual dynamic actions (context specificity). As with any physiological assessment tool, one needs to consider factors such as joint range of motion, metabolic stresses, motor unit recruitment and firing patterns, intermuscular coordination, and movement speed.

Schmidtbleicher (1986) has suggested that to accommodate context specificity, isoinertial strength assessment tools should take into account the continuum of isoinertial strength qualities that seem to characterize the majority of "powerful" dynamic actions. This continuum incorporates high, slow force productions at one end (maximal isoinertial strength) and moves toward progressively faster and lower force productions (isoinertial speed strength or power, isoinertial starting strength, isoinertial explosive strength, and isoinertial SSC strength) (Schmidtbleicher 1986; Vrijens et al. 1990). While this continuum of strength qualities appears to provide a useful framework for assessment of sport-specific isoinertial strength and power, establishing the validity and reliability of the measures remains perhaps the greatest challenge to the advancement of isoinertial strength assessment techniques.

Maximal Isoinertial Strength Assessment

Maximal isoinertial strength itself can be further differentiated according to the muscle action associated with a given phase of force development. The two subcategories are maximal concentric and maximal eccentric strength.

Maximal Concentric Strength

Maximal concentric strength is determined by the greatest load that can be successfully lifted during a movement requiring concentric muscle activity—that is, the maximum load that can be moved through a positive range of motion (i.e., against gravity) (Schmidtbleicher 1986). Clearly, maximal concentric strength is expressed in those actions that involve the movement of the athlete's own body weight or of heavy external loads from a static position (e.g., jumping and shot putting, respectively). In these instances, the agonist muscle groups generally work from an elongated or pre-stretched state.

Maximal Eccentric Strength

Maximal eccentric strength refers to the maximal load that can be resisted during a muscle-lengthening action (i.e., through a negative range of motion or with gravity). This strength quality is required in sporting actions that entail high levels of shock absorption (e.g., long jump, high jump, triple jump) or deceleration (e.g., changing direction and pivoting).

Maximal Isoinertial Strength

Maximal isoinertial strength is usually determined from tests of low-repetition maxima (e.g., one-repetition maximum [1RM] to three-repetition maximum [3RM]). A 1RM score refers to the load that can be lifted only once during a given task; a 3RM score refers to the load that can be lifted no more than three times for a given action. While the 1RM is a more "maximal" effort, 2 or 3RM tests are sometimes preferred over the 1RM test because they may be more reliable and less injurious and yet remain high correlates of the 1RM test (Sewall and Lander 1991; Hoeger et al. 1990; Jacobs et al. 1988). An example of 1-3RM tests of maximal concentric and eccentric strength is the traditional bench press in which the bar is pressed from the chest against a maximal load (elbow extension and shoulder flexion) or resisted to the chest against a maximal load (elbow flexion and shoulder extension).

Isoinertial Speed Strength Assessment

Speed strength refers to the ability of the neuromuscular system to produce the greatest possible impulse in the shortest possible time, or its capacity to overcome resistance with the greatest contraction speed possible. Since power is expressed as the product of force and speed, the term speed strength tends to be used synonymously with power (Vrijens et al. 1990).

Categorization of Speed Strength

One can further categorize speed strength according to the temporal impositions placed on any given powerful contraction (that is, as starting strength or explosive strength), while another strength quality within the broad category of speed strength is isoinertial SSC strength.

- **Starting strength.** The ability of the neuromuscular system to develop the greatest possible strength in the shortest possible time may be referred to as starting strength. This strength quality does not necessarily relate to total movement outcome but rather to rapid force production at the beginning of a muscle contraction. It is essential for optimal performance in sports like boxing and fencing that require great initial speed. It is based on the capacity to involve a great number of motor units at the beginning of the contractions and to develop high initial force.
- **Explosive strength.** Explosive strength refers to the ability of the neuromuscular system to continue developing already initiated tension as quickly as possible. It relates to the capacity to achieve a rapid rate of force development (RFD). In speed strength- or power-oriented tasks in which a load is light (i.e., low inertia), a high level of starting strength is necessary. As the inertial load increases, more explosive strength is required. Ultimately, when the load is very high, maximal strength clearly plays a pivotal role.
- **Isoinertial SSC strength.** Sometimes also referred to as reactive strength, this quality is necessary for the development of force in rapid SSC muscle actions—for example, the contact phase in sprint running. Here the extensor muscles are activated before ground contact and thus show a better stiffness, which magnifies the tension at the tendon. A high level of reactive strength means a more economic and effective eccentric phase and in turn a more powerful concentric phase (Vrijens et al. 1990; Schmidtbleicher 1986). During the stretching of the muscle, reflex activities may provide for a higher activation of motor units than is possible during maximal voluntary contractions (Schmidtbleicher 1986). Again, tension at the tendon is enhanced, resulting in a higher force push-off or concentric muscle action.

General Test Procedures

Isoinertial speed strength and SSC strength may be determined from tests that require short-burst (explosive) efforts against body weight or only small external resistances. Simple vertical jump tests (countermovement jump [CMJ], squat jump), for example, seem to be effective for gauging the speed strength qualities of the lower limb musculature during concentric knee and hip extension (squat jump) and during SSC knee and hip flexion followed by knee and hip extension (CMJ) (Harman et al. 1990; Shetty and Etnyre 1989; Komi and Bosco 1978; Zamparo et al. 1997; Wilson et al. 1995; Cordova and Armstrong 1996; Young and Bilby 1993). Similarly, the seated shot put throw has been used traditionally to determine the speed strength qualities of the upper limb musculature during concentric elbow extension and shoulder flexion (Gillespie and Keenum 1987).

The dependent measures adopted from the more traditional tests of isoinertial speed strength are generally derived "indirectly:" jump height, performance time, or throw displacement is taken as an approximation of mechanical work and/or power. While these tests do involve isoinertial muscle actions and in some cases provide reasonable correlates of power (Dowling and Vamos 1993; Harman et al. 1990), they offer little insight into the isoinertial strength qualities—starting, explosive, and SSC strength—that are central to performance of most isoinertial, high-power sporting actions.

Specific isoinertial strength qualities of both upper and lower limb actions can be accurately assessed using force or torque transducers such as those in most ground reaction force (GRF) plates. Direct force measurement during a given explosive action allows analysis of an individual's capacity to develop force over time (Cordova and Armstrong 1996; Wilson et al. 1995; Dowling and Vamos 1993): a force versus time function is produced that highlights the individual's RFD and impulse for a chosen explosive movement. If this movement is specific to the individual's focal sporting activity, direct force measurement can probably help differentiate specific aspects of the person's capacity for speed strength.

The force-time functions derived from direct force measurement permit one to specify force production capabilities at predetermined stages of high-force muscle activity (e.g., starting strength at 30 ms after the onset of muscle contraction). While little work has so far gauged the sensitivity of these measures to longitudinal training, they show promise for establishing meaningful markers of isoinertial strength adaptation (Cordova and Armstrong 1996; Wilson et al. 1995; Dowling and Vamos 1993).

There are few data on the reliability of either traditional or more recent isoinertial strength assessment techniques. Also, the apparent specificity to athletic performance of the more complex isoinertial speed strength measures has only generally been implied and not rigorously assessed.

Traditional Tests for Isoinertial Strength Qualities

A number of protocols, long used as basic tools for strength assessment, are easily integrated into weight-lifting programs and commonly involve little equipment.

The One-Repetition Maximum

The 1RM for various tasks has been used for many years to assess maximal isoinertial strength. Some have questioned this approach to assessment on the basis that although the activity is more dynamic than an isometric contraction, it bears little resemblance to many athletic tasks in terms of posture or pattern or speed of movement. Most 1RM movements (e.g., squat or bench press) are nonspecific to real movement tasks and occur over a matter of seconds rather than milliseconds. This is in contrast to the majority of isoinertial movements in sporting contexts, which require force development over milliseconds. In fact, some have defined 1RM loads as quasi-isometric (Siff 1993). Again, the majority of sporting tasks involve the execution of forceful contractions in minimal time (e.g., sprint stride, jump takeoff, football kick), emphasizing high-power isoinertial contractions. It is logical that 1RM tests requiring low-repetition, high-load contractions with no time constraint may not be sensitive to isoinertial strength changes resulting from faster execution or higher repetition maxima weights. Information from 1RM tests, relating more to power and/or strength endurance, is perhaps more important for athletic profiling or diagnosis than as an indicator of maximal strength.

Vertical Jumps

Historically, vertical jump tests have been widely used to assess isoinertial strength qualities in the lower limb musculature. Factors such as arm swing (Harman et al. 1990; Shetty and Etnyre 1989), pre-stretch or countermovement, and knee angle at starting position (Komi and Bosco 1978; Zamparo et al. 1997; Wilson et al. 1995; Cordova and Armstrong 1996) can affect vertical jump performance. Thus, although some vertical jump tests may truly reflect the underlying strength qualities of the knee and hip extensor muscles, others may be more influenced by motor skill or peripheral elements.

It is easy to reduce the influence of arm swing and therefore of motor skill in vertical jump performance. Subjects are often required to jump with their hands on their hips (Young and Bilby 1993) or with their arms constantly extended above the head. These protocols require specialized equipment that avoids the need to make a chalk mark on a wall at the peak of the jump. Use of a jump mat, a slat/vane apparatus, or the Smith machine (which allows a barbell to slide in a vertical plane along two upright support rails) also diminishes the skill component of the vertical jump. The protocols presented later in this chapter include information on these devices.

Upper-Body Assessments

One- to three-repetition maximum tests are widely used to assess maximal isoinertial strength of the upper-body musculature. For example, bench press and bench pull movements determine the maximum isoinertial strength of the shoulder/elbow extensor and shoulder/elbow flexor musculature, respectively. However, few protocols are in general use for determining any of the other qualities in the isoinertial strength continuum.

One protocol that presumes to assess speed strength qualities of the shoulder/elbow extensors is the seated shot put throw. Seating the subject in this test appears to limit the contributing muscle activity to the upper body. The distance that the shot is thrown is taken as an index of power of the relevant upper-body musculature. One study has shown that seated shot put performance correlates well with weight-lifting performance (Gillespie and Keenum 1987), but there are no data relating seated shot put performance to performance in other sports.

Recent Tests for Isoinertial Strength Qualities

There is much interest in measuring power through use of isoinertial jumping and weight-lifting tasks as discussed in the next two sections. For example, height (or work) on load curves have been developed using weighted jump tasks with a wide range of loads (Hakkinen et al. 1987, 1988; Viitasalo 1985a, 1985b). Some argue that in comparison to weight-lifting tasks such as the RM squat, these tests have a dynamic accelerative motion more closely approximating that in a range of sports. However, opponents of this form of assessment tend to emphasize the potential for athlete injury and the poor reliability and objectivity due to intertrial, interathlete, and interlaboratory variations.

Force Versus Time Functions

Clear evidence that static tests for RFD are not meaningful correlates of movement-specific strength qualities emerged from a study showing correlations between both RFD and maximum force during isometric, concentric, and SSC actions and sprint running performance (Wilson et al. 1995). In that investigation, all movements involved an upright squat position performed in a Smith machine apparatus. Subjects were required to "attempt" to perform maximum, explosive hip and knee extension movements that in the case of the concentric and SSC actions involved jumping into the air (squat jump and CMJ, respectively), and in the case of the isometric task resulted in no such external movement. For the concentric and isometric movements, subjects adopted a starting knee angle of 110° and 150°; for the SSC movement, they started in an upright standing position and "dipped" to similar knee angles before exploding into the air. From the force-time profiles for these movements, the maximum force, maximum RFD, force at 30 ms, and impulse at 100 ms were used as test variables. Sprint running performance was determined from a 30 m indoor sprint started from a crouched position. Only variables calculated from the concentric test correlated to the running task and were able to discriminate effectively between good and bad sprint performance.

Wilson et al. (1995) showed that although isometric RFD tests lack specificity to dynamic performance, a potentially high correlate of performance is concentric, isoinertial RFD involving common musculature and actions. The study also showed that SSC RFD tests involving the hip and knee extensor musculature may have limited relationship to sprint running performance. This appears to counter the tenet of specificity, as sprint running entails sequential, high-force SSC actions. However, the authors postulated that this finding may have been due to a lack of consistency between the eccentric phase of the SSC RFD test (CMJ) and that of the sprint running action. In the CMJ, the eccentric phase is performed very slowly in comparison to that in a sprint action. The authors further theorized that during a ballistic, SSC activity incorporating both eccentric and concentric muscle actions, the limiting factor in overcoming resistance is the concentric muscle force capability. In this way, then, the concentric RFD test may be the superior testing mode in relation to sprint running performance because it more effectively determined the concentric ability that is perhaps the limiting factor to dynamic SSC performance.

Machine Throws

Only one study has measured force development in various upper-body isoinertial tests and compared these with dynamic measures of upper-body muscle function (Murphy and Wilson 1997). Subjects performed maximal concentric bench throw and eccentric bench press tasks, as well as a series of upper-body performance tests. For the concentric and eccentric tests, subjects lay supine on a bench positioned within the Smith machine. Using loads of 30% and 60% of their 1RM bench press score, they threw the barbell for maximum height as explosively as possible. From these tests, the peak force generated from each throw was used as the test variable. For the eccentric test, a load of 130% of 1RM was adopted. The loaded barbell was held by electromagnets at an elbow angle of 120° prior to test commencement. At a signal, the electromagnets were released and subjects were required to resist the accelerating load with as much force and as quickly as possible until it came to rest just above chest level; the peak force produced throughout this movement was recorded as the dependent variable.

The tests of dynamic upper-body performance involved a 1RM bench press task, a seated shot put throw, and two drop bench press throws from a height of 0.25 m at various loads. All tests except the seated shot put throw were performed on a Smith machine apparatus, used as a gauge of upper-body SSC action under different loads. With the subject lying in the bench press position, a loaded barbell was released from a height of 0.25 m above the subject's extended arms (the two loads used were an absolute load of 10 kg and a relative mass of 30% of subject's 1RM). On release of the load as already described, subjects caught the bar and quickly threw it for maximum height. The total positive work done (mass × gravity × height) was calculated as the dependent variable for this test.

The seated shot put throw was conducted as described by Gillespie and Keenum (1987) with horizontal displacement as the dependent variable. Correlation between the isoinertial and performance measures varied from $r = 0.33$ to $r = 0.94$. The only nonsignificant correlations were between the eccentric 130% test and the shot put throw ($r = 0.33$); the highest correlations were between

(a) the concentric 30% and 60% bench throws and the eccentric 130% bench throw tasks (r = 0.94, r = 0.86 respectively) and (b) the 1RM test and the 130% bench throw task (r = 0.93). This study demonstrated good correlations between tests of upper-body isoinertial strength and upper-body performance. However, it is open to conjecture how well these performance tests reflect sporting actions. Moreover, the investigation did not address RFD. Indeed, to the authors' knowledge, no other work has examined isoinertial tests of RFD in the upper body and their relationship to specific dynamic upper-body performance. This appears to remain an important gap in the literature.

Reliability of Isoinertial Strength Measures

Detailed research on the reliability of isoinertial strength measures appears to be sparse. Because the protocols have been in existence longer, there is more work on the reliability of simple RM-type strength assessments. Certainly, widespread use of the more complex, direct force-related assessment techniques in elite athletes requires substantial research across the gamut of athlete groups.

One-Repetition Maximum

The available data appear to support the reliability of simple strength assessments using the 1RM approach. Research has clearly illuminated several issues relating to the potential for data contamination and ways of minimizing it during 1RM determination.

Effect of Load, Habituation, Preload, and Recovery

It has been demonstrated (see review by Abernethy et al. 1995) that the test-retest reliability of 1RM measurements among experienced male and female lifters is high (r = 0.92 to 0.98) (Sale 1991; Hortobagyi et al. 1989; Hennessy and Watson 1994; Hoeger et al. 1990). But the correlation coefficients for more dynamic isoinertial activities (e.g., weighted squat jumps) tend to be lower. For example, Viitasalo (1985a, 1985b) reported a correlation (interclass) of 0.86 to 0.96 between trials for static squat and CMJs involving loads of 0, 20, 40, 60, and 80 kg. Critically, however, the coefficient of variation (CV) increased for both jumps with loads of above 40 kg (e.g., for the CMJs the CVs were 0 kg = 4.3%, 20 kg = 4.8%, 40 kg = 6.8%, 60 kg = 7.1%, and 80 kg = 9.5%) (Viitasalo, 1985a, 1985b). In fact, the relative changes in the CV for the 40, 60, and 80 kg trials as compared to the CV for the 0 kg trial (i.e., [trial CV/CV for the 0 kg condition] $\times$ 100) of Viitasalo (1985a) were markedly greater (i.e., 158, 165, and 221) than are typically seen for isometric, isokinetic, and 1RM measures. According to Abernethy et al. (1995), therefore, these data suggest that there may be threshold loads in isoinertial lifting tests beyond which reductions in reliability seriously compromise validity. Researchers have yet to determine whether these thresholds differ between individuals according to protocol familiarity and/or training status. It is clear that to maximize their ecological validity, tests like the weighted squat jump and drop jump require stringent control as well as extensive habituation of study participants. Finally, Abernethy et al. (1995) note that hydraulic loading represents yet another form of isoinertial loading. Hortobagyi et al. (1989) and Hortobagyi and Katch (1990) indicated that while this form of loading can be reliable (0.89 to 0.93), one must consider the interactions between the movement pattern being assessed, learning effects, the number of trials, and intertrial recovery periods.

Strength at 1RM may also be affected by several acute variables, including preloading and recovery between 1RM efforts (Abernethy et al. 1995). The performance of isoinertial tasks is enhanced by isometric preloading and SSC (Wilson et al. 1995; Tihanyi et al. 1982). Some practitioners suggest that 1RM test results can be adversely affected by the repetitions of the task required before identification of the 1RM. Research suggests that this may be less of a problem than feared, as Anderson and Kearney (1983) reported that 60% of novice study participants required only four trials to reach their 1RM and that all participants had achieved their 1RM by the sixth trial. Also, repeat 1RM bench press measurements were not compromised when there was rest (1, 3, 5, or 10 min) between each effort (Weir et al. 1994). Another study showed that repeat 1RM determinations for the squat and bench press were no different when separated by 2, 6, or 24 h (Sewall and Lander 1991).

Predicted One-Repetition Maximum

Researchers have developed formulas predicting 1RM values from the number of repetitions completed at lesser loads (Brzycki 1993; Lander 1985; Mayhew et al. 1992). The predicted and actual 1RM values tend to be highly correlated (r = 0.89 to 0.96). These equations are based on the assumption that the number of repetitions completed at a given percentage of 1RM does not change with training. This assumption appears to be true for some lifts (bench and leg press), but not all (arm curls, knee extension, sit-ups, leg curls, and lateral pull-downs) (Hoeger et al. 1990). Additionally, as the number of repetitions increases, the variations between the formulas in the predicted 1RM values increase dramatically. Furthermore, Mayhew et al. (1992) reported that several of the formulas (Brzycki 1993; Lander 1985) tend to overestimate 1RM values by between 1.5 and 2.5 kg. Taken together, these data suggest that equations for predicting 1RM have their place (e.g., with the novice lifter and/or where the risk of injury is deemed to be unacceptably high); however, they are not without their problems.

Ground Reaction Force Parameters

The research is equivocal as to the reliability and/or CV of GRF parameters (Cordova and Armstrong 1996; Wilson et al. 1995; Dowling and Vamos 1993). One study examined GRF data in young male and female subjects performing maximal one-legged CMJs (Cordova and Armstrong 1996). Measurement of peak vertical GRF was shown to be very reliable ($r^2 = 0.94$). However, the same study showed vertical impulse to be somewhat unreliable ($r^2 = 0.22$). Wilson et al. (1995) investigated the reliability of GRF data during squat jump and CMJ tasks in young male athletes, assessing repeatability of isoinertial speed strength measures of peak force, force and impulse at 30 ms and 100 ms, and maximum RFD. Coefficients of variation data ranged between 13.2% (impulse at 100 ms during squat jump) to 70.6% (peak force during CMJ). Dowling and Vamos (1993) had moderately trained individuals perform maximum vertical jumps (CMJs) on a GRF plate and looked at a variety of temporal and kinetic variables associated with the resultant instantaneous force-time curves. Unfortunately, they reported CV data only for takeoff velocity (1.7% within subjects).

It may be that the reliability of GRF assessments is affected mainly by the status of the subjects. Unpublished data from our own laboratory reveal that technical error of measurement (TEM; see page 83) scores changed from a group of elite female cyclists to a group of mixed-sex recreationally active individuals performing a squat jump from a starting knee angle of 90° (table 13.1). While the TEM of peak GRF force, force at 100 ms, and total impulse was reasonably low in both groups, TEM scores varied significantly between the groups for force at 30 ms, maximum RFD, and average relative power.

Collectively, the results indicate that peak force and velocity measures within GRF force-time functions may be more reliable tools than associated impulse measures for assessing isoinertial speed strength. However, the literature lacks any focus on the reliability of force-time parameters and therefore complex measures of isoinertial speed strength within representative athlete groups. Logic would dictate the need for this research prior to any prospective isoinertial strength testing of specific athlete groups.

Table 13.1 Percent Technical Error of Measurement (TEM) of Ground Reaction Force (GRF) Data for a Squat Jump

	Subjects	
Parameter	Elite female cyclists %TEM	Mixed-sex, recreationally active %TEM
Peak GRF	2.27	2.09
Force at 30 ms	0.96	3.28
Force at 100 ms	2.52	4.43
Total impulse	4.67	3.17
Maximum RFD	6.46	9.12
Relative average power	6.76	14.27

Jump started from a knee angle of 90°. Subjects were elite female road cyclists (n = 10) and mixed-sex, recreationally active individuals (n = 13).

RFD = rate of force development.

Correlation With Athletic Performance

There is little research on the sensitivity of isoinertial test measures to the effects of training. Callister et al. (1988) reported that sprint training did not change 60 s jumping power even though there were significant improvements in sprint running performance. Pearson and Costill (1988) reported that eight weeks of isokinetic training of one limb produced significant increments in isokinetic but not 1RM strength. According to Sale (1991), isoinertial training that increased 1RM leg press strength did not increase isometric knee extension strength. Conversely, effect size transformations of data presented by Behm (1991) suggested that isokinetic and isoinertial (1RM) measures were similarly sensitive to training regimens involving a variety of equipment.

Murphy and Wilson (1997) tracked training-induced changes in performance using isoinertial tests of muscular function. They conducted tests using concentric and eccentric actions and a 1RM squat, and also determined performance on a 40 m sprint and 6 s cycle test, before and after training. Surprisingly, they found that each isoinertial strength measure correlated poorly with the two performance measures. Only the chosen measure of isoinertial explosive strength (RFD during eccentric strength) was a correlate of performance (with 5 s cycle time). Even this result is surprising in that it runs counter to the edict of context specificity, as the eccentric component of sprint cycling would be less than that of sprint running. Little other work has addressed the relationship between the more complex measures of isoinertial speed strength and athletic performance (table 13.2).

Table 13.2 Pearson Correlation (r) Data—Isoinertial Strength Versus Performance

Reference	Subjects	Test	Performance	r value, test vs. performance
Bosco et al. (1983)	i) 12 young male basketball players ii) 14 young males	Average relative watts after 60 consecutive rebound CMJs.	30 m sprint	i) 0.58 ii) 0.56
	i) 12 young male basketball players ii) 14 young males	Power in Margaria jump test.	30 m sprint	i) 0.20 ii) 0.29
Callister et al. (1988)	46 college men and women, novice RT	15 s CMJ	100 m sprint	0.68 to 0.70
Mayhew et al. (1989)	53 college football players	1RM bench press	36 m and 9 m sprint	0.11 to 0.16
Mero et al. (1981)		SJ, CMJ, DJ	Running speed	0.62 to 0.68
Murphy and Wilson (1997)	30 active males	1RM squat	i) 40 m sprint time ii) 6 s cycle time	–0.41 0.24
		Peak force during SJ (10 kg).	i) 40 m sprint time ii) 6 s cycle time	–0.07 –0.12
		RFD during SJ (10 kg).	i) 40 m sprint time ii) 6 s cycle time	–0.21 –0.08
		Peak force during eccentric squat.	i) 40 m sprint time ii) 6 s cycle time	–0.15 0.10
		RFD during eccentric squat.	i) 40 m sprint time ii) 6 s cycle time	0.07 0.31
		1RM squat.	i) 40 m sprint time ii) 6 s cycle time	–0.41 0.24

(continued)

Table 13.2 *(continued)*

Reference	Subjects	Test	Performance	r value, test vs. performance
Murphy et al. (1995)	24 male recreational athletes, novice RT	Concentric bench press throw (30% and 60% 1RM). Eccentric bench press (130% 1RM)	1RM bench press. Drop bench press throw (10 kg and 30% 1RM bench press). Shot put throw.	0.33 to 0.94
Viitasalo and Auro (1984)	8 national-level male high jumpers	i) SJ height (m) ii) CMJ iii) DJ height (20, 40, 60, 80, and 100 cm)	High jump (m)	i) 0.22 to 0.63 ii) Unreported iii) Unreported
Wilson et al. (1995)	15 male athletes, novice RT	Concentric RFD. SJ (110° and 150° starting knee angle) force at 30 ms, impulse after 100 ms, max force, max RFD); RFD CMJ (force at 30 ms; impulse after 100 ms; max force; max RFD).	30 m sprint time	Only correlation reported was between sprint time and force at 30 ms during the concentric 150° test: 0.62
Young and Bilby (1993)	18 college-age males	1RM squat (relative to BW)	Vertical jump height	0.47

1RM = one-repetition maximum, CMJ = countermovement jump, DJ = drop jump, RFD = rate of force development, SJ = squat jump.

In contrast to the research of Murphy and Wilson (1997), other work has shown isoinertial strength parameters and performance to be reasonable correlates. For example, 1RM squat performance correlated significantly with sprint running performance (Young and Bilby 1993); average power during repetitive CMJ performance with 30 m sprint time (Bosco et al. 1983); squat jump and CMJ power with sprint running speed (Mero et al. 1981); CMJ with sprint running performance (Callister et al. 1988); and squat jump height with high jump performance (Viitasalo and Auro 1984) (table 13.2). However, these may be considered merely the more traditional among isoinertial strength markers.

As with the issue of reliability, therefore, the literature is equivocal on the relationship between the more recent measures of isoinertial strength qualities and strength performance. Despite evidence that traditional isoinertial strength markers may be meaningful correlates of performance, apart from research involving sprinting there have been few comparisons between specific athletic performance measures and isoinertial strength measures.

Description of Protocols in General Use

The following sections present protocols in current use as tests for isoinertial strength qualities. Some of these may not appear to provide useful information on pure strength capacities—for example, the Wingate test uses several contractions over a relatively long period, and the sprint running test targets highly performance-oriented movements. However, these tests are listed because they are at least context specific and because they purport to relate to isoinertial strength qualities. Earlier sections of this chapter indicated the relevance of the remaining tests.

Vertical Jumps

Table 13.3 presents normative data for expected test scores from the literature. Data on selected Australian athletes appear on page 141 (see table 9.8 in that chapter).

Table 13.3 Vertical Jump Scores for Athletes

References	Subjects	Physical characteristics	Vertical jump (cm)
Baker et al. (1993)	22 experienced weight-trained males	Age 26.0 ± 2.6 yr Height 179.6 ± 5.9 cm Body mass 75.0 ± 8.4 kg	49.7 ± 6.9* 53.3 ± 5.9**
			Vertical jump (W/kg)†
White and Johnson (1991)	61 male and female athletes of international (int), national (nat), and regional (reg) levels	Males: Int (n=12) 21.5 yr, 179 cm, 78.8 kg Nat (n=8) 21.6 yr, 173 cm, 72.6 kg Reg (n=11) 18.1 yr, 174 cm, 68.5 kg	Males: Int 17.4 Nat 16.7 Reg 15.8
		Females: Int (n=17) 20.4 yr, 166 cm, 63.07 kg Nat (n=6) 21.8 yr, 165 cm, 63.64 kg Reg (n=7) 17.5 yr, 161 cm, 58.61 kg	Females: Int 15.7 Nat 14.4 Reg 13.6

* Pretraining; **posttraining, three days per week for 12 weeks.

† Peak power in watts was calculated from jump heights and body mass using the Lewis equation (Lewis et al. 1981).

Values are mean ± SD.

PROTOCOLS: Isoinertial Strength

Vertical Jump Using Slat/ Vane-Style Apparatus

Equipment: Vertical jump scores can be determined from commercially available jumping devices—the Yardstick and the Vertec (see the appendix for detailed information).

Test Procedure: The protocol is similar for the two devices.

- First, to establish the athlete's standing reach height, have the athlete displace as many vanes as possible while standing flat-footed and side on to the apparatus—reaching vertically with his or her preferred hand.
- Ask the athlete to jump as high as possible using an arm swing and a countermovement.
- At the peak of the jump, the athlete moves the vanes out of the way.
- The athlete must perform at least three trials.
- To determine vertical jump score, subtract the athlete's standing reach from the highest absolute height jumped (White and Johnson 1991).

Vertical Jump Using Smith Machine

Rationale: In vertical jumping using a Smith machine, the jumper holds a light barbell across the shoulders and, in the process of jumping, pushes the bar as far up the vertical supports as possible (Wilson et al. 1995). The vertical supports of the Smith machine ensure that the jump can be performed only in a vertical plane. Throughout the effort, upper-body posture and arm position stay constant to minimize the confounding effect of these variables.

Refer to table 13.4 for limited normative data for expected test scores for male subjects.

Equipment:

- Smith machine. Smith machines, or sliding squat racks as they are also called, are widely available from most weight-lifting equipment retailers. Importantly, the Smith machine should be capable of supporting a light barbell (e.g., <10 kg) as well as the more conventional 20 kg barbell, as is the case with the Plyopower.

Or:

- Plyopower Smith machine and vertical jump system.

Test Procedure: Explosive power can be measured via a modification of the vertical jump test on the Smith machine. The test requires the subject to execute either a squat jump or a CMJ while holding a bar on the shoulders.

- Instruct the subject to stand in an upright position with feet approximately shoulder-width apart.
- Record the height of the bar when it is stabilized on the subject's shoulders.
- For calculation of the vertical displacement of the bar (centimeters), the Smith machine must be equipped with a system such as a mechanical friction device or an electronic rotary encoder (Wilson et al. 1993).
- At least three trials are required. The best of all trials is recorded.

One-Legged Vertical Jump

Rationale: The one-legged vertical jump test has been shown to be a reliable estimate of lower-extremity functional strength in individual limbs (Cordova and Armstrong 1996).

Selected normative data for expected test scores from the literature are shown in table 13.5.

Equipment:

- A force platform interfaced through a 12-bit analog-to-digital converter to a personal computer

Test Procedure: Before performing this test, subjects must complete multiple habituation trials until they feel comfortable with the protocol and demonstrate correct technique. The recommendation is that familiarization occur on the day before testing.

- The subject is required to stand on the force platform with the right foot. The left knee is flexed at 90° to prevent the left foot from touching the ground, and the subject's arms must be crossed over in front of the chest to eliminate arm swing.
- A countermovement is permissible, however, and the subject is instructed not to go past 90° of flexion with the right knee during the jump.
- The test begins with the subject in an upright position.

Table 13.4 Vertical Jump Scores Using the Smith Machine

			CMJ* max height (cm)	
Reference	**Subjects**	**Physical characteristics**	**Pretraining**	**Posttraining**
Wilson et al. (1993)	55 weight-trained males split into 4 groups:			
	• Gp 1: Control, n=14	24.1 ± 5.4 yr, 173 ± 8.0 cm, 76.9 ± 15.4 kg	37.2 ± 8.2	38.0 ± 8.2
	• Gp 2: Weight training, n=13	21.9 ± 4.3 yr, 173 ± 8.9 cm, 69.0 ± 8.9 kg	40.0 ± 7.0	41.9 ± 7.3
	• Gp 3: Plyometric training, n=13	22.1 ± 6.8 yr, 174 ± 9.9 cm, 71.6 ± 11.9 kg	35.8 ± 6.7	39.5 ± 9.0
	• Gp 4: Dynamic weight training, n=15	23.7 ± 5.8 yr, 178 ± 9.7 cm, 75.9 ± 17.9 kg	35.8 ± 5.8	41.8 ± 5.8
			SJ† max height (cm)	
	• Group 1		35.9 ± 8.1	35.8 ± 7.6
	• Group 2		38.0 ± 7.5	40.4 ± 6.9
	• Group 3		35.6 ± 7.8	37.9 ± 8.2
	• Group 4		33.8 ± 4.9	38.8 ± 5.6

* CMJ = Max vertical jump height with countermovement.

† SJ = Max vertical movement without countermovement.

Values are mean ± SD.

- On the "Go" signal, the subject performs the jump.
- The subject should execute five maximal trials with 1 min rest intervals.
- Record the peak vertical GRF and vertical impulse for all trials (Cordova and Armstrong 1996).

Vertical Jump Using Timing Mat

Rationale: Limitations to the assessment of vertical jump using a traditional jump-and-reach protocol are obvious. The subject's vertical jump is indicated from a chalk mark made on a wall at the summit of the leap. Apart from the fact that the test requires a suitable wall and a chalk marker, it introduces a skill component in that the subject must coordinate jumping and making the chalk mark; logic would dictate that performance might be impaired. On the other hand, although the Smith machine apparatus probably allows for greater control during the determination of jump performance, the device requires considerable laboratory floor space and can be relatively expensive.

Expected test scores not available.

A recently developed electronic timing mat can be used to measure explosive power based on flight time during a vertical jump. The timing mat system consists of a contact mat connected by a cable to a digital timer. The timer, which measures flight time, is triggered when the subject's feet leave the mat and stops when the subject has landed. This method of calculation assumes that the subject's takeoff position and landing position are the same.

Equipment:

- Contact mat (78 × 52 cm)
- Timing module
- Computer with appropriate software (Suitable system available from Swift Performance Equipment [see appendix for details]).

Test Procedure:

- To minimize horizontal and lateral displacement during the jump, subjects should be instructed to perform the jump with their hands on their hips. Carefully explain identical, or at least similar, takeoff and landing postures to subjects and check these for each test.
- Subjects must start the test with a knee angle of 90° and feet placed comfortably apart.
- For estimation of maximal mechanical power of the leg extensor muscle, subjects must jump with maximal effort.

Calculations: In one common protocol, the subject must jump continuously for a certain period (e.g., 15-60 s). The sum of all flight times for the repeated jumps is recorded and used to calculate average mechanical power in the manner of Bosco et al. (1983):

$$\bar{\dot{W}} = \frac{\bar{W}}{\bar{t}_c} \quad (1)$$

where $\bar{W}$ = the total average work performed during 60 s; $\bar{t}_c$ = average total contact time of vertical jumping.

The total work (W) performed during a vertical jump can be calculated using the following formula:

$$W = m \times g \times h \text{ (2)} \quad (2)$$

where m = mass of the subject; g = acceleration of gravity (9.81 $m \cdot s^{-2}$); h = total displacement of center of gravity (CG).

Now calculate the total displacement of CG during the flight (h_f) and the contact period (h_c). The total displacement of CG is:

Table 13.5 One-Legged Vertical Jump Scores

Reference	Subjects	Physical characteristics (mean ± SD)	Test variable	Score (mean ± SD) Trial 1	Trial 2
Cardova and Armstrong (1996)	Males (n = 12) and females (n = 7). All healthy college students	Males: 21.3 ± 4.6 yr, 70.0 ± 2.3 in., 170.5 ± 28.7 lb.	PVGRF(%BW)	1.90 ± .23	1.92 ± 26
		Females: 23.2 ± 5.3 yr, 66.6 ± 4.3 in., 132.4 ± 25.9 lb.	VI(%BW s)	1.30 ± .50	0.91 ± .22

PVGRF = peak vertical ground reaction force, VI = vertical impulse, BW = body weight.

$$h = h_c + h_f \quad (3)$$

The displacement of CG during flight (h_f) is calculated using recorded flight time (t_f) as follows:

$$h_f = (gt_f^2)/8 \quad (4)$$

The displacement of CG during contact can be estimated assuming that the vertical velocity from the lowest point of the CG to the release is linearly increasing (Asmussen and Bonde-Petersen 1974). If the release velocity is v_v and contact time is t_c, then the rise of CG (h_c) during contact is:

$$h_c = (v_v \cdot t_c)/2 \quad (5)$$

Because the vertical release velocity and impact velocity are equal in a harmonic jump series, the vertical velocity (v_v) at the impact-and-release phase can be written:

$$v_v = (gt_f)/2 \quad (6)$$

where t_f = flight time between jumps.

Substituting formula (6) into (5) we can rewrite (5) as:

$$h_c = (gt_f \cdot t_c)/8 \quad (7)$$

The total displacement of CG (h) can then be obtained as follows:

$$h = (gt_f \cdot t_c)/8 + (gt_f^2)/8 = (gt_f \cdot t_t)/8 \quad (8)$$

where t_f = flight time of one jump; t_c = contact time of one jump; t_t = total time of one jump = $t_f + t_c$; and g = gravitational constant (9.81 m · s^{-2}).

The total work performed as expressed in formula (2) during a vertical jump can now be written as follows:

$$\overline{W} = (mg^2 t_f t_t)/8 \quad (9)$$

Assuming that the time of the positive work phase (t_{pos}) during contact can be half of the total contact time (Asmussen and Bonde-Petersen 1974), then the average mechanical power of the positive work phase (W) per body mass is:

$$\overline{W} = W/(m \cdot t_{pos}) = (g^2 \cdot t_f \cdot t_t)/4t_c \quad (10)$$

In the jump series, the electronic timing mat is summing the total flight time (T_f) of all (n) jumps, and therefore the average flight time (t_f) for one jump is:

$$\bar{t}_f = (T_f)/n \quad (11)$$

If in the jump series (including n jumps) the total performance time (T_t) is 60 s, for example, then the average total time (t_t) of one jump is:

$$\bar{t}_t = T_t/n = 60\ s/n \quad (12)$$

The contact time in a jump series of n jumps is the total performance time minus the total flight times. Then the average contact time (t_c) in one jump is:

$$\bar{t}_c = (T_t - T_f)/n = T_c/n \quad (13)$$

Now substituting the individual flight times for the average flight times, the average mechanical power $\left(\frac{\overline{\dot{W}}}{W}\right)$ in a jump series (formula [10]) can be written using the times of the jump series as follows:

$$\frac{\overline{\dot{W}}}{W} = (g^2 T_f T_t)/4nT_c \quad (14)$$

If the total time (T_t) of the jump series is, for example, 60 s, the formula (14) can be written:

$$\frac{\overline{\dot{W}}}{W} = (g^2 \cdot T_f \cdot 60)/4n(60 - T_f) \quad (15)$$

The unit of mechanical power per unit of mass is then watts per kilogram (W/kg).

Thus, to apply formula (15) we need to know only the sum of the flight time recorded by the timer for each jumping performance and the number of jumps executed during that work time period.

Vertical Jump Using Smith Machine and Ground Reaction Force Assessment

Rationale: A protocol in use at the Australian Institute of Sport is designed to assess the isoinertial strength qualities of the lower limb musculature for track and field athletes.

Table 13.6 provides normative data obtained from elite Australian track and field athletes from a number of disciplines.

Equipment:

- Smith machine and GRF plate

Test Procedure:

- Holding a light aluminum bar across the shoulders (approximately 9 kg), the athlete performs a squat jump from a starting knee angle of 120°.
- Instruct the athlete to perform the squat jump as explosively as possible and to aim to jump as quickly and with as much height as possible.

Table 13.6 Isoinertial Speed Strength Parameters on Elite Australian Track and Field Athletes

Subjects		Body mass (kg)	Maximum dynamic strength (MDS) (kg)	Time to MDS (ms)	Explosive strength (max RFD) (N/s)	Starting strength (F30 ms) (kg)	Force at 100 ms (kg)	Impulse at 100 ms (N/s)	Total impulse (N/s)	Average power (W)
Male sprint (n = 12)	Mean	74.7	230.3	104.5	44524.1	30.0	217.3	100.8	168.0	737.0
	SD	7.3	21.1	17.5	11466.0	10.7	28.1	19.7	19.7	113.9
	Max	86.9	272.1	151.0	63004.0	55.0	267.4	131.0	213.8	929.3
	Min	63.2	204.4	86.0	22271.0	15.6	145.7	53.0	135.3	466.2
Female sprint (n = 14)	Mean	59.7	179.4	109.3	30303.0	28.3	164.2	76.4	123.1	469.6
	SD	7.2	23.6	22.3	8336.1	14.9	35.2	27.2	15.7	108.6
	Max	72.8	220.4	145.0	42735.0	68.6	216.1	128.0	143.1	609.6
	Min	52.3	155.6	79.0	20513.0	15.6	108.2	40.5	93.8	305.1

Table 13.7 Isoinertial Speed Strength Parameters From AIS Track and Field Testing Protocol

Parameter	Definition
Starting strength	Force at 30 ms following start of squat jump (SJ)
Explosive strength	Maximum rate of force development during SJ takeoff
Maximum dynamic strength	Peak force achieved during SJ takeoff
Total impulse	Impulse produced during SJ takeoff
Impulse 100 ms	Impulse after 100 ms following start of SJ
Take-off time	Time taken from start of SJ to finish of takeoff phase (i.e., last contact with GRF plate)
Average power	Average power produced during complete SJ takeoff

GRF = ground reaction force; AIS = Australian Institute of Sport.

- From the force-time curves associated with each jump, a number of isoinertial speed strength parameters are recorded (table 13.7).

Drop Jump

Rationale: A vertical jump performed from a squat position (squat jump), with a countermovement (CMJ) or in a rebound movement (drop jump—the vertical jump is immediately preceded by a drop from a given height), can be used to assess the isoinertial concentric and SSC or reactive speed strength qualities of the hip, knee, and ankle extensor musculature. The specific isoinertial strength qualities required for a given sport should dictate the choice of jump. As the drop height increases, the stretch load on the lower limb musculature also increases. This means that the isoinertial reactive or SSC strength of these muscles is further stressed.

Table 13.8 presents selected normative data for expected test scores from the literature.

Equipment: This test requires a stair or stool of known height and a timing mat apparatus as described for the vertical jump using a timing mat. Stairs can be used to provide an incremental drop height (e.g., 20 to 100 cm by 10-20 cm increments).

Test Procedure:

- Subjects drop from the stair or stool, and immediately on landing (on the timing mat) aim to rebound as high and as quickly as possible.
- The timing mat apparatus is used to measure contact time during the ground-contact phase

of the rebound jump (i.e., the amount of time during which the feet are in contact with the timing mat) and vertical height of the rebound jump as previously described (Bosco et al. 1983).

Calculations: To gauge an individual's SSC abilities, a reactive strength index (RSI) has been proposed for use with incremental drop jump protocols (Young, personal communication). The RSI is given by jump height divided by contact time. Thus, if RSI increases or is at least maintained as the drop height is increased, it is assumed that an individual's reactive strength capabilities are sound.

One-Repetition Maximum Tests

Squat

Rationale: The 1RM test is the maximum load that can be lifted only once.

Table 13.9 presents normative and training data for expected test scores from the literature.

Equipment:

- Smith machine or free weights (barbell). The Smith machine is recommended because it reduces skill demand and has good safety features (Young and Bilby 1993).

Test Procedure:

- Subjects should be well warmed up before starting this test. Completion of a number of submaximal lifts with progressively heavier loads is recommended.
- The procedure for determining the 1RM typically involves four lifts with rest periods of 3-5 min.
- The subject begins the test in an upright position with feet positioned shoulder-width apart.
- In a satisfactory lift, the subject descends to a knee angle of 90° of flexion and then returns to the start position (Baker et al. 1994; Murphy and Wilson 1997).
- Weight is increased progressively by 2.5-10 kg until the subject cannot complete the lift (Murphy and Wilson 1997).
- The final weight lifted successfully is recorded as the absolute 1RM. This score can be divided by body weight to establish relative 1RM (Young and Bilby 1993).

Bench Press

Rationale: The 1RM bench press test is a standard test of upper-body strength. This test reflects the maximum load that an individual can press only once from the chest.

Table 13.10 presents normative and training data for expected test scores from the literature.

Equipment:

- Bench press bench
- Barbell
- Masses of 25, 15, 10, 5, 2.5, and 1.25 kg

Test Procedure:

- Before the test, subjects should warm up by completing a number of submaximal lifts with progressively heavier loads.
- The subject lies supine on the bench with knees flexed at ~90° and feet resting on the bench.

Table 13.8 Drop Jump Vertical Jump Scores

Reference	Subjects	Physical characteristics	Jump height (cm)	Drop height* (cm)
Komi and Bosco (1978)	Healthy physical education students Females (n = 25)	20.6 ± 1.2 yr, 58.2 ± 5.6 kg, 165.6 ± 6.0 cm	27.3 ± 3.6	47.6 ± 19.4
	Males (n = 16)	24 ± 1.4 yr, 75.4 ± 11.2 kg, 176.7 ± 8.3 cm	40.3 ± 6.9	63.0 ± 22.7
	Male volleyball players (n = 16)	24.0 ± 3.5 yr, 82.2 ± 7.9 kg, 185.8 ± 6.7 cm	41.0 ± 14.5	66.0 ± 16.3

* The drop jumps were performed from different heights ranging from 20 to 100 cm. The drop height denotes the height that gave the best jump performance.

Values are mean ± SD.

Table 13.9 One-Repetition Maximum (1RM) Squat Scores

Reference	Subjects	Physical characteristics	Absolute 1RM (kg)	
			Pretraining	Posttraining
Murphy and Wilson (1997)	27 recreationally active males			
	Group 1* (n = 14)	Combined mean of groups 1 and 2:	115 ± 20	139 ± 19
	Group 2† (n = 13)	22.0 ± 4 yr, 179 ± 5 cm, 79.5 ± 11.1 kg	126 ± 41	131 ± 41
Baker et al. (1994)	22 experienced men Trained 3 days a week for 12 weeks	20.0 ± 2.6 yr, 180 ± 6 cm, 75.0 ± 8.4 kg	112.7 ± 24.9	141.0 ± 23.5
Young and Bilby (1993)	18 inexperienced male college students	Age range 19-23 yr		
		Group 1**	174.53 ± 24.16	209.33 ± 24.92
		Group 2‡	166.29 ± 22.34	202.85 ± 23.3

* Group 1 trained twice weekly for eight weeks.

† Group 2 maintained normal training patterns.

** Group 1 trained for seven weeks with fast concentric contractions.

‡ Group 2 trained for seven weeks with slow controlled movements.

Values are mean ± SD.

- The subject achieves the hand position on the bar by abducting the arms to 90° and, with the elbows flexed at 90°, placing the proximal balls of the thumb in contact with the barbell (Gullich and Schmidtbleicher 1996).
- In a satisfactory lift, the subject pushes the bar vertically to the point where the elbows are fully extended (Murphy et al. 1994).
- The procedure for determining the 1RM involves four lifts with rest periods of 3-5 min.
- Weight is increased progressively by 2.5-5 kg until the subject cannot complete the lift.
- The final weight lifted successfully is recorded as the absolute 1RM. This score can be divided by body weight to establish relative 1RM (Young and Bilby 1993).

Isoinertial Force-Mass Test

Rationale: The isoinertial force-mass test assesses the peak force generated from concentric and eccentric contractions.

Selected normative and training data for expected test scores from the literature are shown in table 13.11.

Equipment:

This test can be performed in a bench press movement to reflect upper-body power or in an upright squat position to determine lower limb power. The procedure described here is for a squat test.

- For leg power assessment, a Smith machine placed over a GRF platform

Test Procedure:

- Subjects are instructed to stand on the force platform with feet positioned shoulder-width apart.

Concentric test

- The absolute load is 10 kg.
- The knee angle required for the concentric test is 1.9 radians (~110°).
- On the "Go" signal, the subject jumps as fast and high as possible.
- Peak force, RFD, and vertical displacement are measured and calculated as a percentage of the subject's body weight.

Eccentric test

- The eccentric component of this test is performed at a load of 200% of body mass.
- The subject begins the test in the squat position with a knee angle of 2.8 radians (~160°).
- The bar is held in place by an electromagnetic brake.

Table 13.10 One-Repetition Maximum (1RM) Bench Press Scores

Reference	Subjects	Physical characteristics	Subject selection	Absolute 1RM (kg)	
				Worst (n = 4)	Best (n = 4)
Murphy et al. (1994)	13 experienced recreationally active males	23 ± 4 yr, 178.4 ± 7.3 cm, 81.1 ± 9 kg	Isometric† Concentric‡	90 ± 20 86.2 ± 7.5	114 ± 7.2 119.4 ± 5.2
				Pretraining	Posttraining
Baker et al. (1994)	22 experienced males*	20.0 ± 2.6 yr, 179.6 ± 5.9 cm, 75.0 ± 8.4 kg	Not applicable	82.1 ± 11.7	92.6 ± 11.5

† Subjects completed an isometric bench press with results recorded via a force platform. The four worst and best performers on this test then completed the 1RM bench press.

‡ Subjects completed a barbell throw for maximum height, and the four worst and best performers on this test then completed the 1RM bench press.

* Trained three days a week for 12 weeks.

Values are mean ± SD.

- The subject makes contact with the bar but does not exert any force until the instructor gives an auditory signal that the brake is to be released.
- On the signal, the athlete exerts as much force as possible against the falling load.
- Metal stops on the Smith machine prevent the load from descending past 1.9 radians knee angle.
- Peak force, RFD, and vertical displacement are measured and calculated as a percentage of the subject's body weight (Murphy and Wilson 1997).

Other Upper-Body Tests

Bench Throw

Rationale: Bench press throws in the Smith machine can also be performed against variable loads to provide a closer match to sport-specific demands. Most of the powerful upper-body actions in sporting contexts entail acceleration of the involved limbs until after completion of the movement goal (e.g., the release of an external mass or completion of a contact phase between the limb and an external mass). Conventional 1RM tasks that use weight-lifting movements and require the subject to decelerate the barbell at completion of the movement would not appear to be valid approximates of these dynamic sporting actions.

Table 13.12 contains limited normative data for expected test scores from the Australian Institute of Sport.

To facilitate a closer match with sport movement demands, bench press throw protocols allow the subject to release an external load before the limbs enter a nonspecific decelerative kinematic phase.

Equipment:

- Smith machine apparatus
- Bench press bench and barbell (mass of 10-20 kg)

Bench press throws can also be performed against variable loads to more closely match sport-specific demands. As a general guide, masses from 5 to 100 kg may be used in weighted bench press throw tasks depending upon the age, strength, and experience of the subject.

Test Procedure:

- The subject lies supine on the bench with knees flexed at ~90° and feet resting on the bench.
- As with vertical jumps in the Smith machine, a rotary encoder is used to determine displacement of the barbell, which in this case is thrown up the vertical supports.
- If no countermovement is adopted and the bar is thrown from the chest (chest throw), throw displacement can be used to gauge isoinertial concentric speed strength qualities of the upper limb musculature.
- If a countermovement is adopted (countermovement chest throw), throw displacement will approximate isoinertial concentric SSC speed strength of the same muscles.

Table 13.11 Normative Scores for the Isoinertial Force-Mass Test

Reference	Subjects	Physical characteristics	Test variable	Score (mean ± SD)		
					Pretraining	Posttraining
Murphy and Wilson (1997)	27 recreationally active males Group 1* (n = 14) Group 2† (n = 13)	Combined mean of groups 1 and 2: 22.0 ± 4 yr, 179 ± 5 cm, 79.5 ± 11.1 kg	C10FOR (N)	Gp 1	1340 ± 154	1400 ± 137
				Gp 2	1310 ± 203	1340 ± 168
			C10RFD (N/s)	Gp 1	8820 ± 2758	9980 ± 2780
				Gp 2	9070 ± 3440	9270 ± 2830
			E200FOR (N)	Gp 1	1630 ± 212	1770 ± 354
				Gp 2	1630 ± 433	1570 ± 390
			E200RFD (N/s)	Gp 1	12900 ± 4050	13000 ± 3460
				Gp 2	12400 ± 4330	13900 ± 5860

* Group 1 trained twice weekly for eight weeks.

† Group 2 maintained normal training patterns.

C10 = concentric isoinertial test with an absolute mass of 10 kg, E200 = eccentric isoinertial test at a load of 200% of body mass, FOR = peak force, RFD = rate of force development.

Table 13.12 Normative Scores for the Upper-Body Bench Throw

Reference	Subjects	Physical characteristics (mean ± SD)	Range (mm)
Unpublished data from AIS test protocol	8 national-level male cricketers (fast bowlers)	20.0 ± 2 yr, 180.8 ± 4.0 cm, 78.4 ± 9.1 kg	522–879
Unpublished data from AIS test protocol	9 national-level male boxers	19.2–24.7 yr, 55–91 kg	501–860
Unpublished data from AIS test protocol	8 national-level male canoe polo players	21–28 yr, 72.3–85.8 kg	591–791
	7 national-level female canoe polo players	20.8–36.8 yr, 57.7–81.4 kg	301–460

All bench throw tests performed with 9 kg barbell on Smith machine apparatus. AIS = Australian Institute of Sport.

Drop Bench Press Throw

Rationale: The drop bench press throw is used to assess upper-body performance in a dynamic SSC action (Murphy et al. 1994).

Table 13.13 contains limited normative data for expected test scores from the literature.

Equipment:

- A bench positioned under a Smith machine. In this case, it is advantageous to have a Smith machine equipped with an electromagnetic braking system.
- Murphy et al. (1994) recommend loads with an absolute mass of 10 kg and a relative mass of 30% of maximum.

Test Procedure:

- The subject lies supine on the bench with knees flexed and performs the test with varying loads.
- The loaded bar is held 0.25 m above the extended arms of the subject and is released by the instructor following an auditory signal.
- Subjects are required to catch the bar and throw it quickly for maximal height. Metal stops on the Smith machine are placed between the subject's chest and the bar to minimize the risk of injury.
- To calculate total positive work done, multiply the mass of the bar-weight system by gravity and by the height to which the bar was thrown (mass × gravity × height).

Table 13.13 Normative Scores for the Drop Bench Press Throw

References	Subjects	Physical characteristics	Test format	Joules	
				Worst (n = 4)	Best (n = 4)
Murphy et al. (1994)	13 experienced recreationally active males	23 ± 4 yr, 178.4 ± 7.3 cm, 81.1 ± 9 kg	Isometric*	138.8 ± 20.6	178.2 ± 13.2
			Concentric*	136.3 ± 16.2	175.8 ± 16.9
			Isometric†	207.5 ± 34.7	286.1 ± 48.9
			Concentric†	204.5 ± 30.3	301.1 ± 27.3

* A load of an absolute mass of 10 kg.

† A load of a relative mass of 30% of maximum.

Values are mean ± SD.

Seated Shot Put Throw

Rationale: The seated shot put throw has been shown by Gillespie and Keenum (1987) to be a reliable and valid measure of upper-body power.

Limited normative data for expected test scores from the literature are presented in table 13.14.

Equipment:

- A standard shot put (3.63 kg)

Test Procedure:

- Subjects need to be seated with hips against the back of a seat or wall. If possible, they should be strapped into the chair to prevent use of the lower body.
- The shot put is held with both hands against the center of the chest. The forearms are positioned parallel to the ground.
- After assuming the correct starting position, the subject performs at least three trials.
- Use a tape measure to obtain the distance thrown in centimeters (Gillespie and Keenum 1987).

Table 13.14 Normative Scores for the Seated Shot Put Throw

References	Subjects	Physical characteristics	Test format	Distance (cm)	
				Test 1	Test 2
Gillespie and Keenum (1987)	57 male college students enrolled in beginners weight-training courses		Angle of release controlled (cm)	445.8 ± 63.3	465.8 ± 61.2
			Angle of release uncontrolled	474.2 ± 69.2	495.73 ± 68.6
			Subject selection	Worst (n = 4)	Best (n = 4)
Murphy et al. (1994)	13 experienced recreationally active males	23 ± 4 yr, 178.4 ± 7.3 cm, 81.1 ± 9 kg	Isometric†	456 ± 104	493 ± 67
			Concentric‡	417 ± 53	523 ± 28

† Subjects completed an isometric bench press with results recorded via a force platform. The four worst and best performers on this test then completed a seated shot put.

‡ Subjects completed a barbell throw for maximum height, and the four worst and best performers on this test then completed a seated shot put.

Values are mean ± SD.

Other Lower-Body Tests

30-Second Wingate Cycle Ergometer Test

Rationale: The Wingate anaerobic test was originally developed to assess anaerobic performance in populations ranging from young children and persons with physical disabilities to elite athletes. The test involves a 30 s maximal effort during pedaling (or arm cranking for upper body) against a constant force. The indexes measured are peak power (watts), average power (watts), and rate of fatigue.

Table 13.15 shows selected normative data for expected test scores from the literature.

It is reasonable to assume that maximal isoinertial contractions during short-burst pedaling of 30 s duration will approximate the maximal isoinertial power development characteristics of the lower limb musculature. However, the pedaling movement will accentuate isoinertial power only during hip extension, knee extension, and slight ankle plantarflexion-type movement.

Equipment:

- Any mechanical bicycle ergometer such as the Monark or Fleisch

Electrically braked ergometers are appropriate provided that they have a constant-force mode.

Test Procedure:

- The subject performs a 30 s maximal effort during pedaling or arm cranking against a constant force.
- Pedal revolutions can be counted by a computer or mechanical automated counter.
- The use of toe stirrups increases performance on the Wingate test because force can be applied to the pedal throughout the full cycle. For this reason, toe stirrups should always be used.

Although evidence is equivocal regarding the optimal force setting to yield the highest mean and peak power, the original recommended equation for force settings using the Monark ergometer was 0.075 kp/kg—a force equivalent to mechanical work of 4.41 J per pedal revolution per kilogram body weight. A review of literature by Bar-Or (1987) suggested that the force applied to induce maximal mean and peak power needs to be 20-30% higher. Factors such as training background, age, and sex will influence the optimal force level required. As a general guideline for testing with the Monark, Bar-Or (1987) recommended use of a force of 0.090 kp/kg with adult nonathletes and 0.100 kp/kg with adult athletes.

Table 13.15 Normative Scores for the 30 s Wingate Cycle Ergometer Test

			Test variable			
Reference	**Subjects**	**Physical characteristics**	**Average power (W)**	**Average power (W/kg)**	**Max power (W)**	**Max power (W/kg)**
White and Johnson (1990)	61 male and female athletes of international (int), national (nat), and regional (reg) levels	Males:				
		Int (n=12) 21.5 yr, 179 cm, and 78.8 kg	735	9.32	836	10.6
		Nat (n=8) 21.6 yr, 173 cm, and 72.6 kg	694	9.54	859	11.8
		Reg (n=11) 18.1 yr, 174 cm, and 68.5 kg	616	8.39	772	11.2
		Females:				
		Int (n=17) 20.4 yr, 166 cm, and 63.07 kg	546	8.65	669	10.6
		Nat (n=6) 21.8 yr, 165 cm, and 63.64 kg	495	7.76	650	10.2
		Reg (n=7) 17.5 yr, 161 cm, and 58.61 kg	434	7.43	572	9.8

6-Second Bicycle Test

Rationale: The 6 s bicycle test involves a short maximal-effort sprint on an Exertech front-access cycle ergometer. As with the Wingate test, short-burst pedaling for 6 s will allow for some approximation of the maximal isoinertial power development characteristics of the lower limb musculature. Again, the pedaling movement emphasizes isoinertial power during hip extension, knee extension, and slight ankle plantarflexion-type movement. However, it is logical that the comparatively shorter duration of the 6 s test will afford more accurate determination of absolute power than the longer Wingate test, as this emphasizes alactic anaerobic energy derivation only.

Table 13.16 provides selected normative data for expected test scores from the literature.

Equipment:

- Repco Exertech-10 cycle ergometer (or a highly geared cycle ergometer—46:14/46:14), calibrated in the range of 500-1500 W
- Exertech work monitor unit (or software to acquire data at ~40 Hz)
- Toe clips and heel straps to ensure that the subject's foot cannot pull off pedal
- Two stopwatches

Test Procedure:

- Subjects are instructed to begin in a seated position, pedaling at a power output of ~300 W.
- After a few seconds the instructor gives the "Go" signal for starting the test.
- The subject immediately stands and pedals with maximal effort for 6 s.
- The best of two trials is recorded.
- The best of three efforts is used to determine the peak power output with a minimum of 5 min recovery between each effort.

It is noteworthy that the pretest power output of ~300 W may vary depending on the age group and training status of the individual (Murphy and Wilson 1997; Wilson et al. 1993).

Table 13.16 Normative Scores for the 6 s Bicycle Test

References	Subjects	Physical characteristics	Peak power output (W)	
			Pretraining	Posttraining
Murphy and Wilson (1997)	27 recreationally active males			
	Group 1* (n = 14)	Combined mean of groups 1 and 2:	1120 ± 211	1220 ± 160
	Group 2† (n = 13)	22.0 ± 4 yr, 179 ± 5 cm, 79.5 ± 11.1 kg	1070 ± 122	1120 ± 138
Wilson et al. (1993)	55 weight-trained males split into four groups:			
	• Gp 1: Control, n=14	24.1 ± 5.4 yr, 173 ± 8.0 cm, 76.9 ± 15.4 kg	1028 ± 359	1046 ± 374
	• Gp 2: Weight training, n=13	21.9 ± 4.3 yr, 173 ± 8.9 cm, 69.0 ± 8.9 kg	1022 ± 279	1078 ± 263
	• Gp 3: Plyometric training, n=13	22.1 ± 6.8 yr, 174 ± 9.9 cm, 71.6 ± 11.9 kg	967 ± 266	996 ± 289
	• Gp 4: Dynamic weight training, n=15	23.7 ± 5.8 yr, 178 ± 9.7 cm, 75.9 ± 17.9 kg	975 ± 251	1022 ± 274

*Group 1 trained twice weekly for eight weeks.

†Group 2 maintained normal training patterns.

Values are mean ± SD.

Standing Broad Jump

Rationale: Many use the standing broad jump to identify an individual's level of explosive power of the lower limb musculature. The test emphasizes powerful knee and hip extension from a starting posture marked by deep knee flexion.

Selected normative data for expected test scores are presented in table 13.17.

Test Procedure:

- The subject stands with feet comfortably apart, behind a line.
- Hands must remain on hips throughout the jump in order to eliminate the contribution of the arms.
- Subjects are to jump maximally and are allowed to perform a countermovement prior to takeoff.
- Subjects perform at least three trials.
- Measure the distance (in centimeters) from the takeoff line to the back of the heel closest to the takeoff line (Robertson and Fleming 1987).

Table 13.17 Normative Scores for the Standing Broad Jump

Reference	Subjects	Test variable	Score (mean ± SD)	
Robertson and Fleming (1987)	Females (n=4) and males (n=2)	Average distance	215.2 ± 25.3 cm	
		Peak vertical force	2.05 ± 0.13 × BW	
		Peak horizontal force	0.65 ± 0.07 × BW	
			% contribution of total work done*	
			Joint	Percent
			Hip	45.9 ± 8.5
			Knee	3.9 ± 3.9
			Ankle	50.2 ± 6.3
			Total/total change in energy**	85.0 ± 9.5

* Percentage contributions of the leg joints to the total work done in the propulsive phases of the standing broad jump.

** Percent of work done by the leg joints in relation to the change in total energy of the body.

BW = body weight.

part IV

Protocols for the Physiological Assessment of Players of Specific Sports

chapter 14

Protocols for the Physiological Assessment of Basketball Players

Andrea Stapff

Like any competitive team sport, basketball requires a substantial skill component from both an individual and a team perspective. Physiologically, the requirements of basketball include aerobic and anaerobic energy production, muscular strength and endurance, and flexibility. Physical conditioning is therefore important for enhancing performance, reducing the potential of injury, and extending the career of players at peak level (Stone and Steingard 1993). Physiological testing is a valuable tool that can help coaches and sport scientists

- construct individual position-specific physiological profiles,
- assess and monitor the effectiveness of training programs,
- help identify gains or losses in skill development, and
- aid in talent identification.

The following protocols are currently used for assessing national squads. They are intended to provide routine assessment of physiological status and represent a range of parameters. These tests have been selected on the basis of their ability to allow for direct comparison, their cost and time effectiveness, and their applicability to the game of basketball.

For each of the tests outlined in this chapter, a set of "expected" or target scores is presented for the various squads. These scores were derived from subjective evaluation of data collected in Australia between 1993 and 1996 and after consultation with national team coaches. Summary results for the various squads are also presented for each test. These results were derived from analysis of all test records collected between 1993 and 1999.

Laboratory and Field Environment and Subject Preparation

It is important to control subject preparation and laboratory or field conditions for every testing occasion to ensure that testing is both reliable and meaningful. Refer to chapter 2 for an outline of the pretest conditions that should be observed.

Equipment Checklist

1. Anthropometry

Refer to chapter 5 for a comprehensive description of relevant equipment.

2. Anaerobic power, capacity, and performance tests

For each of the following tests, refer to chapter 9 for a description of relevant equipment. Note that the 20 m sprint must be conducted on a basketball court.

- Countermovement jump test
- Vertical jump test
- 10 s and 5 3 6 s cycle ergometer tests
- 20 m sprint test

3. Aerobic power tests

Refer to chapter 9 for a description of relevant equipment.

4. Maximal oxygen consumption test ($\dot{V}O_2max$)
   - ☐ Treadmill with minimum speed range 0-16 kph, gradient range 0-10%
   - ☐ Gas collection and analysis system according to general recommendation
   - ☐ Heart rate monitor
5. 20 m multistage shuttle run test
   - ☐ Basketball court, and refer to chapter 9.
6. Muscular strength tests

Upper-body strength test:
- ☐ Bench press apparatus
- ☐ Weights in kilograms

Lower-body strength test:
- ☐ Power rack
- ☐ Squat apparatus
- ☐ Weights in kilograms
- ☐ Goniometer
- ☐ Marking pen

PROTOCOLS: Basketball

Forms that may be photocopied and used for data collection are found on pages 233-237. They are used in Australia at the national-athlete level.

Anthropometry

Rationale: Anthropometric measurements of stature, mass, and sum of skinfolds provide a clear appraisal of the structural status of an individual at any given time (Ross and Marfell-Jones 1991). Detailed athletic profiles are also valuable in describing the characteristics of elite athletes across sports (Withers et al. 1987), by position within a sport (Morrow et al. 1980), and at various stages throughout a yearly training cycle.

Practically, anthropometric measurements, particularly sum of skinfolds, are used as a basis for training and dietary interventions. Whether an event is primarily aerobic or anaerobic, increased fat mass will be detrimental to performance. Moreover, in sports requiring speed or explosive power, such as basketball, excess fat, which causes an increase in body mass, will decrease acceleration unless proportional increases in force are applied (Norton et al. 1996a).

Measurements Required: Height, body mass, and sum of seven skinfolds are required (refer to chapter 5 for procedures).

Limitations: Chapter 5 outlines the limitations of anthropometric assessment. Allowances for body types must be made when interpreting skinfolds, especially for females with somatotypes consisting of a high endomorphic component.

Population Data: Table 14.1 provides data for international female as well as junior Australian basketball players.

Normative Data: Tables 14.2 and 14.3 outline normative data for various squads and current results from data collected between 1993 and 1996.

Anaerobic Power Capacity and Performance Tests

Laboratory measurements of anaerobic power and capacity are relevant to those athletes whose sport requires a significant contribution from either or both of the adenosine triphosphate-phosphocreatine and glycolytic pathways. Examples of basketball performance via these two pathways include moves involving speed, acceleration, and explosiveness such as rebounding, jump shooting, driving layups, and shot blocking, in addition to longer-duration, repetition efforts such as a series of fast breaks or high-speed play.

Four laboratory tests and one performance field test have been selected to assess relevant anaerobic power and capacity abilities. They are the countermovement jump, vertical jump, 10 s cycle ergometer test, 5 × 6 s cycle ergometer test, and 20 m sprint.

Countermovement Jump

Rationale: The countermovement jump (CMJ) is a vertical jump without the arm swing. The intent of eliminating the arm swing is to reduce the skill/coordination requirement of the test and focus the effort on the leg extensor muscles. Therefore the CMJ is preferable to the vertical jump for testing leg power and for monitoring the effects of strength-training programs that target the leg muscles (Young 1994a).

Test Procedure: Refer to page 139 for a description of the test procedure.

Table 14.1 Anthropometric Data for Senior Female Basketball Players at the 1994 World Championships and for Junior Australian Players

Group	n	Age (yr)	Height (cm)	Mass (kg)	Σ7* (mm)
Australian senior female					
Center	6	24.8 (5.9)	188.4 (6.3)	79.4 (9.0)	85.2 (28.2)
Forward	2	24.5 (4.9)	188.7 (2.8)	76.2 (2.4)	89.1 (36.6)
Guard	4	26.2 (4.4)	173.5 (6.7)	67.3 (7.4)	73.0 (16.8)
World senior female					
Center	47	24.1 (3.2)	189.0 (6.4)	82.6 (8.2)	88.0 (21.1)
Forward	57	25.1 (3.8)	181.2 (5.9)	73.3 (5.1)	75.8 (20.3)
Guard	64	25.4 (3.3)	171.9 (6.1)	66.1 (6.2)	76.6 (22.4)
Australian junior female	139	–	178.4 (9.6)	69.2 (8.3)	–
Australian junior male	95	–	198.4 (7.7)	94.4 (11.5)	–
AIS** female	362	17.9 (0.9)	180.4 (7.6)	72.5 (8.4)	91.7 (18.9)
AIS** male	261	18.4 (0.9)	198.5 (7.8)	94.6 (10.4)	72.0 (27.0)

*Σ7 = triceps, subscapular, biceps, supraspinale, abdominal, thigh, medial calf.

**AIS = Australian Institute of Sport

Values are mean with SD in parentheses.

Adapted, by permission, from T. Ackland, A. Schreiner, and D. Kerr, 1994, "Anthropometric profiles of world championship female basketball players" (abstract). International Conference of Science and Medicine in Sport. Brisbane: Sports Medicine Australia.

Table 14.2 Normative Skinfold Data, Sum of Seven* (mm)

	Female	Male
Junior	95	85
Senior	75	65

*Σ7 = triceps, subscapular, biceps, supraspinale, abdominal, thigh, medial calf.

Table 14.3 Results of Skinfold Testing, Sum of Seven* (mm)

	Female		Male	
	Senior	Junior	Senior	Junior
n =	134	362	–	261
Mean	81.5	91.7	–	72.0
SD	20.4	18.9	–	27.0
Minimum	45.9	55.0	–	29.5
Maximum	161.5	144.0	–	161.5

*Σ7 = triceps, subscapular, biceps, supraspinale, abdominal, thigh, medial calf.

Vertical Jump Test

Rationale: The vertical jump is a laboratory measure of explosive/anaerobic power of the legs. The protocol outlined in this section is basketball specific, incorporating a one-step movement prior to the jump. Young (1994b) demonstrated the necessity to select jumps that are as specific as possible to the requirements of the sport. While it is true that the game of basketball incorporates a mixture of jump types that can be affected by such factors as playing position, team, and opposition team playing style or tactics (Young 1994b), the vertical jump protocol described here is both familiar to basketball players and representative of a skill used in the game. The proviso is that the test be seen in this light.

Test Procedure: The test procedure is described on page 141. Note, however, that for basketball, one step is used prior to the jump to make it more sport specific.

1. After sufficient warm-up, the subject stands against a wall (either left or right side is permitted) with feet flat on the floor and the body close to the wall.

2. The subject chalks his or her fingertips, elevates the shoulder, and stretches out the arm and

hand closest to the board, leaving a mark at the height of full stretch on the measuring board. The position of the initial mark is recorded.

3. The subject takes a step backward with one foot only, leaving the other firmly planted at the base of the measuring board—assuming a crouch position to whatever depth is preferred, with outstretched arms behind the body.

4. From this stationary starting position, the subject brings both feet together and leaps up as high as possible, leaving a chalk mark on the measuring board with the inner hand.

5. As this test involves some skill, three attempts are allowed with a minimum of 30 s between attempts. The distance between the initial mark and the highest jump is recorded.

Limitations: This test is limited by the ability of the subject to coordinate the various contributing body parts (i.e., skill); therefore it is extremely important that the tester provide coaching to maximize correct technique.

Normative Data: Tables 14.4 and 14.5 outline normative data for various squads and results from data collected between 1993 and 1999.

Table 14.4 Normative Vertical Jump Data (cm)

	Female	Male
Junior	50	70
Senior	55	75

Table 14.5 Results of Vertical Jump Testing (cm)

	Female		Male	
	Senior	Junior	Senior	Junior
n =	42	121	–	86
Mean	47.6	46.2	–	65.5
SD	6.6	5.6	–	7.1
Minimum	31.0	31.0	–	50.0
Maximum	61.0	60.0	–	85.0

10-Second Cycle Ergometer Test

Rationale: See page 136 for discussion of the rationale for this test.

Test Procedure: Refer to page 136 for a comprehensive description of test administration and limitations.

Normative Data: Tables 14.6 and 14.7 outline normative data for various squads and results from data collected from 1997 to 1999.

Table 14.6 Normative 10 s Cycle Ergometer Test Data

Female senior squad	
Work (J/kg)	Peak power (W/kg)
115	15.5

Table 14.7 Results of 10 s Cycle Ergometer Testing

	Female senior squad	
	Work (J/kg)	Peak power (W/kg)
n =	41	41
Mean	111.8	13.8
SD	18.6	2.3
Minimum	58.3	8.1
Maximum	138.3	17.6

5 × 6-Second Cycle Ergometer Test

Rationale: The 5 × 6 s cycle ergometer test involves five repetitions of 6 s of maximal sprinting with 24 s recovery. This test was selected as a laboratory method to assess the requirement in basketball to repeat short-duration sprints over an extended period of time. In a study of the bioenergetics of basketball, Gillam (1985) emphasizes the importance of high-intensity interval work with repeated bouts of recovery as a fundamental aspect of the game. The 5 × 6 s test of repeated sprint ability has been shown to be reliable and valid (Fitzsimons et al. 1993), allowing for a better assessment of the specific match fitness of team sport players.

Test Procedure: Refer to page 137 for a comprehensive description of test administration, data reduction, limitations, and interpretation.

Normative Data: Tables 14.8 and 14.9 outline normative data for various squads and results from data collected between 1997 and 1999.

Table 14.8 Normative 5 × 6 s Cycle Ergometer Test Data

Female senior squad		
Total work (J/kg)	**Work Dec (%)**	**Power Dec (%)**
330	< 7	< 5

Table 14.9 Results of 5 × 6 s Cycle Ergometer Testing

	Female senior squad		
	Total work (J/kg)	**Work Dec (%)**	**Power Dec (%)**
n =	41	41	41
Mean	304.5	7.5	6.4
SD	47.3	3.4	3.2
Minimum	166.5	2.04	1.76
Maximum	392.2	14.6	15.5

20-Meter Sprint Test

Rationale: Basketball players require the ability to sprint at maximal or near-maximal speed. High-speed play and fast breaks are two examples of situations in which the skill of "running the floor" is vital.

The 20 m sprint test was selected as a sport-specific performance measure of sprint ability. The 20 m distance represents the most probable maximal distance a player might run. The test also allows for assessment of acceleration and speed over shorter distances within the 20 m. Practically, this test is useful to assess an athlete's performance in the skill of "running the floor" and adds to the nonspecific laboratory measures of anaerobic power and capacity.

Test Procedure: Refer to page 130 for a comprehensive description of the test procedure.

Limitations: This test is limited by player motivation and environmental conditions.

Normative Data: Tables 14.10 through 14.12 outline normative data for various squads and results from data collected between 1993 and 1999.

Aerobic Power Tests

Aerobic power is the rate at which oxygen is used by the tissues during prolonged bouts of exercise. Stone and Kroll (1991) have identified it as the base of sport conditioning and a major component of

Table 14.10 Normative 20 m Sprint Test Data (seconds)

	Female			Male		
	Acceleration (0–5 m)	**Speed (10–20 m)**	**Combined (0–20 m)**	**Acceleration (0–5 m)**	**Speed (10–20 m)**	**Combined (0–20 m)**
Junior	1.15	1.40	3.30	1.01	1.12	3.00
Senior	1.05	1.30	3.30	0.98	1.10	3.00

Table 14.11 Results of Female Squads 20 m Sprint Testing (seconds)

	Senior			Junior		
	Acceleration (0–5 m)	**Speed (10–20 m)**	**Combined (0–20 m)**	**Acceleration (0–5 m)**	**Speed (10–20 m)**	**Combined (0–20 m)**
n =	30	28	28	59	99	99
Mean	1.18	1.47	3.48	1.11	1.42	3.33
SD	0.09	0.14	0.27	0.05	0.07	0.13
Minimum	1.06	1.31	3.19	1.00	1.31	3.10
Maximum	1.43	1.78	4.05	1.19	1.60	3.64

Table 14.12 Results of Male Squads 20 m Sprint Testing (seconds)

	Senior			Junior		
	Acceleration (0–5 m)	Speed (10–20 m)	Combined (0–20 m)	Acceleration (0–5 m)	Speed (10–20 m)	Combined (0–20 m)
n =	–	–	–	64	84	84
Mean	–	–	–	1.03	1.28	3.04
SD	–	–	–	0.05	0.07	0.10
Minimum	–	–	–	0.91	1.04	2.80
Maximum	–	–	–	1.14	1.48	3.26

conditioning for basketball. Basketball is a game of continuously changing tempo, requiring players to be able to sustain high levels of continuous efforts. A high degree of aerobic power is therefore necessary to meet energy demands within a game and aid in recovery from anaerobic efforts. In addition, aerobic power enables the athlete to play and practice longer and at higher intensities (Stone and Steingard 1993).

One laboratory and one field test have been selected to measure aerobic power: the maximal oxygen consumption ($\dot{V}O_2max$) test and the 20 m multistage shuttle run test. The $\dot{V}O_2max$ test is a well-established and familiar laboratory test. The 20 m multistage shuttle run test has also been shown to be a valid and reliable test for predicting $\dot{V}O_2max$ (Leger and Lambert 1982; Leger et al. 1988; Ramsbottom et al. 1987), so it is a useful field test when circumstances do not allow laboratory testing.

Maximal Oxygen Consumption ($\dot{V}O_2max$)

Test Procedure:

1. Subjects may choose to wear either basketball footwear or running footwear.
2. Subjects should be allowed 2-5 min of warm-up to become familiar with treadmill and gas collection apparatus.
3. Testing protocol: see table 14.13.

Table 14.13 Summary of $\dot{V}O_2max$ Testing Protocol

	Female	Speed (km/h)		Male	Speed (km/h)	
Time (min)	Grade (%)	Guards and forwards	Centers	Grade (%)	Guards and forwards	Centers
0–1	10	8	0	12	10	0
1–2	10	8	0	12	10	0
2–3	12	10	0	14	12	0
3–4	12	10	0	14	12	0
4–5	14	12	0	16	14	0
5–6	14	12	2	16	14	2
6–7	14	12	4	16	14	4
7–8	14	12	6	16	14	6
8–9	14	12	8	16	14	8
9–10	14	12	10	16	14	10

For each extra minute, increase grade 1%

4. Heart rate is measured during the last 10 s of each minute and immediately prior to completion of the test. The test is terminated on signal from the subject.

5. $\dot{V}O_2max$ is recorded as the highest minute score associated with the following criteria:

- The heart rate reaches a plateau with increasing workload.
- $\dot{V}O_2max$ plateaus with increasing workload ($\dot{V}O_2$ increases less than 150 ml over the score obtained in the previous minute).
- Respiratory exchange ratio value is 1.15 or greater.

Limitations: This test is limited by the protocol and the laboratory sample collection method.

Test Scores and Ranges: Table 14.14 outlines results from data collected between 1997 and 1999.

Table 14.14 Results of $\dot{V}O_2max$ Testing

	Female senior squad	
	L/min	$ml \cdot kg^{-1} \cdot min^{-1}$
n =	33	33
Mean	3.64	49.49
SD	0.33	3.84
Minimum	2.93	41.62
Maximum	4.43	56.39

20-Meter Multistage Shuttle Run Test

Test Procedure: Refer to page 130 for a comprehensive description of the test procedure.

Limitations: This test is limited by the motivation of the subject, and it provides only an estimate of $\dot{V}O_2max$.

Normative Data: Tables 14.15 and 14.16 outline normative data for various squads and results from data collected between 1993 and 1999.

Table 14.15 Normative Test Data for the 20 m Multistage Shuttle Run (Decimal Level)

		Female	Male
Junior	(guards and forwards)	11.0	12.0
	(centers)	10.0	11.0
Senior	(guards and forwards)	12.0	13.0
	(centers)	11.0	11.0

Table 14.16 Results of 20 m Multistage Shuttle Run Testing (Decimal Level)

	Female		Male	
	Senior	Junior	Senior	Junior
n =	6	126	44	89
Mean	11.80	9.73	11.61	12.05
SD	0.80	1.31	1.36	1.35
Minimum	10.40	5.78	8.20	8.18
Maximum	12.40	12.75	14.08	14.62

Strength Testing

Muscular strength is an important aspect of basketball. The game involves skills that must be applied dynamically, explosively, and repeatedly. Muscular strength and power of the legs and hips determine how explosively basketball skills are executed (Stone and Steingard 1993). Arm strength for passing and controlling rebounds and total body strength for maintaining position under the basket are also examples of the importance of strength. In addition, strength is a determinant of speed and agility.

While muscular power and endurance are also essential components of the game, from a practical viewpoint strength testing is used most commonly to assess muscular fitness of basketball players. Muscular endurance and power are more often assessed within, or as part of, the physiological laboratory and field tests mentioned previously.

Strength-Testing Protocol

Bench press and squats are used to assess strength of basketball players. Both exercises are multijoint, and both use the major muscle groups of the upper and lower body. Their movement patterns are also similar to those used in the performance of common basketball skills, both in training and in competition; and they are familiar to basketball players.

The following general principles should guide all the strength-testing protocols that follow:

- Ensure that subjects undertake an extensive warm-up and stretching routine prior to any strength testing. A progressive increase from light to maximal weights over six to eight sets of three repetitions on each of the exercises is an effective way of building the subject into the test.

- All testing must be supervised by the team coach or the individual's coach or an accredited strength and conditioning coach. Safety "spotting" of the subject is required for all attempts at all weights.
- Strength testing consists of a three-repetition maximum (3RM) test at 100% and is the maximum or heaviest weight a subject can lift three times with good technique and without external assistance.
- Testing is to be terminated at failure of volitional strength or at the discretion of the tester(s), if technical aspects of the lift are not met.
- A minimum of 4 min recovery must separate 3RM efforts.

Upper-Body Strength—Bench Press:

- Subjects must complete the bench press using free weights as opposed to a machine.
- Subjects may initially choose the width of grip that they prefer, but this must remain consistent over consecutive attempts and subsequent tests.
- The bar must touch the chest between repetitions but is not allowed to bounce. To prevent this, the athlete must lower the bar at a slow and controlled speed during the eccentric (lowering) phase.
- The subject must be in control of the bar at all times if the repetition is to be valid. An uneven bar during the concentric phase, arching of the lower back, raising of feet off the ground, and bouncing the bar off the chest all invalidate the repetition.

Lower-Body Strength—Squats:

- Squats must be performed in a power rack.
- Athletes place hands evenly on the bar using a grip slightly wider than shoulder-width, with thumbs around the bar.
- The upper back is placed at the center of the bar; chest is held up and out, and shoulder blades are pulled together.
- Feet are positioned shoulder-width or wider apart, with toes pointed slightly out.
- Feet remain flat on the floor; the head faces straight forward throughout the motion.
- If the heel is coming off the floor, a heel block may be used.
- The subject's squat depth should be such that the tops of the thighs are parallel to the floor.
- The athlete must be in control of the bar at all times, maintaining proper technique and reaching the correct depth if the repetition is to be valid.

Athletes must avoid:

- Rounding of the back
- Bouncing at the bottom of the squat
- Excessive forward lean at the waist
- Knees knocking together

Limitations: These tests are limited by subject motivation and technique. Technique must be strictly supervised.

Normative Data: Tables 14.17 and 14.18 outline normative data for various squads and results from data collected between 1993 and 1996.

Table 14.17 Normative Data for Three-Repetition Maximum (3RM) Upper-Body Strength Test (kg); 3RM Squat Strength-to-Weight Ratio*

	Female		Male	
	Bench press (kg)	**Squat**	**Bench press (kg)**	**Squat**
Junior	47.5	> 1.0	85	> 1.25
Senior	60	> 1.25	90	> 1.5

*The strength-to-weight ratio is an easy way to compare the amounts lifted to the athlete's body weight. The equation is as follows:

$$\frac{\text{Amount Lifted (kg)}}{\text{Body Weight (kg)}}$$

Table 14.18 Results of Female and Male Squads Upper-Body (Bench Press) Strength Testing, Three-Repetition Maximum (kg)

	Female		Male	
	Senior	**Junior**	**Senior**	**Junior**
n =	66	74	–	91
Mean	50.7	44.7	–	79.6
SD	5.3	6.0	–	13.4
Minimum	34.0	27.5	–	45.0
Maximum	62.5	57.5	–	102.5

Flexibility

Flexibility is an important aspect of basketball performance. Both dynamic flexibility and agility are requirements in a variety of basketball moves. In addition, when included as part of a conditioning program, flexibility can play a role in preventing injuries.

Flexibility is the range of motion in a joint or series of joints and is joint specific. It cannot, therefore, be measured by a single test in a physiological test battery. The major limitations of the sit-and-reach test, for instance, are outlined clearly on page 104. For assessment of flexibility, readers need to consult respective national squad medical and physiotherapy personnel.

Test Administration

The preceding sections of this chapter have covered tests currently used by national squads. Testing occurs three times per year, although occasionally a fourth testing session may take place, in conjunction with preparation for major competitions. Not all squads do all tests. Table 14.19 outlines the tests used by various Australian squads.

The Australian national senior women's squad performs the maximal oxygen consumption test ($\dot{V}O_2max$) once per year in place of the 20 m multistage shuttle run. The $\dot{V}O_2max$ test usually takes place soon after the main phase of domestic competition and in conjunction with preparation for major competition.

Australian Intensive Training Centre (ITC) squads perform a battery of tests; four of these are outlined in this chapter—anthropometry, vertical jump, 20 m sprint, and 20 m multistage shuttle run. The ITC anthropometry protocol also includes arm span. The procedure for measuring this distance is described by Norton et al. (1996b).

Table 14.19 Suggested Order for Test Administration

	National senior female	National senior male	National junior female	National junior male
Day one				
i Anthropometry	✔	✔	✔	✔
ii Countermovement jump	✔			
iii Vertical jump test—minimum 3 min recovery	✔	✔	✔	✔
iv 20 m sprint test—minimum 5 min recovery	✔	✔	✔	✔
v 10 s cycle ergometer test—minimum 5 min recovery	✔			
vi 5 × 6 s cycle ergometer test—minimum 30 min recovery	✔			
vii 20 m multistage shuttle run test	✔	✔	✔	✔
Day two				
i Maximal oxygen consumption test*	✔			
ii Upper- and lower-body strength tests		✔	✔	✔

*Once per year.

Physiological Assessment for Basketball: Laboratory Based

Name ____________________ Date ___/___/___ Time ________

DOB ________ Squad ____________________ Position: C F G

Training phase: Transition Gen. Prep. Spec. Prep. Competition Peak Rehabilitation

Previous 48-hour training load: None Light Medium Heavy

Venue ____________ Tester/s ____________ Data entry completed by ____________

Anthropometry

Height (cm)	______	Supra (mm)	______
Weight (kg)	______	Abdom (mm)	______
Triceps (mm)	______	Thigh (mm)	______
Subscap (mm)	______	Calf (mm)	______
Biceps (mm)	______	**Total (7)**	______

Tester = ____________________

10-Second Ergometer Max Sprint

	Total work output		Peak power output	
	J	**J/kg**	**W**	**W/kg**
6 s	______	______	______	______
10 s	______	______	______	______

95% of 6 s work = ____________________ J

5 × 6-Second Ergometer Maximal Sprints With 24-Second Recovery

	Total work output		Peak power output	
	J	**J/kg**	**W**	**W/kg**
1	______	______	______	______
2	______	______	______	______
3	______	______	______	______
4	______	______	______	______
5	______	______	______	______
Total	______	______	______	______
% Dec	______	______	______	______

Data acquisition method = ____________________

Vertical Jump

Jump and reach ____________

Stand and reach ____________

Vertical jump ____________

Notes

____

____

____

Physiological Assessment for Basketball: Field Based

20-Meter Sprint

Surface ____________________

Temperature ____________________ °C

Humidity ____________________ %

Raw scores

5 m ____________________ s

10 m ____________________ s

20 m ____________________ s

Calculated scores

Acceleration (0-5 m) ____________________ s

Speed (10-20 m) ____________________ s

Combined (0-20 m) ____________________ s

20-Meter Multistage Shuttle Run

Surface ____________________

Temperature ____________________ °C

Humidity ____________________ %

Raw scores

Level ____________________

Shuttle ____________________

Calculated score

Level ____________________

$\dot{V}O_2$max Data Collection Form: Laboratory Based

Name ______________________________ Date ___/___/___ DOB ________

Training phase: Transition Gen. Prep. Spec. Prep. Competition Peak Rehabilitation

Position: C F G

Previous 48-hour training load: None Light Medium Heavy

Venue ____________ Tester/s ______________________________

Data entry completed by ________________

Maximal Oxygen Consumption ($\dot{V}O_2$max)

Temperature ______________________________ °C

Pressure ______________________________ Hpa

$\dot{V}O_2$max ______________________________ L/min

$\dot{V}O_2$max ______________________________ $ml \cdot kg \cdot min^{-1}$

$\dot{V}_E$max (STPD) ______________________________ L/min

Max heart rate ______________________________ beats/min

Strength-Testing Assessment for Basketball: Laboratory Based

Name ______________________________ Date ___/___/___ DOB ________

Training phase: Transition Gen. Prep. Spec. Prep. Competition Peak Rehabilitation

Position: C F G

Previous 48-hour training load: None Light Medium Heavy

Venue ___________ Tester/s ______________________________

Data entry completed by ______________

Body weight	**Upper-body strength—bench press**	**Lower-body strength—squat**
_____ kg	_____ 3RM kg	_____ 3RM kg

Intensive Training Centre Athlete Assessment for Basketball

Name ______________________ Date ___/___/___ Time ________

DOB __________ Squad ______________________ State ______________

Venue __________ Tester/s ______________________________

Data entry completed by ______________

Height	**Arm span**	**Weight**
_____ cm	_____ cm	_____ kg

Vertical jump

Jump and reach _____ cm

Stand and reach _____ cm

Vertical jump _____ cm

10-meter sprint	**20-meter sprint**	**Sit-ups**	**Agility run**	**Suicide run**	**Sit and reach**	**20-meter shuttle**
_____ s	_____ s	_____ reps	_____ s	_____ s	_____ cm	_____ level
_____ s	_____ s		_____ s	_____ s	_____ cm	_____ m

chapter

15

Protocols for the Physiological Assessment of Cricket Players

Pitre Bourdon, Bernard Savage, and Richard Done

Cricket is the most popular organized summer sport in Australia, with approximately 470,200 competitors aged 14 and over (Department of Sport, Recreation and Tourism 1986). Several factors are fundamental to the quality of performance in the game of cricket: skill, correct mental approach, and physical fitness. It is on the third factor that this chapter will focus by outlining a recommended field test battery designed to measure the physiological demands of cricket.

To present a case for physiological testing in cricket, several features of the modern game need to be highlighted. First, the game has changed in recent years. One-day games are now an integral part of most programs, demanding fast scoring rates, quick over rates, and field restrictions requiring players to field athletically in a variety of positions. Secondly, fitness is an integral component of performance; as fatigue develops, skill deteriorates. Being fit delays the onset of fatigue. Therefore, optimal fitness levels will enable fast bowlers to string together longer spells at top pace without losing line and length; batsmen will maintain their timing, placement, and concentration; fielders will hold their physical performance, reaction time, and concentration longer. Thirdly, fitness plays an important role in injury prevention. Those players possessing adequate levels of strength, endurance, and flexibility in the muscle groups used in cricket not only will enhance their potential to perform better, but also may reduce their susceptibility to injury.

Regular fitness testing is now an important part of the preparation of many squads. The following protocols have been documented with the aim of developing greater uniformity in the physiological assessment of the high-performance cricketer. These tests are easy to administer in the field environment and require very little specialized testing equipment. Results provide the coach with an indication of where an individual's strengths and weaknesses lie in relation to the various components of fitness important to cricket, as well as providing a measure of progress in response to prescribed training programs.

Test Environment and Subject Preparation

Refer to page 128 for a discussion of important issues in test environment, subject preparation, and scheduling of tests.

Equipment Checklist

See page 129 for descriptions of the equipment for the following, except where noted:

- Anthropometry (see chapter 5)
- Vertical jump
- Flexibility
- Sprint
- Abdominal stage test
- Multistage fitness test

1. Speed—sprint tests (10, 20, and 40 m)
   - ☐ Preferably a standardized 50 m nonslippery smooth indoor surface (otherwise a flat, even-length grass surface can be used)

- ☐ Light gates accurate to 0.01 s (preferably dual beam)
- ☐ Recommended footwear (normal running shoes, not spikes)
- ☐ 50 m measuring tape

2. Agility—run a three
   - ☐ Preferably a standardized 25 m nonslippery smooth indoor surface (otherwise a flat, even-length grass surface can be used)
   - ☐ Light gates accurate to 0.01 s (preferably dual beam)
   - ☐ Measuring tape
   - ☐ Masking tape and/or cones
   - ☐ Two stopwatches (preferably with split timing available)
   - ☐ Recommended footwear (normal running shoes, not spikes)
   - ☐ Cricket bat

PROTOCOLS: Cricket

Anthropometry

Chapter 5 outlines the procedures for measuring height, mass, and skinfolds.

Normative Data: Table 15.1 shows anthropomorphic data for Australian cricket players.

Leg Power—Vertical Jump (Double Leg Jump)

Rationale: The vertical jump is a widely recognized test of leg extensor power for which both normative and sport-specific data are available (Young 1994). The test is simple to administer and does not require large or expensive pieces of equipment. Furthermore, it allows the assessment of large numbers of players in a relatively short period of time.

Leg strength and power are important for all players. In cricket, these qualities contribute to the speed and agility required for fielding, wicketkeeping, and running between wickets. Leg strength and power are also vital for fast bowling: they provide a solid base, assist in absorbing the large forces experienced in the delivery stride, and assist in the efficient transfer to momentum from the run-up.

Test Procedure: Refer to page 141 for the procedure for this jump.

Normative Data: See table 15.2 for vertical jump scores of Australian cricket players.

Table 15.1 Anthropometric Characteristics of Australian Cricket Players

	Mass (kg)				Height (cm)				Sum of 7 (mm)*			
Squad/position	**n**	**Av**	**SD**	**Range**	**n**	**Av**	**SD**	**Range**	**n**	**Av**	**SD**	**Range**
AIS Academy												
Batsman	63	77.6	7.0	62.9–91.5	45	168.9	5.3	168.9–194.5	63	62.1	18.9	32.5–112.6
Pace bowler	46	86.4	8.3	68.1–107.5	30	189.2	6.3	176.2–208.9	46	56.2	16.3	31.5–98.1
Keeper	13	78.5	4.3	71.7–85.2	8	177.1	3.7	170.0–181.3	13	56.7	10.9	38.8–74.5
Spin bowler	18	74.8	7.1	61.1–85.6	10	182.5	5.6	174.0–192.0	17	65.2	21.8	42.1–125.3
Sheffield Shield	59	79.9	8.8	62.9–103.2	59	182.7	7.2	174.1–208.9	59	66.3	24.2	30.0–145.3
Male Test	27	85.8	8.7	72.8–109.8	27	183.6	7.9	169.8–202.0	27	74.7	25.1	47.4–130.2
Female Test	62	63.2	7.3	53.4–82.2	62	169.7	5.5	155.7–182.1	62	96.1	34.8	47.3–213.3

*Σ7 = triceps, subscapular, biceps, supraspinale, abdominal, thigh, medial calf.

The AIS Academy is the National Development Squad; Sheffield Shield are state-level players; Test are international players.

Table 15.2 Vertical Jump Scores of Australian Cricket Players

Squad/position	Vertical jump (cm) n	Av	SD	Range
AIS Academy				
Batsman	34	61.1	6.5	49–78
Pace bowler	24	60.5	6.3	45–70
Keeper	6	62.6	2.7	58–65
Spin bowler	8	57.4	3.9	50–63
Sheffield Shield	59	54.8	8.3	38–69
Male Test	27	52.6	9.5	22–78
Female Test	62	41.3	5.1	31–53

The AIS Academy is the National Development Squad; Sheffield Shield are state-level players; Test are international players.

Flexibility—Sit-and-Reach Test

Rationale: The accurate assessment of flexibility is joint specific and requires specialized equipment such as goniometers referenced precisely to anthropometric landmarks (see chapter 7). The limitations of the sit-and-reach test as a method to assess flexibility are well described (Jackson and Baker 1986), but a number of sports in Australia continue to use this test because it is simple to administer and because a pool of normative data is available. A more comprehensive rationale for selection of flexibility tests is presented on page 137.

Test Procedure: Refer to page 138.

Normative Data: Table 15.3 presents sit-and-reach data for Australian cricket players.

Speed—Sprint Tests (10, 20, and 40 Meter)

Rationale: The 10, 20, and 40 m sprint tests may be used as tests of an athlete's explosive ability, rate of force development, and maximal running velocity. In cricket, players must accelerate over short distances in the infield and run over longer distances when chasing the ball to the boundary. Therefore, testing speed over these distances is relevant to cricket as it simulates the varying possibilities during match play.

Test Procedure: Refer to page 130.

Normative Data: See table 15.4 for sprint test scores of Australian cricket players.

Table 15.3 Sit-and-Reach Test Scores of Australian Cricket Players

Squad/position	Sit and reach (cm) n	Av	SD	Range
AIS Academy				
Batsman	40	14.6	4.0	5 to 23
Pace bowler	27	17.4	6.3	2 to 29
Keeper	8	18.7	3.1	15 to 23.5
Spin bowler	11	13.1	6.4	-3 to 22
Sheffield Shield	59	12.1	7.0	-5.5 to 28
Male Test	27	11.6	7.6	-12 to 22
Female Test	62	13.4	5.7	3.5 to 29.5

The AIS Academy is the National Development Squad; Sheffield Shield are state-level players; Test are international players.

Table 15.4 Sprint Test Scores of Australian Cricket Players

Squad/position	n	10 m (s) Av	10 m (s) SD	10 m (s) Range	20 m (s) Av	20 m (s) SD	20 m (s) Range	40 m (s) Av	40 m (s) SD	40 m (s) Range
AIS Academy										
Batsman	6	1.78	0.1	1.70–1.86	3.04	0.1	2.82–3.19	5.42	0.2	5.08–5.65
Pace bowler	5	1.76	0.0	1.75–1.79	2.99	0.1	2.88–3.07	5.30	0.1	5.17–5.49
Keeper	1	1.73	–	–	2.93	–	–	5.17	–	–
Spin bowler	2	1.78	0.0	1.76–1.79	3.08	0.0	3.06–3.10	5.49	0.1	5.44–5.54
Sheffield Shield	–	–	–	–	–	–	–	–	–	–
Male Test	–	–	–	–	–	–	–	–	–	–
Female Test	40	2.07	0.1	1.9–2.3	3.52	0.2	3.2–3.9	–	–	–

The AIS Academy is the National Development Squad; Sheffield Shield are state-level players; Test are international players.

Agility—Run a Three

Rationale: Agility is a very important physiological parameter for many team sport players and in the game situation is apparently a combination of neurological processing ("reading the play"), coordination, and speed. The run a three uses the 505 test (see page 132 for a description of this test) and places it into a cricket-specific setting to assess agility in relation to performance in cricket.

Preparation:

- Using the measuring tape, masking tape, and cones, mark out the points of the course according to the diagram in figure 15.1: place markers 17.7 m apart (the length of a cricket pitch). Place another marker 12.7 m from the starting point. This is where the timing gates are to be placed for the measurement of turn speed.
- Set up the timing lights ~2 m apart to form a gate for the players.

Test Procedure:

1. Instruct subject to assume the starting position, with one foot behind the crease line and cricket bat in hand, without hands touching the ground.

2. After a signal that the light gates are set or stopwatch timers are prepared, the subject should start when ready. The hand timing of the run a three should begin when the subject's back foot leaves the ground and should finish as the bat crosses the crease line at the end of the third run.

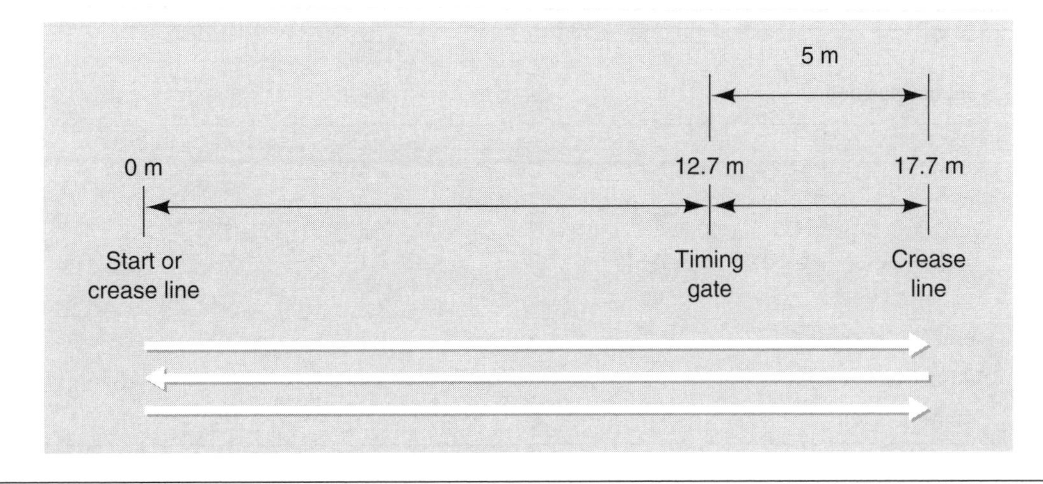

Figure 15.1 Dimensions for the agility test course.

3. The light gates will start timing automatically as the subject passes through the light gate on the first run. After turning at the crease line, the subject will pass again through the timing gate. The time displayed on the screen is recorded as the turning speed. This should be recorded immediately, as the subject will pass again through the gates on the third run and restart the timing. Alternatively, the average time from the two stopwatches should be recorded.

4. The average time from the two stopwatches for the run a three should be recorded to the nearest 0.1 s.

5. Record time on the appropriate recording sheet and reset the timing device.

6. The subject should complete at least two trials.

7. The fastest time for turning speed and run a three is recorded as the best score.

Normative Data: Table 15.5 shows test scores of Australian cricket players.

Muscular Strength: Abdominal Stage Test

The problems and issues of strength assessment are described in chapter 10.

Rationale: See page 142 for a discussion of the rationale for this test. Strength in the abdominal region is important in cricket, in particular for pre-

Table 15.5 Agility Test Scores of Australian Cricket Players

		Run a three (s)*			Turn speed (s)†		
Squad/Position	**n**	**Av**	**SD**	**Range**	**Av**	**SD**	**Range**
AIS Academy							
Batsman	62	9.29	0.4	8.60–10.15	2.05	0.1	1.84–2.60
Pace bowler	45	9.36	0.4	8.68–10.85	2.08	0.1	1.87–2.43
Keeper	13	9.36	0.6	8.67–10.80	2.06	0.2	1.86–2.34
Spin bowler	18	9.53	0.3	8.82–10.37	2.12	0.1	1.85–2.36
Sheffield Shield	59	9.32	0.4	8.68–9.99	2.16	0.1	1.86–2.39
Male Test	27	9.65	0.5	8.61–11.20	2.12	0.1	1.79–2.48
Female Test	62	10.62	0.5	9.80–11.90	2.33	0.1	2.06–2.78

* Timing completed by handheld stopwatch.

† Timing completed with light gates.

The AIS Academy is the National Development Squad; Sheffield Shield are state-level players; Test are international players.

Table 15.6 Abdominal Stage Test Scores of Australian Cricket Players

		Stage		
Squad/position	**n**	**Av**	**SD**	**Range**
AIS Academy				
Batsman	11	3	1.3	2–6
Pace bowler	9	4	0.9	4–6
Keeper	2	3	2.1	2–5
Spin bowler	11	4	6.4	4–5
Sheffield Shield	59	4	1.1	1–5
Male Test	27	4	0.4	3–5
Female Test	20	5	1.0	2–7

The AIS Academy is the National Development Squad; Sheffield Shield are state-level players; Test are international players.

venting injury in fast bowlers. Abdominal strength assists in stabilization of the trunk and also provides a strong and powerful torso to assist in power production during the delivery. Batsmen and fielders will also benefit from good abdominal strength during long periods in the field or at the crease.

Test Procedure: Refer to page 143.

Normative Data: See table 15.6 for scores of Australian cricket players.

Aerobic Power—Multistage Fitness Test

Rationale: The multistage fitness test has been selected as the test of aerobic power for a number of reasons. The test is easy to administer both in the lab and on the road; it is inexpensive; and it is similar to activities in many team sports with respect to the stop, start, and change-of-direction movement patterns. This 20 m progressive shuttle run test has been found to be a sufficiently accurate estimate of aerobic power (Brewer et al. 1988; Leger and Lambert 1982).

Test Procedure: Refer to page 130.

Normative Data: See table 15.7 for multistage fitness test scores of Australian cricket players.

Table 15.7 Multistage Fitness Test Scores of Australian Cricket Players

		Laps		
Squad/position	**n**	**Av**	**SD**	**Range**
AIS Academy				
Batsman	46	112.6	24.0	54–149
Pace bowler	34	104.3	26.8	45–139
Keeper	10	111.4	34.0	45–147
Spin bowler	15	105.7	34.4	51–158
Sheffield Shield	59	112.0	17.3	70–147
Male Test	27	97.0	14.4	78–122
Female Test	62	85.1	14.6	41–111

The AIS Academy is the National Development Squad; Sheffield Shield are state-level players; Test are international players.

chapter 16

Protocols for the Physiological Assessment of Cross-Country Skiers and Biathletes

David T. Martin, Christer Skog, Ian Gillam, Lindsay Ellis, and Kate Cameron

Successful cross-country skiers possess high levels of cardiovascular fitness and upper-body endurance (Eisenman et al. 1989; Sharkey 1984). Because physical fitness is important for success, ski coaches have long sought the advice of exercise physiologists to help them select and train champion skiers. Laboratory and field testing is administered primarily to quantify initial sport-specific fitness in a standard environment. These data enable evaluation of the effects of a training program or an ergogenic aid and in some cases the identification of individual strengths and weaknesses. Laboratory and field testing can also help one to better understand the physiological stress imposed by different types of training. Finally, talent identification programs may benefit from laboratory data that characterize the physiological traits of successful competitors.

Before testing an athlete it is important to recognize the demands of competition. In the 1997 world championships for cross-country skiing, men and women participated in four individual events and a relay. Race distances were 10, 15, 30, and 50 km for men and 5, 10, 20, and 30 km for women. Skiing technique for these races was evenly divided between freestyle (skating) and classic (diagonal). As in other endurance sports, the objective in cross-country skiing is to cover a fixed distance in the shortest time possible.

In biathlon, athletes attempt to combine skiing speed with shooting accuracy. Intermittent periods of shooting (five shots) lasting between 30 and 50 s are included in the race between segments of skiing. Thus, the skiing portion of the race is divided into intervals. Depending upon the race distance, biathletes will shoot on either one or two occasions from both the prone and standing positions. Five events take place at the World Biathlon Championships over distances ranging from 7.5 to 15 km for women and from 10 to 20 km for men.

New waxing and skiing techniques combined with sophisticated trail-preparation machinery are allowing skiers to reach high speeds during competition. Using the freestyle technique (skating), the winner of the women's 15 km event at the 1997 world championships in cross-country skiing averaged 2:25 (min:s) per kilometer. In the men's 30 km freestyle event, the winner averaged 2:12 (min:s) per kilometer. Thus, internationally competitive cross-country skiers can now ski more than 25 kph over hilly terrain for distances between 15 and 30 km.

Although skiing technique and equipment will influence a skier's competitiveness, at the elite level there are certain sport-specific physical fitness requirements. In groups of skiers with a wide range of skiing ability, aerobic power as indicated by maximum oxygen consumption ($\dot{V}O_2max$) may be one of the most important laboratory predictors of success (Bergh 1987). Additionally, a skier's lactate threshold and upper-body endurance should be well developed. Laboratory assessment of fitness in cross-country skiers and biathletes should include movement patterns that require use of the upper body in a ski-specific manner (e.g., ski walking on a treadmill). A 1-9 min maximal upper-body effort using the double-pole technique will also provide useful information. Because it is difficult to duplicate skiing precisely in the laboratory, many international ski programs use field testing to stratify heart rates into training zones.

Laboratory Testing of Cross-Country Skiers and Biathletes

Sport scientists have employed numerous laboratory tests to simulate the movement patterns used by cross-country skiers. Other selected fitness tests evaluate the state of conditioning in those muscles primarily recruited during cross-country skiing.

Testing $\dot{V}O_2max$

Scandinavian countries have a rich tradition of testing aerobic power in cross-country skiers using treadmills. Ski striding with poles on a rather large treadmill has been used in Norway, Sweden, Finland, and more recently in the United States. A variety of arm ergometers have been modified to be used in conjunction with the treadmill in laboratories that do not have a treadmill large enough to accommodate walking with poles. Unfortunately, many upper-body ergometers (e.g., Exergeni) allow athletes to hang by their arms during the final work stages of a $\dot{V}O_2max$ test. This results in a decrease in the overall work performed and therefore may not elicit maximal rates of oxygen consumption. For this reason, ski walking with poles is generally the preferred test modality when testing cross-country skiers (figure 16.1).

Figure 16.1 An Australian cross-country skier during a maximal treadmill ski walking test at the Australian Institute of Sport in Canberra.

The ski pole-treadmill interface is a problematic aspect of treadmill walking with poles. The treadmill belt needs to be modified with two strips of dense rubber that allows the metal ski-pole tips to stick without damaging the treadmill belt. Alternatively, the ski-pole tips can be covered with rubber stoppers so that the ski poles can stick to an unmodified treadmill belt. It is possible to roller ski on treadmills if the belt surface and the size of the treadmill are appropriate. Previous research indicates that this mode of testing results in a lower $\dot{V}O_2max$ than achieved during treadmill running (Rundell 1996). Both freestyle and classic techniques can be incorporated during roller skiing on a treadmill; but it is important to use a safety harness, since even experienced cross-country skiers have been known to fall during the final stages of testing.

An air-braked arm-leg bicycle ergometer has been used extensively for testing cross-country skiers at the Australian Institute of Sport (AIS) primarily because a suitable treadmill has only recently become available for ski striding with poles. Also, the arm-leg cycling activity is relatively easy to learn. $\dot{V}O_2max$ was similar during treadmill running and arm-leg cycling in nine Australian junior skiers at the AIS in June 1985. The mean ± SD $\dot{V}O_2max$ for the arm-leg bike was 64.3 ± 5.7 ml · kg^{-1} · min^{-1}; for treadmill running it was 64.1 ± 6.1 ml · kg^{-1} · min^{-1}. More recently these data have been confirmed using two members of the 1997 national cross-country ski team. However, athletes performed these tests preseason before they had developed a high degree of sport-specific fitness. Because ski striding on a treadmill with poles can be learned quickly and is performed at relatively slow treadmill speeds, this type of testing has been adopted for assessment of $\dot{V}O_2max$ in cross-country skiers and biathletes. Laboratory testing of this type should allow for the maximum recruitment of specifically trained muscle mass, resulting in the highest possible $\dot{V}O_2max$.

Testing Upper-Body Fitness

A modified Repco kayak ergometer (see appendix) can be used to assess upper-body power specific to cross-country skiing. With the Repco kayak ergometer mounted to a bench with the handles just below shoulder height, the skier can engage in a sport-specific series of contractions simulating double poling. (Note: A Repco swim bench may be used if a kayak ergometer is not available. However, our experience indicates that the cords connecting the handguns to the turbine rub on the housing, causing them to fray quite quickly. Ski-pole straps are attached to the cords from the ergometer so that the normal poling action can be simulated. The standard ergometer handguns may also be modified so that they closely mimic the action of a ski pole held vertically.)

Researchers in Sweden have been successful in modifying a Concept II rowing ergometer to allow skiers to simulate double poling (see figure 16.2). A metal plate is attached to the ergometer where the seat would be positioned during the "catch phase" of the rowing stroke. During the test the skier stands on the plate. The seat (under the standing plate) is then modified so that ski poles can be attached. The rowing handle is removed, and the chain is attached to the seat. This allows the skiers to double-pole by standing on the plate and poling the seat behind them. One of the advantages of this system is that the control panel of the Concept II can be used to quantify power output.

Figure 16.2 A Concept II rowing ergometer modified to allow testing of cross-country skiers using the double-pole technique.

Other commercially available ergometers have been constructed that allow for a skier to replicate the double-poling movement pattern. Some are simply a modification of a slide board. The skier lies or kneels on a board that can slide or roll up and down an incline as the skier double-poles. Although these ergometers can be useful for training and general muscle-endurance tests, most lack the ability to precisely quantify work and power output. Systems that do allow work to be quantified tend to be expensive.

Laboratory Environment and Subject Preparation

Follow the guidelines presented in chapter 2 regarding the laboratory environment, subject preparation, and calibration of equipment. Additional guidelines more specific to skiers are the following:

- **Laboratory environment.** During treadmill walking and upper-body poling tests it is important to provide a consistent airflow to aid in thermoregulation. Because even minimal airflow can reduce heart rate for a given power output, the fan should be adjusted to consistently generate between 10 and 15 kph wind speed. These conditions should not change in subsequent testing sessions.
- **Scheduling testing.** In situations in which skiers will be tested multiple times throughout the year, testing should take place at the same time of day on each occasion. At the AIS, the battery of tests is completed in the following order:

1. A light day of training with some competition-specific intensity (<15 min). The, on the day prior to testing:
2. 7:00 A.M.—assessment of skinfolds and a resting, overnight-fasted, venous blood draw
3. Later that day, a treadmill $\dot{V}O_2max$ test (with a large squad, the first test starts at 9:00 A.M., and one test takes place each hour thereafter)
4. The following day, field lactate testing in the morning (~10:00 A.M.)

• **Subject preparation.** When testing a squad of athletes, ensure that all athletes know their testing time and location. Prior to the test day, it is advisable that a physician examine any athletes who have an existing health or injury complaint.

On test day, athletes should refrain from eating or training prior to morning blood tests and skinfold assessment. Details of any recent injuries, illness, and training problems should be recorded, as this may help to explain any aberrant findings (see the pretest questionnaire in chapter 2). Along with explaining test procedures and obtaining informed consent, clearly explain to athletes that their participation in testing is voluntary and that they can stop a test for any reason they choose. For subjects who have not engaged in a ski walking test on a treadmill using poles, a 5-15 min familiarization period may be helpful.

• **Equipment calibration.** It is particularly important that the treadmill speed and grade are accurate and reliable prior to testing. Also, lactate controls (approximately 4 and 8 mmol/L) and calibration gases should be used to confirm accuracy of the lactate, CO_2, and O_2 analyzers.

• **Training.** Ask subjects to refrain from physical training on the day of testing. It is also suggested that athletes should not have raced in the previous two days. Since most elite skiers compete in one or two races every weekend during the ski season, neither Monday nor Tuesday is an advisable test day. Also, since many skiers have to travel 6 to 8 h to the testing location, it is useful to schedule a rest day prior to testing. One desirable testing scenario incorporates light training with 5-15 min of high-intensity exercise (5 × 3 min intervals above threshold) on the day before the test. If possible, this training session should be standardized.

• **Diet.** A typical high-carbohydrate diet (>6 g · kg body mass^{-1} · day^{-1}) including liberal intake of fluids is suggested for the two days leading up to testing. Athletes may drink water or a sport drink leading up to the $\dot{V}O_2$max test; however, they should drink sparingly during the 30 min prior in order to avoid the need for a toilet break during the initial portion of the test.

• **Warm-up.** No warm-up is required before either the $\dot{V}O_2$max or the field lactate test, since the initial 12 min of the protocol involves very low exercise intensities. Athletes may engage in some light stretching prior to testing, especially if they have any muscle groups that are particularly prone to becoming tight during exercise. As already mentioned, athletes unfamiliar with ski striding on a treadmill may require 5-15 min familiarization. It is also useful to allow subjects to briefly walk on the treadmill before starting the test in order to check the position of the mouthpiece, the length of the ski pole, and the operation of the treadmill and heart rate monitor.

• **Immediately prior to testing.** Complete the recording forms for all tests as detailed in chapter 2; also, barometric pressure should be recorded along with temperature and humidity. Immediately before conducting the test, briefly reiterate to the athlete what will be required and ask if he or she has any questions or concerns. Emphasize how the test will terminate (e.g., athlete grabs handrails and treadmill belt will stop). Record all data using a pencil, and cross out with a single line any data recorded incorrectly.

Equipment Checklist

1. Blood testing
   - ☐ Alcohol swabs
   - ☐ Elastic arm cuff to enhance venous blood pressure
   - ☐ Cotton balls or gauze pads
   - ☐ 21-gauge Vacuette needles (see appendix for details about this product)
   - ☐ Vacuette needle holder
   - ☐ 8 ml serum separator Vacuette tube (speckled top)
   - ☐ 2-4 ml EDTA Vacuette (lavender top)
   - ☐ Labeling pen
   - ☐ Rubber gloves
   - ☐ Sharps container
   - ☐ Biohazard bag for contaminated soft materials
2. Anthropometry
   - ☐ Stadiometer
   - ☐ Electronic platform scale (calibrated 20-120 kg)
   - ☐ Harpenden skinfold calipers (dynamically calibrated)
   - ☐ Marking pen
   - ☐ Anthropometric tape (Lufkin model W606PM)
   - ☐ Bone calipers
   - ☐ Recording forms
3. $\dot{V}O_2$max test
   - ☐ Calibrated treadmill (0.9 × 1.6 m suggested)

- ☐ Ski poles with rubber stoppers on tips (~40 cm less than athlete height)
- ☐ Pretest questionnaire for illness, injury, and training (see chapter 2)
- ☐ Expired gas analysis system, including Hans Rudolph (Kansas City, Missouri) headset for Hans Rudolph R2700 respiratory valve (a noseclip is also required)
- ☐ Heart rate monitor
- ☐ Electric fan capable of 10-15 kph wind speed
- ☐ Stopwatch
- ☐ Recording sheet with two pencils
- ☐ Perceived exertion scale (Borg 6-20)
- ☐ Thermometer (for air temperature)
- ☐ Testing protocol
- ☐ Rubefacient cream, such as Finalgon (Boehringer Mannheim), to warm the sampling site (fingertip or earlobe)
- ☐ Blood lactate analyzer capable of 1-20 mmol/L lactate measurements
- ☐ Capillary blood collection supplies:
  - __ Autolancet such as Autolet (Owen Mumford), lancets, platforms
  - __ Alcohol swabs
  - __ Tissues
  - __ 100 ml capillary tubes (heparinized)
  - __ Biohazard bags
  - __ Sharps container
  - __ Latex gloves
  - __ Lab coat
  - __ Protective glasses

4. Field lactate test
   - ☐ Polar heart rate monitor (capable of logging and downloading data; e.g., Polar Sport Tester, Vantage, Nevada)
   - ☐ Roller skis and ski poles
   - ☐ Meter wheel for measuring distance
   - ☐ Paint or marking tape for selecting testing loop
   - ☐ Stopwatch
   - ☐ Recording sheet with two pencils
   - ☐ Perceived exertion scale (Borg 6-20)
   - ☐ Thermometer (for air temperature)
   - ☐ Blood lactate analyzer (e.g., Accusport [Boehringer Mannheim] or Yellow Springs Instruments [YSI 2300stat; Yellow Springs, Ohio])
   - ☐ Capillary blood collection supplies as for $\dot{V}O_2$max test, plus:
     - __ Yellow Springs Instruments pipette
     - __ Microcentrifuge tube with YSI lysing cocktail for preserving sample

PROTOCOLS: Cross-Country and Biathlons

Venous Blood Analysis

Rationale: It is recommended that cross-country skiers undergo regular blood screening before and after the competitive season. Monitoring iron status is of particular interest because cross-country skiers have been shown to have the highest incidence of suboptimal iron status among all winter sport athletes. In one study of competitive female cross-country skiers, 50% were observed to suffer from prelatent iron deficiency as indicated by plasma ferritin values below 30 ng/ml, and 7% were anemic (Clement et al. 1987). Although it is common to observe a decrease in ferritin independent of changes in hemoglobin mass in well-trained endurance athletes, decreases in ferritin below 20 ng/ml require medical attention. Blood testing can also be used to screen overall health status (white blood cell enumeration) and to provide insight into how well training loads are being tolerated (creatine kinase activity, urea, and uric acid).

In 1996, the international governing body for skiing, Fédération Internationale de Ski, incorporated random blood testing for measurements of hemoglobin concentration prior to World Cup ski races. Ski officials hoped that imposing a limit on hemoglobin concentration of 16.5 and 18.5 g/dl for women and men, respectively, would discourage the suspected illegal use of human recombinant erythropoietin. Routine blood testing can be used to establish whether skiers normally exceed this limit.

In addition, many athletes have their blood tested when they feel sick or run-down. Their results are compared to large reference ranges established from normal, relatively unfit individuals. It may be more relevant to use data collected from athletes on a number of occasions when they feel healthy in order to establish individual reference ranges that may help in the detection of slight changes for that athlete. Blood tests following successful training periods and competition can be used to highlight an acceptable range for many blood parameters. Extensive research is required to clarify how subtle changes in blood parameters can be used to best guide future training efforts.

Test Procedure:

1. The athlete, in a fasted state, should report to the laboratory for routine venous blood collection in the morning after a light day of training.

2. Blood should be collected only by a person who has been trained in standard phlebotomy procedures (refer to chapter 6 for more details about venepuncture procedures).

3. Blood samples should be collected from a superficial forearm vein while the athlete is in a supine position. The athlete should remain supine for 10-20 min prior to the blood test so that plasma volume can completely equilibrate to the new body position.

4. The amount of blood collected into Vacuette tubes depends upon the hematology and biochemistry equipment available. Typically, 2 ml of whole blood and 2 ml of serum are adequate for a complete whole blood count and a full blood chemistry panel analysis. Whole blood needs to be collected and mixed in an EDTA Vacuette tube, and the blood for serum needs to be centrifuged in a Vacuette serum separator tube after the blood is allowed to clot.

5. After collection of the blood sample, pressure should be applied to the venepuncture site and the athlete should be monitored for 5-10 min afterward as a precaution.

6. It can be educational to tell the athlete exactly how much blood was collected (approximately 12 ml) and to explain that this amount of blood loss is minimal (in relation to 4-6 L of total blood volume) and will not detrimentally affect the ability to train or compete.

Standard laboratory procedures should be followed for whole blood count and serum biochemistry analysis. Only laboratories that participate in routine quality-assurance programs (such as that conducted by the Royal College of Pathologists of Australia) should be used for blood analysis. When possible, blood samples should be analyzed within 1-3 h of the blood collection and in the same laboratory.

Data From Australian Skiers: As just mentioned in connection with venous blood analysis, normal clinical ranges are available for most standard hematology and biochemistry assays. For extreme variation these normal limits can provide a useful reference range, but previous data from an individual athlete are likely to be the most useful for comparisons. With frequent blood monitoring of a well-trained athlete it is possible to identify reference ranges of hematological and biochemical parameters during heavy training, tapering, and racing phases of the year. These reference values are then available for reference at times when there are unexplained decrements in performance.

When ferritin values are less than 20 ng/ml but hemoglobin values are within a normal range, a dietary review is warranted. If inadequate amounts of iron are consumed, or in cases in which an iron absorption problem is suspected, the athlete may be prescribed an iron supplement (ferrous sulfate) by a physician. An increase in hemoglobin concentration after 1-2 weeks of iron supplementation strongly suggests that the dietary iron intake was inadequate. However, if hemoglobin concentration (or more importantly, total hemoglobin mass) does not change, then inadequate iron intake is usually not the problem. In rare instances, folic acid or vitamin B12 deficiency may be the cause of the low hemoglobin. It is important to note that moderately low ferritin values (<45 ng/ml) are frequently observed in endurance female athletes who are very aerobically fit and capable of successful performances. When these athletes receive an iron supplement either orally or via an injection, ferritin values generally increase but hemoglobin concentrations rarely increase—suggesting that iron intake was not inadequate.

Results from blood tests performed on a group of Australian junior cross-country skiers are presented in table 16.1. Note the large individual variation in many of the parameters. A more extensive hematology and biochemistry analysis was performed on blood collected from three male members of the 1997 Australian cross-country ski team the morning before and after they had competed in the Kangaroo Hoppet ski marathon (42 km; figure 16.3). Data demonstrate normal variation that can occur in blood results following cross-country ski racing. Hemoglobin concentration was 0.5-2.0 g/dl lower the morning after the ski race. Decreases in hemoglobin and hematocrit are likely the result of a plasma volume expansion and not a large decrease in red blood cell mass. Even after more than 15 h of recovery, urea and uric acid were elevated by 15-20%, creatine kinase was elevated by 160%, and aspartamine amino transferase (AST) and bilirubin were 40% higher. Changes that take place in healthy fit skiers following successful competition need to be recognized before changes in blood data can be used to diagnose training problems. It is

Table 16.1 Summary of Venous Blood Data from Australian Junior Cross-Country Skiers

Skier category	n	White blood cell count (× 10^9/L)	Red blood cell count (× 10^{12}/L)	Hb (g/dl)	Hct (%)	Ferritin (ng/ml)
Junior men	9	5.5 ± 0.23 (4.3–7.2)	4.09 ± 0.8 (4.5–5.2)	15.5 ± 2.5 (14.0–16.5)	45.3 ± 3.0 (41.1–48.6)	51 ± 1.1 (21–107)
Junior women	4	5.2 ± 1.2 (4.2–6.9)	4.75 ± 0.8 (4.23–5.94)	13.8 ± 1.1 (12.5–15.1)	41.4 ± 3.3 (37.6–45.1)	69 ± 1.2 (18–170)

Skiers participated in a training camp at the Australian Institute of Sport in 1990. Hb=hemoglobin, Hct=hematocrit.

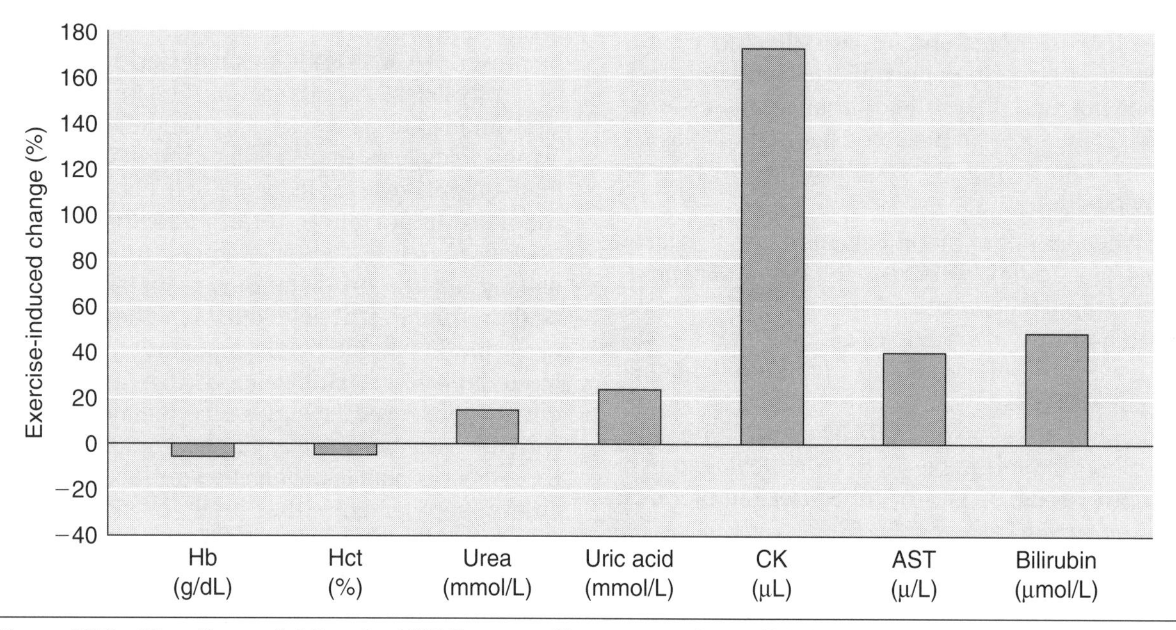

Figure 16.3 The effects of skiing the 1997 Kangaroo Hoppet marathon (42 km) on a selection of hematological and biochemical parameters. (Mean data from 3 skiers placing in the top 11 [including the race winner] are presented as a % change from prerace values. Venous blood was collected from a forearm vein the morning before and the morning after the race.)

possible that blood testing may help a coach better understand how the relative overload imposed by racing is affecting athletes. However, further research is required to clarify the specific blood parameters that should be used for monitoring fatigue.

Anthropometry

Rationale: Skinfold calipers can be used to measure the thickness of the skin at specifically identified anatomical sites (see chapter 5). Because the thickness of the skin is proportional to the amount of subcutaneous fat, the sum of a series of skinfolds in conjunction with body weight can be used to estimate changes in body composition (see table 16.2). The primary objective of monitoring skinfolds is for the coach and the athlete to better understand the balance between caloric expenditure (exercise) and caloric intake (eating). Whereas a gradual decrease in skinfolds over a competitive season is normal and should be expected, rapid gains or losses in skinfolds are not desired, as they tend to reflect radical changes in diet, training, or both.

Cross-country skiers and biathletes do not want to be excessively fat; however, it is important to remember that performances can suffer when an athlete is too lean (possibly due to the prolonged period of insufficient energy intake and not as a direct consequence of being lean). Thus, the primary aim in monitoring skinfolds is to understand

Table 16.2 Possible Changes in Skinfolds and Body Mass and Corresponding Interpretation of Changes in Body Composition

Body mass	Skinfolds	Interpretation
→	→	→ Muscle or fat mass
→	↑	↓ Muscle mass, ↑ fat mass
→	↓	↑ Muscle mass, ↓ fat mass
↑	→	↑ Muscle mass or fluid, ↓ fat mass
↑	↑	↑ Fat mass, maybe muscle mass
↑	↓	↓Fat mass, ↑ muscle mass
↓	→	↓ Muscle mass, → fat mass
↓	↑	↓ Muscle mass, ↑ fat mass
↓	↓	↓ Fat mass, maybe ↓ muscle mass

→ = No change, ↑ = increase, ↓ = decrease.

the effects of training and also to identify skinfolds associated with optimal results.

Research using twins demonstrates that genetic factors strongly influence how rapidly and to what extent body fat changes in response to a given caloric imbalance (Bouchard et al. 1990). These data explain why some athletes struggle to lose body fat although others who are eating and training in the same manner quickly become lean. It may be beneficial for coaches and athletes to focus on behavior issues (i.e., diet and exercise) instead of outcome issues (i.e., skinfolds and body weight). Most coaches and sport scientists criticize the athlete with the highest skinfolds regardless of whether that person is eating and training well. In contrast, lean athletes who eat poorly and train inconsistently can leave skinfold tests pleased with their fitness status. Again, because the ultimate goal is for the skier to perform well, skinfold data should be interpreted in conjunction with a general knowledge of eating and training practices. Those who have monitored skinfolds regularly in a group of athletes will recognize that each athlete can have a different level of body fat when performing his or her best. For this reason it is useful to establish individual reference ranges for each athlete.

Measurements Required: With the skier in underwear, the following measurements are made:

- Height
- Body mass
- Skinfold thickness for biceps, triceps, subscapular, supraspinale, abdominal, thigh, and calf sites (see chapter 5 for descriptions of these sites)

Data should be appropriately identified and entered into a database that allows comparison of an individual's results with those from previous testing sessions.

Interpretation of Data: Anthropometric data collected from world-class cross-country skiers can be used to establish appropriate reference ranges. Although anthropometric data are not available for current internationally successful cross-country skiers, data have been collected from top Australian cross-country skiers during the competition phase of their domestic ski season. Table 16.3 presents the mean data collected from a group of 4 Australian cross-country skiers who all placed within the top 10 in the 1997 Kangaroo Hoppet ski marathon. Data, collected within two weeks of competition, illustrate a "typical" anthropometric profile of an elite Australian cross-country skier. Although these data

Table 16.3 Mean Anthropometric Data Collected From Four Elite Australian Cross-Country Skiers

Variable	Mean	SD
Age (yr)	28.3	0.9
Height (cm)	179.0	4.7
Mass (kg)	72.0	2.8
Biceps (mm)	2.6	0.2
Triceps (mm)	5.5	1.1
Subscapular (mm)	7.1	0.8
Iliac crest (mm)*	6.2	1.6
Supraspinale (mm)	3.8	0.4
Abdominal (mm)	7.7	2.7
Thigh (mm)	8.2	1.7
Calf (mm)	5.2	1.5
Sum of 8 sites (mm)	46.4	8.1
Sum of 7 sites (mm)	40.2	6.7

*Not included in sum of seven sites.

Data collected within two weeks of the 1997 Kangaroo Hoppet ski race (42 km). These skiers placed first, third, sixth, and ninth.

can be useful for evaluating results collected from developing athletes, it is important to tell younger athletes (especially women) that low skinfolds are the result of prolonged periods of consistent training in combination with a healthy diet. When monitoring skinfolds and body mass over time, it becomes clear that there are only a limited combination of changes in body mass and skinfolds. Table 16.2 provides a guide for interpreting changes in body mass and sum of skinfolds.

Aerobic Power

Rationale for Testing Maximal Oxygen Consumption: The maximum oxygen consumption ($\dot{V}O_2max$) values of elite cross-country skiers are some of the highest documented in humans (Bergh 1982). Bergh (1982) has reported an average $\dot{V}O_2max$ of 83 ml · kg^{-1} · min^{-1} for the Swedish national team, and one world-class skier recorded 94 ml · kg^{-1} · min^{-1}. Additionally, data collected from both successful and unsuccessful cross-country skiers at the international and national level have been published (Bergh 1987). Analysis of these data indicates that regardless of how $\dot{V}O_2max$ was expressed, world-class cross-country skiers possess significantly higher $\dot{V}O_2max$ scores than less successful competitors. This relationship was observed for both the men and the women. On the basis of these data, Dr. Bergh concluded that "there is . . . very little chance of winning gold medals in the Olympic Games or world championships with $\dot{V}O_2max$ more than a few per cent below 350 and 290 ml · kg^{-1} · min$^{-2/3}$ for male and female skiers, respectively" (330).

Swedish researcher Arthur Forsberg has also commented on the relationship between absolute oxygen consumption and performance in cross-country ski races. On the basis of laboratory measures of $\dot{V}O_2max$ and skiing performance in male cross-country skiers, Forsberg has concluded that Swedish team members who never progress beyond the national team have values between 5.0 and 5.3 L/min; Swedish team members qualifying for world championships or Olympics have values between 5.6 and 5.8 L/min; and Swedish skiers winning international medals consistently have values above 6.0 L/min.

Rationale for Assessing the Lactate Threshold: The observation that exercise intensity ranges between 80% and 90% $\dot{V}O_2max$ during races (Bergh 1982; Jette et al. 1976) indicates that cross-country skiers are competing at an oxygen consumption equal to, or in some cases just above, their "anaerobic threshold." It is therefore important that the "anaerobic threshold" occur at a rapid skiing velocity. Results of laboratory testing (treadmill walking in conjunction with arm ergometry) have shown that the onset of blood lactate accumulation (>4 mmol/L lactate) expressed as a percentage of $\dot{V}O_2max$ can be as high as 90% in top U.S. biathletes (Martin 1989). Published reports have indicated that the "anaerobic threshold" of elite endurance athletes expressed as a percentage of $\dot{V}O_2max$ rarely exceeds 90% (Kindermann et al. 1979). Thus, it is possible that an "anaerobic threshold" occurring at 90% of $\dot{V}O_2max$ reflects an athlete who is nearing an adaptive limit. It has been suggested that the absolute workload at which the "anaerobic threshold" occurs is more critical for success in endurance sports than is the relative relationship to $\dot{V}O_2max$ (Jacobs 1986).

An evaluation of Australian cross-country skiers by sport scientists at the AIS showed that the power output in watts (W), W/kg, and W · kg$^{-2/3}$ on an arm-leg ergometer at the "anaerobic threshold" (identified using ventilatory parameters) was the best predictor of skiing performance in both junior and senior male and female skiers (Gillam, routine testing of Australian skiers, unpublished, 1990). Readers should note that the large variability in fitness of these junior skiers increases the ability of laboratory parameters to predict performance.

Successful cross-country skiers typically engage in high training loads (600-1000 h/year) and incorporate a wide variety of training modalities. The threshold training heart rate can be used as a reference point allowing more precise quantification of relative training intensity. This provides coaches and athletes increased resolution of relative training intensity as they evaluate successful and nonsuccessful training programs.

Testing Modality: Cross-country skiing techniques place heavy demands on the upper body for forward propulsion. Biomechanists at Pennsylvania State University have estimated that 66% of the forward propulsive forces generated during VI skating (i.e., alternate skate) up an 8° slope are generated from the poles (Smith 1989). In a published abstract, Smith emphasized the importance of the upper body for cross-country skiing, stating, "The legs, while important for support of the body, do little to propel the skier uphill" (14-15). In Smith's study the forces applied by each pole were slightly greater than half of the skier's body weight. These data illustrate the importance of upper-body power and endurance for cross-country skiing and highlight the need for a test designed to determine the skier's upper-body $\dot{V}O_2max$ using ski-specific poling movements.

Maximal oxygen uptake in cross-country skiers should be determined using an incremental exercise test that includes both upper- and lower-body exercise (Rundell 1996). The principle of sport specificity suggests that $\dot{V}O_2max$ will be higher when the selected movement patterns for testing mimic those predominantly used during training. However, it is not clear whether an upper-body poling motion is required for cross-country skiers to achieve their highest oxygen uptake. Higher $\dot{V}O_2max$ scores have been recorded during treadmill running as compared to snow skiing (Niinimaa et al. 1979). Others have reported that the highest $\dot{V}O_2max$ values occur in trained skiers during skiing (Stromme et al. 1977) and treadmill walking (Hermansen 1973). A direct comparison between running and roller skiing during an incremental-intensity treadmill test showed a higher $\dot{V}O_2max$ during running (Rundell 1996). It is likely that the phase of training and the ski-specific fitness influence which testing modality results in the highest $\dot{V}O_2max$.

Test Procedure: A ski walking treadmill protocol developed to test cross-country skiers in Finland has been modified for testing Australian skiers. It requires athletes to walk on a treadmill using ski poles that have rubber stoppers covering the metal tips of the poles. Skiers should spend 5-15 min familiarizing themselves with ski walking before attempting the protocol. Men and women are tested using the same protocol (see table 16.4). The use of one protocol for testing all skiers has the advantage of allowing comparisons across age groups and gender.

1. After the skier is familiar with walking on the treadmill using ski poles, the following equipment is attached to the subject: a headset and Hans-Rudolph valve, a heart rate monitor, rubefacient cream for ear or fingers, and a noseclip.

2. The first stage lasts for 4 min and then stops for 45 s. Just prior to the rest, the heart rate is recorded.

3. During the rest period skiers are asked to rate their perception of effort using a Borg 6-20 scale, and a capillary blood sample is collected from a fingertip or earlobe.

4. The subsequent stages last for 3:15 (min:s).

5. Skiers are encouraged to walk for as long as possible using long strides. Eventually subjects may want to engage in hill-bounding as they attempt to achieve a maximal performance time.

6. Regardless of heart rate values, skiers should be encouraged to stay on the treadmill protocol for as long as possible. Once the skier can go no further, the treadmill is stopped; maximal heart rate and performance time are recorded, and the headset is removed.

7. Speed and grade decrease to 6% and 4 kph for a brief period (approximately 5 min) of active recovery.

8. After 1 min, ask the skier to rate his or her perceived exertion; at 2 min post-maximal effort, stop the treadmill and collect a final capillary blood sample.

9. The subject may want to walk during recovery for another 2-10 min.

Interpretation of Data: One of the most important variables tested is performance time. Assuming that treadmill speed and grade are calibrated correctly and that the protocol was administered accurately, the maximal time achieved should be a good indicator of current fitness. Because ski striding is most similar to the classic (diagonal) technique, treadmill times should be greatest after training blocks that emphasize hill striding, classic skiing, or roller skiing. Members of the 1997 Australian National Cross-Country Ski Team, tested at the end of the Australian ski season, recorded times between 31:30 and 34:00 min. In Canberra (600 m altitude), these performance times resulted in a $\dot{V}O_2max$ between 71 and 80 ml $\cdot$ kg^{-1} $\cdot$ min^{-1}. Figure 16.4 reflects the rate of oxygen uptake for each stage of the protocol for nine Australian cross-country skiers who had recently raced in the 1997 Kangaroo Hoppet ski marathon. Submaximal oxygen consumption expressed as ml $\cdot$ kg^{-1} $\cdot$ min^{-1} should not vary by more than 5 ml $\cdot$ kg^{-1} $\cdot$ min^{-1} above or below the values identified in figure 16.4. If values are outside this limit, it is possible that the treadmill or the gas analysis equipment is not calibrated properly.

Both heart rate and lactate can be plotted against the treadmill speed and grade (figure 16.5). Although there are many techniques for detecting a lactate threshold, the Dmax method has recently become popular because of the objective manner in which a point on the lactate curve can be reliably identified (Cheng et al. 1992; Zhou and Weston 1997). Using the peak heart rate, a series of four training zones can be identified: active recovery/ long slow distance, 65-75% HRmax; aerobic training, 75-85% HRmax; threshold training, 85-92% HRmax; and $\dot{V}O_2max$ training, 92-100% HRmax. As exercise intensity increases, a greater percentage of slow-twitch fibers are recruited. When the skier

Table 16.4 Australian Ski-Striding Treadmill Protocol for Assessment of the Lactate Threshold and Maximal Oxygen Consumption

Stage (#)	Time (min)	Grade (%)	Speed (km/h)	Estimated $\dot{V}O_2$ ($ml \cdot kg^{-1} \cdot min^{-1}$)
	Entry stage for juniors	04.0	05.5	
1	00:00–04:00	06.0	06.0	26.2
	04:01–04:45	06.0	00.0	
2	04:46–08:00	08.0	06.5	31.6
	08:01–08:45	08.0	00.0	
3	08:46–12:00	10.0	07.0	39.8
	12:01–12:45	10.0	00.0	
4	12:46–16:00	12.0	07.5	48.8
	16:01–16:45	12.0	00.0	
5	16:46–20:00	14.0	08.0	58.8
	20:01–20:45	14.0	00.0	
6	20:46–24:00	16.0	08.5	66.3
	24:01–24:45	16.0	00.0	
7	24:46–28:00	18.0	09.0	70.4
	28:01–28:45	18.0	00.0	
8	28:46–32:00	20.0	09.5	
	32:01–32:45	20.0	00.0	
9	32:46–36:00	22.0	10.0	
	36:01–36:45	22.0	00.0	
10	36:46–40:00	22.0	11.0	
	40:01–40:45	22.0	00.0	
11	40:46–44:00	22.0	12.0	

Current record is 34:00 min (September 1997).

Each stage is 3:15 min and is followed by a 45 s break during which blood samples are collected. Subsequent stages increase by 2% grade and 0.5 km/h.

begins to exceed a threshold training intensity, some fast-twitch fibers are called into action. The change in muscle fiber recruitment is also associated with a shift from oxygen-dependent (aerobic) to oxygen-independent (anaerobic) energy-yielding systems. An "active recovery/long slow distance intensity" recruits primarily slow-twitch muscle fibers and is fueled by the aerobic breakdown of fats. Heart rates are low, and blood lactate is typically less than 2 mmol/L.

During the "aerobic training intensity," more slow-twitch muscle fibers are recruited, and a shift to aerobic metabolism of carbohydrates occurs. The exercise session can last between 1 and 3 h, and blood lactate is usually between 2 and 4 mmol/L. Threshold training promotes recruitment of fast oxidative fibers and begins to incorporate oxygen-independent (anaerobic) metabolism of carbohydrate. Blood lactate accumulates to between 4-6 mmol/L, and exercise sessions tend to be less than 30-40 min. $\dot{V}O_2max$ training intensity uses some fast-twitch muscle fibers and relies heavily upon the oxygen-independent breakdown of carbohydrate, which results in the rapid accumulation of lactic acid (>6.0 mmol/L). $\dot{V}O_2max$ intervals may last 5-10 min, at the end of which the heart rate is within 5 beats/min of maximal. These training zones are sometimes referred to as E1, E2, E3, and E4. One can use training zones to ensure that athletes are training at the desired relative intensities and also to more accurately quantify the specific training that has been completed.

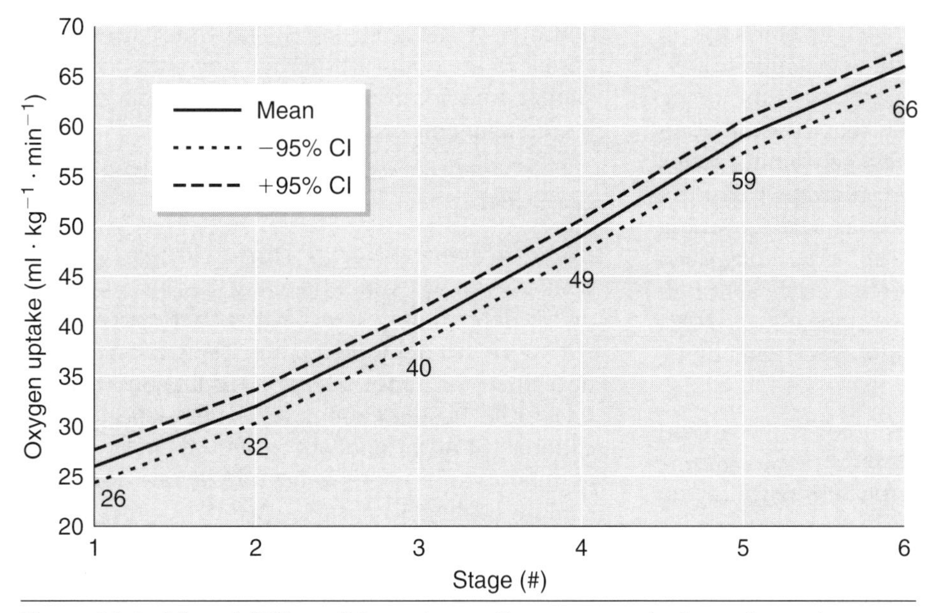

Figure 16.4 Mean (±95% confidence interval) oxygen uptake from nine male cross-country skiers during the Australian ski-striding protocol. (Data were collected in Canberra, Australia (600 m altitude), under simulated sea level conditions.)

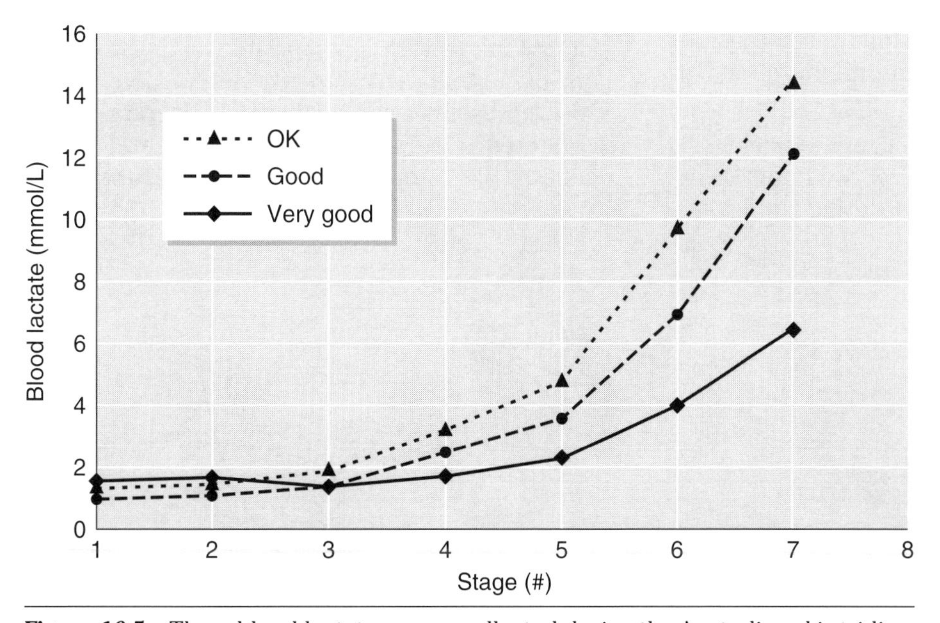

Figure 16.5 Three blood lactate curves collected during the Australian ski-striding protocol. (Each curve represents an Australian cross-country skier tested one month after the Australian ski season. One skier is relatively new to the sport [OK]; one skier can place top 10 in some domestic races [Good]; and one skier has won multiple national titles in skiing [Very good]. Data were collected in Canberra, Australia [600 m altitude], under simulated sea level conditions.)

Upper-Body Fitness Tests

Researchers have compared a skier's $\dot{V}O_2max$ as determined using just the arms and as determined using arms and legs together (Sharkey 1984; Sharkey and Heidel 1981). Elite cross-country skiers are able to achieve a high upper-body $\dot{V}O_2max$ expressed as a percentage of the treadmill running $\dot{V}O_2max$ (Millerhagen et al. 1983). Upper-body $\dot{V}O_2max$ expressed as a percentage of treadmill $\dot{V}O_2max$ ranged from a low of 60% in an untrained female to as high as 85% in the elite male cross-country skiers. Other tests of upper-body endurance have incorporated short time-trial efforts lasting between 1 and 9 min. During this time the skier uses the double-pole technique, and the total work accumulated on a specialized upper-body ergometer is recorded. Results have shown a significant correlation between "short-duration double-poling power" and ski performance in top U.S. female biathletes (Rundell and Bacharach 1995). In the same study, the researchers reported that a 1 km uphill double-pole time-trial on snow could significantly predict the overall U.S. biathlon rankings. Clearly, upper-body fitness is important for both the biathlete and the cross-country skier.

Field Lactate Threshold Test

Rationale: Roller skiing and ski striding with poles on a treadmill produce movement patterns similar to those observed during actual skiing or roller skiing. However, the physiological response to simulated skiing in a laboratory is not always similar to that observed in the field (e.g., heart rates for a given oxygen consumption tend to be higher). With the advent of portable heart rate and lactate monitors and "preserving cocktails" that allow storage of whole blood for up to three days at 4-10° C, it is now possible to perform maximal incremental-intensity tests in the field. The general assumption is that the evaluation of fitness and the establishment of training zones from field testing will be more relevant when the exercise modality closely matches actual training and racing modalities. Whereas field testing does increase the specificity of exercise movement patterns, one limitation is that environmental variables such as temperature and wind speed cannot be strictly controlled.

Test Procedure: This protocol is for a field maximal incremental-intensity test in current use by the Australian cross-country and biathlon ski teams.

1. A section of gradual uphill terrain is identified, and a 1.25-1.50 km distance is measured.

2. Skiers complete five repeat intervals of increasing intensity over the prescribed distance.

3. Using a Borg 6-20 scale, skiers are told to perform the five intervals at the following intensities: easy, moderately easy, moderate, hard, and extremely hard.

4. To help achieve the appropriate pacing, skiers are told to perform the first two intervals at a pace similar to that selected during a long, slow distance-training session. The third interval is approximately marathon race pace; the fourth is similar to a 5 km race pace; and the last is completed using maximal intensity.

5. A heart rate-lactate curve is constructed for each skier.

Data Interpretation: Threshold heart rates in the field can be estimated using the Dmax method or by determining heart rates that correspond with 3-4 mmol/L blood lactate. Using the methods described earlier for interpreting data on $\dot{V}O_2max$, heart rate training zones for roller skiing can be calculated. Additionally, the lactate curve can be compared to previous heart rate-lactate curves to evaluate changes in fitness. A shift downward and rightward in the lactate curve indicates that for a given heart rate there is less lactate accumulation. Examining the heart rate and lactate response to skiing speed can also enhance the evaluation of fitness. However, these comparisons may be influenced by weather conditions and type of roller skis. Figure 16.6 illustrates a typical heart rate-lactate curve with appropriate training zones in one of Australia's top cross-country skiers. Figure 16.7 represents a series of velocity-lactate curves collected during a 1997 developmental cross-country ski camp held in Jindabyne, New South Wales.

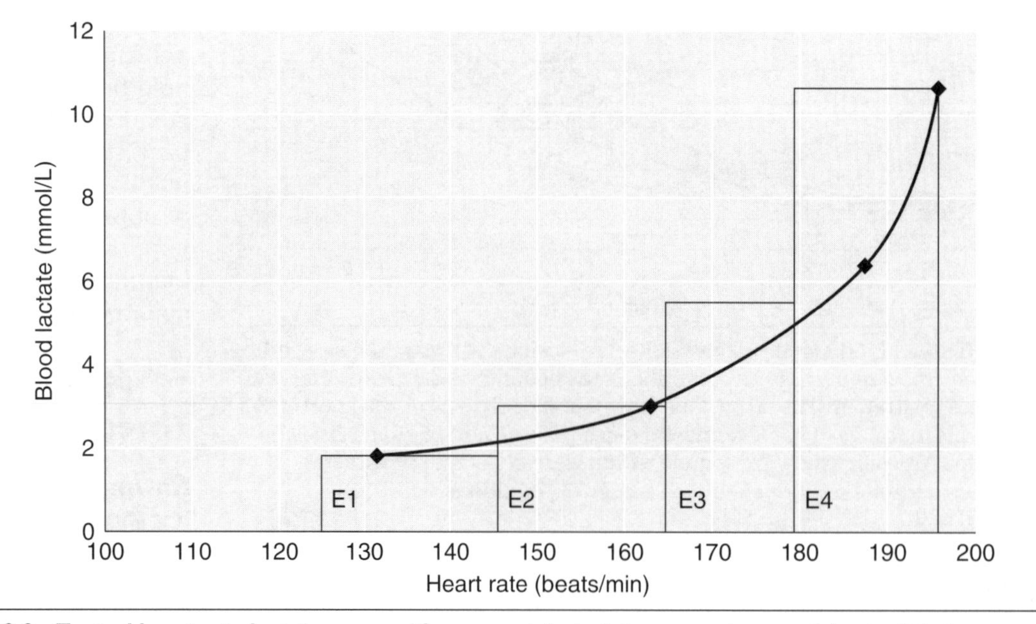

Figure 16.6 Typical heart rate-lactate curve with appropriate training zones in one of Australia's top cross-country skiers.

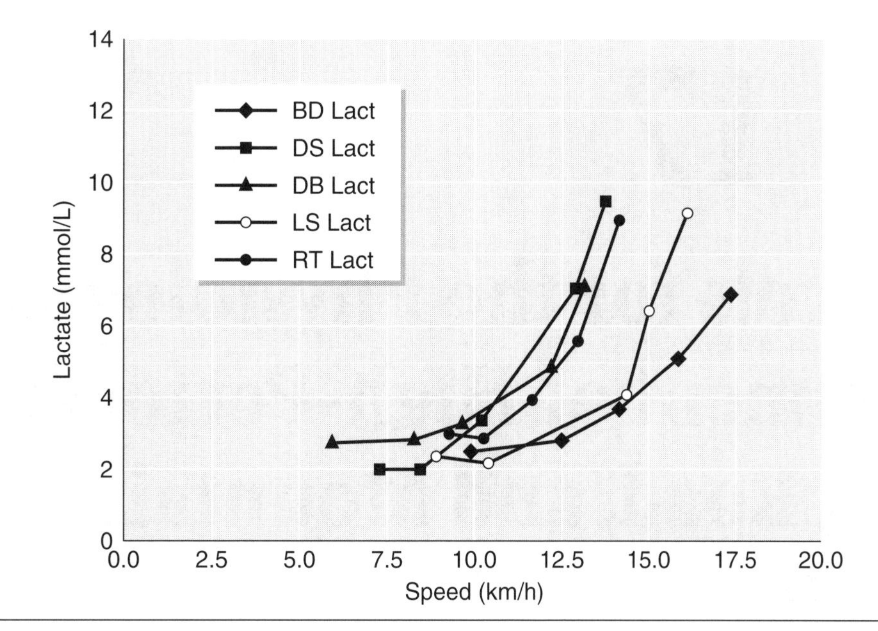

Figure 16.7 Velocity-lactate curves collected from five male cross-country skiers during a field "step test."

chapter 17

Protocols for the Physiological Assessment of High-Performance Track, Road, and Mountain Cyclists

Neil Craig, Charlie Walsh, David T. Martin, Sarah Woolford, Pitre Bourdon, Tom Stanef, Peter Barnes, and Bernard Savage

The sports of competitive track and road cycling are physiologically demanding. In the majority of track events the cyclist must maximally tax both the aerobic and anaerobic metabolic pathways, while the majority of road events are completed at submaximal-intensity levels. Although changes in riding position and cycling equipment can result in substantial increases in cycling performance, the greatest increases are achieved by changing physiological parameters, that is by training. Thus the engine or the physiological makeup of the cyclist plays a major role in any successful cycling performance, and coaches have used physiology testing for the following purposes:

- To monitor and assess the effectiveness of their training program and examine alternative preparation techniques
- To prescribe and implement varying training intensities
- To construct an event-specific physiological profile
- To construct a cyclist's individual profile of the key physiological indexes contributing to track and road cycling success
- To augment information that is already available (e.g., track and road performances, field observations) when making decisions on team selection
- To aid in talent identification
- To predict cycling performance
- To help detect and/or confirm any acute or chronic overtraining syndrome

The following protocols have been documented with the aim of developing greater uniformity in the physiological assessment of high-performance track and road cyclists. One problem in any attempt to standardize test procedures will be a disagreement among sport scientists as to what the standard method should be. Notwithstanding this, recent discussions between cycling coaches and sport scientists in Australia resulted in agreement with respect to the following protocols and assessment conditions there.

Cycle Ergometers

Without doubt in high-performance track and road cycling, the cyclist adapts specifically to the physiological, biomechanical, and psychological stresses imposed by training and competition. Therefore, performance on any test will be a function of physiological components, neuromuscular skill, and motivation. For information on calibrating cycle ergometers, see chapter 8.

Types of Ergometers

While both the physiology and the skill component argue for specificity, the psychological component will be higher when the test ergometer is similar to the bicycle used in training and competition. These considerations have led to the construction of three specific cycle ergometers for assessing elite track and road cyclists. Each ergometer has a fully adjustable seat pillar with quick-release seat clamps and is fitted with a Turbo racing saddle. A magnetic sensor is mounted adjacent to a 60-tooth steel sprocket allowing use of the Repco Ex50 Exertech work monitor unit. Handlebars are of the pursuit style. A Polar Cyclovantage (Polar Electro Oy, model no. 45910) is mounted on the handlebar in a position easily viewed by the cyclist. Crank lengths are 170 mm; cyclists provide their own pedals.

- **Track ergometer.** This is an ergometer designed specifically for assessing cyclists involved in track training and competition events. It is an air-braked cycle ergometer with a single 48-tooth chain ring. The intermediate drive consists of a 14-tooth sprocket and a 40-tooth chain ring. The cluster on the flywheel consists of 23-, 21-, 19-, 17-, 15-, 14-, and 13-tooth sprockets. The gearing ratio is readily selected by means of a control lever connected to a conventional derailleur.
- **Road ergometer (1).** This is an ergometer designed specifically for assessing cyclists involved in road training and competition events. It is an air-braked cycle ergometer with a 48-tooth chain ring on the crank to a 14-tooth sprocket on the intermediate drive. The intermediate drive also consists of a double chain ring (44/35 tooth). The cluster on the flywheel consists of 23-, 22-, 21-, 20-, and 19-tooth sprockets. The sprocket and chain ring are readily selected by means of control levers connected to a conventional derailleur.
- **Road ergometer (2).** This road ergometer was designed specifically for assessing junior road cyclists (e.g., road-team time trials) involved in road training and competition events. It is also used for assessing track cyclists when their road training involves large volumes (km) completed at high cadences (120-130 rpm) and when there is a need to develop a cadence-specific $\dot{V}O_2$ versus power output regression line for assessing the maximal accumulated oxygen deficit. It is an air-braked cycle ergometer with a 48-tooth chain ring on the crank to a 20-tooth sprocket on the intermediate drive. The intermediate drive also consists of a double chain ring (44/35 tooth). The cluster on the flywheel consists of 23-, 22-, 21-, 20-, 19-, and 18-tooth sprockets. The sprocket and chain ring are readily selected by means of control levers connected to a conventional derailleur.
- **Lode Excalibur Sport.** The Lode Excalibur Sport (Lode BV, Groningen, The Netherlands) is an electromagnetically braked cycle ergometer that can provide a constant power output through a range of cadences (80-120 rpm; hyperbolic mode) or a power output that changes in proportion to cadence (linear mode). This ergometer has the advantage of allowing cyclists to self-select their preferred cadence during maximal oxygen uptake tests using an incremental power output protocol. Testing protocols can be programmed using the Work Load Programmer so that cyclists experience a resistance that is precise and reproducible over a series of testing sessions. The ergometer is constructed to allow the use of racing pedals, aerobars, and drop handlebars and a racing saddle. Additionally, adjustments of the seat height, seat forward-backward position, and handlebar position allow for a wide range of cyclists to simulate their normal riding position. A computer interface and software developed at the Australian Institute of Sport enable cadence, power output, and work to be recorded every second during ergometer tests.

Bicycle Tool Kit

The following bicycle tools will be useful for setting riding positions and performing general maintenance procedures:

- 5 and 6 mm allen keys
- 15 mm pedal spanner
- Cloth/Metal measuring tape
- Chain breaker
- Chain lubricant
- Screwdriver
- Moly-based grease for head stem and seat pillar
- Spare seat bolts

Laboratory Environment and Subject Preparation

Guidelines for preparation for testing, in addition to those presented in chapter 2, are as follows:

- **Yearly testing program.** For any structured yearly training program, the coach and sport scientist should decide and plan in advance all testing dates. This will help replicate all testing conditions.

• **Health status.** Where possible, the cyclist should be tested under consistent nutritional status and health condition.

• **Testing.** Where possible, time of day and order of testing should be standardized.

• **Training.** Where possible, type of training sessions within 24 h of testing should be standardized. These training sessions should be of a light, recovery nature (less than 50 km at less than 75% maximal heart rate).

• **Familiarity.** Familiarity is particularly important with any open-ended performance test that will include a pacing strategy.

• **Equipment.** The gearing ratios should be exactly as described above for each air-braked ergometer. For calibration guidelines, refer to chapter 8.

• **Laboratory environment.** The laboratory setting should be controlled for conditions such as observers and noise. An electric fan is useful to help cool cyclists completing protracted tests. The coach should be encouraged to be present at all testing sessions.

• **Test procedures.** The following procedures are common for all tests:

a. Record cyclist's information on an appropriate data sheet (see "Test Data Sheet" as an example).
b. Calibrate all equipment prior to the testing session, and record calibration results on data sheet (see "Calibration Data Sheet").
c. Have the cyclist set up the ergometer (seat height, head stem). Record setup position (head stem, seat pillar) for future reference.
d. Place the Repco Ex50 Exertech work monitor unit where the cyclist can easily see it. Place the electric fan in a position acceptable to the cyclist.
e. Fully explain the test procedure and have the cyclist complete the appropriate warm-up.
f. If micro blood samples are required (e.g., for blood lactate measurement), place a small quantity of Finalgon on one of the cyclist's fingers to ensure that the capillary blood is arterialized prior to sampling. The Finalgon should be in place for at least 10 min and then wiped clean.

Equipment Checklist

1. Anthropometry
   - ☐ Stadiometer mounted on wall
   - ☐ Electronic platform scale accurate to ±0.05 kg
   - ☐ Harpenden skinfold calipers
   - ☐ Marking pen
   - ☐ Anthropometric tape (Lufkin, model W606PM)
   - ☐ Bone calipers
   - ☐ Anthropometer
2. Aerobic power ($\dot{V}O_2$max) and maximal heart rate (HRmax)
   - ☐ Modified air-braked track or road cycle ergometer
   - ☐ Repco Ex50 Exertech work monitor
   - ☐ Gas collection and analysis system according to recommendation
   - ☐ Heart rate and cadence monitor
   - ☐ Electric fan
   - ☐ Pulse oximeter
   - ☐ Blood lactate analyzer
   - ☐ Supplies for preparation of tray for fingertip blood sampling:
     - __ Autolets, lancets, platforms
     - __ Sterile alcohol swabs
     - __ Tissues
     - __ 50 μL tubes (appropriately labeled and prepared)
     - __ Hazard bags
     - __ Sharps container
3. Blood lactate transition threshold test
   - ☐ Modified air-braked track or road cycle ergometer
   - ☐ Repco Ex50 Exertech work monitor
   - ☐ Gas collection and analysis system according to recommendation
   - ☐ Heart rate and cadence monitor
   - ☐ Pulse oximeter
   - ☐ Blood lactate analyzer
   - ☐ Blood gas and acid-base analyzer
   - ☐ Supplies for preparation of tray for fingertip blood sampling:
     - __ Autolets, lancets, platforms
     - __ Sterile alcohol swabs

Test Data Sheet

Name: ______________________________ Date/time: ________ /________

Age : ______ yr Date of birth: __________ Height: ______ cm Weight: ______ kg BSA: ______ m^2

FSA: ______ m^2 Ergometer: ______________ Sport event: ______________

Temperature: ______ °C pH_2O: __________ Relative humidity: ________ % $\dot{V}O_2max$: ________

Barometric pressure: ______________ mm Hg Lactate threshold: ______________

Anthropometry: ______

Tricep	______	______	______ mm
Subscapular	______	______	______ mm
Iliac crest	______	______	______ mm
Bicep	______	______	______ mm
Front thigh	______	______	______ mm
Medial calf	______	______	______ mm
Pectoral	______	______	______ mm
Abdominal	______	______	______ mm
Supraspinale	______	______	______ mm
Midaxillary	______	______	______ mm

Sum 6 ______ mm

Other tests:

Sum 7 ______ mm

Flexibility ______ cm

Sum 8 ______ mm

Vertical jump ______cm

Body fat ______ %

Previous exercise:

(1) Before test ______________

(2) Day before ______________

Medication ______________

Other information:

Diet

(1) Breakfast ______________

(2) Lunch ______________

Calibration Data Sheet

Lactate:

La– standard: ______ mmol/L

Measured ______

Oxygen Consumption:

	Pre	Post
Ventilometer	______	______
O_2	______	______
CO_2	______	______

$\dot{V}O_2$max criteria:

HR ______

$\dot{V}_E/\dot{V}O_2$ ______

R ______

Hla– ______

Protocol:

Time (min)	Pre	Exercise duration	Recovery
Parameters	0	__ __ __ __ __ __ __ __ __ __ __ __ __	__ __ __ __ __ __
Heart rate (bpm)	__	__ __ __ __ __ __ __ __ __ __ __ __ __	__ __ __ __ __ __
Power output (W)	__	__ __ __ __ __ __ __ __ __ __ __ __ __	__ __ __ __ __ __
Power (W/FSA)	__	__ __ __ __ __ __ __ __ __ __ __ __ __	__ __ __ __ __ __
rpm	__	__ __ __ __ __ __ __ __ __ __ __ __ __	__ __ __ __ __ __
Running speed (km/h)	__	__ __ __ __ __ __ __ __ __ __ __ __ __	__ __ __ __ __ __
Gradient (%)	__	__ __ __ __ __ __ __ __ __ __ __ __ __	__ __ __ __ __ __
% Saturation	__	__ __ __ __ __ __ __ __ __ __ __ __ __	__ __ __ __ __ __
pH	__	__ __ __ __ __ __ __ __ __ __ __ __ __	__ __ __ __ __ __
HCO_3^- (mmol/L)	__	__ __ __ __ __ __ __ __ __ __ __ __ __	__ __ __ __ __ __
Blood lactate (mmol/L)	__	__ __ __ __ __ __ __ __ __ __ __ __ __	__ __ __ __ __ __
Log (La–) mmol/L	__	__ __ __ __ __ __ __ __ __ __ __ __ __	__ __ __ __ __ __
$\dot{V}O_2$ (L/min)	__	__ __ __ __ __ __ __ __ __ __ __ __ __	Comments:
$\dot{V}O_2$ ($ml \cdot kg^{-1} \cdot min^{-1}$)	__	__ __ __ __ __ __ __ __ __ __ __ __ __	
Log $\dot{V}O_2$ (L/min)	__	__ __ __ __ __ __ __ __ __ __ __ __ __	
% $\dot{V}O_2$max	__	__ __ __ __ __ __ __ __ __ __ __ __ __	

- _ Tissues
- _ 50 to 125 μL tubes (appropriately labeled and prepared)
- _ Hazard bags
- _ Sharps container

4. Alactic anaerobic power and capacity test
   - ☐ Modified air-braked track cycle ergometer
   - ☐ Repco Ex50 Exertech work monitor and data collection software
   - ☐ Cadence monitor

Note: The data collection software was developed and written at the South Australian Sports Institute (see appendix for details).

PROTOCOLS: Cycling

Anthropometry

Table 17.1 provides normative anthropometric data for male and female track and road cyclists. These are specific to the corresponding training phase. Tables 17.2 and 17.3 illustrate the large degree of seasonal variation that can occur with skinfold measurements, both within an individual and within a training squad. Technical error of measurement data are also provided (see page 83 for details about the calculation procedures for this statistic).

Rationale: Additional nonfunctional weight, of either the rider or the bike, has a quadruple effect in slowing a bicycle down. First, because added weight increases the inertia, it will slow the rate of acceleration. In track cycling, where the acceleration phase is a large part of the total event time, acceleration is a highly important factor. Therefore, although a small amount of added weight does not appreciably change top speed, it will lengthen the time to reach this speed. Secondly, added weight slows speed in hill climbing. Lifting extra weight over a hill takes energy that is never recovered on the descent (Kyle 1991). Therefore, any benefit of added weight on downhill courses is far outweighed by the detriment of the same added weight on uphill courses—a consideration pertinent to out-and-back courses. Thirdly, weight adds to the rolling resistance of tires, with additional weight causing added deformation in the tread and sidewalls. Finally, increased fat mass has a significant impact on the cyclist's frontal surface area, which is an important consideration for cycling at high velocities.

From a practical point of view, Olds et al. (1993, 1995b) estimated that an increased body fat mass of 2 kg would increase a 4000 m individual-pursuit cycling performance time by about 1.5 s (20 m) and a 40 km individual time trial by about 15 s (180 m). From this estimation it is obvious that under any circumstances, nonfunctional weight is detrimental to competitive cycling performance—hence the importance of routinely monitoring the cyclist's body composition. In relation to sprint events, McLean and Ellis (1992) have demonstrated that body mass is a good predictor of peak power and capacity in an all-out 15 s cycle ergometer test. This simple anthropometric measure and its strong correlation with cycling power events have implications for the process of talent identification.

Measurements Required: Age, height, body mass, and skinfolds are required (refer to chapter 5). Those coaches or cyclists wishing to experiment with the calculation of relative body fat can find a summary of body density prediction equations by Norton (1996) in *Anthropometrica*. Be sure to select a population-specific regression equation.

Wind resistance is the primary resistance to motion faced by the cyclist and is determined by the cyclist's projected frontal surface area (FSA), velocity, equipment, clothing, and environmental conditions. In cycling, many physiological parameters are normalized to FSA. The FSA is calculated according to the equation of McLean (1993):

$$\text{Frontal surface area (m}^2) = 0.00215 \times \text{Wt (kg)} + 0.18964 \times \text{Ht (m)} - 0.07961$$

where:

Wt = body mass in kilograms
Ht = standing stretch height in meters

Limitations: For discussion of the limitations of these measurements, refer to chapter 5.

Aerobic Power ($\dot{V}O_2$max) and HRmax

Rationale: With use of aerodynamic equipment, streamlined body positioning, and specialized training techniques, average speeds of 50 km/h or

Table 17.1 Anthropometric Data for Male and Female High-Performance Cyclists

Group	n	Age (yr)	Height (cm)	Mass (kg)	Σ6 (mm)	Σ7 or Σ8 (mm)	% BF	Training phase
Aust. senior male track endurance	11	21.5 2.3	178.1 4.3	74.40 4.81	45.4 8.8	57.9 12.1	8.4 1.6	After 12 weeks of aerobic training
Aust. senior male track endurance	10	20.8 3.1	179.5 2.9	73.30 4.90	43.8 9.4	54.8 13.0	8.0 1.5	1 week before national championships
Aust. senior male track endurance	9	20.1 3.4	178.3 4.4	74.70 7.50	38.7 8.0	n/a n/a	7.4 1.1	2 days before world championships
Aust. senior male track sprint	5	19.7 2.7	176.0 3.1	76.15 5.85	46.3 12.3	60.2 17.1	8.7 1.8	General preparation
Aust. senior male road squad	14	21.2 1.7	181.7 6.8	73.85 7.60	48.0 9.0	59.4 12.0	8.7 1.6	General preparation
Aust. junior male road squad	13	17.2 0.7	176.2 7.9	66.50 7.30	42.9 5.3	53.3 6.3	8.0 0.9	General preparation
State senior male road squad	4	19.7 1.3	177.7 4.0	74.45 8.30	49.6 6.8	62.7 9.5	8.9 1.0	1 week before national championships
State junior male road squad	4	16.8 0.3	178.0 8.5	65.75 4.30	43.2 3.0	55.1 2.7	7.7 0.3	1 week before national championships
Aust. senior female road squad	8	24.0 3.1	168.2 6.4	57.55 4.7	58.0 6.2	65.4 6.9	15.7 6.2	General preparation
Aust. senior female olympic road squad	8	26.8 2.8	163.4 6.9	50.95 3.65	43.0 7.6	48.4 8.8	11.5 2.5	End of intensive and extensive aerobic road training phase (16 weeks)
State senior female road squad	6	27.2 4.4	165.2 4.5	58.65 2.8	68.7 10.5	73.8 9.7	16.9 1.3	1 week before national championships
Talent identification male	11	15.3 0.8	172.9 6.8	60.50 6.7	39.7 5.7	47.0 6.3	n/a n/a	Selection camp
Talent Identification female	9	14.9 1.0	156.5 5.7	46.65 5.45	57.6 14.0	60.2 12.2	n/a n/a	Selection camp

TEM = technical error of measurement (chapter 5); BF = body fat; ICC = interclass correlation coefficient; HR = heart rate; FSA = frontal surface area (p. 263).

1. Mean value listed above and standard deviation below.
2. Σ6 skinfolds males and females = triceps + subscapular + iliac crest + biceps + front thigh + medial calf
   Males: TEM = 0.4 mm %TEM = 0.90% ICC = 0.999 (Harpenden caliper)
   Females: TEM = 0.7 mm %TEM = 0.90% ICC = 0.999 (Harpenden caliper)
3. Σ7 skinfolds females = triceps + subscapular + supraspinale + biceps + abdominal + front thigh + medial calf
   TEM = 0.7 mm %TEM = 0.80% ICC = 0.999 (Harpenden caliper)
4. Σ8 skinfolds male = triceps + subscapular + supraspinale + biceps + abdominal + front thigh + medial calf + mid-axilla
   TEM = 0.6 mm %TEM = 1.10% ICC = 0.998 (Harpenden caliper)
5. %BF females: TEM = 0.22% %TEM = 1.24
   %BF males: TEM = 0.08% %TEM = 0.96
6. %BF males calculated by regression equation of Withers et al. (1987a).
7. %BF females calculated by regression equation of Withers et al. (1987b).

Table 17.2 Seasonal Variation in Skinfold Measurements of a High-Performance (Olympic Gold Medallist) Track Endurance Cyclist

Date	Σ6 (mm)	Training phase
11/16/87	54.8	End of 4-week transition/recovery phase
11/25/87	48.5	
12/2/87	45.8	2 weeks of low-intensity volume training
12/9/87	44.0	
1/13/88	38.7	
1/27/88	32.9	
2/2/88	34.7	3 months of intensive and extensive aerobic training
2/17/88	32.7	
3/2/88	30.6	
3/16/88	30.3	High-intensity aerobic and anaerobic track training
5/18/88	34.7	
6/2/88	33.7	Major competition (world championships)

1. Σ6 skinfolds = triceps + subscapular + biceps + iliac crest + front thigh + medial calf.
2. TEM = 0.4 mm or 0.90%.
3. ICC = 0.999 (Harpenden caliper).

Table 17.3 Seasonal Variation in Skinfold Measurements for the Australian Female Olympic Road Cycling Squad

Date	Mass (kg)	Σ7 (mm)	% BF
January 1992	52.40	56.7	13.4
February 1992	51.70	53.8	12.8
April 1992	51.95	48.4	11.5

1. n = 7.
2. Σ7 skinfolds = triceps + subscapular + supraspinale + biceps + abdominal + front thigh + medial calf.
3. TEM = 0.7 mm or 0.80%, ICC = 0.999 (Harpenden caliper).

greater are now being achieved in track endurance events. Cycling at such speeds would require a steady-state $\dot{V}O_2$ ranging from 90 to 100 ml · kg^{-1} · min^{-1}. Assuming that the track cyclist has a $\dot{V}O_2$max of 76 ml · kg^{-1} · min^{-1}, this would mean that he or she would be operating at approximately 120-130% $\dot{V}O_2$max with a concomitantly large contribution from the anaerobic metabolic pathways. Thus, a high $\dot{V}O_2$max, together with the ability to achieve it quickly and maintain it, would enable a large, rapid, and sustained aerobic energy release and would reduce premature reliance upon a large proportion of the finite oxygen deficit. In view of this, it is not surprising that $\dot{V}O_2$max and its associated indexes are significantly correlated with track cycling performance (Craig et al. 1993).

From a practical point of view, Olds et al. (1993) predict that a 15% improvement in $\dot{V}O_2$max (5.14-5.91 L/min) would enable the track cyclist to complete a 4000 m individual pursuit approximately 15.5 s faster. In addition, $\dot{V}O_2$max is one of the main physiological parameters contributing to road cycling performance (Olds et al. 1995a, 1995b). In modeling road cycling performance, Olds et al. (1995b) predicted that a 20% change in $\dot{V}O_2$max (e.g., 4.53 ± 0.91 L/min) would result in a change in predicted time of approximately ±7-10% over a 26 km individual time-trial course.

Test Procedure:

1. Complete common test procedures as outlined earlier in this chapter, and connect the cyclist to all measuring equipment (e.g., mouthpiece, heart rate monitor, pulse oximeter).

2. Have the cyclist complete the warm-up outlined in table 17.4 while checking final calibration.

3. The cyclist is required to pedal at a constant workload for 1 min work durations. The test is continuous, with cyclists stopping only on a full-minute or 30 s interval when they can no longer maintain the required workload. Once cyclists commit to a 60 or 30 s interval, they must complete the interval even though they may be unable to sustain the required workload.

4. Consult table 17.4 for information regarding type of ergometer, starting workloads and ramp increases, cadence range, and sprocket selections. If the tester wishes to assess ventilatory equivalents for estimation of a ventilatory threshold, the starting workload may need to be lower for both females and males.

Data Collection:

- Record age, height, mass, and calculated FSA.
- Record $\dot{V}O_2$, $\dot{V}CO_2$, $\%SaO_2$, respiratory exchange ratio (RER), $\dot{V}_E$, $\dot{V}_E/\dot{V}O_2$, $\dot{V}_E/\dot{V}CO_2$, and power output for each 30 s of the test.
- The $\dot{V}O_2$max is the average of the highest two consecutive readings.
- The respiratory valve mouthpiece and noseclip must be worn throughout the whole test.
- Record heart rate during the last 10 s of each minute and/or last 30 s of each work interval.
- If peak blood lactate is required, blood samples should be taken at completion of the test and then at 1, 2, 5, and 7 min postexercise. The cyclist has stopped pedaling and remains in a seated position; no warm-down is allowed for the first 7 min postexercise.

Data Analysis and Practical Application: $\dot{V}O_2$max should be presented in the following units: L/min, $ml \cdot kg^{-1} \cdot min^{-1}$, and L/FSA.

Recently, several studies have indicated the importance of maximal muscle power as a predictor of cycling performance (Craig et al. 1993; Hawley and Noakes 1992). Craig et al. (1993) reported a significant correlation ($r = -0.79$) between the power output at $\dot{V}O_2$max and 4000 m individual-pursuit cycling time, while Hawley and Noakes (1992) found that peak power during progressive exercise to exhaustion explained 82% of the variability in cycle time over 20 km. Hence, when assessing cycling performance, one should not interpret the $\dot{V}O_2$max without reference to the power output corresponding to $\dot{V}O_2$max. This power output should always be reported in conjunction with $\dot{V}O_2$max. In addition, this power output can be utilized for prescribing $\dot{V}O_2$max aerobic intervals on the cycle ergometer.

Having recorded HRmax, one can calculate the following heart rate training zones:

- Endurance 1 (E1): <75% HRmax
- Endurance 2 (E2): 75-85% HRmax
- Endurance 3 (E3): 85-92% HRmax
- Endurance 4 (E4): >92% HRmax

Table 17.4 Protocol Selection for Assessing the $\dot{V}O_2$max of Male and Female Road and Track Cyclists

	Ergometer (type and no.)	Warm-up	Starting workload and ramp increase (W)	Cadence range selection (rpm)	Sprocket (tooth no.)
Road					
Male senior	Road (1)	5 min at 150 W	175 with 25 increase	90-100	As required
Male junior	Road (1)	5 min at 125 W	150 with 25 increase	90-105	As required
Female senior	Road (1)	5 min at 75 W	100 with 25 increase	80-90	As required
Female junior	Road (1)	5 min at 50 W	75 with 25 increase	75-85	As required
Track					
Male senior	Track	5 min at 150 W	175 with 25 increase	n/a	23
Male junior	Track	5 min at 125 W	150 with 25 increase	n/a	23
Female senior	Track	5 min at 75 W	100 with 25 increase	n/a	23
Female junior	Track	5 min at 50 W	75 with 25 increase	n/a	23

The protocols suggested here are designed to assess the high-performance cyclist. Starting workloads, workload increments, and cadence ranges may not be applicable to the sub-state-level cyclist.

Note: Although these estimated heart rate training zones lack the precision of those derived from a blood lactate transition threshold test, they can provide a valuable training control for the cyclist undertaking only a $\dot{V}O_2max$ test.

Limitations: The cyclist must be fully cooperative and motivated to cycle to exhaustion. When testing young cyclists (e.g., in talent identification programs), be aware that their possible poor attention span and motivation may decrease the likelihood of achieving maximal data. In addition, type of equipment (e.g., mouthpiece size) and protocol may need to be modified.

Normative Data: Table 17.5 provides normative aerobic power/power output (PO) data for male and female track and road cyclists. These are specific to the corresponding training phases. Table 17.6 illustrates the seasonal variation that can occur with differing training and competition phases. Technical error of measurement data are

Table 17.5 Aerobic Power Data for Female and Male High-Performance Cyclists

Group	n	Mass (kg)	$\dot{V}O_2max$ (L/min)	$\dot{V}O_2max$ ($ml \cdot kg^{-1} \cdot min^{-1}$)	$\dot{V}O_2max$ (L/FSA)	PO (W)	HRmax (bpm)	Training phase
Aust. senior male track endurance 1990	12	74.70 4.90	5.57 0.33	74.5 5.0		459 27	202 7	4 weeks into general preparation phase
Aust. senior male track endurance 1991	12	73.15 4.75	5.85 0.35	80.1 4.9		440 24	196 11	Specific preparation 5 days after 4 weeks altitude training (2200 m)
Aust. senior male track endurance 1993-1995	16	73.25 3.65	5.80 0.26	79.3 3.8	13.74 0.63	453 27	198 8	4 weeks into general preparation phase
Aust. senior male track sprint 1991	4	78.45 3.20	5.11 0.06	65.3 3.5		417 30	190 8	4 weeks into general preparation phase
Aust. junior male track endurance 1991	8	70.25 8.10	5.54 0.55	79.1 6.2	13.51 0.78	439 28	200 11	General preparation
Aust. junior male track endurance 1995	12	68.20 8.65	5.15 0.44	76.1 6.1	12.68 0.66	397 33	201 5	General preparation
Aust. junior female track endurance 1995	5	60.45 5.70	3.74 0.31	62.0 3.0	10.05 0.79	296 22	194 4	General preparation
Aust. senior male road 1987	10	72.10 6.30	5.10 0.46	70.8 4.6	12.44 0.68	419 43	202 9	General preparation
Aust. senior male road 1990	14	73.85 7.60	5.50 0.41	74.5 3.5	n/a	487 36	200 7	General preparation
Aust. senior female road 1987	8	56.30 3.10	3.58 0.27	63.8 5.4	n/a	309 24	191 8	General preparation
Aust. senior female road 1992	8	50.95 3.65	3.68 0.18	72.6 6.1	10.82 0.63	344 28	195 5	Specific preparation

1. Mean value listed above and standard deviation below.

2. Absolute $\dot{V}O_2max$	TEM = 0.12 L/min	%TEM = 2.6	ICC = 0.986 (n = 10)
3. Relative $\dot{V}O_2max$	TEM = 2.0 $ml \cdot kg^{-1} \cdot min^{-1}$	%TEM = 3.1	ICC = 0.971 (n = 10)
4. Power output at $\dot{V}O_2max$	TEM = 14 W	%TEM = 3.5	ICC = 0.965 (n = 10)
5. HRmax	TEM = 2 bpm	%TEM = 1.2	ICC = 0.941 (n = 10)

Table 17.6 Seasonal Variation in Maximal Oxygen Uptake for the Australian Female Olympic Road Cycling Squad

Date	Mass (kg)	$\dot{V}O_2$max (L/min)	$\dot{V}O_2$max (ml · kg^{-1} · min^{-1})	$\dot{V}O_2$max (L/FSA)	PO (W)	HRmax (bpm)
January 1992	53.05	3.52	66.4	10.35	333	197
	3.90	0.22	4.7	0.61	26	7
February 1992	52.30	3.62	69.5	10.65	349	196
	3.70	0.17	5.0	0.61	21	5
April 1992	50.95	3.68	72.6	10.82	344	195
	3.65	0.18	6.1	0.63	28	5

1. Mean value listed above and standard deviation below.
2. n = 7
3. Absolute $\dot{V}O_2$max TEM = 0.12 L/min %TEM = 2.6 ICC = 0.986 (n = 10)
4. Relative $\dot{V}O_2$max TEM = 2.0 ml · kg^{-1} · min^{-1} %TEM = 3.1 ICC = 0.971 (n = 10)
5. Power output at $\dot{V}O_2$max TEM = 14 W %TEM = 3.5 ICC = 0.965 (n = 10)
6. HRmax TEM = 2 bpm %TEM = 1.2 ICC = 0.941 (n = 10)

also provided (see page 83 for details about the calculation procedures).

Future Tests Relating to Aerobic Power: In cycling events that depend on the highest rate of energy release over a period of approximately 60 to 300 s, the time taken to achieve $\dot{V}O_2$max is extremely important (Craig et al. 1993; Thoden 1991). As the duration of the performance increases, the ability to maintain $\dot{V}O_2$max, as indicated by time to exhaustion, becomes progressively important (Billat et al. 1994). Hence, routine tests in the future may include measurement of $\dot{V}O_2$ kinetics and cycling time to exhaustion at $\dot{V}O_2$max.

Blood Lactate Transition Thresholds

Rationale: Success in many cycling events depends not only on $\dot{V}O_2$max but also on the percentage of $\dot{V}O_2$max (%$\dot{V}O_2$max) that the cyclist can sustain. Therefore blood lactate transition thresholds are now routinely assessed in order to predict, monitor, and prescribe cycling performance. In relation to road cycling, several studies have reported highly significant correlations between individual time-trial performance and measured thresholds (Coyle at al. 1988, 1991; Craig and Conyers 1988; Miller and Manfredi 1987). These blood lactate transition thresholds may also be important in track endurance cycling, with Craig et al. (1993) reporting significant correlations between lactate, individual anaerobic threshold indexes, and 4000 m individual-pursuit time. In relation to training and adaptation, Olds et al. (1995a, 1995b) predict that a 10% improvement in anaerobic threshold would decrease a 26 km individual time-trial time by approximately 4%. Finally, there is little argument that the concept embodied by the term threshold is valuable to training and competition: that there is a specific intensity of exercise below which endurance is mainly a function of fuel supply and/or body temperature, and above which there is a significant reduction in endurance time that is probably dependent on metabolic disturbances related to acidosis (Thoden 1991). Identification of this point(s) can be extremely useful for planning training and competition strategies.

Test Procedure:

1. Complete common procedures as outlined earlier in this chapter, and connect the cyclist to all measuring equipment (e.g., mouthpiece, heart rate monitor, pulse oximeter).

2. Have the cyclist complete the warm-up outlined in table 17.7 while checking final calibration.

3. The cyclist is required to pedal at a constant workload for 5 min work durations. The test is continuous, with cyclists stopping only when they can no longer maintain the required power output. Strongly encourage the cyclist to complete a final full-minute workload.

4. Consult table 17.7 for information regarding type of ergometer, starting workloads and ramp increases, cadence range, and sprocket selections.

Use the ergometer's gearing mechanism to ensure that the cyclist's pedaling cadence stays within the required range. Note: In assessment of junior males, after completion of the 250 W workload, a 25 W increase may be a more appropriate workload increase. You will need to use your judgment.

5. As the cyclist progresses through the test protocol, provide continual information concerning cadence ranges and target power outputs. The cyclist is permitted to drink during the first 2.5 min of each workload.

6. The test should be completed within 25-40 min.

Data Collection:

- Record age, height, mass, and calculated FSA.
- Record $\dot{V}O_2$, RER, $\%SaO_2$, $\dot{V}_E$, and power output. Power output is averaged over the 5 min work duration; the respiratory variables are the average of the last 2 min of each 5 min work duration.

Note: This procedure requires the mouthpiece and noseclip to be in place only during the last 2.5 min of any 5 min work duration. Toward the end of the test (i.e., the last 6-10 min), the cyclist should have the mouthpiece and noseclip in place continuously.

- Record heart rate during the last 15 s of each 5 min work duration and heart rate at the completion of the test.

Blood Collection:

- Collect a pre-warm-up blood sample from the fingertip and analyze the sample for lactate concentration and (if required) pH, blood gases, and bicarbonate concentration.
- Throughout the test, blood is collected from the fingertip during the last 30 s of each 5 min work duration and analyzed in the same manner as the pre-exercise blood sample.

Data Analysis and Practical Application: Detection of the lactate threshold (LT) and modified individual anaerobic threshold (IAT) is derived using the method and software described in chapter 21. Oxygen uptake and heart rate at these thresholds can be calculated using linear regression techniques (for this linear regression procedure, use only data that come from completed 5 min workloads).

The following indexes should be reported for each threshold: $\dot{V}O_2$, $\%\dot{V}O_2max$, heart rate (HR), %HRmax, power output, and pH (if measured).

As most cyclists have access to a heart rate monitor to help monitor training intensities, the following heart rates can be prescribed from the test just outlined:

- Endurance 1 (E1): $< \text{LTHR} \pm \sqrt{2} \times \text{TEM}$;
- Endurance 2 (E2): $\text{LTHR} = (\sqrt{2} \times \text{TEM}) - (\text{IATHR} - \sqrt{2} \times \text{TEM})$;
- Endurance 3 (E3): $\text{IATHR} \pm \sqrt{2} \times \text{TEM}$;
- Endurance 4 (E4): $> \text{IATHR} + \sqrt{2} \times \text{TEM}$.

Note: TEM = technical error of measurement (Pederson and Gore 1996).

Table 17.8 provides some approximate guidelines (such as duration of effort, blood lactate levels, perceived exertion, power output) for each of these training zones.

Limitations: The method or methods used to identify the blood lactate transition thresholds may affect the index values associated with each threshold. When comparing test data, be sure to employ the same identification method.

Table 17.7 Protocol Selection for Assessing the Blood Lactate Transition Thresholds of Road and Track Cyclists

Group	Ergometer (type and no.)	Warm-up	Starting workload and ramp increase (W)	Cadence range selection (rpm)	Sprocket (tooth no.)
Male senior	Road (1)	5 min at 75-100 W	100 with 50 W increase	90-100	As required
Male junior	Road (1)	5 min at 75-100 W	100 with 50 W increase	90-105	As required
Female senior	Road (1)	5 min at 75-100 W	100 with 25 W increase	80-90	As required
Female junior	Road (1)	5 min at 50-75 W	75 with 25 W increase	75-85	As required

Table 17.8 Approximate Guidelines (for Males) for the Four Aerobic Training Zones

Training zone	Heart rate (% HRmax)	Duration of effort (min)	Blood lactate (mmol/L)	Power output (W)	Cycling speed (km/h)	Perceived exertion
Endurance 1 (E1)	< 75%	30–120	< 1.5	< 260	< 28	Recovery, easy
Endurance 2 (E2)	75–85%	120–450	1.5–3.5	260–350	28–37	Comfortable
Endurance 3 (E3)	85–92%	15–60	3.5–6.0	350-370	37–45	Uncomfortable
Endurance 4 (E4)	> 92%	3–5	> 6.0	> 370	45–50	Stressful

Normative Data: Table 17.9 provides normative data for the LT and IAT thresholds of male and female track and road cyclists. These data are specific to the corresponding training phases. Figure 17.1 illustrates the seasonal variation that can occur with the blood lactate curve and its associated thresholds for an individual cyclist. For both table 17.9 and figure 17.1, TEM data are provided (see page 83 for details of the calculation procedures).

Laboratory 30-Minute Time Trial

Rationale: For the road cyclist, the time trial is sometimes referred to as the "race of truth." In the time trial, tactical decisions and drafting are not important as the cyclist simply attempts to cover the prescribed distance in the shortest time possible. For this reason a laboratory assessment of time-trial fitness can be used as an important dependent variable for monitoring the effects of training, diet, psychological preparation, and ergogenic aids (Bishop 1997).

The laboratory 30 min time trial serves as a standardized performance task that is not influenced by environmental conditions (heat or altitude) or aerodynamics. Although many have adopted ride time until exhaustion at a fixed percentage of $\dot{V}O_2max$ as the cycling performance test of choice (Jeukendrup et al. 1996), recent research has demonstrated that the average power output during a fixed-duration test may be better for reflecting fitness because this parameter is reliable (Jeukendrup et al. 1996; Bishop 1997), correlates with actual time-trial performance in the field (Coyle et al. 1991), and more closely replicates the demands of actual competition. Assuming a maximal, well-paced effort, results from this test reliably and accurately reflect time-trial cycling-specific fitness.

The heart rate and power output that a cyclist can sustain for 30 min can be used as a point of reference to guide future training sessions. A cyclist who can sustain 300 W for 30 min should be able to complete the following interval-training session: 4 × 10 min at 305 W with 5 min recovery. This type of interval training represents an overload that is consistent with the cyclist's current level of fitness. Over time, this type of interval session should stimulate adaptations that will allow the cyclist to improve his or her 30 min maximal power output.

Test Procedure:

1. The 30 min laboratory time trial is typically completed after assessment of aerobic power using a 5 min incremental-intensity protocol to establish lactate transition thresholds (LTT5-min) as described earlier in this chapter. Although one recovery day (<40 km E1 and E2) is commonly prescribed between the LTT5-min test and the 30 min time-trial test, it is important to note that in well-trained road cyclists, performance during the 30 min time trial does not appear to be detrimentally affected by an LTT5-min test performed 24 h earlier. In fact, some road cyclists have commented that they actually felt better during the 30 min time trial when they had completed an LTT5-min test the day before. As with most laboratory tests, it is important to be consistent. For historical reasons, one day of recovery is typically programmed into the testing week.

2. One hour is scheduled for testing each cyclist, and two physiologists assist with administration of the test. Cyclists are told to report to the laboratory 20 min early to complete a pretest questionnaire (see chapter 2, page 16) as well as a 10-15 min warm-up using their bicycle attached to a wind-trainer. After the warm-up, adhere to the common procedures outlined earlier in the chapter, and connect the cyclist to all measuring equipment.

Table 17.9 Blood Lactate Transition Threshold Data for High-Performance Cyclists

Group	n	Lactate threshold					Individual anaerobic threshold					Training Phase
		PO (W)	HR (bpm)	% HRmax	$\dot{V}O_2max$ (L/min)	$\dot{V}O_2max$	PO (W)	HR (bpm)	% HRmax	$\dot{V}O_2max$ (L/min)	$\dot{V}O_2max$	
Aust. junior male track endurance 1994	7	249 29	147 8	74.6 4.5	3.36 0.36	61.4 3.5	341 24	179 5	90.2 2.5	4.48 0.28	81.9 4.1	Specific preparation
Aust. senior male track endurance 1990	11	213 17	150 7	74.4 4.2	2.98 0.16	53.6 3.6	342 21	179 6	88.8 2.8	4.52 0.27	81.2 2.5	End of transition
Aust. senior male track endurance 1994	11	263 20	152 10	75.4 3.5	3.66 0.23	6.25 2.8	358 27	178 12	91.3 2.7	4.92 0.36	85.0 4.3	End of transition
Aust. senior male road 1987	10	199 43	139 12	67.5 9.1	2.98 0.45	56.4 6.7	299 41	172 6	85.9 2.8	4.01 0.41	77.8 4.2	General preparation
Aust. senior male road 1990	7	240 18	148 11	73.4 4.7	3.52 0.22	63.8 4.6	352 28	179 8	88.7 3.1	4.72 0.39	85.1 3.6	General preparation
Aust. senior female road 1987	8	156 24	142 11	74.7 3.6	2.11 0.36	58.7 7.4	223 21	170 8	89.4 1.6	2.87 0.3	79.9 4.5	General preparation
Aust. senior female road 1992	8	176 18	155 8	79.0 2.8	2.33 0.20	66.0 7.0	228 22	179 8	91.0 1.8	2.87 0.21	82.0 7.0	General preparation

1. Mean value listed above and standard deviation below.

2. Lactate threshold

PO	TEM = 7 W	%TEM = 3.4
HR	TEM = 1 bpm	%TEM = 1.0
Absolute $\dot{V}O_2$	TEM = 0.13 L/min	%TEM = 4.2

3. Individual anaerobic threshold

PO	TEM = 5 W	%TEM = 1.6
HR	TEM = 2 bpm	%TEM = 1.4
Absolute $\dot{V}O_2$	TEM = 0.08 L/min	%TEM = 2.0

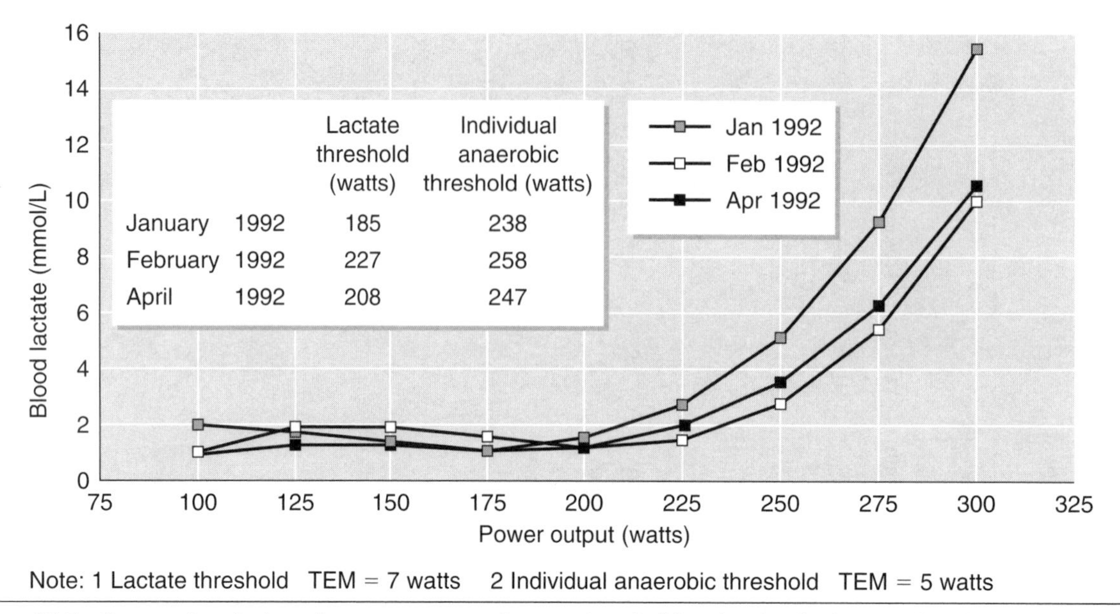

Figure 17.1 Seasonal variation of power output at lactate threshold and individual anaerobic threshold for an Australian female road cyclist.

3. Before the start of the test, inform cyclists of their average power output during their previous 30 min test and/or their power output at threshold (IAT or Dmax) from the recent LTT5-min test. This power output serves as a reference point so cyclists can quickly settle into an appropriate pacing strategy. Knowing a previous test score can also help motivate the athlete.

4. The 30 min time trial can be performed using wind-braked, electromagnetically braked, or friction-braked cycle ergometers. For all types of ergometers it is important that the cyclist be able to replicate a typical racing position (i.e., aerobars, bike position) and select a comfortable cadence during the test. In the case of the Lode Excalibur Sport described previously, it is necessary to select a linear factor that results in the appropriate power output using a desired cadence. The following relationship allows selection of the correct linear factor:

$$\text{Power Output (watts)} = \text{LF} \times (\text{Cadence})^2$$

where LF refers to the linear factor and cadence to revolutions per minute. Thus, at a cadence of 100, a linear factor of 0.015 would result in a power output of 150 W. Refer to the Lode Excalibur Sport operating manual for details on how to operate the ergometer in the linear testing mode.

5. Five minutes before starting the test, the cyclist is allowed to become familiar with the desired power output and cadence by completing one to two 20 s efforts at the target power output. During this period athletes are asked whether they would like any further adjustments of the bike, the heart rate monitor, or mouthpiece. Within 2 min of the start of the test, a pretest capillary blood sample is collected. Cyclists are reminded to treat this test as they would a real competitive effort. Because many athletes during their first test tend to start too aggressively, it is useful to remind them to pace themselves to achieve the highest average power output over 30 min.

6. Two watches are started when the test begins, and the display of accumulated work needs to be easy to view or recorded continuously. It is important that cyclists receive feedback on both time and power output throughout the test.

Data Collection:

1. Expired air analysis is performed from 0 to 5, 10 to 15, and 20 to 25 min.

2. Heart rate, accumulated work (kJ), and cadence (rpm) are recorded every minute.

3. Every 5 min and within 2 min of finishing the test, athletes are asked to rate their perception of effort using the Borg 6-20 scale; and a capillary blood sample is collected for measurement of lactate, glucose, bicarbonate, and pH.

4. Based on accumulated work, an average power output is calculated (e.g., 1 min accumulated work in kilojoules multiplied by 1000 divided by 60 based on the relationship 1 watt = 1 J/s). Power output is reflected as an average absolute and relative power output for 30 min (W and W/kg).

5. Oxygen uptake is presented as an average of all readings made over the 15 min of monitoring and is expressed as L/min, ml · kg^{-1} · min^{-1}, and $\%\dot{V}O_2$peak using the previous LTT5-min data.

6. Other parameters collected during the 30 min test are reflected as an average score and a maximum (minimum in the case of pH) score.

Data Analysis and Practical Application: Of primary concern is the cyclist's average power output expressed as both watts and watts per kilogram body mass. Whereas average power output during a 1 h laboratory time trial has been correlated to 40 km time-trial performance (Coyle et al. 1991), it is important to recognize that changes in laboratory time-trial performance may not precisely parallel changes in time-trial performance in the field. Additionally, it is also assumed, but not well documented, that improvements in the power-to-weight ratio (W/kg) reflect an increased ability to ride up hills. One can use other associated physiological parameters to better understand changes in performance. Generally, improved performance reflects improved cycling-specific fitness. However, it is necessary to consider several other aspects of training and testing:

1. Performance has improved: This may result from a better pacing strategy, better motivation, increased aerobic fitness, increased anaerobic fitness, or less residual fatigue. It is also possible that what the athlete does the day before and even hours before the test may affect performance. Recently collected data from Australian road cyclists demonstrate that even one day of high-intensity training can promote a 5-10% increase in plasma volume. However, at this time the best advice is to keep the pretesting protocol as consistent as possible.

2. Performance has decreased: The first thing to check is the pacing strategy. Many times an extremely aggressive start leads to a big decrease in power output during the middle third of the performance trial. This obviously results in a decrease in the average power output. Other possibilities include a decrease in aerobic fitness, a decrease in anaerobic fitness, a decrease in motivation, and a large degree of residual fatigue. Fatigue can usually be distinguished from a loss of fitness because fatigued cyclists display a lower heart rate and sometimes a lower lactate for a given workload and perception of effort. A detrained athlete, on the other hand, displays elevated heart rates and lactates for a given power output. A careful evaluation of training history prior to the performance test can help refine the explanation of a poor performance test.

3. Performance is unchanged: Even when performance is not dramatically different from that in previous tests, it is of interest to examine the RER, lactate, and pH to gain insights into anaerobic fitness. In some cases performance is unchanged despite large increases in average heart rate, perceived exertion, and lactate. It is possible that this scenario reflects a detrained but very fresh and very motivated athlete.

Limitations: As with any performance test, it is very important for the athlete to be sufficiently motivated to give a maximal effort and for pacing to be appropriate. The respiratory exchange ratio (R) and the rating of perceived exertion (RPE) can be used as a rough guide to evaluate the relative exercise intensity. When R is close to 1.00 and the perception of effort is >17, it is likely that the athlete is giving a near-maximal, well-paced effort. When the R is less than 0.95 and the RPE is <16, one should interpret the data with caution. It may be more important to understand why the athlete was not sufficiently motivated. With regard to pacing, the best individual results tend to display a fairly flat power output profile over time with a gradual increase in power over the last 5-10 min (see figure 17.2a). If there is a large power output decrease (more than 30-50 W) after 5-10 min, the cyclist may have been too aggressive; and if power output increases by more than 100 W during the last 5 min of the test, the cyclist was likely too conservative.

Other limitations associated with testing athletes (i.e., calibration of bicycle and equipment), consistent environmental conditions, and similar time of day for intraindividual testing sessions should also be recognized.

Normative Data: Figures 17.2a and 17.2b, as well as figure 17.3, give typical results for elite road cyclists. Note that for 11 Australian male cyclists the mean heart rate, power output, and lactate concentration were 183 beats/min, 354 W, and 6.6 mmol/L, respectively. Figure 17.3 illustrates that the average power during a 30 min time-trial output (380 W) can be well above that identified at Dmax threshold (320 W).

Alactic Power and Capacity

Rationale:

Sprint track cycling requires a maximal rate of energy expenditure that must be matched by a rapid rate of energy resynthesis. In particular, the flying 200 m and matched sprint race are track

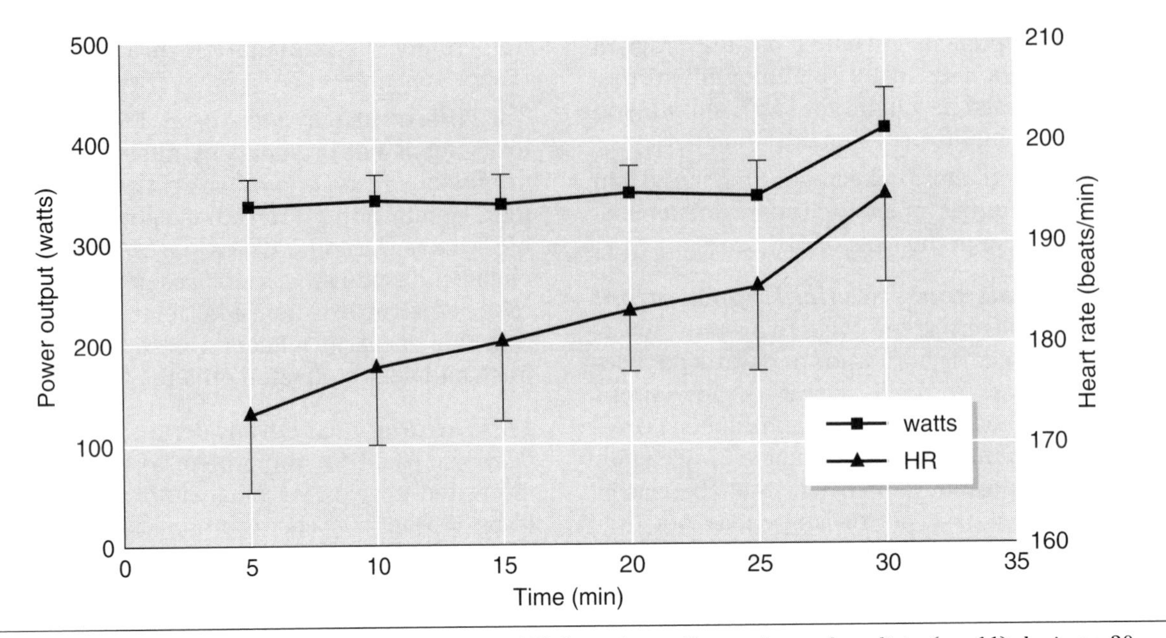

Figure 17.2a Power output and heart rate (mean ± SD) from Australian male road cyclists (n = 11) during a 30 min laboratory time trial.

All cyclists were either European professionals (N = 4) or Australian Institute of Sport scholarship holders (n = 7). Data were collected in December 1995 using a Lode Excalibur Sport cycle ergometer. For the group, average heart rate and power output for the 30 min test were 183 beats/min and 354 W.

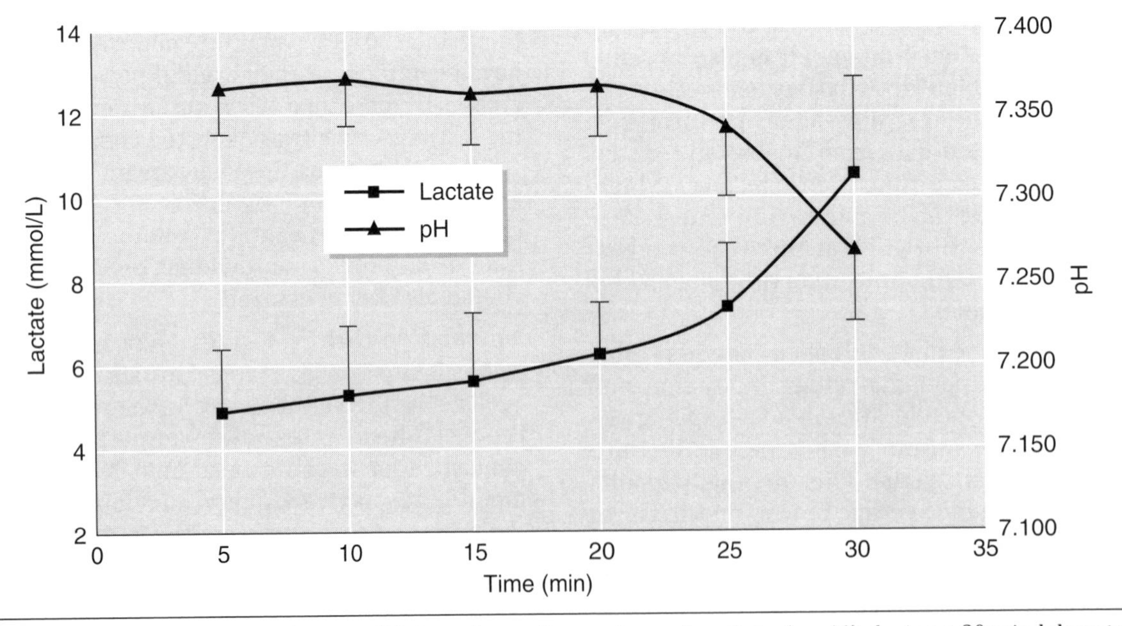

Figure 17.2b Lactate and pH (mean ± SD) from Australian male road cyclists (n = 11) during a 30 min laboratory time trial.

All cyclists were either European professionals (n = 4) or Australian Institute of Sport scholarship holders (n = 7). Data were collected in December 1995 using a Lode Excalibur Sport cycle ergometer. For the group, average lactate and pH for the 30 min test were 6.6 mmol/L and 7.35 pH.

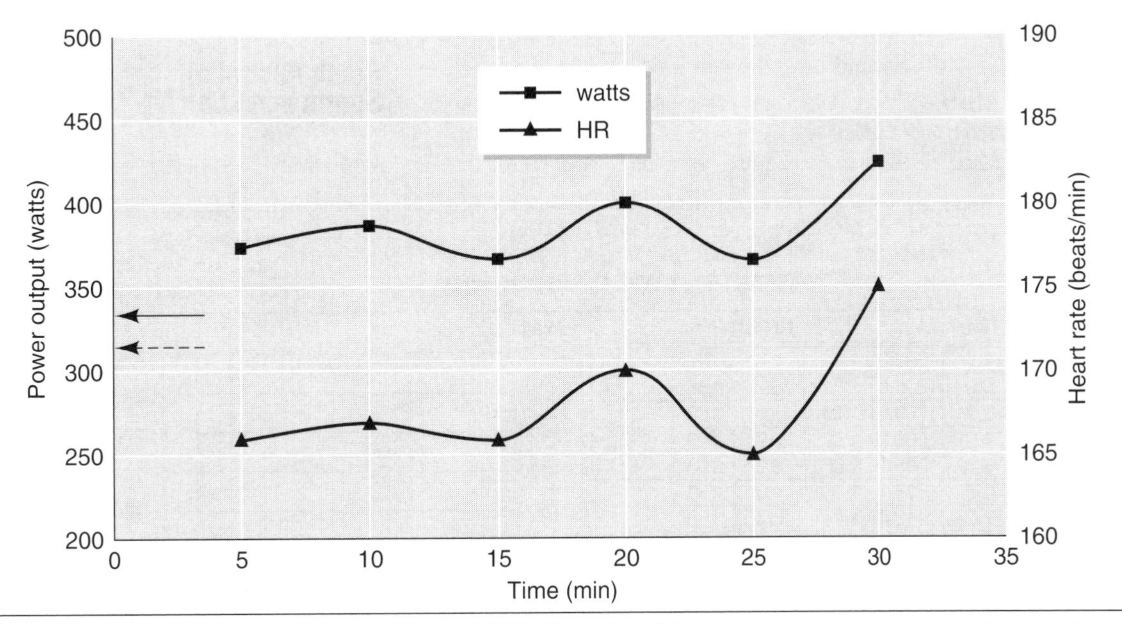

Figure 17.3 Power output and heart rate from a 1996 Olympian (SH) during a 30 min laboratory time trial.

An LTT5-min was performed one week prior to the time trial, allowing calculation of power output at 4 mmol/L lactate (solid arrow—343 W) and Dmax threshold (dotted arrow—320 W). Data were collected in December 1995 using a Lode Excalibur Sport cycle ergometer. Of the group data shown in figures 17.2a and 17.2b, the highest average power output (386 W), with an average heart rate of 168 beats/min, was recorded by SH.

cycling events lasting approximately 10-15 s, with the alactic anaerobic energy system being the dominant metabolic energy pathway (Bouchard et al. 1991; Burke 1986). This energy system is also relevant to other track and road cycling events, such as the start of the 1000 m time trial, individual and team pursuit, and the sprint to the finish line at the end of a road race. The following protocol purports to be a measure of alactic anaerobic performance. Capacity is defined as the total work output during the 10 s all-out effort, with the highest performance output per second (peak power) being a measure of alactic power (Bouchard et al. 1991). Finally, the relationship between cycling performance and cycle ergometer performance has been clearly established. International-class cyclists produce more work in a short-term ergometer test than non-international-level racing cyclists (White and Al-Dawalibi 1986), and higher maximum average power outputs have been produced by sprint cyclists than by pursuit cyclists (Davies and Sandstrom 1989).

Test Procedure:

1. Complete common procedures as outlined previously. If appropriate, connect the track ergometer to the interface unit and enter the appropriate information in the computer software (figure 17.4 and the equipment checklist presented earlier in this chapter).

2. The test utilizes the track cycle ergometer. Select the 19- and 21-tooth sprocket when testing males and females, respectively.

3. Cyclists are allowed a warm-up of their choice. Record the length and content of the warm-up.

4. The test is an all-out maximum effort (no pacing technique allowed) with the cyclist remaining in the saddle for the full length of the test.

Data Collection:

- Record age, height, mass, and calculated FSA.
- The computer software records work done, accumulated work done, average power output, and average cadence for each 1 s sampling period. From this information the software will produce a summary table depicting peak cadence, peak power (absolute and relative), minimum power (absolute and relative), % drop-off in power, total work (absolute and relative), and average power (absolute and relative). The Alactic Power Data Sheet is an example of such a summary.

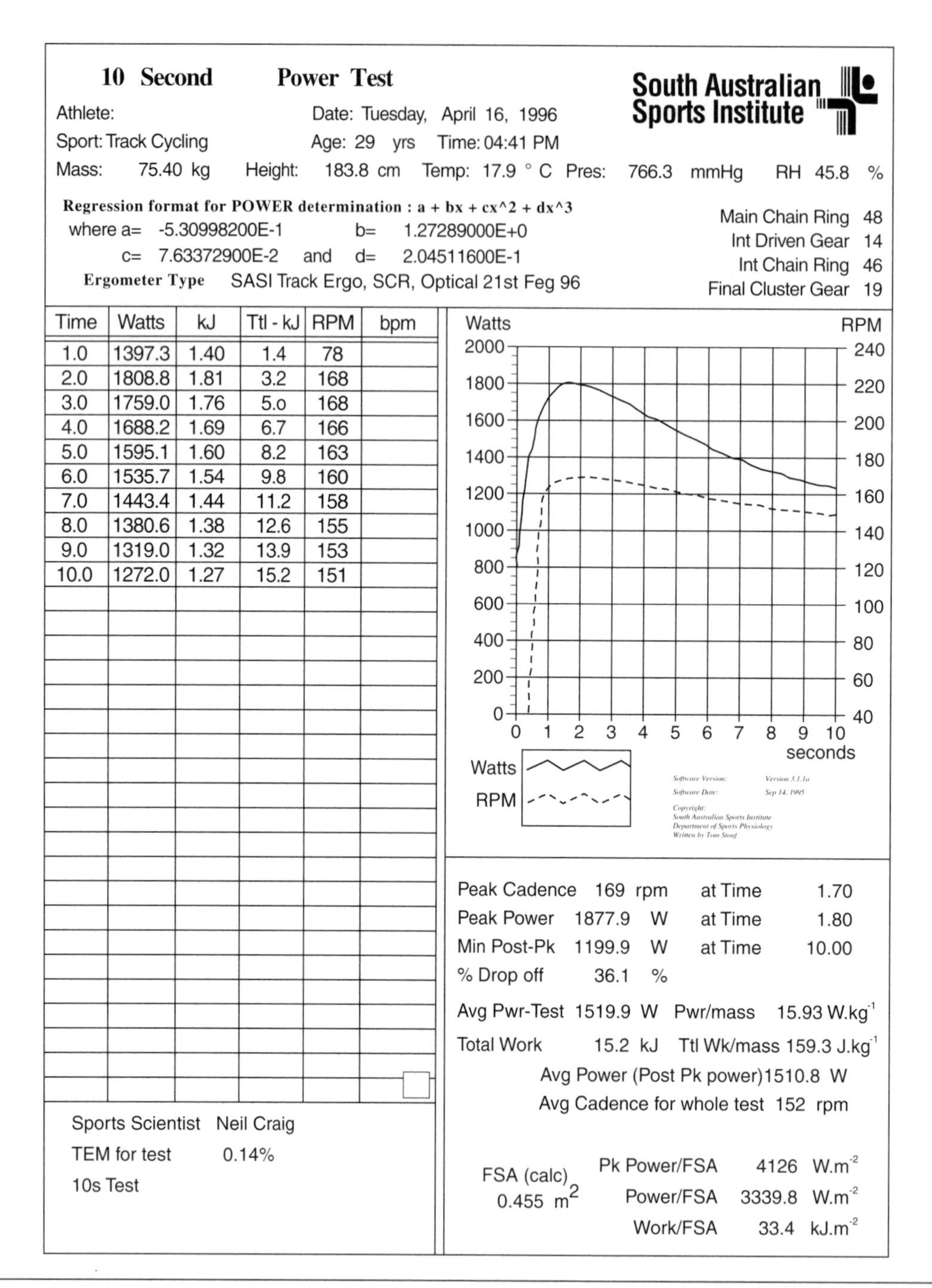

10 Second Power Test

South Australian Sports Institute

Athlete:
Date: Tuesday, April 16, 1996
Sport: Track Cycling
Age: 29 yrs Time: 04:41 PM
Mass: 75.40 kg Height: 183.8 cm Temp: 17.9 ° C Pres: 766.3 mmHg RH 45.8 %

Regression format for POWER determination : a + bx + cx^2 + dx^3
where a= -5.30998200E-1 b= 1.27289000E+0
c= 7.63372900E-2 and d= 2.04511600E-1
Ergometer Type SASI Track Ergo, SCR, Optical 21st Feg 96

Main Chain Ring 48
Int Driven Gear 14
Int Chain Ring 46
Final Cluster Gear 19

Time	Watts	kJ	Ttl - kJ	RPM	bpm
1.0	1397.3	1.40	1.4	78	
2.0	1808.8	1.81	3.2	168	
3.0	1759.0	1.76	5.0	168	
4.0	1688.2	1.69	6.7	166	
5.0	1595.1	1.60	8.2	163	
6.0	1535.7	1.54	9.8	160	
7.0	1443.4	1.44	11.2	158	
8.0	1380.6	1.38	12.6	155	
9.0	1319.0	1.32	13.9	153	
10.0	1272.0	1.27	15.2	151	

Peak Cadence 169 rpm at Time 1.70
Peak Power 1877.9 W at Time 1.80
Min Post-Pk 1199.9 W at Time 10.00
% Drop off 36.1 %

Avg Pwr-Test 1519.9 W Pwr/mass 15.93 $W.kg^{-1}$
Total Work 15.2 kJ Ttl Wk/mass 159.3 $J.kg^{-1}$
Avg Power (Post Pk power) 1510.8 W
Avg Cadence for whole test 152 rpm

Sports Scientist Neil Craig
TEM for test 0.14%
10s Test

FSA (calc) 0.455 m^2
Pk Power/FSA 4126 $W.m^{-2}$
Power/FSA 3339.8 $W.m^{-2}$
Work/FSA 33.4 $kJ.m^{-2}$

Figure 17.4 Alactic Power Data Sheet.

Data Analysis and Practical Application:

- Data should be expressed as an absolute score and relative to body weight and FSA.
- The 100% test effort can be used to prescribe alactic interval training on the ergometer.

Normative Data: Table 17.10 provides normative data for senior male and female sprint cyclists.

Table 17.10 Alactic Power and Capacity Data for High-Performance Sprint Cyclists

Variable	Senior male (n = 7)	Senior female (n = 5)
Peak power (W)	1701 163	1072 66
Peak power (W/kg)	18.56 2.28	14.63 0.42
Peak power (W/FSA)	3805 256	2668 106
Peak cadence (rpm)	165 6	155 3
Percent drop-off in power (%)	27.5 4.1	21.5 2.8
Total work done (kJ)	13.9 1.3	8.8 0.6
Total work done (J/kg)	157.1 31.1	119.6 5.0
Total work done (kJ/FSA)	31.1 1.9	21.8 1.2
Average power (W)	1393 137	876 62
Average power (W/kg)	15.71 0.66	11.95 0.49
Average power (W/FSA)	3112 198	2178 124

Mean value listed above and standard deviation below.

chapter **18**

Protocols for the Physiological Assessment of Golfers

Aaron Russell and David Owies

Unlike many other sports, golf does not require high levels of aerobic power or muscular strength and power for success; but it does require an extremely high level of skill. To maintain these skill levels for prolonged periods, golfers need to develop several fitness components. The important ones are cardiorespiratory fitness, body composition, strength, and flexibility. Although physiological testing for golf is relatively new, it can be used to identify specific fitness areas that the golfer may need to develop as a basis for overall improved golf performance.

As in other sports, fatigue is among the major factors that decreases performance of golfers. The high skill level required, combined with the small margin for error with each shot, means that even a small level of fatigue may greatly reduce performance. Fatigue will influence factors such as concentration, neuromuscular coordination, and muscular strength (Wilmore and Costill 1994) that are integral to the precise execution of many golf skills. An increased level of cardiorespiratory fitness enables the golfer to complete a round of golf feeling less exhausted and thus better able to continue to execute skills at a high level. In addition, a high level of cardiovascular fitness gives the golfer the ability to cope more efficiently with hot or humid conditions and to recover more effectively from competition and training.

The ability to withstand fatigue over several rounds of golf (and in hot or humid conditions) is further enhanced with a low level of body fat. A low level of body fat also causes less stress on the lower back and may improve the ability to swing more freely (Chim and McMaster 1996).

In amateur golfers, the most common sites of injury due to overuse (68%) and poor swing technique are the back, shoulder, elbow, and knee (Jobe and Yocum 1988; Batt 1992). These areas, as well as the wrist, have also been observed to be the most common sites of injury for professional golfers (McCarroll and Gioe 1982). Cahalan et al. (1991) observed that wrist injuries occurred from overuse and were associated with a lack of forearm strength, while the degeneration with age and weakness in the shoulder rotator cuff muscles increases the risk of overuse injury (Jobe et al. 1989).

Abdominal strength is associated with efficient trunk rotation in conjunction with spinal flexibility. Increased abdominal muscle group strength has been observed to increase the power of the golf swing and increase club head speed (Brendecke 1990). Improved flexibility and strength in the abdominal group have been recommended for injury prevention (Jobe et al. 1994). Abdominal strength is measured using a progressive seven-stage abdominal test.

Good flexibility is essential for playing golf. Golfers need a good range of spinal rotation and shoulder mobility to successfully execute the golf swing (Chim and McMaster 1996). In addition, agonist and antagonist muscles should be well balanced to prevent injuries and maintain posture. Specific golf flexibility should be tested by a physiotherapist familiar with the requirements of golf.

Equipment and suggested protocols follow. If you wish, you may photocopy and use the Golf—Physiological Assessment form as needed to record your assessment results.

Golf—Physiological Assessment

Name: ______________________________ Date : __________

Age: ____________ (yr) Height: ________ (cm) Mass: ________________ (kg)

Room temperature: ______________ Barometric pressure: ______________

Parameter	Result
Aerobic power	
W/kg	
Sum of seven skinfolds (mm)	
Biceps	
Triceps	
Subscapular	
Supraspinale	
Mid-abdominal	
Front thigh	
Medial calf	Sum of seven SF: mm
Grip strength (kg)	
Right hand	
Left hand	
Knee flexion/extension (N · m)	Peak torque
Right knee flexion	
Left knee flexion	
Left/Right flexion (deficit)	
Right knee extension	
Left knee extension	
Left/Right extension (deficit)	
Right flexion/extension	
Left flexion/extension	

(continued)

Parameter	Result
Shoulder internal and external rotator cuff (N · m)	Peak torque
Right external rotator cuff	
Left external rotator cuff	
Left/Right external rotator cuff (deficit)	
Right internal rotator cuff	
Left internal rotator cuff	
Left/Right internal rotator cuff (deficit)	
Right external/internal	
Left external/internal	
Forearm pronation and supination (N · m)	Peak torque
Right forearm supination	
Left forearm supination	
Left/Right supination (deficit)	
Right forearm pronation	
Left forearm pronation	
Left/Right pronation (deficit)	
Right supination/pronation	
Left supination/pronation	

Laboratory Environment and Subject Preparation

For laboratory and subject preparation, see chapter 2.

Equipment Checklist

1. Anthropometry
   - ☐ Stadiometer accurate to 0.1 cm
   - ☐ Weight scales accurate to ±0.1 kg
   - ☐ Harpenden skinfold calipers (dynamically calibrated)
   - ☐ Steel anthropometric girth tape
2. Cardiorespiratory fitness
   - ☐ Monark 814E cycle ergometer
   - ☐ Polar Electro heart rate monitor
3. Strength
   - ☐ Hand-grip dynamometer
   - ☐ Cybex isokinetic dynamometer (model 770)

PROTOCOLS: Golf

Anthropometry

See table 18.1 for normative data.

Test Procedure: Stretch height, body mass, and the sum of seven skinfolds are assessed according to the techniques and procedures outlined in chapter 5.

Considerations for the Coach: An increase in the sum of skinfolds indicates that the golfer is carrying excess body fat. The effect is similar to that of having to swing with a weight strapped to the body. The extra weight also makes it more difficult for the golfer to tolerate hot and humid playing conditions. Excess body fat may reduce the golfer's concentration because it will increase fatigue. It may also reduce the range of movement during the swing.

Table 18.1 Sum of Seven Skinfolds* (mm)

	Mean	Range	Desired standard
Male	95.1	44–172	< 80
Female	121.0	62–181	< 100

*The seven skinfold sites are biceps, triceps, subscapular, supraspinale, mid-abdominal, front thigh, and medial calf.

Normative values have been established from test results obtained from Australian Institute of Sport and Victorian Institute of Sport golf scholarship holders between 1993 and 1997. These are state- and national-level players.

Cardiorespiratory Fitness

For normative data, see table 18.2.

Test Procedure:

1. Calculate the target heart rates for the subject at 55%, 65%, and 75% of the age-predicted maximum using the formula:

Maximum Heart Rate = 220 (beats/min) – Age (yr)

2. Adjust the seat height to suit the subject, and record the number of visible seat holes. The seat height should be set so that the subject's leg is slightly flexed when the pedal is at the bottom of the revolution.

3. Place the heart rate monitor approximately 5-10 cm below the subject's nipple. Moisten the electrode pads to ensure transmission of the heart rate signal. Place the receiver within ~1 m of the subject to obtain a heart rate signal.

4. The subject is to complete at least three workloads, each lasting 3 to 4 min. The initial workload is 1.0 kp (kilopond) for men and 0.5 kp for women. The subject starts cycling at 60 rpm. Read the heart rate after the first minute and record it on the data sheet. Repeat this step for the second and the third minute. If the heart rate change between minute 2 and 3 is not less than or equal to 3 beats/min, another minute at that work rate is required until there is no more than a 3-beat difference.

5. Repeat step 4 for the second and third workloads. The increments should be either 0.5 or 1.0 kp. If the target heart rate for the third workload is not reached, another workload is required. Also, if the heart rate exceeds 90% of the age-predicted

heart rate, the subject is working too hard and the test should be stopped.

6. To cool down, the subject should continue cycling for several minutes with a light workload. The heart rate should return to below 60% of the age-predicted maximum heart rate.

7. Calculate the subject's physical work capacity (PWC) according to the following example. An 80 kg man cycles for 3 min to achieve his 75% age-predicted heart rate of 150 beats/min at a workload of 3 kp at a pedal rate of 60 rpm. To establish the PWC, use the following equation:

$$PWC = 3 \text{ kp} \times 6 \text{ m/rev} \times 60 \text{ rev/min}$$
$$= 18 \text{ kpm/rev} \times 60 \text{ rev/min}$$
$$= 1080 \text{ kpm/min}$$

Given that 1 watt equals 6.12 kpm/min,
1080/6.12 = 176.5 W

To establish the PWC in watts per kilogram, divide the 176.5 by the subject's body weight:

$$176.5/80 = 2.21 \text{ W/kg}$$

Considerations for the Coach: An increased aerobic power indicates that the golfer is able to produce more energy aerobically per kilogram of body weight. This may help the golfer to concentrate for longer periods during a round as well as decrease physical and mental fatigue during a more demanding three- or four-day tournament.

Table 18.2 Normative and Target Physical Work Capacity at 75% of Maximum Heart Rate

	Mean (W/kg)	Range	Desired standard
Male	1.86	1.31-2.84	2.55
Female	1.43	0.71-2.28	2.02

Normative values are for subjects between the ages of 19 and 29 years and are taken from *Australian Fitness Norms* (Gore and Edwards 1992).

Flexibility Testing

A complete musculoskeletal screening should also be performed to identify any weaknesses in the spine, shoulders, hips, wrist, and elbows. Melbourne's Golf Injury Clinic has developed an extensive flexibility assessment (see appendix for more information). In addition, chapter 7 provides a theoretical consideration of flexibility as well as specific assessment protocols. The muscles screened include the following:

- Trapezius
- Levator scapulae
- Rhomboid
- Deltoid
- Pectoralis major
- Teres major
- Teres minor
- Supraspinatus
- Infraspinatus
- Long extensor compartment
- Short extensor compartment
- Long flexor compartment
- Short flexor compartment
- Tibialis posterior
- Opponens pollicis
- Flexor pollicis
- Quadratus lumborum
- Latissimus dorsi
- Glutei group
- Piriformis psoas major
- Adductor muscle group
- Tensor facia latae
- Iliotibial band
- Hamstrings, quadriceps
- Gastrocnemius

The muscles are tested on the right and left sides of the body and are rated as either passive or active.

Strength-Testing Protocols

The Cybex dynamometer is to be set up according to the manufacturer's instructions and should be calibrated prior to the testing session. Experience indicates that the Cybex 770 has ±1 N · m measurement error. However, there may be a significant biological error of measurement that should be reduced through adequate warm-up and familiarization with the test pattern prior to the testing session. Note that chapter 12, "Strength Assessment by Isokinetic Dynamometry," provides detailed protocols as well as a critique of isokinetic strength testing, while chapter 10, "Introduction to the Assessment of Strength and Power," covers other issues of strength and power assessment.

Grip Strength

For normative data, see table 18.3.

Test Procedure:

1. Ensure that the grip dynamometer is set at zero before starting the test.

2. Fit the size of the subject's hand to the dynamometer by adjusting the base and the handle of the dynamometer. The handle should rest on the middle phalange, and the base should rest on the first metacarpal.

3. In the starting position, subjects stand with their heels, buttocks, and back resting against a wall. The arm is raised vertically above the head with the palms facing inward.

4. The subject performs the task by gripping the dynamometer as hard as possible while moving the arm through a 180° arc for 3 s. The arm must remain straight (that is, fully extended at the elbow) throughout the movement.

5. Record the score to the nearest 0.5 kg.

6. Repeat the procedure for the other hand. Repeat twice for each hand and record the best score.

Considerations for the Coach: Both left and right grip strength need to be well developed to control the forces that are generated from the rotation of the body. A weak grip may result in the golfer's trying to hold the club too tight in order to control the shot. In so doing the golfer may tighten up during the swing, which can result in a misdirected shot.

Table 18.3 Normative and Target Grip Strength Scores

		Mean (kg)	Range	Desired standard
Male	(Right)	51	25–70	> 55
	(Left)	54	23–74	> 55
Female	(Right)	40	28–54	> 45
	(Left)	42	28–54	> 45

Normative values have been established from test results obtained from Australian Institute of Sport and Victorian Institute of Sport golf scholarship holders between 1993 and 1997. These are state- and national-level players.

Knee Extension and Flexion 60° Per Second

For normative data, see table 18.4. Figure 18.1 shows knee extension and flexion.

Considerations for the Coach: Individuals should have no more than a ±10% difference in strength between the right and left legs. A difference greater than ±10% may make it more difficult to control weight transfer from one leg to the other during the swing. In addition, it is desirable for the hamstrings to possess approximately 55-75% the strength of the quadriceps. With increased hamstring strength, the golfer may play with "stiff legs" and be unable to comfortably hold the knees in the slightly flexed position. On the other hand, too much quadriceps strength may mean that the golfer squats down too far during the swing.

Table 18.4 Peak Normative and Target Scores for Isokinetic Knee Extension and Flexion

			Mean (N · m)	Range	Desired standard
Male	Quadriceps	(right)	183	98–268	> 200
		(left)	199	81–283	> 200
	Hamstrings	(right)	126	75–200	> 130
		(left)	120	66–176	> 130
Female	Quadriceps	(right)	144	94–207	> 160
		(left)	133	82–191	> 160
	Hamstrings	(right)	93	63–143	> 104
		(left)	91	56–134	> 104

Normative values have been established from test results obtained from Australian Institute of Sport and Victorian Institute of Sport golf scholarship holders between 1993 and 1997. These are state- and national-level players. Dynamometer set at 60°/s.

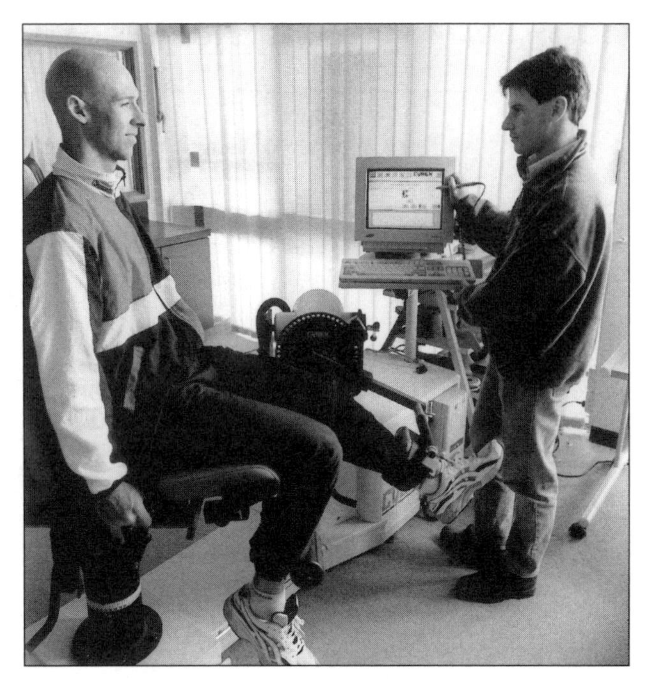

Figure 18.1 Knee extension and flexion at 60°/s.

Shoulder Internal and External Rotation 60° Per Second

For normative data, see table 18.5. Figure 18.2 illustrates internal and external rotation of the shoulder.

Considerations for the Coach: In a right-handed golfer performing external rotation, it is desirable for the rotator cuff muscles to be able to develop approximately 60-70% of the strength developed during internal rotation. A left external rotator cuff with too much strength may lead to hooking of the ball. There should be no more than a 10% difference in strength between the right and left sides.

Figure 18.2 Shoulder internal and external rotation at 60°/s.

Table 18.5 Peak Normative and Target Scores for Isokinetic Shoulder Internal and External Rotation

			Mean (N · m)	Range	Desired standard
Male	Internal	(right)	64	40–87	> 70
		(left)	62	34–83	> 70
	External	(right)	47	25–64	> 50
		(left)	43	24–60	> 50
Female	Internal	(right)	44	24–56	> 50
		(left)	40	22–54	> 50
	External	(right)	33	16–49	> 40
		(left)	33	15–50	> 40

Normative values have been established from test results obtained from Australian Institute of Sport and Victorian Institute of Sport golf scholarship holders between 1993 and 1997. These are state- and national-level players. Dynamometer set at 60°/s.

Forearm Pronation and Supination 30° Per Second

For normative data, see table 18.6. Figure 18.3 illustrates forearm pronation and supination.

Considerations for the Coach: For control of the club head, as well as for injury prevention, equal development of forearm strength during pronation and supination is desirable. A right-handed golfer with too much strength during left forearm supination may block the ball on impact.

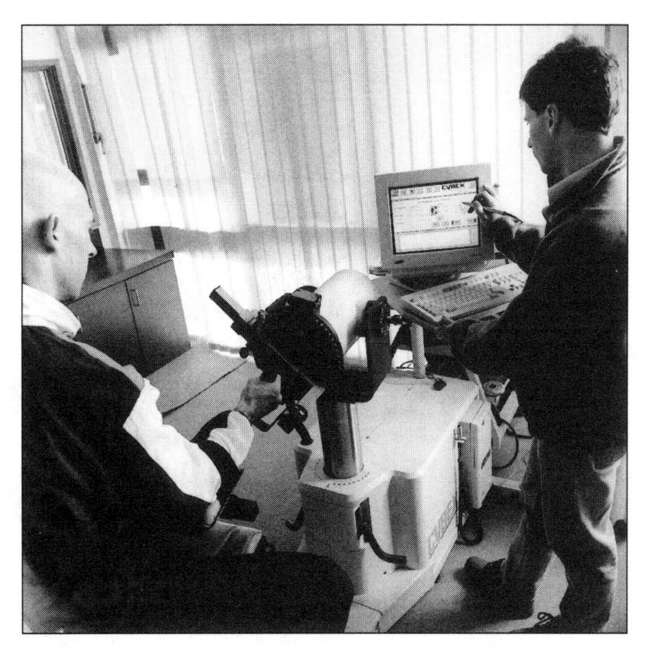

Figure 18.3 Forearm pronation and supination at 60°/s.

Table 18.6 Peak Normative and Target Scores for Isokinetic Forearm Pronation and Supination

			Mean (N · m)	Range	Desired standard
Male	Pronation	(right)	16	10–25	> 20
		(left)	15	8–25	> 20
	Supination	(right)	15	7–28	> 20
		(left)	15	7–28	> 20
Female	Pronation	(right)	10	0–15	> 10
		(left)	10	0–15	> 10
	Supination	(right)	9	0–20	> 12
		(left)	9	0–16	> 12

Normative values have been established from test results obtained from Australian Institute of Sport and Victorian Institute of Sport golf scholarship holders between 1993 and 1997. These are state- and national-level players. Dynamometer set at 30°/s.

chapter 19

Protocols for the Physiological Assessment of Male and Female Field Hockey Players

Steve Lawrence and Ted Polglaze

At the international level, field hockey is fast and free-flowing, with players continually changing their motion (on average every 5 s). Heart rate and lactate analysis of international competition indicates that players use on average 75-85% of their maximum oxygen uptake and can accumulate, during play, blood lactate levels in excess of 10 mmol/L on a regular basis. The tests in the following pages have been chosen to measure the power or capacity of energy systems deemed relevant to the physical performance of field hockey. Time, resource, and labor costs have also been considered in the selection of tests.

These tests will probably be modified as new knowledge and technology become available. For instance, the use of a cycle ergometer for measuring anaerobic power and capacity is not considered ideal because the action of cycling is nonspecific to a running-based sport. Therefore an attempt is currently being made to develop a valid and accurate laboratory measure of power during sprint running.

The test scores in this chapter represent 12 years of regular assessment (over 4000 individual records) of Australia's elite hockey players and therefore should provide a useful reference for comparison.

Pretest Protocols and Subject Preparation

The following specific pretest conditions should be observed in addition to those recommended in chapter 2.

- **Training.** Athletes should undertake no training inducing severe physiological or neural fatigue in the 24 h before testing. This will preclude any maximal strength training, high-intensity skill, or physical training.
- **Field environment.** Environmental conditions will influence field testing; therefore documentation of ambient temperature, humidity, light, and wind is essential. In addition, all field testing for hockey players should occur on a wetted artificial grass surface with the players wearing their normal playing footwear.

Equipment Checklist

1. Anthropometry

See chapter 5 for a description of appropriate equipment.

2. Anaerobic power and capacity tests

☐ Vertical jump

_ Calibrated board (1 cm increments) firmly attached to a high vertical wall

_ Gymnastics chalk dust (magnesium carbonate)

☐ 10 s maximal cycle ergometer test

Refer to chapter 9, "Testing Protocols for the Physiological Assessment of Team Sport Players," for a description of relevant equipment.

☐ Test of repeatability (5 × 6 s cycle ergometer test)

Refer to chapter 9 for a description of relevant equipment.

☐ Acceleration and speed test

__ 60 m of even surface

Refer to chapter 9 for a description of relevant equipment.

☐ Agility

__ Surface as for acceleration and speed test

Refer to chapter 9 for a description of relevant equipment.

3. Aerobic power tests

☐ Maximal oxygen consumption ($\dot{V}O_2max$)

__ a. Treadmill: minimum speed range 0-16 km/h, gradient range 0-10%

__ b. Expired gas analysis system according to general recommendations in chapter 8

__ c. Heart rate monitoring system

__ d. Stopwatch

__ e. Data sheet

☐ Shuttle run test

__ 10 × 25 m area of even surface

Refer to chapter 9 for a description of relevant equipment.

See sample data collection forms for individual and team assessments at the end of this chapter.

PROTOCOLS: Field Hockey

Anthropometry

See table 19.1 for normative data.

Rationale: The rationale for anthropometric measurements of stature, mass, and sum of skinfolds is the same as for basketball (page 225).

In hockey, which requires speed and explosive power, excess fat will decrease acceleration unless proportional increases in force are applied (Norton et al. 1996). Excess fat will also increase the load the musculoskeletal system must absorb during movement, increasing the risk of injury.

Measurements Required: Height, body mass, and sum of skinfolds (refer to chapter 5) are needed.

Limitations: Refer to chapter 5. One must make allowances for body types in interpreting skinfolds, especially for females with somatotypes containing a high endomorphic component.

Anaerobic Power Capacity and Performance Tests

Examples of hockey performance via the adenosine triphosphate-phosphocreatine and glycolytic pathways include moves that involve speed, acceleration, and explosiveness such as tackling, making space to receive the ball, or moving to cover the ball carrier, as well as longer-duration, repetition efforts such as a series of fast attacks and/or counterattacks. Three laboratory tests have been selected to measure the power and capacity of the anaerobic energy systems. Two field-based performance tests have also been included to assess functional performance utilizing anaerobic power and capacity.

Vertical Jump

See table 19.2 for normative data.

Rationale: The vertical jump is an established measure of explosive/anaerobic power of the lower limbs and hips that is easy to perform, requires limited equipment, and is common to many power-related sports, allowing easy comparison.

Procedures: The test procedure is described on page 141.

Limitations: Because this test is limited by technique, it is extremely important that the tester provide coaching to maximize technique: for example, the subject jumps up, not forward, and does not extend arm forward or backward of the vertical when reaching to mark the top of the jump.

Table 19.1 Anthropometric Characteristics of Elite and Subelite Players

Sex	Squad	Position	n	Mean	SD	Range
				Height (cm)		
Female	Senior	Field	385	166.7	5.1	155.6–178.0
		Goalkeeper	78	170.8	4.4	162.8–179.4
	Youth	Field	171	165.7	5.2	156.6–178.2
		Goalkeeper	25	166.3	3.2	162.5–174.1
	Subelite	Field	93	164.6	5.9	154.5–175.5
		Goalkeeper	12	166.9	2.5	162.1–171.7
Male	Senior	Field	287	178.7	5.5	165.9–191.0
		Goalkeeper	28	179.2	2.9	170.6–183.6
	Youth	Field	111	177.9	7.3	165.5–196.6
		Goalkeeper	16	183.9	3.9	177.7–188.3
	Subelite	Field	25	181.2	6.0	169.0–190.2
		Goalkeeper	9	182.8	5.5	175.2–188.6
				Mass (kg)		
Female	Senior	Field	791	61.1	5.5	46.8–75.3
		Goalkeeper	159	68.0	4.3	56.6–75.4
	Youth	Field	304	62.2	5.5	48.0–77.2
		Goalkeeper	39	69.9	6.4	53.9–83.4
	Subelite	Field	153	59.7	6.3	49.8–85.9
		Goalkeeper	15	69.2	4.2	63.5–74.8
Male	Senior	Field	709	76.8	5.6	58.8–93.0
		Goalkeeper	78	81.0	4.9	72.1–89.7
	Youth	Field	300	74.8	7.1	58.2–96.7
		Goalkeeper	41	78.3	3.8	66.3–84.9
	Subelite	Field	69	75.0	8.5	62.5–98.2
		Goalkeeper	11	77.8	6.2	68.2–86.1
				Skinfolds Σ7* (mm)		
Female	Senior	Field	521	68.2	13.6	36.6–126.4
		Goalkeeper	89	84.4	13.4	56.4–116.9
	Youth	Field	220	78.0	16.0	26.0–126.0
		Goalkeeper	30	98.9	21.9	65.2–166.9
	Subelite	Field	137	80.7	15.0	48.5–130.7
		Goalkeeper	13	103.3	14.5	87.0–136.6
Male	Senior	Field	416	56.2	15.1	28.7–123.1
		Goalkeeper	46	51.9	13.5	36.8–88.9
	Youth	Field	219	52.1	13.9	27.9–111.2
		Goalkeeper	39	59.6	11.1	41.2–80.6
	Subelite	Field	63	51.1	10.8	28.6–82.3
		Goalkeeper	9	54.2	25.0	33.1–107.1

*Σ7 = triceps, subscapular, biceps, supraspinale, mid-abdominal, front thigh, and medial calf.

Table 19.2 Vertical Jump Scores of Elite and Subelite Players

			Vertical jump (cm)			
Sex	**Squad**	**Position**	**n**	**Mean**	**SD**	**Range**
Female	Senior	Field	491	46	4	33–56
		Goalkeeper	94	42	6	31–60
	Youth	Field	185	43	4	29–53
		Goalkeeper	25	38	5	27–47
	Subelite	Field	68	42	4	34–50
		Goalkeeper	8	41	4	36–47
Male	Senior	Field	363	56	5	40–77
		Goalkeeper	44	59	4	50–68
	Youth	Field	133	57	7	38–72
		Goalkeeper	19	53	5	42–60
	Subelite	Field	21	56	8	40–68
		Goalkeeper	5	54	8	49–68

10-Second Maximal Ergometer Sprint

See table 19.3 for normative data.

Rationale: Refer to page 136.

Procedures: Refer to page 136 for a comprehensive description of test administration and limitations.

5 × 6-Second Repeat-Effort Test (Cycle Ergometer)

See table 19.4 for normative data.

Rationale: For the rationale for this test, refer to page 137.

Procedures: Refer to page 137 for a comprehensive description of test administration, data reduction, limitations, and interpretation.

Acceleration and Speed Test

See table 19.5 for normative data.

Rationale: This test provides an objective measure of a highly relevant functional physical requirement of the game—to accelerate to make a position nearby and/or accelerate and sprint to make a position over an extended distance. Performance analyses of international men's and women's hockey indicate that field players make between 600 and 800 changes of velocity per game. On average, a field player can be expected to sprint farther than 30 m more than 13 times during a game.

Procedures:

1. The start gate is positioned on the start line; gates are also set at 10, 30, and 40 m from the start line on either side of the sprint path. The preferred sprint path is set down the back line of a hockey field with the sideline being utilized as start line, as this sprint path has a zero slope on most fields. Sprint paths along the center line or sidelines have increasing gradients, which will influence performance.
2. Subjects position themselves in a crouch "ready" position with front foot toeing the start line.
3. In their own time, subjects sprint through and past the 40 m line.
4. Record time to reach the 10, 30, and 40 m gates.
5. Athletes should do a minimum of two trials with at least 2 min rest between each one.

Data Reduction:

- Fastest measured 0-10 m time is recorded as the acceleration score.
- Fastest measured 30-40 m time is recorded as the speed score.
- Fastest measured 0-40 m time is recorded as a combined acceleration and speed score.

Table 19.3 Ergometer Sprint Scores of Elite and Subelite Players

Sex	Squad	Position	n	Mean	SD	Range
				10 s work (J/kg)		
Female	Senior	Field	465	129.7	10.4	106.0–168.9
		Goalkeeper	84	120.5	8.0	106.0–136.9
	Youth	Field	135	122.3	7.7	106.0–148.1
		Goalkeeper	8	118.5	4.0	111.9–124.9
	Subelite	Field	35	126.0	10.3	105.1–153.2
		Goalkeeper	2	118.8	9.7	111.9–125.6
Male	Senior	Field	372	155.1	13.8	106.1–187.0
		Goalkeeper	44	154.5	10.8	124.0–175.7
	Youth	Field	128	149.7	15.9	110.2–196.9
		Goalkeeper	16	143.7	8.5	129.3–161.0
	Subelite	Field	21	148.2	13.4	129.6–176.9
		Goalkeeper	5	145.0	7.3	135.0–155.0
				Peak power (W/kg)		
Female	Senior	Field	474	16.08	1.42	12.11–19.10
		Goalkeeper	87	14.85	1.17	12.11–16.70
	Youth	Field	137	15.21	1.05	12.60–18.01
		Goalkeeper	15	13.92	1.24	12.58–15.96
	Subelite	Field	44	15.23	1.72	12.04–17.77
		Goalkeeper	4	13.80	1.47	12.70–15.93
Male	Senior	Field	355	18.93	1.75	12.21–23.13
		Goalkeeper	43	19.65	1.53	14.29–22.48
	Youth	Field	128	17.95	1.87	13.10–24.30
		Goalkeeper	16	18.02	1.26	15.59–19.89
	Subelite	Field	21	18.31	1.80	15.79–22.40
		Goalkeeper	5	18.34	0.86	17.29–19.40

Limitations: The test is affected by player motivation, technique, and environmental conditions.

Agility Test—505 Test

See table 19.6 for normative data.

Rationale: The ability to change direction quickly is extremely important for the hockey player. The 505 test of agility is a valid and reliable test of this quality (Draper and Lancaster 1985).

Procedures: The general procedure for this test is described on page 132. Note, however, that this test must be conducted on a hockey field.

1. A zero line is marked on the sprint surface (sideline at junction of back line).
2. The start marker is placed on the back line 15 m from the zero line.
3. The timing gates are placed either side of the sprint path 5 m from the zero line.

Limitations: Footwear, in particular, will influence performance; therefore only playing shoes must be allowed. Player motivation will also influence performance, as will the quality of the test surface.

Aerobic Power Tests

Aerobic power is the rate at which oxygen is utilized by the tissues during prolonged bouts of exercise. Hockey is a game of constantly changing tempo, requiring players to sustain high levels of

Table 19.4 Ergometer Total Work and Fatigue Index of Elite and Subelite Players

Sex	Squad	Position	n	Mean	SD	Range
				5 × 6 s total work (J/kg)		
Female	Senior	Field	428	349.0	27.0	282.2–427.3
		Goalkeeper	80	326.5	26.0	282.2–388.6
	Youth	Field	124	328.9	21.1	282.0–373.3
		Goalkeeper	8	305.6	18.6	283.5–326.1
	Subelite	Field	39	329.7	24.5	281.3–373.1
		Goalkeeper	2	329.9	24.1	312.8–346.9
Male	Senior	Field	337	412.6	32.4	297.2–471.0
		Goalkeeper	44	399.1	31.5	339.3–468.9
	Youth	Field	125	400.4	38.5	293.8–520.7
		Goalkeeper	16	374.1	22.7	337.2–414.4
	Subelite	Field	21	404.2	31.3	300.3–445.9
		Goalkeeper	4	369.3	27.0	340.3–404.7
				Work decrement (%)		
Female	Senior	Field	413	9.4	3.6	3.0–23.8
		Goalkeeper	82	9.6	3.0	4.0–16.9
	Youth	Field	111	8.8	3.4	3.1–22.3
		Goalkeeper	12	12.4	4.8	5.1–23.6
	Subelite	Field	40	9.8	4.4	3.2–25.0
		Goalkeeper	5	9.6	2.9	6.9–14.0
Male	Senior	Field	333	10.1	4.6	3.4–28.3
		Goalkeeper	43	15.3	4.3	5.3–23.6
	Youth	Field	124	11.2	4.7	3.1–24.1
		Goalkeeper	16	11.7	4.6	4.5–18.9
	Subelite	Field	20	10.9	5.0	3.4–22.4
		Goalkeeper	4	11.4	2.4	7.8–12.9
				Power decrement (%)		
Female	Senior	Field	398	8.0	2.9	3.0–20.8
		Goalkeeper	82	8.4	2.6	3.1–13.8
	Youth	Field	107	7.1	3.0	3.2–17.8
		Goalkeeper	12	9.9	4.4	4.1–18.9
	Subelite	Field	43	7.7	3.9	3.0–24.7
		Goalkeeper	5	8.5	3.1	5.2–12.2
Male	Senior	Field	330	9.7	3.9	3.0–24.1
		Goalkeeper	42	12.4	4.0	4.4–18.7
	Youth	Field	113	9.6	4.0	3.2–19.7
		Goalkeeper	16	9.7	3.8	5.0–16.6
	Subelite	Field	19	9.9	5.2	3.2–25.2
		Goalkeeper	4	10.8	0.7	10.0–11.7

Table 19.5 Sprint Scores of Elite and Subelite Players

Sex	Squad	Position	n	Mean	SD	Range
				Sprint 0–10 m (s)		
Female	Senior	Field	409	1.98	0.07	1.80–2.24
		Goalkeeper	75	2.07	0.07	1.90–2.21
	Youth	Field	193	1.99	0.08	1.82–2.18
		Goalkeeper	28	2.11	0.08	1.97–2.30
	Subelite	Field	132	1.99	0.07	1.78–2.28
		Goalkeeper	14	2.06	0.09	1.91–2.20
Male	Senior	Field	131	1.81	0.07	1.61–2.00
		Goalkeeper	13	1.84	0.11	1.70–2.06
	Youth	Field	152	1.82	0.07	1.63–2.09
		Goalkeeper	22	1.90	0.07	1.70–1.99
	Subelite	Field	63	1.85	0.07	1.67–2.03
		Goalkeeper	15	1.85	0.12	1.65–2.06
				Sprint 0–40 m (s)		
Female	Senior	Field	406	6.04	0.22	5.38–6.74
		Goalkeeper	67	6.41	0.22	6.00–7.06
	Youth	Field	193	6.10	0.21	5.63–6.76
		Goalkeeper	28	6.61	0.23	6.20–7.19
	Subelite	Field	132	6.11	0.21	5.66–6.84
		Goalkeeper	14	6.42	0.30	5.96–7.05
Male	Senior	Field	118	5.38	0.17	4.97–5.80
		Goalkeeper	13	5.45	0.18	5.14–5.78
	Youth	Field	140	5.40	0.17	4.98–5.89
		Goalkeeper	19	5.61	0.18	5.23–5.92
	Subelite	Field	63	5.48	0.21	5.05–6.16
		Goalkeeper	15	5.58	0.35	5.02–6.18
				Sprint 30–40 m (s)		
Female	Senior	Field	346	1.32	0.07	1.11–1.55
		Goalkeeper	57	1.43	0.07	1.30–1.62
	Youth	Field	163	1.34	0.06	1.10–1.52
		Goalkeeper	22	1.49	0.06	1.40–1.66
	Subelite	Field	116	1.34	0.06	1.21–1.52
		Goalkeeper	12	1.47	0.09	1.36–1.63
Male	Senior	Field	111	1.15	0.05	1.05–1.25
		Goalkeeper	11	1.16	0.03	1.10–1.21
	Youth	Field	130	1.15	0.04	1.07–1.27
		Goalkeeper	19	1.19	0.04	1.14–1.25
	Subelite	Field	57	1.17	0.06	1.08–1.38
		Goalkeeper	11	1.21	0.08	1.09–1.37

Table 19.6 505 Agility Scores of Elite and Subelite Players

Sex	Squad	Position	Agility (s) n	Mean	SD	Range
Female	Senior	Field	284	2.47	0.13	2.25–2.96
		Goalkeeper	46	2.53	0.10	2.34–2.77
	Youth	Field	173	2.45	0.10	2.17–2.79
		Goalkeeper	27	2.54	0.11	2.29–2.72
	Subelite	Field	124	2.49	0.12	2.22–2.99
		Goalkeeper	14	2.54	0.11	2.43–2.79
Male	Senior	Field	74	2.30	0.11	2.10–2.78
		Goalkeeper	8	2.30	0.09	2.14–2.46
	Youth	Field	96	2.28	0.10	2.06–2.57
		Goalkeeper	14	2.28	0.06	2.28–2.48
	Subelite	Field	52	2.31	0.12	2.03–2.54
		Goalkeeper	15	2.40	0.13	2.11–2.56

continuous effort. A high degree of aerobic power is therefore necessary to meet energy demands within a game, whether it directly contributes to energy requirements or aids recovery from anaerobic efforts. In addition, aerobic power enables the athlete to play and practice longer and at higher intensities.

Maximal Oxygen Consumption Test

See table 19.7 for normative data.

Rationale: Determination of oxygen consumption by open-circuit spirometry during a suitably structured progressive test to volitional exhaustion is the accepted criterion method for determining maximal aerobic power.

Procedures:

1. The test is conducted on a motorized treadmill. Two to five minutes of warm-up are required to familiarize athletes with treadmill and gas collection apparatus. The warm-up speed should be one that allows athletes to run using their normal gait. Table 19.8 provides the exact test protocol.

2. Heart rate is measured during the last 10 s of each minute and immediately prior to completion of the test.

3. Termination of test is on signal from the subject.

4. $\dot{V}O_2$max is the highest minute score (may include the cumulative addition of consecutive shorter periods adding up to 1 min) associated with the following criteria:

- Heart rate reaching a plateau with increasing workload
- Plateau in $\dot{V}O_2$max with increasing workload ($\dot{V}O_2$ increases less than 150 ml over the score obtained during the previous minute)
- An r value of 1.15 or greater

Limitations: Scores obtained for maximum oxygen consumption are influenced by the test protocol and the method of sample collection that

Table 19.7 $\dot{V}O_2$max Scores of Junior and Senior Players

Sex	Squad	Position	$\dot{V}O_2$max ($ml \cdot kg^{-1} \cdot min^{-1}$) n	Mean	SD	Range
Female	Senior	Field	126	51.25	4.72	39.38–63.61
	Youth	Field	29	50.49	4.89	44.35–66.71
Male	Senior	Field	110	59.56	4.30	49.32–71.71
	Youth	Field	52	59.72	4.22	51.91–70.07

Table 19.8 Testing Protocol for Maximal Oxygen Consumption

	Males		Females	
Time (min)	Speed (km/h)	Grade (%)	Speed (km/h)	Grade (%)
0–1	12	0	10	0
1–2	12	0	10	0
2–3	14	0	12	0
3–4	14	0	12	0
4–5	16	0	14	0
5–6	16	2	14	2
6–7	16	4	14	4
7–8	16	6	14	6
8–9	16	8	14	8
9–10	16	10	14	10

For each extra minute beyond 10 min, grade increased by 1%.

the laboratory employs, as well by as subject motivation.

20-Meter Multistage Shuttle Run

See table 19.9 for normative data.

Rationale: It is often neither practical nor cost effective to utilize gas analysis to determine maximal aerobic power. The 20 m multistage shuttle run test allows multiple players to be tested at once, for little cost.

The 20 m multistage shuttle run test has been shown to be a valid and reliable test for the prediction of $\dot{V}O_2$max (Leger and Lambert 1982; Leger et al. 1988; Ramsbottom et al. 1987), making it a useful field test when circumstances do not allow for laboratory testing.

Procedures: Refer to page 130. Note that the test area should be located in a corner of the hockey field to ensure that it is as flat as possible.

Limitations: The test score has a high degree of relationship to $\dot{V}O_2$max ($r \approx 0.9$); however, it must be acknowledged that a degree of error exists (20% of variance in tests is not explained by $\dot{V}O_2$max).

Administration of Tests

Although the timing of the test will depend largely on the training schedule, it is important to administer them at a consistent time of day to reduce circadian rhythm influence on performance (Winget et al. 1985). The following order of testing should be adhered to:

Laboratory tests

1. Anthropometry
2. Warm-up and stretch
3. Vertical jump
4. 10 s ergometer sprint: minimum of 3 min recovery
5. 5 × 6 s ergometer sprint: minimum of 30 min recovery
6. Maximal aerobic power test

Field tests

1. Acceleration and speed test
2. Agility test
3. 20 m multistage shuttle run

Laboratory testing (excluding $\dot{V}O_2$max) is normally carried out at the end of each training macrocycle to help determine the effectiveness of the program completed and the structure of the next program. The $\dot{V}O_2$max test is normally carried out twice per year, at the end of general preparation and immediately preceding major competition.

Field tests of speed and agility are conducted before and after those training macrocycles involving speed/agility development. The field aerobic power tests (20 m multistage shuttle run) are conducted on all test occasions that do not require laboratory assessment of $\dot{V}O_2$max.

Table 19.9 20 m Multistage Shuttle Run Scores of Elite and Subelite Players

			Shuttle			
Sex	Squad	Position	n	Mean	SD	Range
Female	Senior	Field	277	11.81	0.97	8.54–13.46
	Youth	Field	151	11.33	1.37	8.11–15.31
	Subelite	Field	98	11.04	0.97	9.06–13.38
Male	Senior	Field	221	13.34	1.03	9.91–15.5
	Youth	Field	113	13.37	1.01	11.00–15.69
	Subelite	Field	48	13.08	0.98	8.82–15.08

Physiological Assessment

Team Sport (Laboratory Based)

PHYSIOLOGICAL ASSESSMENT—HOCKEY:

Name ______________________________ Date ___/___/___ Time ________

DOB ______________ Squad ______________________ Position: ST MF FB GK

Training phase: Transition Gen. prep. Spec. prep. Competition Peak Rehabilitation

Previous 48-hour training load: None Light Medium Heavy

Venue ______________ Tester/s ____________ Data entry completed by __________

Anthropometry

Height (cm)	______	Supra (mm)	______
Weight (kg)	______	Abdom (mm)	______
Tricep (mm)	______	Thigh (mm)	______
Subscap (mm)	______	Calf (mm)	______
Bicep (mm)	______	**Total (7)**	______

Tester = ______________________________

10-Second Ergometer Max Sprint

	Total work output		**Peak power output**	
	J	**J/kg**	**W**	**W/kg**
6 s	______	______	______	______
10 s	______	______	______	______

95% of 6 s work = ____________ J

5 × 6-Second Ergometer Maximal Sprints With 24-Second Recovery

	Total work output		**Peak power output**	
	J	**J/kg**	**W**	**W/kg**
1	______	______	______	______
2	______	______	______	______
3	______	______	______	______
4	______	______	______	______
5	______	______	______	______
Total	______	______	______	______
% Dec	______	______	______	______

Data acquisition method = ______________________

Vertical Jump

Jump and reach ______________

Stand and reach ______________

Vertical jump ______________

Notes

____

____

____

$\dot{V}O_2$max Assessment (Female)

Team Sport (Laboratory Based)

$\dot{V}O_2$max ASSESSMENT: HOCKEY (Female)

Name ______________________ Date ___/___/___ Time ________

DOB ____________ Squad ______________ Position: ST MF FB GK

Training phase: Transition Gen. prep. Spec. prep. Competition Peak Rehabilitation

Previous 48-hour training load: None Light Medium Heavy

Venue ____________ Tester/s ____________ Data entry completed by ____________

Time (min)	Speed (km/h)	Slope (%)	Heart rate
1 ________	10 ________	0 ________	________
2 ________	10 ________	0 ________	________
3 ________	12 ________	0 ________	________
4 ________	12 ________	0 ________	________
5 ________	14 ________	0 ________	________
6 ________	14 ________	2 ________	________
7 ________	14 ________	4 ________	________
8 ________	14 ________	6 ________	________
9 ________	14 ________	8 ________	________
10 ________	14 ________	10 ________	________
11 ________	14 ________	11 ________	________

$\dot{V}O_2$max Assessment (Male)

Team Sport (Laboratory Based)

$\dot{V}O_2$max ASSESSMENT: HOCKEY (Male)

Name ______________________ Date ___/___/___ Time ________

DOB ____________ Squad ______________________ Position: ST MF FB GK

Training phase: Transition Gen. prep. Spec. prep. Competition Peak Rehabilitation

Previous 48-hour training load: None Light Medium Heavy

Venue ____________ Tester/s ____________ Data entry completed by ____________

Time (min)	Speed (km/h)	Slope (%)	Heart rate
1 ________	12 ________	0 ________	________
2 ________	12 ________	0 ________	________
3 ________	14 ________	0 ________	________
4 ________	14 ________	0 ________	________
5 ________	16 ________	0 ________	________
6 ________	16 ________	2 ________	________
7 ________	16 ________	4 ________	________
8 ________	16 ________	6 ________	________
9 ________	16 ________	8 ________	________
10 ________	16 ________	10 ________	________
11 ________	16 ________	11 ________	________

Running Acceleration Speed and Agility

Squad ______________________________ Date ___/___/___ Time ________

Training phase: Transition Gen. prep. Spec. prep. Competition Peak Rehabilitation

Surface ______________________________ Venue ______________

Environmental conditions:

Temperature ________ Humidity ________ Wind speed ________ Direction ________

Tester/s ______________________ Data entry completed by ______________

Name	**DOB**	**40 m sprint**						**Agility (505)**		
		Trial 1			Trial 2					
		10 m	30 m	40 m	10 m	30 m	40 m	Trial 1	Trial 2	Trial 3

20-Meter Multistage Shuttle Test

20-Meter Multistage Shuttle Test (English version)

Squad ______________________ Date ___/___/___ Time ________

Training phase: Transition Gen. prep. Spec. prep. Competition Peak Rehabilitation

Surface ______________________ Venue ________

Environmental conditions:

Temperature ________ Humidity ________ Wind speed ________ Direction ________

Tester/s ______________________ Data entry completed by ________

Name	DOB	Level	Shuttle	Name	DOB	Level	Shuttle

Skinfold Data Collection

Squad: ____

Tester: ____ Date: ____/____/____

Calipers used: ____________________________ Position ID: ST MF FB GK

Data entry completed by: ________________

Name	DOB	Posit.	Ht (cm)	Wt (kg)	Tricep (mm)	Subscap (mm)	Bicep (mm)	Supra (mm)	Abdom (mm)	Thigh (mm)	Calf (mm	Total

Data Collection Proforma—Field Test Proforma

Team Sport

FIELD TEST PROFORMA (INDIVIDUAL): HOCKEY

Name ______________________________ Date ___/___/___ Time ________

DOB ______________ Squad ______________________ Position: ST MF FB GK

Training phase: Transition Gen. prep. Spec. prep. Competition Peak Rehabilitation

Previous 48-hour training load: None Light Medium Heavy

Surface ______________________________ Venue ______________

Environmental conditions:

Temperature ________ Humidity ________ Wind speed ________ Direction ________

Tester/s ______________________ Data entry completed by ______________

Anthropometry

Height (cm)	______	Supra (mm)	______
Weight (kg)	______	Abdom (mm)	______
Tricep (mm)	______	Thigh (mm)	______
Subscap (mm)	______	Calf (mm)	______
Bicep (mm)	______	**Total (7)**	______

40 m sprint						**Agility (505)**		
Trial 1			Trial 2					
10 m	30 m	40 m	10 m	30 m	40 m	Trial 1	Trial 2	Trial 3

20 m Multistage Shuttle (English Version)	
Level	
Shuttles	

chapter 20

Protocols for the Physiological Assessment of Netball Players

Lindsay Ellis and Paul Smith

Success in netball at an elite level requires players to possess an exceptional level of skill in combination with high levels of aerobic and anaerobic fitness (Chad and Steele 1990). The game itself is an interval team sport requiring repeated short, high-intensity efforts. As activity patterns vary between playing positions, the physical and physiological requirements will vary accordingly (Chad and Steele 1990; Otago 1983).

The tests in this chapter are those currently in use by Australian national squads and comprise field tests that players can perform on court at the training venue, as well as strength tests that require access to a weight-training facility. In use for six years in both field and training settings, they have been found to be sensitive to training and also time and cost effective. The field tests and strength tests are two separate procedures and must be conducted on separate occasions. The field tests measure some of the major physical abilities required for playing netball—namely, agility or footwork, acceleration, endurance, flexibility, and "explosive" leg power. Resistance training is an important component of the netball player's physical preparation, and the selected strength tests enable basic diagnosis of strength and changes related to training.

In general, field tests and strength tests should be carried out at the end of each appropriate macrocycle to help determine the effectiveness of the program completed and the structure of the next program. Therefore the majority of the tests will be carried out four or five times per year.

Athlete Preparation and Test Conditions

For details concerning athlete preparation and conditions and environment for test sessions, refer to chapter 2.

Equipment Checklist

1. Field tests

The equipment required for field tests is described in chapter 9.

2. Strength tests
   - ☐ Power rack (particularly relevant rather than a squat rack if an athlete has poor technique)
   - ☐ Olympic bar
   - ☐ Weights (20 kg disks down to 1.25 kg disks)
   - ☐ Weight belt
   - ☐ Blocks or disks that can be used for heel lift
   - ☐ Bench press rack

Order of Tests

It is important that the tests be completed in the same order to control the interference between tests. This order also allows valid comparison of different test occasions. The order is as follows:

- Field tests:
  1. Height, mass, and skinfolds
  2. Warm-up (10 min easy running and stretching; six "stride-throughs" over 30 m, building in intensity from 85% to 100% maximum with jog recovery; and four sprints from stationary starts over 10 m, building in intensity from 90% to 100%)
  3. Sprint test
  4. Agility test
  5. Vertical jump
  6. Sit and reach
  7. Multistage fitness test
- Strength tests:
  1. Squat
  2. Bench press
  3. Abdominal strength test

PROTOCOLS: Netball

Anthropometry

For normative data, see table 20.1.

Rationale: Stature, body mass, and sum of skinfolds provide objective measures of the athlete's body structure. These measurements are important in quantifying differential growth and training influences (Ross and Marfell-Jones 1987). In particular, one can use skinfold measurements to monitor and understand the effect of diet and training intervention on body composition. For example, skinfold measurements may be used to determine whether changes in body mass are due to changes in body fat levels or lean body mass (Norton et al. 1994). Increased body fat levels may have an adverse effect on netball performance. Excess fat has no functional role in activities on the court and can be regarded as dead weight. Excess fat will decrease acceleration and is detrimental to the player's ability to rebound or leap for the ball (Norton et al. 1994).

The development of anthropometric profiles of elite netball players is also important. Analysis of such information helps increase understanding of the physical requirements of particular netball positions (Bale and Hunt 1986).

Test Procedure: The equipment and techniques for measuring height, body mass, and seven skinfold thicknesses are presented in chapter 5. The seven sites are triceps, subscapular, biceps, supraspinale, abdominal, front thigh, and medial calf.

Although the description of skinfold measurement procedures seems simple, a high degree of technical skill is essential for consistent results, especially when the procedures are carried out under field test conditions (see page 71). It is therefore important that these measurements be taken by an experienced tester who has been trained in these techniques. It is also important that the same tester be used for each retest to ensure reliability.

The individual skinfold measures as well as the sum of the seven should be recorded.

Speed and Acceleration

See table 20.2 for normative data.

Rationale: While netball players rarely sprint at maximum speed during the game, play entails a large number of efforts that involve acceleration from a jog, shuffle, or stationary position (Steele and Chad 1992). These running efforts average less than 2 s; but the acceleration involved in making position, evading an opponent, or intercepting a pass is an important requirement of the game. Therefore, the ability of the netball player to accelerate is best indicated by the times to 5 and 10 m, with 20 m being the maximum distance any player might travel.

20-Meter Sprint Test Procedure: The equipment and test procedure are described on pages 129 and 130.

Ideally the speed tests should take place on a wooden floor that is used for indoor court play. It is important to ensure that the surface used is consistent for each test: results can be affected by dust or dirt on the floorboards or with a totally different playing surface. It is also important that athletes complete the tests in their competition shoes.

Athletes must warm up adequately in preparation for the sprint and agility tests. A poor warm-up may result in unreliable or slower results. The warm-up described in the earlier section on test

Table 20.1 Anthropometric Characteristics of State- and National-Level Players

Squad	Time	n	Mean	SD	Range
			Height (cm)		
Australian Open	April 1997	15	178.4	5.2	169.0–184.5
	June 1995	16	178.8	4.3	169.7–184.5
	August 1992	15	178.4	4.0	170.4–184.5
SASI state	May 1995	23	175.9	3.9	169.5–182.8
AIS U/21	1997	15	177.3	4.3	167.5–182.8
	1996	12	177.8	5.6	171.2–187.0
U/17 state players	1997	76	175.6	6.2	159.8–190.5
			Mass (kg)		
Australian Open	April 1997	16	69.7	5.9	59.9–82.5
	June 1995	16	71.5	5.3	63.0–79.5
	August 1992	17	71.1	4.9	78.6–69.4
SASI state	May 1995	23	68.8	5.3	60.3–78.0
AIS U/21	1997	15	68.5	8.1	56.4–85.4
	1996	12	72.2	5.1	64.9–78.1
U/17 state players	1997	76	67.3	7.8	51.6–89.3
			Skinfolds Σ7* (mm)		
SASI state	May 1995	23	103.1	23.5	66.5–164.0
AIS U/21	1997	15	82.4	29.4	33.0–158.5
	1996	12	102.7	28.2	41.5–163.5

*Σ7 = triceps, subscapular, biceps, supraspinale, mid-abdominal, front thigh, and medial calf.

AIS = Australian Institute of Sport, SASI = South Australian Sports Institute.

AIS and SASI are state-level players, one level below the national Australian Open players.

order is recommended; however, any warm-up used should be similar, should be of at least this length and intensity, and must include several maximal sprints and start efforts.

Between each sprint trial there should be a minimum of 2 min recovery.

Agility

For normative data, see table 20.3.

Rationale: The basic movement patterns of netball players involve numerous sideways movements, sudden changes in direction, and quick stops and starts. A player will perform shuffling or sideways movement of the body at speed and full effort between 100 and 300 times a game, depending on the playing position (Steele and Chad 1992).

The 505 test has been selected as a measure of netball agility or footwork. The 505 test is a valid and reliable measure of the individual's ability to make a single change of direction after reaching a high speed (Draper and Lancaster 1985). The test is not meant to simulate the specific movements of the game but was designed to place greatest emphasis on measuring the player's ability to decelerate, change direction, and accelerate from changes of direction.

505 Agility Test Procedure: The equipment and test procedure are described on pages 129 and 132.

Power and Jump Ability

See table 20.4 for normative data.

Rationale and Interpretation: The vertical jump test was selected as the method for evaluating both explosive leg power and jumping ability.

Table 20.2 20 m Sprint Times of State- and National-Level Players

Squad	Time	n	Mean	SD	Range
				5 m (s)	
Australian Open	April 1997	16	1.16	0.06	1.09–1.27
	June 1995	16	1.15	0.09	1.05–1.32
	August 1992	17	1.07	0.05	0.93–1.15
SASI state	May 1995	23	1.15	0.10	1.01–1.37
AIS U/21	1997	15	1.15	0.07	1.06–1.30
	1996	12	1.08	0.03	1.06–1.15
QAS U/19	1995–97	22	1.16	0.05	1.00–1.30
U/17 state players	1997	76	1.23	0.08	1.08–1.52
				10 m (s)	
Australian Open	April 1997	16	1.98	0.08	1.87–2.13
	June 1995	16	1.97	0.09	1.84–2.16
	August 1992	17	1.87	0.06	1.72–1.96
SASI state	May 1995	23	1.94	0.11	1.80–2.17
AIS U/21	1997	15	1.96	0.10	1.85–2.16
	1996	12	1.89	0.04	1.84–1.96
QAS U/19	1995–97	22	1.99	0.05	1.88–2.18
U/17 state players	1997	76	2.05	0.09	1.89–2.37
				20 m (s)	
Australian Open	April 1997	16	3.40	0.12	3.19–3.60
	June 1995	16	3.36	0.12	3.19–3.56
	August 1992	17	3.28	0.09	3.06–3.41
SASI state	May 1995	23	3.36	0.14	3.12–3.65
AIS U/21	1997	15	3.37	0.15	3.17–3.61
	1996	12	3.29	0.08	3.18–3.42
QAS U/19	1995–97	22	3.41	0.09	3.25–3.62
U/17 state players	1997	76	3.49	0.15	3.04–3.93

AIS = Australian Institute of Sport, QAS = Queensland Academy of Sport, SASI = South Australian Sports Institute. AIS, QAS, and SASI are state-level players, one level below the national Australian Open players.

The test is easy to perform in the field and requires limited equipment; in addition, substantial sport-specific normative data are available.

The test provides two jump heights: the relative vertical jump height and absolute jump height. The absolute jump height is the highest point reached or touched when jumping, while the relative jump height is the difference between the standing reach and absolute jump height. Absolute jump height provides an objective measure of the specific skill and physical ability to rebound, with the height reached indicating a component of the performance of goalers and defenders. Relative jump height can be used to monitor the effects of strength-training programs that target the legs and to make comparisons between players.

Readers should note that the vertical jump test result can be affected by the coordination of trunk and arms and that the interpretation of any changes due to strength training may be obscured by changes in the jump action used (Young 1994). It is assumed that the jump action of elite-level netball players will not change, although junior netball players may improve their vertical jump

Table 20.3 505 Run Times of State- and National-Level Players

Squad	Time	n	Mean	SD	Range
			Right foot 505 time (s)		
Australian Open	April 1997	16	2.47	0.13	2.22–2.65
	June 1995	16	2.41	0.09	2.24–2.55
	August 1992	17	2.32	0.08	2.18–2.46
SASI state	May 1995	23	2.50	0.12	2.35–2.84
AIS U/21	1996	12	2.40	0.10	2.33–2.54
QAS U/19	1995–97	22	2.47	0.11	2.32–3.65
			Left foot 505 time (s)		
Australian Open	April 1997	16	2.48	0.18	2.18–2.74
	June 1995	16	2.41	0.08	2.26–2.54
	August 1992	17	2.32	0.07	2.21–2.43
SASI state	May 1995	23	2.48	0.08	2.29–2.63
AIS U/21	1996	12	2.44	0.10	2.36–2.60
QAS U/19	1995–97	22	2.48	0.09	2.30–3.71

AIS = Australian Institute of Sport, QAS = Queensland Academy of Sport, SASI = South Australian Sports Institute. AIS, QAS, and SASI are state-level players, one level below the national Australian Open players.

result due to skill and growth-related changes. Therefore, it may be prudent when one is testing younger netball players to also use the countermovement jump, described on page 139, which eliminates the arm action.

The portable jump measurement device, the Yardstick, is the preferred instrument for testing the vertical jump. It is also possible to test the countermovement jump using this device. While it is acceptable to use a wall-mounted vertical jump board, higher jumps are attained with use of the Yardstick (see page 139).

Vertical Jump Test Procedure: The equipment and test procedure are described on pages 130 and 141. In addition, the following points apply to the measurement of relative and absolute jump height:

- Reach height must be recorded to help ensure test reliability for relative height.
- The absolute jump height is recorded to the nearest 1 cm.
- The relative jump height is the difference between the two points (reach height and jump height) recorded to the nearest 1 cm.
- The jump measurement device used to conduct the test should be recorded.

Flexibility

See table 20.5 for normative data.

Rationale: The sit-and-reach test requires a combined joint action movement that provides a crude approximation of flexibility around the hip joint. As a test of specific flexibility it is generally recognized as having major limitations (Jackson and Baker 1986). Despite these, however, the test does have some specific relation to the physical requirements of netball such as the need for goalers or defenders to field or intercept low bounce passes. The test is extremely easy to administer, requiring little equipment; and substantial sport-specific normative data are available. Readers may refer to page 104 for a more comprehensive rationale for selection of flexibility measures.

Sit-and-Reach Test Procedure: The equipment and test procedure are described on pages 129 and 138.

Endurance

See table 20.6 for normative data.

Rationale: The contribution of the aerobic energy system to netball play will depend on factors such as playing position, the closeness of the match between the competing teams, and the dis-

Table 20.4 Vertical Jump Scores of State- and National-Level Players

Squad	Time	n	Mean	SD	Range
			Relative jump (cm)		
Australian Open[Y]	April 1997	16	53.4	4.6	45–61
	June 1995	16	48.4	6.7	40–59
	August 1992	17	50.7	5.9	43–62
SASI state[W]	May 1995	23	42.6	4.9	34–52
AIS U/21[Y]	1997	15	49.0	7.0	39–63
	1996	12	45.8	4.0	39–54
QAS U/19[Y]	1995–97	22	46.0	4.0	41–55
U/17 state players[Y]	1997	76	44.5	5.5	35–56
			Absolute jump (cm)		
Australian Open[Y]	April 1997	16	298	9	284–312

[Y] = Yardstick protocol (recommended), [W] = wall-mounted board protocol (not recommended).
AIS = Australian Institute of Sport, QAS = Queensland Academy of Sport, SASI = South Australian Sports Institute.
AIS, QAS, and SASI are state-level players, one level below the national Australian Open players.

Table 20.5 Sit-and-Reach Scores of State- and National-Level Players

Squad	Time	n	Mean	SD	Range
			Sit and reach (cm)		
Australian Open	April 1997	13	15.5	5.8	4–25
	June 1995	9	16.1	6.0	4.5–24
	August 1992	17	15.4	5.8	4.5–25
AIS U/21	1996	12	17.3	6.2	4.5–24
U/17 state players	1997	76	11.5	6.7	–15.0 to 28

AIS = Australian Institute of Sport are state-level players, one level below the national Australian Open players.

tribution or pattern of court play (Chad and Steele 1990; Woolford and Angove 1991). Because players may need to sustain high levels of intensity with little opportunity for substantial recovery, the level of aerobic power required is high (Chad and Steele 1990). Furthermore, a high level of aerobic power will enable the netball athlete to play and practice longer at higher intensities.

The multistage fitness test has been selected because it is a valid and reliable estimate of aerobic power, is simple and efficient to administer, and uses movement patterns that incorporate changes of direction similar to those in the game of netball.

Multistage Fitness Test Procedure: The equipment and test procedure are described on pages 129 and 130.

Strength Tests

The coach or instructor who administers strength tests should be qualified and experienced in teaching strength-training methods. The tester should closely scrutinize the athlete's technique during the warm-up, particularly for the squat, and should give instruction and coaching tips if needed. It is also the tester's responsibility to ensure that the

Table 20.6 Multistage Fitness Test Scores of State- and National-Level Players

Squad	Time	n	Mean	SD	Range
			Level/Shuttle (as a decimal)		
Australian Open	April 1997	16	11.18	1.12	9.10–13.38
	June 1995	12	11.21	1.25	8.60–13.10
	August 1992	17	11.52	1.19	8.82–13.38
SASI state	May 1995	23	10.73	1.33	8.27–13.00
QAS U/19	1995–97	18	10.33	0.90	9.60–11.40
U/17 state players	1997	76	9.99	1.31	7.10–13.07
			Estimated $\dot{V}O_2max$ ($ml \cdot kg^{-1} \cdot min^{-1}$)		
Australian Open	April 1997	16	50.6	3.9	43.6–58.6
	June 1995	12	50.9	4.3	41.8–57.4
	August 1992	17	51.9	4.3	42.7–58.6
SASI state	May 1995	23	49.3	4.6	40.8–57.1
AIS U/21	1997	15	50.8	5.6	39.5–62.4
	1996	12	53.1	5.8	44.6–64.7
QAS U/19	1995–97	18	47.8	3.1	43.6–54.6
U/17 state players	1997	76	46.5	–	36.6–57.3

AIS = Australian Institute of Sport, QAS = Queensland Academy of Sport, SASI = South Australian Sports Institute. AIS, QAS, and SASI are state-level players, one level below the national Australian Open players.

lifting area is clear and safe and that equipment is in good working order.

For the squat and bench press tests, the bar should be loaded such that no more than five sets are performed in achieving the three-repetition maximum (3RM) or six-repetition maximum (6RM). A 6RM protocol is used for athletes who have not been regularly exposed to the higher-intensity loads (i.e., lower than five reps) in training. The athlete should have a minimum of 3 min recovery between sets.

The athlete's 3RM or 6RM best result will be converted to a one-repetition maximum (1RM) prediction using the conversion factors of 1.08 and 1.2, respectively (Baker 1995).

If the tester believes that the athlete's technique is so poor that he or she cannot perform the squat test safely, it is imperative that the test not be conducted. Those athletes should perform a split squat as described further on. The split squat load is not recorded for the test, but is an indicator to the coach of basal leg strength levels. Before leaving the weight-training facility the athlete should perform a number of light sets of squats (barbell only) under instruction from the tester. Follow-up technical coaching or testing, or both, should be arranged at this time.

For normative data, see table 20.7.

Rationale: Netball play involves explosive actions such as leaping for intercepts, rebounding, and hard changes of direction. These explosive actions require the athlete to exert high levels of force over a relatively short period of time. Various investigators have demonstrated that maximum force and rate of force development are important to the performance of such activities (Hakkinen et al. 1986; Viitasalo and Aura 1984; Schmidtbliecher 1992). Furthermore, resistance training and plyometric training effectively develop the strength qualities of maximum force and rate of force development, respectively (Hakkinen et al. 1985; Schmidtbliecher 1992). Resistance training, therefore, should be an important component of the netball player's physical preparation, and the selected strength tests enable basic diagnosis of strength levels and changes related to training. The bench press and squat lifts have been employed because they both test multijoint function, use the major muscle groups of the upper and lower body, and are common training exercises familiar to the high-level player.

Squat

Equipment:

- A power rack rather than squat rack should be used if there is access to one, particularly if the athlete has poor technique.
- Olympic bar, weights (from 20 kg disks down to 1.25 kg disks).
- Weight belt.
- Blocks or weight disks that may be used as a lift under the athlete's heels.

Procedure:

1. A 3RM or 6RM test is to be used, as already stated.

2. The eccentric phase of the test should be fairly slow and well controlled.

3. The squat should be a "half squat" with the top of the thighs parallel with the floor at the bottom of the movement. The feet should be positioned shoulder-width or slightly wider apart.

4. During the test the heels should be in contact with the floor or a block at all times.

5. It is important to maintain correct posture with the back flat and chest out, and to avoid excessive forward lean.

6. It is the athlete's choice whether to use a weight belt or not.

Split Squat

Equipment:

- A power rack rather than squat rack should be used if there is access to one, particularly if the athlete has poor technique.
- Olympic bar, weights (from 20 kg disks down to 1.25 kg disks).
- Weight belt.

Procedure:

1. The athlete stands erect with the barbell resting on the shoulders, or grasping the dumbbells at his or her side with the feet shoulder-width apart. The head is upright and the shoulders back (i.e., good posture).

2. The athlete takes an exaggerated step forward with one leg. The forward ankle should be directly under the forward knee with the rear knee directly or slightly behind the hips. The trunk should be straight and upright. The shoulder-width stance should be maintained with both feet pointing forward in line with the knees. The back foot rests on the ball of the foot. The weight is evenly distributed over both feet.

3. The movement is initiated by bending the back knee with the front knee following in concert. This allows the body to be lowered without forward movement (as opposed to the lunge, where the back leg remains straight and the body rocks forward with the load taken on the front quads).

4. The body is lowered until the knee of the back leg just touches the floor.

5. Pausing at the bottom, the lifter extends both legs and rises to the start position. The feet remain in the split position throughout the entire exercise.

6. On completion of the 3RM or 6RM, have the athlete alternate the legs and retest. It is the forward leg that designates left or right. Both legs should be tested, and the score of the weakest leg recorded.

Safety is an important aspect of this test. There should be a spotter on each side of the athlete. Ensure that the athlete's forward step is the correct length; this will avoid undue torque around the forward knee joint. The eccentric phase should be slow and controlled, with a pause at the bottom to avoid bouncing through the bottom phase and injuring the knee of the back leg. Avoid dumbbells greater than 15 kg (use a barbell instead) so that the shoulders or hands do not fatigue.

Bench Press

Equipment:

- Bench press rack
- Olympic bar, weights (from 20 kg disks down to 1.25 kg disks)

Procedure:

1. A 3RM or 6RM test is to be used.

2. The eccentric phase of the test should be fairly slow and well controlled.

3. The bottom of the movement occurs when the bar touches the chest. There should be no bouncing of the bar off the chest. The hips and shoulders must remain in contact with the bench at all times. The back should not be arched excessively during or before the press. The hands should be positioned on the bar such that they are directly above the elbow when the bar touches the chest.

Seven-Level Sit-Up Abdominal Strength Test: One Attempt

See table 20.8 for normative data.

The equipment and test procedure are described on pages 130 and 143.

Table 20.7 Strength Scores of Australian Senior National-Level Players

	Body mass (kg)	Estimated 1RM bench press (kg)	Estimated 1RM squat (kg)	Estimated 1RM bench press (kg/kg body mass)	Estimated 1RM squat (kg/kg body mass)
n	12	12	10	12	12
Mean	70.6	51.3	82.1	0.7	1.2
Minimum	82.5	43.2	70.2	0.6	1.0
Maximum	59.5	62.1	102.6	0.9	1.5

1RM scores are estimated from 3RM results.

All tests completed in May 1997.

Table 20.8 Abdominal Stage Test Scores of State- and National-Level Players

Squad	Time	n	Mean	SD	Range
			Number of stages		
Australian Open	April 1997	12	5.6	–	4–7
	May 1997	9	5.3	1.2	4–7
QAS U/19	1995–97	18	3	1	2–7

QAS = Queensland Academy of Sport.

QAS are state-level players, one level below national Australian Open players.

chapter 21

Protocols for the Physiological Assessment of Rowers

Allan Hahn, Pitre Bourdon, and Rebecca Tanner

All major national- and international-level rowing races are over a distance of 2000 m. Depending on boat category, weather conditions, and such factors as the depth and temperature of the water, the time taken to complete the distance typically varies between 5 min 30 s and 9 min.

Purposes of Testing Rowers

The physiological testing of rowers has three major aims:

- To assess characteristics believed to be the primary determinants of rowing performance
- To produce guidelines for determining appropriate training intensities
- To ensure that rowers are coping with and adapting to the prescribed training programs

The required information should come from the least possible number of tests so that coaches and athletes do not have to spend too many hours in the laboratory at the expense of training. During a single visit to the laboratory, rowers can undergo routine blood screening, anthropometric assessment, and a rowing ergometer test. This is sufficient to provide information on all of the aims just mentioned. Normally, the rowers will require full assessment only four to five times per year.

- **Determinants of rowing performance.** Performance has been found to correlate highly with maximum oxygen uptake (Secher et al. 1982), although anaerobic capacity may also be influential. Much evidence indicates that anthropometric characteristics may also have some influence on rowing performance. Champion rowers tend to be taller (Hahn 1990) and heavier (Secher and Vaage 1983; Hahn 1990) than their less successful counterparts. In addition, they typically have lower subcutaneous skinfold readings (Hahn 1990). Many have long limbs, not only in absolute terms but also in proportion to their height (Ross et al. 1982; Hahn 1990). However, Ross et al. (1982) believe that factors relating to body proportionality are probably less important than sheer size in determining success. It is possible that, for mechanical reasons, anthropometric factors have a greater influence on performance in sweep-oar rowing than in sculling. Anthropometric studies of lightweight rowers competing at the 1985 world championships showed that they, too, tended to be tall, lean, and long-limbed when compared with a reference population of university students (Rodriguez 1986). It seems clear, then, that the physiological assessment of rowers should include measurement of maximum oxygen uptake, anaerobic capacity, specific strength or power, and various anthropometric characteristics.

- **Establishing guidelines for determining training intensities.** Rowing has an accepted training classification in which work intensities are divided into six categories. The basis for this division is the identification of two specific points on the lactate-workload curve. The first is the point beyond which the lactate concentration begins to rise above resting levels (i.e., the lactate threshold [LT]). The second is the point beyond which lactate begins to accumulate very rapidly (i.e., anaerobic threshold [AT]). Training performed below AT is regarded as promoting muscle "utilization" of oxygen. The utilization category is divided into three subcategories. Work carried out below LT is termed U3; that performed between LT and AT is divided into two halves, with the lower half termed U2 and the upper half U1. Anaerobic threshold work (i.e., work performed right at AT) constitutes a category in itself. Training performed above AT is classified as "transport" work. The sixth training category comprises "anaerobic" work. This classification is not based on the lactate curve, but includes all efforts performed at or above race pace.

- **Monitoring responses to training.** Responses to training can be assessed largely in terms of change in some of the performance determinants just described. However, during certain phases of training the emphasis may be, for example, on improving the aerobic capabilities of skeletal muscle. Change in maximum oxygen uptake may not necessarily reflect such an improvement. A shift in the curve indicating the relationship of blood lactate or blood pH to workload could be a more sensitive indicator. Furthermore, one can often obtain valuable information by monitoring routine hematological parameters and serum ferritin concentrations. Reduction of ferritin concentrations is a common response to endurance training and is believed to reflect a decrease in body iron stores. If iron reserves are depleted beyond a certain critical point, ability to adapt to training is affected. Resting blood characteristics, and blood lactate and/or pH responses to incremental exercise, should therefore be monitored for elite rowers.

Laboratory Environment and Subject Preparation

To standardize test conditions, the following specific guidelines—in addition to the general pretest guidelines recommended in chapter 2—should be met:

- **Training.** The athlete must not train at all in the 15 h preceding the test. On the day before the test, the afternoon training session should consist of no more than 12 km on the water, and should be of low intensity. There should be no heavy weight training and no exercise to which the athlete is not accustomed.

- **Diet.** Athletes should eat a normal meal (incorporating a high carbohydrate component) on the evening preceding the test and, if scheduling allows, also on the day of the test. However, they should eat nothing for at least 2 h preceding the test. They should give special attention to ensuring good hydration in the lead-up to the test.

Equipment Checklist

1. Routine hematology
   - ☐ Disposable rubber gloves
   - ☐ Tourniquet
   - ☐ Sterile alcohol swabs
   - ☐ Cotton wool
   - ☐ 10 ml syringe
   - ☐ 21-gauge disposable needle
   - ☐ EDTA tube
   - ☐ Serum tube
   - ☐ Tissues
   - ☐ Band-Aids
   - ☐ Sharps container
   - ☐ Biohazard bag
2. Anthropometry

The relevant equipment is described in chapter 5.

3. Ergometer tests
   - ☐ Concept IIc rowing ergometer
   - ☐ Heart rate monitoring system
   - ☐ Expired gas analysis system (according to general recommendations)
   - ☐ Lactate analyzer (and blood gas analyzer if possible)
   - ☐ Blood sampling equipment:
     - __ Finalgon ointment or cream
     - __ Autolet, lancets, platforms
     - __ Sterile alcohol swab
     - __ Tissues
     - __ Heparinized capillary tubes
     - __ Pipette (if required)
     - __ Disposable rubber gloves
     - __ Sharps container
     - __ Biohazard bag

Routine Hematology

A 5 ml sample is drawn from an antecubital vein with the athlete in a supine position. Approximately 3 ml is injected into an EDTA tube, thoroughly mixed, and subsequently analyzed for erythrocyte count, hemoglobin concentration, hematocrit, mean cell volume, and leukocyte count using a standard hematology analyzer. The remaining blood is placed in a serum tube and allowed to clot. Serum is then extracted to determine ferritin concentration. All blood sampling and analysis must be performed by appropriately qualified staff.

After blood collection, athletes should be advised to apply pressure to the sampling site for 2-3 min to ensure that bleeding has stopped. Those responsible should attempt to ensure that for an individual rower, all blood samples taken during a season are collected at the same time of day.

Normative values for the various hematological parameters are listed in table 21.1.

Table 21.1 Normative Hematological Data of Australian Rowers*

	No. of rowers	Mean	SD	Range
Red blood cell count ($\times 10^{12}$/L)				
HW male	296	5.17	0.35	4.47–5.87
HW female	286	4.61	0.29	4.03–5.19
LW male	159	4.92	0.33	4.26–5.58
LW female	87	4.54	0.31	3.92–5.16
Hematocrit (%)				
HW male	296	46.3	2.7	40.9–51.7
HW female	286	41.5	2.4	36.7–46.3
LW male	159	44.9	2.7	39.5–50.3
LW female	87	41.3	2.3	36.7–45.9
Hemoglobin concentration (g/dl)				
HW male	296	15.4	0.8	13.8–17.0
HW female	286	13.8	0.9	12.0–15.6
LW male	159	14.9	0.9	13.1–16.7
LW female	87	13.6	0.8	12.0–15.2
Mean cell volume (fl)				
HW male	296	89.5	3.7	82.1–96.9
HW female	286	90.2	3.3	83.6–96.8
LW male	159	91.4	3.1	85.2–97.6
LW female	87	91.3	4.3	82.7–99.9
White blood cell count ($\times 10^{9}$/L)				
HW male	296	6.1	1.6	2.9–9.3
HW female	286	6.7	1.8	3.1–10.3
LW male	159	6.2	1.6	3.0–9.4
LW female	87	6.3	1.3	3.7–8.9
Serum ferritin concentration (ng/ml)				
HW male	296	139	80	30–300
HW female	286	62	37	20–140
LW male	159	156	74	30–300
LW female	87	65	33	20–140

*All rowers at least of national standard.

HW = heavyweight, LW = lightweight.

Anthropometry

Body mass and skinfold thicknesses are routinely measured when rowers undergo ergometer testing. A full anthropometric profile including height, sitting height, arm span, girths, breadths, and lengths is generally completed only once or twice per season. Individual anthropometric characteristics may be measured more often if required, particularly in the case of lightweight athletes or athletes who may still be growing. Normative values for the various anthropometric characteristics are presented in table 21.2. Following are guidelines for performance of all anthropometry tests.

Height: Standing height or stature is measured using the method described in chapter 5.

Body Mass: The body mass of the rower, who is clothed only in underwear or in a rowing suit, is measured using scales that have recently been calibrated against certified masses. The reading is taken to the nearest 0.10, 0.05, or 0.02 kg, according to the resolution of the scales.

Sitting Height: A specialized apparatus is used for measurement of sitting height. This consists of a bench (approximately 80 cm in height) backing on to a stadiometer, with the zero point at the bench surface. The athlete sits on the bench, with the lower legs hanging over the front edge at right angles to the thighs, and is instructed to sit as tall as possible, without tilting the head backward. The tester orients the head in the Frankfort plane and applies gentle traction to the mastoid processes. The buttocks and back of the athlete must be firmly against the stadiometer. Arching of the lower back should be avoided. The athlete is instructed to take a deep breath, and the measurement is made before exhalation. The measurement involves bringing the horizontal arm of the stadiometer downward until it is in firm contact with the vertex of the athlete's head. The stadiometer scale is read to the nearest 0.1 cm.

Arm Span: The athlete stands facing a wall on which an appropriate scale has been placed. The feet must be 3-4 cm apart. The arms are stretched out parallel to the floor. The tip of the middle finger of one hand is placed on the zero point of the scale. The athlete is then asked to stretch out to obtain the greatest possible span. The span is read to the nearest 0.5 cm.

Girths: All girth measurements are made using the methods described in chapter 5.

Skinfold Measurements: Skinfold measurements are taken at seven different sites (triceps, subscapular, biceps, supraspinale, abdominal, front thigh, and medial calf) for both males and females. The measurements are made using the methods described in chapter 5.

Other Anthropometric Measures: Occasionally there may be reason to conduct more detailed anthropometric assessment, such as somatotyping or determination of O-scale profiles. These assessments should be made only by staff who have achieved appropriate accreditation as at least Level 2 anthropometrists. The measurements should also be made using the methods outlined in chapter 5 or as described by Norton et al. (1996).

Ergometer Tests

A Concept IIc rowing ergometer is the preferred ergometer for use during all tests. This ergometer is simple in design, easy to maintain, fairly portable, and relatively inexpensive. If it is not feasible to use a Concept IIc rowing ergometer, a Concept IIb rowing ergometer should serve as the alternative.

All ergometer tests must be completed according to the following guidelines.

Concept IIc:

Category	Lever setting
Lightweight female	2
Heavyweight female	3
Lightweight male	3
Heavyweight male	4

Concept IIb: All tests must be conducted using the small gear (i.e., chain on large cog) and with the vent closed.

The requirements listed are based on biomechanical studies suggesting that these settings are associated with the least discrepancy between forces generated on the ergometer and those generated on the water.

There may be minor variations between ergometers with regard to flywheel characteristics and resistance to flywheel rotation. Therefore, all tests on a particular rower during a single season should ideally be conducted on the same well-maintained ergometer. It is important to take precautions to ensure that the ergometer cannot slip during the test. These precautions may consist of bolting the

Table 21.2 Normative Anthropometric Data of Australian Rowers*

	No. of rowers	Mean	SD	Range
Height (cm)				
HW male	49	191.9	5.0	181.5–203.2
HW female	43	179.2	3.6	170.3–186.9
LW male	27	182.7	5.5	170.0–191.5
LW female	28	169.3	6.2	157.0–178.0
Weight (kg)				
HW male	49	90.2	4.7	76.6–99.2
HW female	43	74.0	5.4	64.3–89.0
LW male	27	72.6	2.1	67.4–75.0
LW female	28	61.6	1.7	55.1–64.6
Skinfolds (mm)—Σ7 sites				
HW male	49	47.5	8.3	33.2–67.4
HW female	43	86.3	21.1	40.7–152.2
LW male	27	41.9	7.0	31.8–61.7
LW female	28	67.9	13.2	50.8–101.6
Sitting height (cm)				
HW male	23	101.8	3.1	95.2–110.0
HW female	24	94.5	2.3	89.7–98.6
LW male	12	96.6	2.5	92.1–101.7
LW female	10	90.8	2.4	85.2–93.5
Arm span (cm)				
HW male	17	199.8	5.6	184.5–206.0
HW female	22	182.9	4.5	174.5–194.5
LW male	12	188.9	4.2	181.0–195.5
LW female	28	174.0	2.9	169.0–180.0

*All rowers at least of national standard.

HW = heavyweight, LW = lightweight.

ergometer to the floor, taping it down (duct tape is effective in this regard), or placing it on a nonslip surface. Additionally, the ergometer should be set up so that it is no closer than 1 m to the nearest wall or barrier.

Progressive Ergometer Test

The progressive ergometer test is designed to allow determination of maximum oxygen uptake, the workloads and heart rates corresponding to LT and AT, and the relationship of blood lactate and blood pH to workload. It may also provide a basis for estimating anaerobic capacity. Normative values for the major parameters measured during the progressive ergometer test are presented in table 21.3. The values are based on tests conducted on rowers of at least national standard. However, many of the tests were performed in Canberra, at an altitude of 610 m. Recent studies (Gore et al. 1996) suggest that even such mild altitude may result in some depression (average of about 7%) in the maximum oxygen uptake readings of elite athletes.

The progressive ergometer test is based on 4 min increments, with a maximum of seven stages. The stages are separated by 1 min recovery intervals. The test protocol is individualized on the basis of work capacity, according to the steps outlined in the next section.

Table 21.3 Normative Data for Progressive Ergometer Tests of Australian Rowers*

	No. of rowers	Mean	SD	Range
Maximum oxygen uptake (L/min)				
HW male	14	5.48	0.32	4.91–6.16
HW female	19	3.93	0.32	3.52–4.64
LW male	12	4.82	0.25	4.26–5.14
LW female	6	3.44	0.24	3.14–3.84
Maximum oxygen uptake ($ml \cdot kg^{-1} \cdot min^{-1}$)				
HW male	14	59.5	4.4	52.6–68.5
HW female	19	52.5	3.8	46.3–60.1
LW male	12	65.0	3.4	58.5–69.9
LW female	6	55.6	2.5	52.8–59.4
Workload (m/min) at lactate threshold				
HW male	14	254	9	237–269
HW female	19	226	6	215–236
LW male	13	245	11	230–269
LW female	6	215	3	210–218
Heart rate (beats/min) at lactate threshold				
HW male	14	150	7	137–164
HW female	19	155	8	142–175
LW male	13	155	9	140–174
LW female	6	146	13	123–161
Heart rate at lactate threshold (as %HR max)				
HW male	14	77.6	3.4	72.0–82.1
HW female	19	78.1	3.9	68.9–86.2
LW male	13	79.8	4.8	71.4–85.8
LW female	6	76.8	5.0	69.1–83.0
Oxygen uptake (L/min) at lactate threshold				
HW male	14	3.56	0.26	3.20–4.05
HW female	19	2.52	0.20	2.20–2.81
LW male	13	3.20	0.30	2.69–3.59
LW female	6	2.16	0.12	1.97–2.29
Oxygen uptake at lactate threshold (% $\dot{V}O_2max$)				
HW male	14	65	4	57–72
HW female	19	64	5	58–74
LW male	13	65	6	55–72
LW female	6	63	5	56–69
Blood lactate concentration (mmol/L) at lactate threshold				
HW male	14	1.3	0.2	1.0–1.8
HW female	19	1.0	0.2	0.6–1.5
LW male	13	1.2	0.3	0.7–2.1
LW female	6	0.9	0.4	0.5–1.6

	No. of rowers	Mean	SD	Range
Workload (m/min) at anaerobic threshold				
HW male	14	289	6	276–300
HW female	19	256	6	241–268
LW male	13	278	7	266–287
LW female	6	241	5	234–250
Heart rate (beats/min) at anaerobic threshold				
HW male	14	178	8	164–190
HW female	19	182	6	171–192
LW male	13	180	7	166–192
LW female	6	173	10	157–182
Heart rate at anaerobic threshold (as %HRmax)				
HW male	14	91.9	2.0	88.7–95.6
HW female	19	92.0	2.3	86.4–94.5
LW male	13	92.6	2.7	89.6–98.0
LW female	6	91.0	1.8	88.2–93.0
Oxygen uptake (L/min) at anaerobic threshold				
HW male	14	4.69	0.28	4.35–5.18
HW female	19	3.40	0.21	3.06–3.79
LW male	13	4.11	0.21	3.77–4.63
LW female	6	2.96	0.17	2.76–3.27
Oxygen uptake at anaerobic threshold (% $\dot{V}O_2max$)				
HW male	14	86	2	82–88
HW female	19	87	3	81–92
LW male	13	84	3	78–91
LW female	6	86	2	83–88
Blood lactate concentration (mmol/L) at anaerobic threshold				
HW male	14	3.9	0.5	3.0–5.0
HW female	19	3.5	0.7	2.5–5.0
LW male	13	3.8	0.8	2.6–5.5
LW female	6	3.5	0.7	2.9–4.8

*All rowers at least of national standard.

HW = heavyweight, LW = lightweight.

Determination of Test Protocol:

1. Ascertain the best time recorded by the subject for a race simulation test during the preceding season. In most cases, the test will have been performed over a "distance" of 2000 m. Convert the time to a pace for each 500 m.

Example:

If the best time for 2000 m is 6 min 20 s (6:20), the pace per 500 m would be

$$\frac{380\text{ s}}{(2000\text{ m}/500\text{ m})} = \frac{380}{4} = 95.0\text{ s}$$

or approximately 1:35 per 500 m.

2. Add 4 s per 500 m to the pace derived in step 1. This gives the pace the rower should be required to maintain in the sixth stage of the test.

Example:

1:35 per 500 m + 0:04 per 500 m = 1:39 per 500 m

3. Add successive amounts of 6 s per 500 m to the pace derived in step 2 in order to calculate the required pace for the earlier workloads.

Example:

Workload 6 = 1:39 per 500 m

Workload 5 = 1:45 per 500 m

Workload 4 = 1:51 per 500 m

Workload 3 = 1:57 per 500 m

Workload 2 = 2:03 per 500 m

Workload 1 = 2:09 per 500 m

Note: Workload 7 (i.e., the final workload) is maximal. The rower is asked to maintain the fastest possible pace for the 4 min period.

4. Calculate the target meters for each minute of the first six test stages. In so doing, it is important to account for the fact that the digital display of pace per 500 m does not have resolution to tenths of a second. Therefore, a rower who appears to be maintaining a 2:09 pace may actually be averaging 2:09.8, and consequently may fail to attain the target meters calculated on the basis of 2:09 pace. To overcome this problem, the target meters should be calculated using a pace 1 s per 500 m slower than the designated target pace.

Example:

Workload 1	= 2:09 per 500 m
+ 1 s per 500 m	= 2:10 per 500 m
Convert to m/s	= 500 m per 130 s
	= 3.846 m/s
Convert to m/s	= 3.846 m/s × 60 s
	= 230.8 m/min

Thus, minute-by-minute targets for the first 4 min workload would be 231, 462, 692, and 923 m.

Alternatively, the Concept IIc ergometer allows the user to display the "average split" on the display unit. If preferred, this display may be used as opposed to time per 500 m with rowers required to maintain the overall average split for each workload. With use of the average split display, final meters for each workload must be obtained from the display unit in the rest periods between workloads.

5. Make a record sheet detailing both the target pace and the target meters for each workload. Allow space for recording actual meters and stroke ratings. A record sheet based on the example just presented appears on page 325, "Data Recording Sheet." The very slight differences in the minute-by-minute targets as compared to those listed in the example are due to rounding errors associated with manual as opposed to automated computation.

All the steps listed can be easily computerized. However, for convenience, table 21.4 makes it possible to determine test protocols and the target meters for each workload simply by reference to the best 2000 m (or 2500 m) time of the rower to be tested. The 2000 m and/or 2500 m times appear in increments of 4 and 5 s, respectively, since differences of lesser magnitude have no influence on the test protocol. Where the time for a particular rower falls halfway between two increments, the less demanding targets should be selected.

Note: Once the protocol for a particular rower is established at the beginning of the season, it should not be varied during the season, even if the 2000 m time improves. For senior rowers, some coaches even prefer to keep the protocol constant for the 4-year period of an Olympiad.

Equipment Preparation: The equipment for the progressive test must be prepared as follows:

1. Check that the ergometer chain is clean and well oiled. If it is not, pour one teaspoon of Concept oil onto a paper towel and rub the towel gently along the length of the chain. An assistant will need to draw the ergometer handle back to make the whole of the chain accessible.

2. Ensure that the ergometer slide is clean and that the seat can be moved freely. Check that the ergometer display is operational. If it is not, replace the batteries in the display unit.

3. Ensure that the lactate and (if required) blood gas analyzers are turned on at least 45 min before the scheduled start of the test. Calibrate the analyzers against reference standards shortly before the commencement of testing.

4. Ensure that oxygen and carbon dioxide analyzers are turned on at least 45 min before the first test. Shortly before commencement of the test, calibrate the analyzers against at least three reference gases that span the physiological range.

5. Calibrate the device to be used for measurement of pulmonary ventilation (unless using a reference device, such as a Collins gasometer). Use a syringe of known volume to check the accuracy of the calibration across a number of different levels of simulated ventilation.

Some laboratories may not be able to include respiratory gas analysis with all progressive tests. In these cases, steps 4 and 5 will not apply.

Test Administration: When the rower arrives at the laboratory, the tester should check to ensure that he or she has met the guidelines for athlete preparation. Explain to the athlete that there must be no warm-up before the test as this could influence heart rate and lactate readings for the early workloads. Administration of the test should then proceed as follows:

1. Attach the heart rate monitoring device to the athlete and ensure that the device is operating correctly.

2. Place a small quantity of Finalgon on one of the athlete's earlobes, or on some fingertips, to ensure that the capillary blood is arterialized prior to sampling.

3. If necessary, the athlete should adjust the position of the ergometer foot stretcher according to preference.

4. Ask the athlete to be seated on the ergometer. If the test will include gas analysis, position the breathing apparatus (respiratory valve, etc.) and ensure that the athlete is as comfortable as possible. Have the athlete take several light strokes and make any necessary adjustment to respiratory hoses or other apparatus. Ensure that the hose is not pulling on the mouthpiece at any stage during the stroke.

5. Collect a pre-exercise blood sample from the earlobe or fingertip using a capillary tube, and analyze the sample for lactate concentration and (if required) pH, blood gases, and bicarbonate concentration.

Note: With use of a fingertip as a sampling site, it is especially important to ensure that there is no subsequent bleeding onto the ergometer handle. Absorption of blood into the material covering the handle poses a risk of contaminating other subjects.

6. Press the "Reset" button on the ergometer display; then select "Work Time." Using the arrow keys, set the work time to 4:00. Then select "Rest Time" and use the arrow keys to set to 0:59. Finally, press the "Ready" button.

Note: Setting "Rest Time" at 59 s rather than 1 min allows for the fact that there is typically a slight delay in the activation of the ergometer clock at the onset of work, and facilitates synchronization between ergometer time and computer time.

7. Put the noseclip in place (if required) and ensure that the athlete's feet are secured in position on the foot stretcher. Ask the athlete to take hold of the ergometer handle in preparation for the start of the test.

8. Activate the gas analysis system for test commencement and instruct the athlete to begin rowing when appropriate.

9. As the athlete works through the test protocol, provide continual information concerning target pace and target meters. Stroke ratings, heart rates, and the number of meters completed must be recorded for each workload. Blood is collected from the earlobe or fingertip during all rest periods and is treated and analyzed in the same way as the pre-exercise blood sample. During the rests, permit the athlete to remove the noseclip and the mouthpiece or face mask if using these and to have a drink. However, it is important to ensure that the breathing apparatus and the noseclip are back in position well before the start of the next work bout (approximately 10-15 s).

10. If the athlete manages to complete the six set workloads, advise him or her that the aim during the seventh is to complete as much work as possible in the 4 min period. Emphasize the importance of finishing the whole 4 min, preferably holding a fairly constant pace.

11. At the end of the test, remove the noseclip and breathing apparatus as rapidly as possible.

12. An earlobe or fingertip blood sample should be collected as soon as possible after the end of the test. The samples must be analyzed as indicated in step 5.

13. Check the calibration of the gas analyzers.

14. If there are no more tests, ensure that the gas analysis system, the lactate analyzers, and any other equipment used for the testing are shut down in accordance with normal procedures. Ensure that the respiratory apparatus is thoroughly sterilized, rinsed, and hung up to dry.

Analysis of Test Results: The maximum oxygen uptake is recorded as the highest value actually attained over a period of a full minute. Thus, if the gas analysis system is based on 30 s sampling periods, the maximum oxygen uptake is the sum—or, if all results are expressed in L/min, the average—of the highest two consecutive readings. With 15 s sampling periods, the maximum oxygen uptake is the highest value obtained on the basis of any four consecutive readings.

Submaximal oxygen uptakes are calculated by averaging the readings recorded during the final

Table 21.4 Rowing 4 min Step Test Protocols—Look-Up Chart for Determining Progressive Test Protocols

Best time		Step 1 targets		Step 2 targets		Step 3 targets		Step 4 targets		Step 5 targets		Step 6 targets		Step 7 targets	
2000 m (min:s)	**2500 m (min:s)**	**Pace time/ 500 m (min:s)**	**Distance final meters (m)**	**Pace time/ 500 m (min:s)**	**Distance final meters (m)**	**Pace time/ 500 m (min:s)**	**Distance final meters (m)**	**Pace time/ 500 m (min:s)**	**Distance final meters (m)**	**Pace time/ 500 m (min:s)**	**Distance final meters (m)**	**Pace time/ 500 m (min:s)**	**Distance final meters (m)**	**Pace time/ 500 m (min:s)**	**Distance final meters (m)**
5:40	7:05	1:59	1000	1:53	1053	1:47	1111	1:41	1176	1:35	1250	1:29	1333		
5:44	7:10	2:00	992	1:54	1043	1:48	1101	1:42	1165	1:36	1237	1:30	1319		
5:48	7:15	2:01	984	1:55	1034	1:49	1091	1:43	1154	1:37	1224	1:31	1304	Maintain	
5:52	7:20	2:02	976	1:56	1026	1:50	1081	1:44	1143	1:38	1212	1:32	1290		
5:56	7:25	2:03	968	1:57	1017	1:51	1071	1:45	1132	1:39	1200	1:33	1277		
6:00	7:30	2:04	960	1:58	1008	1:52	1062	1:46	1121	1:40	1188	1:34	1263	lowest	
6:04	7:35	2:05	952	1:59	1000	1:53	1053	1:47	1111	1:41	1176	1:35	1250		
6:08	7:40	2:06	945	2:00	992	1:54	1043	1:48	1101	1:42	1165	1:36	1237		
6:12	7:45	2:07	938	2:01	984	1:55	1034	1:49	1091	1:43	1154	1:37	1224	split	
6:16	7:50	2:08	930	2:02	976	1:56	1026	1:50	1081	1:44	1143	1:38	1212		
6:20	7:55	2:09	923	2:03	968	1.57	1017	1:51	1071	1:45	1132	1:39	1200		
6:24	8:00	2:10	916	2:04	960	1:58	1008	1:52	1062	1:46	1121	1:40	1188	possible	
6:28	8:05	2:11	909	2:05	952	1:59	1000	1:53	1053	1:47	1111	1:41	1176		
6:32	8:10	2:12	902	2:06	945	2:00	992	1:54	1043	1:48	1101	1:42	1165		
6:36	8:15	2:13	896	2:07	938	2:01	984	1:55	1034	1:49	1091	1:43	1154	for full	
6:40	8:20	2:14	889	2:08	930	2:02	976	1:56	1026	1:50	1081	1:44	1143		
6:44	8:25	2:15	882	2:09	923	2:03	968	1:57	1017	1:51	1071	1:45	1132		
6:48	8:30	2:16	876	2:10	916	2:04	960	1:58	1008	1:52	1062	1:46	1121	four (4)	
6:52	8:35	2:17	870	2:11	909	2:05	952	1:59	1000	1:53	1053	1:47	1111		
6:56	8:40	2:18	863	2:12	902	2:06	945	2:00	992	1:54	1043	1:48	1101		
7:00	8:45	2:19	857	2:13	896	2:07	938	2:01	984	1:55	1034	1:49	1091	minutes	
7:04	8:50	2:20	851	2:14	889	2:08	930	2:02	976	1:56	1026	1:50	1081		
7:08	8:55	2:21	845	2:15	882	2:09	923	2:03	968	1:57	1017	1:51	1071		

7:12	9:00	2:22	839	2:16	876	2:10	916	2:04	960	1:58	1008	1:52	1062	
7:16	9:05	2:23	833	2:17	870	2:11	909	2:05	952	1:59	1000	1:53	1053	
7:20	9:10	2:24	828	2:18	863	2:12	902	2:06	945	2:00	992	1:54	1043	Maintain
7:24	9:15	2:25	822	2:19	857	2:13	896	2:07	938	2:01	984	1:55	1034	
7:28	9:20	2:26	816	2:20	851	2:14	889	2:08	930	2:02	976	1:56	1026	
7:32	9:25	2:27	811	2:21	845	2:15	882	2:09	923	2:03	968	1:57	1017	lowest
7:36	9:30	2:28	805	2:22	839	2:16	876	2:10	916	2:04	960	1:58	1008	
7:40	9:35	2:29	800	2:23	833	2:17	870	2:11	909	2:05	952	1:59	1000	
7:44	9:40	2:30	795	2:24	828	2:18	863	2:12	902	2:06	945	2:00	992	split
7:48	9:45	2:31	789	2:25	822	2:19	857	2:13	896	2:07	938	2:01	984	
7:52	9:50	2:32	784	2:26	816	2:20	851	2:14	889	2:08	930	2:02	976	
7:56	9:55	2:33	779	2:27	811	2:21	845	2:15	882	2:09	923	2:03	968	possible
8:00	10:00	2:34	774	2:28	805	2:22	839	2:16	876	2:10	916	2:04	960	
8:04	10:05	2:35	769	2:29	800	2:23	833	2:17	870	2:11	909	2:05	952	
8:08	10:10	2:36	764	2:30	795	2:24	828	2:18	863	2:12	902	2:06	945	for full
8:12	10:15	2:37	759	2:31	789	2:25	822	2:19	857	2:13	896	2:07	938	
8:16	10:20	2:38	755	2:32	784	2:26	816	2:20	851	2:14	889	2:08	930	
8:20	10:25	2:39	750	2:33	779	2:27	811	2:21	845	2:15	882	2:09	923	four (4)
8:24	10:30	2:40	745	2:34	774	2:28	805	2:22	839	2:16	876	2:10	916	
8:28	10:35	2:41	741	2:35	769	2:29	800	2:23	833	2:17	870	2:11	909	
8:32	10:40	2:42	736	2:36	764	2:30	795	2:24	828	2:18	863	2:12	902	minutes
8:36	10:45	2:43	732	2:37	759	2:31	789	2:25	822	2:19	857	2:13	896	
8:40	10:50	2:44	727	2:38	755	2:32	784	2:26	816	2:20	851	2:14	889	

2 min of each submaximal workload. Similarly, submaximal heart rates are the average values for the final 2 min of each submaximal workload.

Data collected from Australian laboratories participating in the national Lactate Quality Assurance Program indicate that, at low levels of lactate concentration, measurement errors exceeding ±0.2 mmol/L are rare. Thus, a measured rise of more than 0.4 mmol/L during the course of a progressive test is likely to represent a real increase in lactate concentration. Lactate threshold is determined as the real data point preceding that point on the lactate curve (i.e., third-order polynomial fit describing the relationship between lactate and workload) where the lactate concentration reaches a level 0.4 mmol/L above the minimum recorded lactate reading. The minimum may sometimes be the pre-exercise value.

Once the workload at LT has been determined, the corresponding oxygen uptakes and heart rates can be calculated by linear regression of these parameters against workload.

Detection of AT is achieved using a modification of the Dmax method developed by Cheng et al. (1992). This involves determining the maximum perpendicular distance (Dmax) from a straight line joining LT and the peak of the fitted lactate-workload curve, to the curve itself. The workload corresponding to the Dmax is the AT workload. The oxygen uptake and the heart rate at this workload can be calculated using the regression technique just described.

Computerized analysis allows for quite simple determination of the Dmax and associated threshold points (ADAPT—Automatic Data Analysis for Progressive Tests; see appendix for details). Regression equations can be easily established for both the curve and the straight lines.

2000-Meter Ergometer Test

This test involves simulation of a race. The rower is required to complete a "distance" of 2000 m in the shortest possible time. Whenever possible, the test should be performed with gas analysis.

Equipment Preparation: Equipment preparation is the same as for the progressive test.

Test Administration: When the rower arrives at the laboratory, check to ensure that the individual has met the guidelines for athlete preparation. Advise the athlete that an extensive pretest warm-up is recommended. Administration of the test should then proceed as for the Progressive Ergometer Test, with the following exceptions:

- After completing steps 1-5, press the "Reset" button on the ergometer display and then select "Meters." Use the arrow keys to set the display to 2000. Then press the "Ready" button.
- During the test, record the number of meters completed at the end of each 30 s period. Stroke ratings and heart rates should be recorded immediately after each 250 m distance reading. Ensure that the time taken to complete the entire 2000 m is recorded to the nearest 0.1 s.
- An earlobe or fingertip blood sample should be collected as soon as possible after the end of the test, and at 2, 4, and 6 min of inactive recovery. The samples should be analyzed as in step 5 of the progressive test.

Analysis of Test Results: The peak oxygen uptake is recorded as the highest value actually attained over a period of a full minute, as for the progressive test described earlier. Peak values for heart rate, ventilation (both STPD [standard temperature of 0° C, standard pressure of one atmosphere or 760 mm Hg, and dry, indicating the absence of water vapor] and BTPS [body temperature, ambient pressure, and saturated with water vapor]), and respiratory exchange ratio are recorded as the highest values achieved during any minute of the test.

Anaerobic capacity can be estimated by calculating the splits (in meters) for each 30 s period of the test and, on the basis of extrapolation from submaximal data collected during the most recent progressive test, calculating the oxygen requirement for that split. Measured oxygen uptakes are then subtracted from the estimated oxygen requirement to give the oxygen deficit for each 30 s. The oxygen deficits over the entire period of the test (up to the end of the 30 s collection period preceding test completion) are then summed to obtain the maximal accumulated oxygen deficit. This is considered an index of anaerobic capacity (Medbø et al. 1988).

Biomechanical instrumentation of the ergometer, if available, can provide valuable information concerning performance in the 2000 m test, particularly with regard to specific strength and power. It enables accurate determination of the total work done, and this can be expressed relative to oxygen uptake. It can also reveal considerable variation between athletes with regard to the peak forces generated, work per stroke, the shape of the force curve, and the extent to which rating contributes to the total work output. When considered in conjunction with other known characteristics of the athlete, such as performance in weight-training activities, this can be of great assistance in developing individualized training programs.

Normative Data: Normative values for the time taken to complete the 2000 m test are presented in table 21.5. The values are based on athletes of at least national standard.

To date, few measurements of maximum accumulated oxygen deficit have been made on high-performance rowers using the combination of data obtained from a 4 min incremental test and a 2000 m test. Consequently, it is not yet possible to present appropriate normative data. However, it is already evident that the accumulated oxygen deficit can vary considerably between individuals. Values ranging from 2.07 to 6.01 L, and from 28.5 to 76.5 ml/kg body weight, have been observed in preliminary studies conducted at the South Australian Sports Institute.

6000-Meter Ergometer Test

During this "endurance" test, the rower is required to complete a "distance" of 6000 m in the shortest possible time. This test is generally performed without gas analysis.

Equipment Preparation: Equipment preparation is the same as for the progressive test.

Test Administration: When the rower arrives at the laboratory, check to ensure that he or she has met the guidelines for athlete preparation. After advising the athlete that an extensive pre-test warm-up is recommended, proceed with administration of the test:

1. Apply Finalgon as for the progressive test.
2. Have the athlete adjust the foot stretcher as described for the progressive test.
3. Ask the athlete to be seated, then collect and analyze the pre-exercise blood sample as for the progressive test.
4. Before commencement of the test, press the "Reset" button on the ergometer display, and then select "Meters." Use the arrow keys to set the display to 6000. Finally, press the "Ready" button.
5. Ensure that the feet are secured in position on the foot stretcher as detailed for the progressive test. Ask the athlete to take hold of the ergometer handle in preparation for the start of the test.
6. Instruct the athlete to begin rowing when appropriate.
7. During the test, record the time taken to complete each 500 m split. Stroke ratings and heart rates should be recorded immediately after each 500 m distance reading. Ensure that the time taken to complete the entire 6000 m is recorded to the nearest 0.1 s.
8. An earlobe or fingertip blood sample should be collected as soon as possible after the end of the test, and at 2, 4, and 6 min of inactive recovery. The samples must be analyzed as described in step 5 of the progressive test.

Normative Data: Normative values for the time taken to complete the 6000 m test are presented in table 21.6. The values are based on athletes of at least national standard.

Table 21.5 Normative Data for 2000 m Ergometer Test of Australian Rowers*

	No. of rowers	Time for 2000 m (min:s)		
		Mean	SD	Range
HW male	123	6:06	0:26	5:54–6:24
HW female	78	7:03	0:20	6:40–7:35
LW male	82	6:27	0:04	6:11–6:56
LW female	36	7:27	0:04	7:19–7:53

*All rowers at least of national standard.

HW = heavyweight, LW = lightweight.

Table 21.6 Normative Data for 6000 m Ergometer Test of Australian Rowers*

	No. of rowers	Time for 6000 m (min:s)		
		Mean	SD	Range
HW male	260	20:06	0:42	18:53–23:14
HW female	139	22:55	0:57	21:09–25:59
LW male	170	20:50	0:29	19:36–22:15
LW female	83	23:51	0:36	22:41–25:40

*All rowers at least of national standard.

HW = heavyweight, LW = lightweight.

Data Recording Sheet—Progressive Ergometer Test

Physiological Assessment of Rowers
Progressive Test—Concept IIc Rowing Ergometer

Name: ______________________________ Date: ____________

Height: ____________ Weight: ____________ Date of birth: ______________________

Target pace	Target m/min	Cumulative meters	Stroke rating	Actual meters	Heart rate
2:09	231	231	______	______	______
		462	______	______	______
		692	______	______	______
		923	______	______	______
2:03	242	242	______	______	______
		484	______	______	______
		726	______	______	______
		968	______	______	______
1:57	254	254	______	______	______
		508	______	______	______
		763	______	______	______
		1017	______	______	______
1:51	268	268	______	______	______
		536	______	______	______
		804	______	______	______
		1071	______	______	______

(continued)

Physiological Assessment of Rowers
Progressive Test—Concept IIc Rowing Ergometer *(continued)*

Name: ______________________________ Date: ____________

Target pace	Target m/min	Cumulative meters	Stroke rating	Actual meters	Heart rate
1:45	283	283	______	______	______
		566	______	______	______
		849	______	______	______
		1132	______	______	______
1:39	300	300	______	______	______
		600	______	______	______
		900	______	______	______
		1200	______	______	______
Maximum	n/a	n/a	______	______	______
		n/a	______	______	______
		n/a	______	______	______
		n/a	______	______	______

chapter 22

Protocols for the Physiological Assessment of Rugby Union Players

David Jenkins and Peter Reaburn

Rugby union is a game played between two teams of 15 players over two 40 min periods. Although there is considerable variation in the fitness demands relative to each playing position, all players require a high degree of strength, speed, power, and endurance fitness.

Few studies have comprehensively assessed the physiological demands of rugby union. However, recent work by Deutsch et al. (1998) has confirmed that players can cover 6 km during a game. The average distance that players sprint is approximately 15 m, and the total distance sprinted during competition ranges from 100 m (forwards) to 350 m (backs). Players run at near-maximal pace for an additional total of 370 m (forwards) and 550 m (backs).

In 1989, David Clark and Brian O'Shea from the Australian Institute of Sport (AIS) Rugby Unit developed a sport science program to support Australian rugby into the 1990s. Results from each of four testing sessions throughout the season allow modification of the player's training and a change in emphasis for a particular component of fitness. Once test results are analyzed, individual goals are set that each player is encouraged to achieve at the next set of tests.

The tests currently used for rugby union in Australia are specific, valid, and reliable field tests that most rugby clubs can administer independently of sport science facilities and without expensive or complicated equipment. A booklet by Jenkins (1995) available from the Australian Rugby Union (ARU, P.O. Box 188, North Sydney, 2059, Australia) provides a comprehensive rationale for the tests described in this chapter.

Laboratory Environment and Subject Preparation

The chapters on quality assurance (chapter 1) and pretest preparation (chapter 2) address factors that influence the precision and accuracy of test data. Many of the considerations relevant to laboratory testing are equally important for field testing. The following is a brief list of those factors that the ARU has identified as needing particular attention. Bear in mind that our athletes are often tested at several different locations around the country and that the data are normally pooled for analysis.

- Players should have at least one day of rest before being tested. Unless players have time to recover from the previous series of training sessions, their test results will not be valid.
- Test at the same time of day on each testing occasion. Marked differences in air temperature should be avoided if at all possible (e.g., midday testing in Queensland may yield data that are disparate from those for the same

tests conducted at 7:00 A.M. in Canberra). Consistency between the testing centers and across testing occasions is vital.

- Ensure that the players warm up thoroughly (by running and stretching) before all tests and that they have sufficient recovery between tests.
- Keep the order of testing the same for each athlete across each testing occasion. Ideally, all players should begin with the same test (strength tests followed by the sprints and finally the endurance test). However, if numbers are high and time is short, players may have to be tested in a "circuit," with different groups beginning with different tests.
- Take care to achieve consistency in running surfaces—again between centers and across testing occasions. Footwear is significant and should be consistent between testing occasions.

The results of any test have most value when athletes are retested. Progress to a particular training program can be assessed and the training modified if necessary. Whereas scores from just one testing session allow comparison only between different players, retest scores allow monitoring of a player's change in fitness.

The following sections present details of the test equipment and protocols used to evaluate rugby union players in Australia. For each test, a brief rationale is given and the methods are then described. Most of the normative data cited in this chapter were compiled using test results recorded by AIS (Rugby) scholarship holders between 1989 and 1995.

Equipment Checklist

1. Aerobic fitness (multistage fitness test)

Refer to chapter 9 for a description of relevant equipment.

2. Speed (10 m sprint, 35 m sprint, and the phosphate recovery test)

Refer to chapter 9 for a description of relevant equipment.

- ☐ Stopwatches (at least two)

3. Body fat (skinfold sum)

Chapter 5 includes a description of appropriate equipment.

4. Strength tests
   - ☐ Weight-training equipment for bench press
   - ☐ Pull-up (chin-up) bar
   - ☐ Pressure Biofeedback Stabilizer (see appendix for details)

PROTOCOLS: Rugby

Aerobic Fitness

See table 22.1 for normative data.

Rationale: Depending on their playing position, rugby union players will cover up to 6 km during a match. Moreover, the importance of aerobic fitness for multiple-sprint athletes is now well documented (Jenkins 1993). Rugby players, irrespective of their playing position, must have an acceptable level of endurance ability.

There are literally dozens of field and laboratory tests that can provide an index of endurance fitness. Over the past six years the ARU has used a 5 km run, a 3 km run, and variations of the multistage fitness test.

Test Procedure: Refer to page 130 for a comprehensive description of the test procedure.

Speed (10 Meters)

See table 22.2 for normative data.

Rationale: Recent analysis of rugby union match play has shown that the average distance that players sprint ranges from 10 to 20 m (Deutsch et al. 1998). Acceleration is therefore far more dominant than "maximal" speed when "quickness" is both measured and trained. Speed over 10 m provides a good index of a player's sprint ability specific to the distances typical of a game.

Test Procedure:

1. Measure a 10 m distance and ensure that any wind is a crosswind.
2. Set up the electronic timing lights exactly 10 m apart.

Table 22.1 Normative Data for the 20 m Multistage Fitness Test

	Playing position			
	Tight 4	Backrow and hookers	Inside backs	Outside backs
	Levels completed (number)			
October	9.7 ± 0.9	10.5 ± 0.8	10.2 ± 0.6	10.2 ± 0.6
November	10.0 ± 0.9	11.0 ± 0.7	10.2 ± 0.6	11.0 ± 0.5
December	10.5 ± 0.8	11.8 ± 0.7	11.5 ± 0.5	11.5 ± 0.4
February	11.0 ± 0.8	13.0 ± 0.8	12.5 ± 0.6	12.5 ± 0.4

Scores are based on data collected from Australian Institute of Sport (Rugby) scholarship holders between 1989 and 1995.

Table 22.2 Normative Data for the 10 m Sprint

	Playing position			
	Tight 4	Backrow and hookers	Inside backs	Outside backs
	Time (s)			
October	1.80 ± 0.08	1.80 ± 0.05	1.74 ± 0.07	1.74 ± 0.06
November	1.78 ± 0.06	1.77 ± 0.06	1.70 ± 0.08	1.70 ± 0.05
December	1.75 ± 0.07	1.70 ± 0.04	1.65 ± 0.05	1.65 ± 0.03
February	1.70 ± 0.04	1.65 ± 0.03	1.65 ± 0.07	1.65 ± 0.05

Scores are based on data collected from Australian Institute of Sport (Rugby) scholarship holders between 1989 and 1995.

3. Allow the player to warm up thoroughly by running and stretching.

4. Sprints must start from a standing, stationary position; the player's feet must be as close as possible to the starting line but far enough behind to prevent the player from breaking the light beam.

5. A player is allowed three sprints, and the best time is recorded for later analysis.

Phosphate Recovery Anaerobic Capacity Test

See table 22.3 for normative data.

Rationale: This test is a modified version of a protocol originally called the phosphate decrement test (reviewed by Dawson et al. 1991). A player is timed over 35 m on eight consecutive occasions; each sprint is separated by a recovery period of approximately 24 s.

Rugby union players must have the ability to recover quickly following short bouts of high-intensity exercise. Players rarely get sufficient time between sprints to fully recover (i.e., to achieve complete resynthesis of creatine phosphate). The phosphate recovery test is ideal for evaluating a player's ability to resist fatigue while enduring demands (times and distances) similar to those experienced in a game.

Test Procedure:

1. Place timing lights, ideally, at three positions (35 m apart), in a straight line (as shown in figure 22.1).

2. Brief players on the test procedures, and stress the importance of a maximum effort throughout the test.

3. On the instruction "Go," the player sprints the 35 m from start line 1 to cross the timing line and

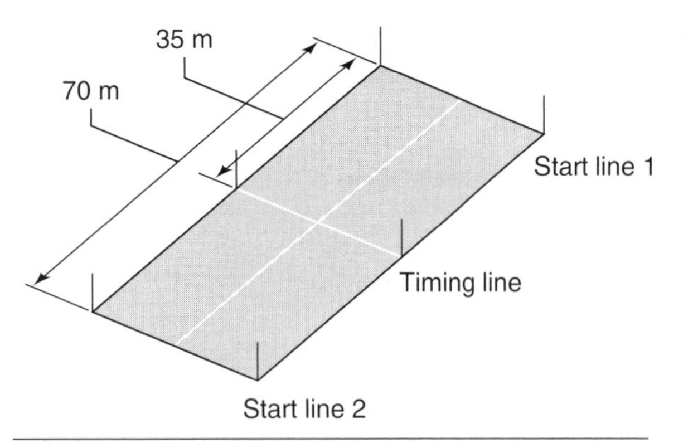

Figure 22.1 Modified version of the phosphate recovery test. Timing lights are placed at the two start lines and at the 35 m mark, labeled "timing line" on this diagram.

then slows down and jogs or walks to start line 2. The player then turns, rests, and prepares to sprint in the opposite direction (toward start line 1). For the next sprint, players sprint the 35 m to the timing line, slow down, and walk or jog to start line 1.

4. Players repeat this procedure until they have completed eight 35 m sprints.

5. Time between the start of each 35 m sprint is exactly 30 s. A 5 s countdown is given prior to the start of each sprint.

6. Each sprint is recorded to 0.01 s.

Data Analysis: In the following sample calculation, the means of sprints 3 and 4, 5 and 6, and 7 and 8 are each compared to the mean of sprints 1 and 2. The differences are shown in the column on the far right. These differences are summed and expressed as a percentage relative to the mean speed of sprints 1 and 2.

Sample calculation:

Sprint	Time (s)		
1	5.25		
2	5.26	5.25 (mean 1 and 2)	
3	5.32		
4	5.48	5.40 (mean 3 and 4)	0.15 (mean 1 and 2 – mean 3 and 4)
5	5.62		
6	5.78	5.70 (mean 5 and 6)	0.45 (mean 1 and 2 – mean 5 and 6)
7	5.81		
8	5.81	5.81 (mean 7 and 8)	0.56 (mean 1 and 2 – mean 7 and 8)
		Total 1.16s	

$$\frac{1.16 \text{ (total fatigue)}}{5.25 \text{ (mean time, sprints 1 and 2)}} \times 100 = 22.1\% \text{ fatigue}$$

Provided that the player has given 100% effort during the test, two measurements can be used

Table 22.3 Normative Data for the Best 35 m Sprint and for the Phosphate Recovery Test

	Playing position			
	Tight 4	Backrow and hookers	Inside backs	Outside backs
	35 m sprint times (s)			
October	5.20 ± 0.18	5.10 ± 0.09	5.10 ± 0.15	5.00 ± 0.08
November	5.10 ± 0.10	5.00 ± 0.12	5.00 ± 0.08	4.95 ± 0.11
December	5.00 ± 0.06	4.90 ± 0.19	4.80 ± 0.08	4.80 ± 0.13
February	5.00 ± 0.05	4.85 ± 0.08	4.75 ± 0.11	4.70 ± 0.07
	Phosphate recovery test (fatigue %)			
October	23 ± 3.2	20 ± 5.1	20 ± 6.2	20 ± 5.8
November	18 ± 4.9	17 ± 5.8	17 ± 3.6	17 ± 4.1
December	15 ± 3.0	13 ± 3.7	13 ± 6.4	13 ± 2.1
February	12 ± 3.2	10 ± 4.1	10 ± 2.3	10 ± 3.3

Scores are based on data collected from Australian Institute of Sport (Rugby) scholarship holders between 1989 and 1995.

to assess the individual's fitness. First, speed over 35 m is available. Second, the ability to maintain initial sprint speed is given by the fatigue index (%). These two figures are not necessarily related. For example, a player can be very fast over the initial sprints but fatigue quickly.

Body Composition

See table 22.4 for normative data.

Rationale: Estimating body composition enables a coach or conditioner to monitor changes in a player's body fat, muscle bulk, or both. No player wants to carry excess bulk around the playing field for 60-80 min. However, certain playing positions (e.g., tight five) may require greater muscle mass for power generation and/or protection in tackles, rucks, and mauls.

A coach or conditioner can measure changes in body composition using a set of scales, a pair of skinfold calipers, and a girth tape. For example, a drop in total skinfolds accompanied by increases in both body mass and certain girths would indicate an increase in muscle bulk.

To ensure that body composition tests are accurate and repeatable from one test to another, it is essential that the same measurer use the same methods consistently each time a team or player is measured. Listed here are the recommended sites and methods for measuring skinfolds, girths, and height and body mass.

Skinfold Sites: Historically, we have used the "Telford" method, which was developed and used at the Australian Institute of Sport in the mid-1980s. This method used a sum of seven sites for females (biceps, triceps, subscapular, suprailiac, abdominal, thigh, and medial calf) and sum of eight sites for males (female sites plus mid-axilla). Details on each of these sites appear in chapter 5.

It is now recommended that a sum of seven sites be used for both female and male athletes. These seven skinfold sites allow the coach or conditioner to use most of the available methods of body fat estimation. The seven skinfold sites are biceps, triceps, subscapular, suprailiac, abdominal, thigh, and medial calf. Skinfold measures appear simple, but only experienced persons should conduct them.

Girth Sites: Although, at least to date, girths have not been measured with rugby players, possible sites for girth measures are discussed in detail in chapter 5. Use of a number of sites is suggested for monitoring changes in body composition in rugby players. These sites are arm relaxed, chest, waist, gluteal (hip), midthigh, and calf. Chapter 5 provides descriptions of four of these sites. The additional sites of chest and midthigh girth, as well as the technique for taking girth measures, are described by Norton et al. (1996). As with skinfold measures, precise girths should be measured only by experienced persons.

Height and Mass: The equipment and procedures for measuring height and body mass are described in detail in chapter 5.

Three-Repetition Maximum Bench Press and Chin-Ups Strength Tests

See table 22.5 for normative data.

Rationale: Rugby union players need upper-body strength for both attack and defense during

Table 22.4 Normative Data for Sum of Eight Skinfold Sites

	Playing position			
	Tight 4	Backrow and hookers	Inside backs	Outside backs
	Skinfold sum (mm)			
October	130 ± 43	100 ± 33	83 ± 17	80 ± 11
November	124 ± 39	95 ± 19	75 ± 14	72 ± 12
December	112 ± 35	91 ± 16	65 ± 15	63 ± 10
February	110 ± 23	90 ± 15	62 ± 10	65 ± 10

Scores are based on data collected from Australian Institute of Sport (Rugby) scholarship holders between 1989 and 1995. Eight sites (rather than the now recommended seven) were used during this period. The eight sites were triceps, biceps, subscapular, suprailiac, abdominal, mid-axilla, front thigh, and medial calf.

Table 22.5 Normative Data for 3RM Bench Press, Total Chin-Ups, and Mass Lifted for 3RM Chin-Ups

Playing position			
Tight 4	**Backrow and hookers**	**Inside backs**	**Outside backs**
3RM bench press mass (kg)			
118.9 ± 16.8	112.7 ± 11.1	90.6 ± 11.8	117.4 ± 23.3
Total chin-ups (number)			
9.1 ± 6	15.5 ± 10.5	14.6 ± 5.9	23.7 ± 10.5
3RM chin-up mass (kg)			
3.5 ± 3.7	10.0 ± 10.2	11.3 ± 4.8	15.0 ± 4.5

Scores are based on data collected from senior representative Australian Rugby Union players in February 1996.

a game. The bench press is used as an index of the strength of the elbow extensors and the muscles that laterally flex the arm; chin-ups are used to assess the muscular strength/endurance of the elbow flexors and shoulder extensors. Taken together, these two tests provide a reasonable index of upper-body strength.

Bench Press Test Procedure:

1. Players warm up and stretch shoulders and chest:

- 10 repetitions using bar only
- 8-10 repetitions at approximately 30% of three-repetition maximum (3RM)
- Stretching of shoulders and chest
- 6 repetitions at approximately 60% of 3RM

2. Players should use overhand grip, and hands must be approximately shoulder-width apart.

3. Players must use 2 s in the lowering (eccentric) phase, hold the bar on the chest for 1 s, and then use 2 s to fully extend the bar.

4. Players then bench press 3 repetitions of their maximal possible weight.

5. Players take 5 min of rest between attempts.

Chin-Ups Test Procedure:

1. Players warm up and stretch shoulders and latissimus muscles:

- 8-10 repetitions of lat pull-downs at approximately 30% of 3RM
- Stretching of shoulders and lats
- 6 repetitions of lat pull-downs at approximately 60% of 3RM

2. Players then perform chin-ups using an overhand grip (shoulder-width apart) and achieving full arm extension.

3. Hips and knees must be extended throughout the test, and the body must remain in a near-vertical position.

4. As a variation on the total number of chin-ups achieved, players can repeat the task with a belt (and added weight) attached to their waist.

Abdominal Strength

Chapter 10 describes the problems and issues of strength assessment.

Rationale: Abdominal strength is now recognized as central to the maintenance of correct posture during most dynamic and static exercises. In addition, the incidence of many injuries to the trunk and hamstring muscles has been attributed to poor "core" strength (which is reflected by weak abdominal control).

Abdominal strength is tested using an instrument called the Pressure Biofeedback Stabilizer. Strength is related to the extent to which pressure can be maintained on an inflatable pouch that lies between the floor and the arch of a player's lower back (while the player lies on his or her back). Two tests are used.

Test 1 Procedure:

1. The player lies supine, and the pouch is placed on the floor under the arch of the lower back.

2. The pouch is inflated until it makes contact with the player's back. A pressure gauge indicates when the pouch has made contact with both the floor and lower back.

3. The player then brings the knees to above the hips. The legs are then flexed so that there is a 90° angle at the knee.

4. The player maintains this body position while isometrically contracting the abdominal muscles so that pressure is exerted downward onto the inflated pouch.

5. The player must hold this contraction for 15 s while breathing normally. The pressure gauge should remain in the same position during the 15 s of abdominal contraction.

Test 2 Procedure:

1. The player assumes the same starting position as in Test 1, and the pouch is again inflated.

2. The player isometrically contracts the abdominal muscles while slowly moving the feet above the floor and away from the body. Both legs are lowered to the floor in an extended position while the player maintains his or her breathing and abdominal contractions.

3. The position of the legs (relative to the floor and body) is noted when the pressure gauge begins to fall. This indicates a significant loss of abdominal control and is used as an index of abdominal strength.

Normative Data: Because this test has just recently been introduced, no data are available for compilation of expected values.

chapter 23

Protocols for the Physiological Assessment of High-Performance Runners

Darren Smith, Richard Telford, Esa Peltola, and Douglas Tumilty

Competitive running is demanding, with events ranging from the 100 m sprint, lasting approximately 10 s, to the marathon, which takes longer than 2 h to complete. The majority of the events will demand high percentages of athletes' aerobic and anaerobic capabilities. Testing of track athletes should therefore be used to

- monitor and assess effectiveness of training programs,
- prescribe and implement training intensities,
- construct event- and athlete-specific physiological profiles,
- aid in talent identification and team selections, and
- help detect acute or chronic overreaching or overtraining syndromes.

The following protocols have been documented in an attempt to increase uniformity in the physiological assessment of elite track athletes. The current and proposed battery of tests includes blood testing, anthropometry, progressive running test, submaximal 60 s test, maximal short-duration test, and sprint-based field tests.

In the past, little attention has been paid to giving track and field coaches and athletes guidelines for training intensities based on the results of physiological testing. Prescription of training intensities from physiological testing has been well established for other sports (e.g., cycling, rowing); the present modification to the aerobic running test will provide endurance runners with similar data. Furthermore, research is currently under way using the sprint test described here (i.e., 3 × 300 m) to determine training loads for sprint to middle-distance athletes.

Laboratory Environment and Athlete Preparation

Standardize test conditions according to the checklists presented in chapter 2. The following additional guidelines are more specific to the testing of runners.

- **Training.** Where possible, standardize time and type of training sessions within the previous 36 h. These training sessions should be familiar and of the light recovery type. For example, an easy day might consist of 16 km of slow running for a marathoner, or some drills and easy accelerations followed by a jog for a 400 m runner.
- **Diet and health status.** There should be no radical changes to diet in the days prior to testing.

Acknowledgments: The authors would like to acknowledge the valuable contribution that selected high-performance track and field coaches and other members of the track and field sport sciences committees have made to this chapter.

However, all athletes should eat high-carbohydrate meals for the two meals preceding the test. They should eat no food for 3 h before the test, but it is important that they consume adequate fluid in the 12 h prior to testing. Athletes should be tested under a consistent health status (preferably healthy).

• **Footwear.** The athlete should wear the same or a similar pair of shoes for each test session. Shoes of varying mass may influence the economy of the runner and perhaps the time to exhaustion during the progressive treadmill test. The type of shoes used during testing should be recorded.

• **Familiarity.** Familiarity is crucial for performance tests and those in which submaximal heart rates (HR) are recorded. Where possible, schedule one or more practice sessions on a separate day.

• **Laboratory environment.** The laboratory should be controlled for noise and onlookers. An electric fan should be used to maintain thermoregulation during the protracted tests. Finally, the athlete's coach should be encouraged to attend.

• **Questionnaire.** Encourage athletes to fill out the standard pretest questionnaire on pages 16-18 in chapter 2, which will help them meet the pretest conditions and will help the test coordinator interpret the results.

Equipment Checklist

1. Routine hematology

See chapter 21 for routine hematology checklist.

2. Anthropometry
   - ☐ Wall-mounted stadiometer
   - ☐ Scales (accurate to ±0.05 kg)
   - ☐ Skinfold calipers (dynamically calibrated)
   - ☐ Marker pen
   - ☐ Steel anthropometric girth tape
   - ☐ Recording sheet
   - ☐ Bench
3. Blood sampling
   - ☐ Recording sheet(s)
   - ☐ Lactate and blood gas analyzers
   - ☐ Blood sampling equipment:
     - __ Rubefacient cream (e.g., Finalgon)
     - __ Sharps container
     - __ Sample tray or rack
     - __ Tissues
     - __ Eppendorf tubes
     - __ Heparinized capillary tubes (100 μl)
     - __ Pipette
     - __ Sterile alcohol swabs
     - __ Biohazard bag
     - __ Autolet (Owen Mumford), lancets, and platforms
4. Progressive running test
   - ☐ Treadmill (calibrated for velocity and elevation)
   - ☐ Gas analysis system
   - ☐ Lactate and blood gas analyzers
   - ☐ Heart rate monitoring system (electrocardiogram or other, plus associated consumables, e.g., electrode gel, razors)
   - ☐ Electric fan
   - ☐ Stopwatch
   - ☐ Blood sampling equipment (see list for blood sampling)
5. Submaximal 60 s short-duration and maximal short-duration tests
   - ☐ Treadmill (calibrated for velocity and elevation)
   - ☐ Heart rate monitoring system (electrocardiogram or other, plus associated consumables, e.g., electrode gel, razor)
   - ☐ Stopwatch
   - ☐ Lactate and blood gas analyzers
   - ☐ Blood sampling equipment (see list for blood sampling)
6. 3×300 m short-duration running test
   - ☐ 400 m running track
   - ☐ Lactate and blood gas (optional) analyzer
   - ☐ Portable HR monitoring system
   - ☐ Stopwatch
   - ☐ Blood sampling equipment (see list for blood sampling)
7. 60 m sprint test
   - ☐ Synthetic running track or equivalent (>90 m)
   - ☐ Timing lights (at least four pairs of lights, preferably dual beam)

PROTOCOLS: Running

Routine Hematology

Blood Testing: A full blood profile and assessment of iron status should be conducted prior to a block of running tests. Runners must be in a rested state (no exercise in the previous 12 h) before giving the blood sample. Additionally, it is necessary to standardize posture and the time athletes have been in that position before taking the sample. A 5 ml sample should therefore be collected from supine athletes after they have rested in that position for at least 5 min. The sample is further divided and analyzed according to the procedure outlined on page 90.

Throughout the training year, blood samples from each athlete should be obtained at the same time of day, preferably in the morning following a 12 h fast.

Normative Data: Comparison to previous data for each athlete will aid in interpretation of blood profiles. Comparisons with normative values (reference ranges) established in the laboratory and elsewhere for elite athletes may also have some value. Making these comparisons is advisable because reference ranges of the normal population may not always be appropriate for elite athletes in heavy training.

Anthropometry

Testing: Measurement of height (centimeters), body mass (kilograms), and sum of skinfolds (millimeters) should take place prior to further testing. Skinfold measurements are taken at seven sites (biceps, triceps, subscapular, supraspinale, abdomen, midthigh, and calf) for both males and females using the techniques endorsed by the International Society for the Advancement of Kinanthropometry (see chapter 5). Depending on the requirements of the athlete, coach, and program, anthropometric measurements can be obtained quite frequently. These tests should ideally be conducted by the same accredited anthropometrist, standardizing the technical error of measurement (TEM; see page 83). Results given to athletes and coaches should take into account these inherent measurement errors by showing the range corresponding to a 68% confidence interval (value $\pm\sqrt{2}\times$TEM) or 95% confidence interval (value $\pm 2\times\sqrt{2}\times$TEM). Additionally, one needs to make allowances for measurements collected during the menstrual phase of female athletes.

Normative Data: Height and body mass results are evaluated in terms of the chronological and biological age of the runner. However, the most appropriate evaluation of an individual's anthropometric results is a comparison to that person's previous results. Comparisons to group means and normal ranges may also be of interest (see tables 23.1 and 23.2, showing data from elite runners visiting the Australian Institute of Sport from 1985 to 1996).

The sum of skinfold values for runners tends to show a degree of variation. It is therefore necessary to consider the phase of the training cycle, as skinfold readings can vary markedly throughout the training season. Note that there is individual variation in the minimal level of SF (sum of skinfolds) that can be maintained with good health.

Table 23.1 Anthropometric Data of Elite Female Runners

	Sprints/Hurdles (100 m-400 m)			Middle distance (800 m-3000 m steeplechase)			Distance (5000 m-marathon)		
	Height (cm)	Mass (kg)	SF (mm)	Height (cm)	Mass (kg)	SF (mm)	Height (cm)	Mass (kg)	SF (mm)
Mean	168.2	57.6	58.2	166.2	58.5	63.4	160.7	52.8	61.0
SD	4.1	4.3	12.7	4.3	5.2	16.1	n/a	4.7	20.5
Minimum	160.9	49.05	37.5	157.4	44.3	29.5	160.7	43.85	31.3
Maximum	175.4	72.9	99.8	172.2	64.8	92	160.7	58.5	96.7
n =	35	49	60	14	30	30	1	16	16

SF = sum of seven skinfolds: biceps, triceps, subscapular, supraspinale, abdomen, midthigh, and calf.

SD = standard deviation.

Table 23.2 Anthropometric Data of Elite Male Runners

	Sprints/Hurdles (100 m-400 m)			Middle distance (800 m-3000 m steeplechase)			Distance (5000 m-marathon)		
	Height (cm)	Mass (kg)	SF (mm)	Height (cm)	Mass (kg)	SF (mm)	Height (cm)	Mass (kg)	SF (mm)
Mean	181.6	75.8	40.2	180.0	64.8	36.1	178.3	64.0	36.0
SD	6.0	6.2	9.6	4.5	5.1	7.4	4.4	5.3	3.7
Minimum	162	57.2	25.0	171.6	53.8	27.3	172.8	58.3	32.2
Maximum	191	94.7	70.9	189.0	75.8	55.5	182.7	71.9	43.9
n =	47	106	106	15	46	47	5	13	13

SF = sum of seven skinfolds: biceps, triceps, subscapular, supraspinale, abdomen, midthigh, and calf.

SD = standard deviation.

Progressive Maximal Test

Rationale: The aerobic power test is designed to allow determination of the running velocity, HR, lactate, and pH corresponding to the blood lactate transitions (lactate [LT] and anaerobic [AT]) and $\dot{V}O_2$peak. It also allows estimates of running economy (i.e., the $\dot{V}O_2$ at a given running speed) and the construction of HR/lactate/pH versus velocity curves. The athlete exercises for 3 min at progressively increasing workloads (note that it is far preferable to complete six workloads). Each workload is separated by a 1 min rest interval. The initial workload, run on a 1% gradient, is typically 8-10 km/h (female athletes and juniors) or 10-12 km/h (male athletes), and successive workloads are 2 km/h faster than the previously completed stage. Previous training and diet will affect economy and HR and lactate curves, so adherence to pretest conditions is essential.

Equipment Preparation: The equipment must be prepared according to the following guidelines:

1. Check that the treadmill is clean and has been recently calibrated for velocity and elevation. To verify proper operation before the athlete arrives, run the treadmill through the range of velocities required in the test.

2. Ensure that the lactate and blood gas analyzers are sufficiently warmed up by switching them on at least 45 min prior to the first test. Calibrate the analyzers against reference standards immediately before the first test.

3. Prior to the first test, ensure that the gas analysis system is also adequately warmed up and calibrated for ventilation and gas composition (using reference gases spanning the physiological range).

4. Adjust the slope of the treadmill to 1% elevation.

Test Procedure: When athletes arrive, check to see whether or not they have met the guidelines for athlete preparation, and have them complete the pretest questionnaire (pages 16-18). Administration of the test should then proceed as follows:

1. Measure the athlete's body mass (±0.05 kg) using balance scales; also measure the mass of one of the athlete's running shoes and record the result on the pretest questionnaire (pages 16-18).

2. Prepare the athlete's skin for placement of the HR monitoring equipment by shaving any excess body hair, cleaning the skin with alcohol swabs, and drying with a tissue. Electrode gel will help with electrical conduction.

3. Place Finalgon on the athlete's right ear or on some fingertips to ensure good blood flow to that area. Finalgon should be applied at least 5 min prior to commencement of the test and removed just before the start.

4. Place a piece of Micropore tape across the bridge of the athlete's nose. This will help keep the noseclip from slipping during the test.

5. Instruct the athlete to take position at the front of the treadmill; when ready, the athlete should have a brief warm-up and familiarization period on the treadmill. During this period, outline the test procedures to the athlete (i.e., 3 min at each workload; 1 min rest with each workload increasing in velocity by 2 km/h; blood samples to be taken after each workload).

6. After the warm-up, obtain a pre-exercise blood sample.

7. Position the headset and mouthpiece on the athlete, ensuring maximum comfort. Instruct the athlete to take hold of the noseclip and position it comfortably to block nasal airflow.

8. Before the test begins, have the athlete hold the handrails and straddle the treadmill. Start the treadmill; when it is operating at the desired speed, have the athlete start running while you are simultaneously starting the gas analysis and time devices. During each workload, indicate the time remaining and check with the athlete about his or her ability to continue. Stop the treadmill at the end of the 3 min, giving the athlete prior warning with a countdown over the last 5 s.

9. Immediately on cessation of exercise, a blood sample should be drawn from the earlobe (or finger) and analyzed for lactate, pH, and bicarbonate levels. Have the athlete identify on the rating of perceived exertion (RPE) chart (Borg 1962) his or her subjective assessment of exertion, and record this value on the data sheet (page 341).

10. During the rest interval, the athlete may remove the noseclip and mouthpiece, but the headset should remain in place. At approximately 30 s before the start of the next workload, instruct the athlete to replace the mouthpiece and noseclip and to assume the starting position. The treadmill should be started at least 15 s before the beginning of the next workload to allow it to reach the speed required for that workload (i.e., 2 km/h faster). After the 60 s of rest, have the athlete start running for a further 3 min.

11. Proceed as outlined above, with the athlete completing at least five (preferably six) different workloads.

12. To assess the appropriateness of a further stage, observe athletes' HR, RPE, and some cardiorespiratory values (i.e., $\dot{V}O_2$, R value) in response to the previous level, and ask them whether they feel that they could attempt to run another 3 min at a higher speed. Note that it is not imperative that the athlete complete all of the last 3 min interval. Meaningful data can be obtained from 1 to 2 min at the workload. This would be better than stopping prematurely.

13. Athletes should continue exercising until exhaustion. With a few seconds left in each 30 s period during the expected final workload, the tester should ask athletes if they are capable of running for at least a further 30 s. Athletes should indicate their decision with hand signals (i.e., thumbs up or thumbs down).

14. The test should be stopped when the athlete voluntarily presses the emergency "Stop" button or indicates to the tester that he or she cannot complete the next 30 s period.

15. At the end of the test, remove the noseclip and headset and immediately obtain the blood sample from the last workload. Blood samples are also collected at 2 and 4 min posttest with the athlete seated and not in an active recovery.

16. If no more tests are scheduled, ensure that the gas analysis system, the blood analyzers, and all other equipment used for testing are shut down and packed away in accordance with normal procedures. Ensure that the respiratory apparatus of the $\dot{V}O_2$ system is thoroughly sterilized, rinsed, and hung up to dry and that all analyzed blood samples are disposed of into a biohazard bag.

Analysis of Test Results: Enter the running velocity (km/h), HR, RPE, lactate, pH, bicarbonate, and $\dot{V}O_2$ data into the ADAPT program (see appendix for information about this program) for determination of the blood lactate thresholds (LT and AT) using the Dmax method (Cheng et al. 1992).

The following data should then be reported for each of LT, AT, and max: velocity (km/h), $\dot{V}O_2$ (L/min and $ml \cdot kg^{-1} \cdot min^{-1}$), %$\dot{V}O_2$max, HR (beats/min), %HRmax, lactate (mmol/L), and pH.

Furthermore, the $\dot{V}O_2$max ($ml \cdot kg^{-1} \cdot min^{-1}$) at 14 and 16 km/h for females and males, respectively, has been used as an index of running economy and performance when expressed as a percentage of $\dot{V}O_2$max.

For prescription of training intensities, HR and lactate zones should be provided using the classifications in table 23.3 based on the prediction of the LT and AT points. See table 23.4 for normative data.

Submaximal 60-Second Run Test

Rationale: This is a 60 s supra-anaerobic threshold running test to assess the accumulation of lactate (and the reduction of pH and bicarbonate) and peak HR. This novel approach to laboratory-based assessment of mechanisms of middle-distance running performance was originally described by Telford (1991). See table 23.5 for normative data.

In theory, as the athlete becomes better prepared for 400 m to middle distance performance, the HR and lactate levels following the test should be reduced. The pH and bicarbonate levels may be higher, depending on the specific nature of the physiological adaptation of the athlete. The authors hope that use of this test will facilitate a

Table 23.3 Training Heart Rate Zones as a Function of Anaerobic Threshold (AT)

Name	Description	HR zones	Perceived exertion	Lactate (mmol/L)
A1	Recovery	Below 75% AT	Easy	< 1.5
A2	Aerobic endurance	76–85% AT	Comfortable	1.5–3.0
A3	Intense aerobic	86–95% AT	Uncomfortable	3.5–5.0
A4	Threshold	96–102% AT	Stressful	4.0–7.0
A5	Lactate tolerance	103% AT–Max	Very stressful	> 7.0
A6	$\dot{V}O_2max$	Max	Maximal	> 7.0

Table 23.4 Target $\dot{V}O_2max$ and Running Economy at 16 km/h

Gender	Economy at 16 km/h ($ml \cdot kg^{-1} \cdot min^{-1}$)	$\dot{V}O_2max$ ($ml \cdot kg^{-1} \cdot min^{-1}$) 800 m-5000 m	$\dot{V}O_2max$ ($ml \cdot kg^{-1} \cdot min^{-1}$) 10,000 m-marathon
Female	45–61	55–65	55–70
Male	45–61	65–80	65–80

Table 23.5 Normative Data for Submaximal 60 s Run

	HR data				Lactate (mmol/L)		pH		HCO_3 (mmol/L)	
	15 s	30 s	45 s	60 s	Pre	Peak-post	Pre	Min-post	Pre	Min-post
Mean	150	165	172	183	3.2	8.7	7.376	7.275	21.2	14.7
SD	9	5	5	10	1.7	1.9	0.020	0.039	1.6	2.2
Min	134	158	164	170	1.2	5.4	7.334	7.214	18.8	11.5
Max	159	170	178	194	6.2	12.8	7.403	7.348	24.1	18.4

better understanding of the mechanisms of improved performance in 400 m to middle distance events in particular. The test might be considered one of physiological/biochemical control during supra-anaerobic threshold efforts.

The submaximal 60 s short-duration test is performed on a treadmill at a speed of 22 km/h for males and 20 km/h for females and at a treadmill gradient of 4%. This represents a change from previous editions of this book (the earlier test used 10% slope and a speed of 18 km/h and 16 km/h for males and females, respectively). These modifications are designed to maintain the metabolic requirements of the test (i.e., to involve similar energy demands), but they place more emphasis on the neuromuscular requirements specific to running fast on the flat. On this basis, this protocol is considered an improvement over the previous "hill" running protocol.

The test is 60 s in duration and thus is terminated before exhaustion for 400 m to middle distance runners, but may bring others close to exhaustion.

Equipment Preparation: The treadmill, blood analyzers, and HR monitoring equipment should be prepared according to the guidelines set out earlier for the progressive maximal test.

Test Administration:

1. Instruct the athlete to complete the pretest questionnaire (pages 16-18), and outline to the athlete the procedures of the test (i.e., 60 s at 20 or 22 km/h for females and males, respectively, at a gradient of 4%; blood samples: pretest and three times posttest).

2. Start the treadmill running and allow the athlete a warm-up period of approximately 5-10 min.

Within this period the athlete should endeavor to complete four efforts of approximately 10-15 s duration, with each separated by at least 1 min of jogging. The first two efforts should be at the test velocity, but with zero gradient; the final two efforts should be at both test speed and elevation.

3. After the standardized warm-up, stop the treadmill and allow the athlete a 5 min rest interval.

4. During the 5 min rest period, prepare the HR monitoring equipment and place Finalgon on the athlete's ear or fingertips. With approximately 2 min remaining, remove the Finalgon with a tissue and obtain a pretest blood sample from either the earlobe or finger for determination of lactate, pH, and bicarbonate.

5. Instruct the athlete to straddle the treadmill in preparation for the start of the test. During this test a "safety spotter" should stand at the side of the treadmill and watch the athlete carefully.

6. Start the treadmill and have the athlete commence running. When the treadmill reaches the required speed, begin timing the test. Provide athletes with 15 s split times throughout the 60 s test. With a few seconds remaining, give athletes a countdown to the end of the test. In preparation to stop running, the athlete should be instructed to take hold of the handrails and straddle the treadmill once again. Stop the treadmill and lower the treadmill gradient and velocity.

7. The athlete may walk on the treadmill or around the laboratory between the posttest blood samples, but should not perform any running activities. Posttest blood samples are collected at 2, 4, and 6 min after completion of exercise.

Analysis of Test Results: Collate the submaximal test data on the data sheet (next page) and present the following in tabulated form: peak HR and pretest and peak lactate, pH, and bicarbonate values.

Maximal Short-Duration Test

Rationale: This test is similar to the submaximal 60 s short-duration test (i.e., 20 or 22 km/h and 4% slope), except that the athlete is required to exercise until exhaustion. The protocol is an adaptation of that first described by Telford (1991). Blood samples are drawn from a prewarmed earlobe prior to the commencement of the test and at 4, 6, 8, and 10 min after the test. These samples are analyzed for lactate, pH, and bicarbonate levels. The test involves running to exhaustion and requires very careful supervision. The inexperienced or highly motivated athlete is at risk of falling, and use of a safety harness is highly recommended. It is clear that neither this test nor any other short-duration treadmill test will measure anaerobic capacity exclusively. The 1.0-2.5 min duration achieved by elite runners on this test also demands significant aerobic energy contribution, muscle and blood buffering, and tolerance to lactate and H^+. See table 23.6 for normative data.

Equipment Preparation: The treadmill, blood analyzers, and HR monitoring equipment should be prepared according to the guidelines presented for the progressive maximal test. Because of the increased chances of injury when the athlete nears exhaustion at the high running speed and elevation this test involves, use of a secure safety harness is highly recommended.

Test Administration: The athlete should be allowed a 15 min rest between the preceding submaximal short-duration test and the maximal short-duration test. Outline the procedures for the test to the athlete (i.e., maximal effort at a speed of 20 or 22 km/h and a gradient of 4%). Communicate to athletes that the test will be stopped when they voluntarily press the emergency "Stop" button, when they indicate to the tester to stop the treadmill, when they grasp the handrails, or when the tester feels that the test should be stopped.

1. Approximately 2 min before the test begins, obtain a pretest blood sample from the earlobe or finger for determination of lactate, pH, and bicarbonate.

2. Instruct the athlete to straddle the treadmill in preparation for the beginning of the test.

3. Start the treadmill and have the athlete begin running. When the treadmill reaches the required speed, begin timing the test. The athlete should be encouraged to exercise for as long as possible. Do not provide split times during this trial. At the completion of the test, press the "Stop" button on the treadmill control (after instruction from the athlete) and have the athlete hold the handrails until the treadmill belt stops.

4. The athlete is permitted to walk on the treadmill or around the laboratory between the posttest blood samples, but should not perform any running. Blood samples are collected at 4, 6, 8, and 10 min after the completion of exercise.

Analysis of Test Results: As in the submaximal test, collate the maximal test data on the data sheet (next page) and present the following in

Progressive Maximal Running Test—Data Sheet

Athlete: ______________________________ Event: ______________

Treadmill slope: 1% ______________ Date: ____________ Time: ______________

Body mass (kg): ________ Environmental conditions (°C and % room humidity): ______________

Raw data

Velocity (km/h)	HR (beats/min)	RPE	Blood lactate (mmol/L)	pH	HCO_3^- (mmol/L)	$\dot{V}O_2$ (L/min)	$\dot{V}O_2$ ($ml \cdot kg^{-1} \cdot min^{-1}$)
Pretest values							

Comments: ______________________________

____

____

____

____

Short-Duration Running Tests—Data Sheet

Athlete: ______________________ Event: ______________

Treadmill speed and slope (km/h and %) ______________ Date: __________ Time: ________

Body mass (kg): ________ Environmental conditions (°C and % room humidity): __________

Raw data—60 s submaximal run test

Collection time	HR (beats/min)	Blood lactate (mmol/L)	pH	HCO_3^- (mmol/L)
Pretest values				
2 min post				
4 min post				
6 min post				
Peak values				

Comments: ______________

Raw Data—Maximal short-duration run test

Collection time	HR (beats/min)	Blood lactate (mmol/L)	pH	HCO_3^- (mmol/L)
Pretest values				
4 min post				
6 min post				
8 min post				
10 min post				
Peak values				

Run duration: ______________

Comments: ______________

Table 23.6 Normative Data for Maximal Run Test

	Time to exhaustion (s)	Lactate (mmol/L) Pre	Lactate (mmol/L) Peak-post	pH Pre	pH Min-post	HCO_3^- (mmol/L) Pre	HCO_3^- (mmol/L) Min-post
Mean	91	5.9	13.9	7.346	7.152	17.4	8.7
SD	24	2.1	3.0	0.022	0.054	2.6	2.2
Min	56	3.6	9.6	7.311	7.080	13.0	5.4
Max	122	10.5	19.7	7.377	7.232	20.6	12.2

tabulated form: maximum run duration; peak HR; and the pretest and peak lactate, pH, and bicarbonate values.

Short-Duration Track Test (3 × 300-Meter Field Test)

Rationale: This test is a modified version of the tests designed in the 1980s by Fohrenbach and colleagues from the Federal Republic of Germany (Fohrenbach et al. 1986) to measure anaerobic running efficiency at different sprinting velocities.

Pretest and Equipment Preparation:

1. A 400 m synthetic running track in good condition is required.

2. At least two stopwatches are needed: one is used to time the 300 m repetitions, and the other is used to coordinate the timing of the various post-300 m repetition blood samplings.

3. At least eight field blood samples will be obtained per athlete; therefore a portable bloods kit with enough consumables (plus spares) should be prepared by standard methods. The lactate samples may be processed immediately after testing or stored for later analysis. The lactate analyzer should be calibrated and warmed up using standard methods.

Test Administration: When the athlete arrives at the training track, check to make sure that the guidelines for athlete preparation have been met. Advise athletes that an extensive pretest warm-up is recommended (i.e., equivalent to their normal individual track-session warm-up). Administration of the test should then proceed as follows:

1. The athlete runs 3 × 300 m intervals from a standing start with full recovery between each.

2. Timing starts when the athlete's back foot leaves the track and stops when the athlete crosses the finish line.

Table 23.7 Target Times for Progressive 300 m Intervals

300 m pb (s)	Approx 80% pb (s)	Approx 90% pb (s)	Approx 95–100% pb (s)
40.0	48.0–50.0	44.0–45.0	< 42.0
37.0	45.0–47.0	41.0–42.0	< 39.0
33.0	41.0–43.0	36.5–38.0	< 35.5

3. The first 300 m effort should be run at about 80% of maximum 300 m running velocity. The second and third 300 m runs should be at 90% and between 95% and 100% of maximum 300 m running velocity, respectively. Ideally, each 300 m should be evenly paced. For example, an athlete with a recent 300 m personal best (pb) performance of 33.0 s should run the three intervals in approximately 42.0, 37.0 and 34.0 s, respectively (see table 23.7).

4. The recovery intervals after the first and second 300 m runs are 10 and 20 min, respectively.

5. Blood samples should be obtained at the following times:

- 2 and 5 min after the first 300 m run
- 2, 5, and 8 min after the second 300 m
- 5, 8, and 12 min after the final run

Analysis of Test Results: Tabulate exact running time (to 0.01 s), running velocity (m/s), peak blood lactates, and HR for each repetition. Graphs of velocity (m/s, x-axis) against the dependent variables (y-axis) of HR (beats/min) and blood lactate (mmol/L) will aid in evaluation of results. Comparisons of the slope and shifts (to the right or left) of running velocity versus blood lactate curves for the same athlete across time will indicate whether or not the athlete has adapted to training.

60-Meter Sprint Test (Maximal Running Velocity)

Rationale: This test measures the ability of an athlete to start (0 to 10 m time; optional), accelerate (10 to 30 m time), and sprint (fastest 10 m interval between 30 and 60 m).

Pretest and Equipment Preparation:

1. A 400 m synthetic running track in good condition is required. Mark 10 m intervals from 0 to 60 m on a straight part of the track.

2. Ideally, place light gates at each 10 m interval. If only four sets of light gates are available, place these at 0, 10, 30, and 60 m.

Test Administration: When the athlete arrives, check to make sure he or she has adhered to the athlete preparation guidelines. Advise athletes that an extensive pretest warm-up is recommended (i.e., equivalent to their normal individual track-session warm-up). Administration of the test should then proceed as follows:

1. Position the athlete so that his or her body is a set distance (e.g., 30 cm) behind the first (starting) light gate.

2. Have the athlete complete three to four trials from either a standing (endurance) or blocks (sprint) starting position.

3. Record the split times for each trial to the nearest 0.01 second.

Analysis of Test Results: Compile the split times and calculate the following intervals: 0 m to each light gate (i.e., 0-10 m, 0-20 m, 0-30 m, etc.); 10 m to 30 m; and the 10 m splits between 30 and 60 m (i.e., from 30 to 40 m, 40 to 50 m and 50 to 60 m, if enough light gates are available; alternatively, divide the 30 to 60 m split by 3 to derive maximum running velocity, if only four light gates are being used). A graph of running velocity (m/s) versus running distance can be compiled from the best splits.

chapter 24

Protocols for the Physiological Assessment of Sailors

Michael Blackburn

Sailing is primarily a tactical, strategic, and technical sport. However, as wind strength increases, so too do the physiological demands of the sport as a result of the greater effort required to control the boat proficiently. Fitness requirements vary slightly with the class of boat sailed, but the primary actions for the Olympic dinghy and keelboat classes include steering, hiking (leaning over the side of the boat), and sheeting (pulling on the rope or ropes controlling the sails) and for the Olympic sailboard class, pumping (repeated fanning on the sail).

In a typical five- to seven-day championship sailing regatta, crews sail two 60 min races per day with a 20-60 min break between races. For every race, roughly 2 h are spent on the water in sailing to the course area, warming up, racing, and sailing home (i.e., 3-6 h a day on the water).

Although the variability in wind and water conditions makes it impractical to conduct regular on-water physiological testing of sailors to monitor changes in fitness levels, studies of the physiological demands of dinghy sailing have provided useful reference data (e.g., Gallozzi et al. 1993; Vogiatzis et al. 1995).

The results obtained from the following laboratory-based tests will provide a profile of the sailor's physical strengths and weaknesses. These can be compared with normative data that are based on athletes of at least national standard and are provided for each of the Olympic classes. One can then modify the training program considering the importance of each physical capacity in determining on-water performance in specific events.

Fitness test protocols vary among countries and from time to time. Often these have not met specificity and relevance requirements for the physiological testing of sailors. Although the protocols presented here are not as well refined as protocols in physiologically based sports, it is felt that they provide for a good assessment of the capacities important in Olympic-class sailors.

Subject Preparation

Athletes should follow the pretest preparation checklist in chapter 2.

Equipment Checklist

1. Anthropometry
   - ☐ Wall-mounted stadiometer
   - ☐ Electronic scales accurate to ±0.05 kg
   - ☐ Harpenden skinfold calipers
   - ☐ Felt marking pen
   - ☐ Metal anthropometric tape
2. Aerobic power
   - ☐ Repco air-braked cycle ergometer and Exertech work meter (see appendix for

Acknowledgments: Appreciation is extended to the following members of the Sailing Sport Sciences Advisory Committee for their contributions to these protocols: Lex Bertrand, Paul Gastin, Ken Graham, and Justin Walls.

information) or electronically braked ergometer
- ☐ Heart rate monitor (e.g., Polar Electro)
- ☐ Stopwatch

3. Muscle function

Flexibility:
- ☐ Sit-and-reach apparatus
- ☐ Acromat or similar gymnastic mat

Abdominal endurance:
- ☐ Metronome

Grip strength:
- ☐ Grip strength dynamometer (e.g., Baseline Hydraulic Hand Dynamometer; see appendix for information). A calibration service for this device is available through Swift Performance Equipment (see appendix).

4. Ergometer tests of sailing

Sheeting power:
- ☐ Concept II rowing ergometer

Sailboard pumping:
- ☐ Rope to secure ergometer: 3 m of 6 mm pre-stretch
- ☐ Rope to lengthen chain: 3 m of 4 mm pre-stretch
- ☐ Spanner to remove and refit ergometer handle

Hiking endurance:
- ☐ Steel bucket with hiking strap and 15 kg attached
- ☐ Bench (padded)
- ☐ Goniometer or similar equipment
- ☐ Towel, cloth, or paper towel
- ☐ 75 kg in disk weights (e.g., 13 × 5 kg and 1 × 10 kg)

Order and Administration of Tests Across Classes

Tests should proceed in the order shown in table 24.1. Subjects complete only a specific set of tests relating to the demands of their crew position. The approximate duration of testing is 50-60 min.

Table 24.1 Test Order for Crews in the Various Sailing Classes

	Crew position		
	470 skipper; Europe; Laser; Star; Soling crew; Tornado skipper	**470 crew; 49er skipper and crew; Soling skipper; Tornado crew**	**Sailboarder**
Height	1	1	1
Weight	2	2	2
Skinfolds	3	3	3
Aerobic power	4	4	4
Flexibility	5	5	5
Abdominal endurance	6	6	6
Grip strength	7	7	7
Sheeting power	8	8	–
Hiking endurance	9	–	–
Sailboard pumping	–	–	8

PROTOCOLS: Sailing

Anthropometry

See table 24.2 for normative data.

Rationale: Sailing is a weight-supported and weight-dependent activity, so for each Olympic class there are height and weight ranges that have been associated with success. These have emerged naturally as a function of the design of the equipment, particularly the power that can be generated by the sails. Depending on the class of boat sailed, crews deliberately try to gain or lose weight, which often involves adjustment of body fat levels. Therefore, assessment of sailors' body size and composition is important.

Test Procedure: The techniques for measurement of stretch height, mass, and sum of seven skinfolds (biceps, triceps, subscapular, supraspinale, abdomen, front thigh, and medial calf) are described in chapter 5.

Tri-Level Aerobic Power Test

See table 24.3 for normative data.

Rationale: Hiking elicits a high heart rate and blood pressure response but imposes only a moderate aerobic demand as the strength of the static contractions of the quadriceps and other active muscles impinge on local blood vessels and restrict flow (Gallozzi et al. 1993; Vogiatzis et al. 1995). For trapezing crews, aerobic power is less heavily taxed. Hence, in a single fitness-testing session, only a submaximal test of aerobic power is warranted as a general assessment of sailors' fitness. However, specific maximal tests of hiking endurance, sheeting power, and sailboard pumping are included.

In the test presented here, maximal heart rate is predicted for each individual on the basis of age; however, there is considerable variation among individuals' maximums. Consequently, this test should be used primarily for the longitudinal monitoring of individuals rather than for cross-sectional comparison between sailors or between sailors and other athletes.

The submaximal tri-level test is conducted on a calibrated cycle ergometer (the Exertech front-access ergometer is recommended). The test result indicates how much power (watts) subjects can produce at 75% of their maximum heart rate, which is an index of their aerobic power.

Test Procedure:

- Connect the Exertech work monitor (if available) to the Repco cycle ergometer and position it so that the subject is looking directly at the meter needle, thus avoiding parallax error. The work monitor should be set to the low range.
- The subject's feet are placed in the pedals, and he or she is asked to pedal slowly initially so that the seat's height can be set. The leg should be slightly flexed at the knee, with the ball of the foot on the pedal at the bottom of the downstroke.
- The transmitter of a heart rate monitor is fitted and a resting reading obtained on the receiver, which is held by the tester. If the subject's heart rate while sitting on the bike at rest is over 100 beats/min (due to anxiety), the subject should be given time to relax, and the start of the test should be delayed until a heart rate under 100 beats/min is achieved.
- The test begins at a workload of 25 W ("Begin pedaling slowly"), increasing by 25 W each minute. Heart rate is recorded at the end of each minute, before the next workload is started. The subject should be encouraged to pedal smoothly exactly at the required workload. To save time, it is suggested that male subjects who are familiar with the test and are known to take more than 7 min to reach their target heart rate start the test at 50 or 75 W. This starting point must be consistently applied to a given subject.
- The test is stopped at the end of the minute during which the subject's heart rate equals or exceeds the target heart rate (THR) where THR = (220 – age) × 0.75. If possible, leave the heart rate monitor on the subject, as it is required in later tests.
- The workload at which the target heart rate was reached is calculated as described next.

Data Analysis: Data analysis is based on a linear relationship between heart rate and workload.

Sample calculation:

- Target heart rate was 151.
- After 7 min, the workload was 175 W and heart rate was 160. At the end of the 6th min (150 W), heart rate was 148.
- Since 151 represents 3/12 of the difference between 148 and 160, 150 W (6th min) + 3/12 of 25 W (7th min increase) = 156.25.
- A correction factor is applied for temperature and pressure differences:

$$\text{Correction factor} = \frac{P}{760} \times \frac{295}{273 + T}$$

where P is ambient pressure (mm Hg) and T is ambient temperature (°C).

For example, pressure = 745 mm Hg, temperature = 24° C.

Correction factor = 0.974.

- Then, corrected W = 156.25 × 0.974 = 152 (round to nearest whole number).
- If body mass was 75 kg, the aerobic power index is 152 divided by 75 = 2.03 W/kg.

Table 24.2 Target and Mean Anthropometric Characteristics of Olympic-Level Male and Female Sailors

Event	Crew position	n	Target: Height (cm)	Target: Weight (kg)	Mean: Skinfold sum (mm)
470	Skipper	6	170–180	60–65	61.9
470	Crew	6	175–186	67–76	60.8
470 (female)	Skipper	8	155–175	52–60	79.8
470 (female)	Crew	8	170–180	65–70	83.2
49er	Skipper and crew	8	170–190	65–80	63.8
Europe (female)	Skipper	4	165–180	64–71	93.3
Finn	Skipper	2	175–190	88–96	92.3
Laser	Skipper	38	176–188	80–84	54.6
Mistral	Skipper	15	174–183	65–70	37.3
Mistral (female)	Skipper	6	164–175	55–63	86.1
Soling	Skipper	3	170–190	70–90	67.4
Soling	Middleman	3	175–196	90–100	93.9
Soling	Forward hand	3	175–190	80–95	95.1
Tornado	Skipper	6	170–185	70–78	83.1
Tornado	Crew	6	170–185	72–80	63.9
		n	**Mean**	**SD**	**Range**
				Height (cm)	
All classes	Female	26	168.6	6.9	159.6–176.8
	Male	96	178.9	6.3	176.8–194.1
				Weight (kg)	
All classes	Female	26	64.3	9.4	52.1–80.1
	Male	96	77.8	8.5	64.2–113.6
				Skinfold sum (mm)	
All classes	Female	26	104.9	31.6	67.3–156.0
	Male	96	61.9	20.0	34.2–130.7

All sailors are male unless indicated otherwise. Height and weight ranges established from top-performing 1996 Olympians; skinfold values derived from tests on international- or Olympic-level Australian sailors.

Muscle Function

See table 24.4 for normative data.

Rationale: Three simple tests are performed to identify any functional weaknesses in areas of relevance to sailing performance. Sailors often reach in to adjust the sail's controls while seated with the legs extended in front, so the sit-and-reach test is a natural choice to assess functional flexibility. However, this test is known to have major limitations as described on page 104. The abdominal muscles are active in hiking, sheeting, and pumping actions and are important for stabilizing the spine through hours of sitting unsupported. Lastly, a reasonable level of grip strength is required to hold and pull on ropes.

Readers may refer to chapter 10 for a discussion of the problems and issues of strength assessment.

Table 24.3 Mean Aerobic Power (W/kg) of Olympic-Level Male and Female Sailors

			Aerobic power (W/kg)		
Event	**Crew position**	**n**	**Mean**	**SD**	**Range**
470	Skipper	3	2.19	0.49	1.57–3.08
470	Crew	3	2.14	0.49	1.68–2.73
470 (female)	Skipper	4	2.09	0.19	1.70–2.86
470 (female)	Crew	4	2.11	0.20	1.71–3.00
49er	Skipper and crew	8	2.61	0.41	1.95–3.27
Europe (female)	Skipper	4	1.92	0.38	1.46–2.46
Finn	Skipper	2	2.52	0.04	2.48–2.54
Laser	Skipper	34	3.11	0.53	2.18–3.90
Mistral	Skipper	5	3.04	0.26	2.85–3.22
Mistral (female)	Skipper	4	2.32	0.32	2.13–2.69
Soling	Skipper	3	2.29	0.35	1.89–2.49
Soling	Middleman	3	1.65	0.42	1.37–2.09
Soling	Forward hand	3	1.74	0.34	1.48–2.29
Tornado	Skipper	6	2.16	0.40	1.77–2.61
Tornado	Crew	6	2.52	0.54	1.76–3.05
All classes	Female	16	1.96	0.4	1.36–2.69
	Male	76	2.39	0.6	1.26–3.90

All sailors are male unless indicated otherwise. Data derived from tests on international- or Olympic-level Australian sailors.

Sit and Reach Test Procedure for Flexibility:

- The subject stretches the legs, back, arms, and so on after finishing the aerobic test in preparation for the sit-and-reach test and subsequent physical tests.
- The subject sits on the floor in front of the sit-and-reach box/apparatus with the legs straight in front, placing the feet against the vertical surface of the apparatus—no shoes.
- One hand should be placed over the top of the other with the palms facing down, fingertips overlapping, and fingers outstretched; the elbows should be straight.
- The subject leans forward as far as possible, sliding the hands along the ruler of the sit-and-reach box. Full stretch must be held for 3 s to avoid bouncing.
- The distance between the fingers and the vertical surface of the sit-and-reach box at full stretch is measured (if the subject cannot reach the toes, record a negative score). The knees must remain straight (the tester may place his or her hand just above the knees of the subject to hold the legs straight). The best of three trials is recorded (cm).

Abdominal Endurance Test Procedure:

- The subject lies on a mat with knees bent at 90°. The subject starts with the arms straight, hands resting on top of the thighs, and head on mat. The feet are not held.
- The sailor slowly curls up so that the fingertips touch the distal edge of the kneecap, then curls back down with the head returning to the mat and hands sliding back to the starting position. To provide a guide as to how far the subject should sit up for some of the early repetitions, and as necessary to correct form during the test, testers can place their hand at the distal edge of the patella for subjects to touch.
- These crunches are completed to fatigue (to a maximum of 80—but don't tell the subject), at a rate of one every 3 s. There is no rest between repetitions. A metronome is used to help the subject maintain the correct pace.

Table 24.4 Mean Muscle Function Scores of Olympic-Level Male and Female Sailors

Event	Crew position	n	Mean	SD	Range
		Sit and reach (cm)			
470	Skipper	3	15.1	5.83	8–21
470	Crew	3	13.0	4.2	7–15
470 (female)	Skipper	4	19.0	3.8	14–22
470 (female)	Crew	4	16.1	3.0	13–19
49er	Skipper and crew	8	13.8	10.1	–8 to 27
Europe (female)	Skipper	4	11.0	2.1	9–13
Finn	Skipper	2	6.2	2.5	4.5–8
Laser	Skipper	35	23.9	9.3	7–24.5
Mistral	Skipper	15	17.4	2.3	13–22
Mistral (female)	Skipper	6	16.2	6.2	10–21.5
Soling	Skipper	3	13.0	2.3	11.5–16
Soling	Middleman	3	8.0	2.0	6–10
Soling	Forward hand	3	5.6	4.0	2.5–10
Tornado	Skipper	6	5.0	4.6	1–10
Tornado	Crew	6	16.1	4.6	12–21
All Classes	Female	18	14.7	5.9	8–23
	Male	87	13.7	8.5	–10 to 27
		Abdominal endurance (no.) – (max 80)			
470	Skipper	3	59	12.3	45–80
470	Crew	3	49	17.4	28–80
470 (female)	Skipper	4	58	10.1	45–80
470 (female)	Crew	4	50	14.0	32–80
49er	Skipper and crew	8	65	20.4	31–80
Europe (female)	Skipper	4	70	11.8	61–80
Finn	Skipper	2	71	12.7	62–80
Laser	Skipper	35	75	12.3	41–80
Mistral	Skipper	15	80	0	80–80
Mistral (female)	Skipper	6	63	11.2	44–80
Soling	Skipper	3	38	13.0	25–51
Soling	Middleman	3	28	10.0	18–38
Soling	Forward hand	3	50	12.1	37–61
Tornado	Skipper	6	44	15.5	28–59
Tornado	Crew	6	50	13.1	38–64
All classes	Female	18	61.9	17.7	30–80+
	Male	87	60.4	22.5	14–80+

Table 24.4 *(continued)*

Event	Crew position	n	Mean	SD	Range
			Grip strength (kg)		
470	Skipper	3	52	6.8	47–60
470	Crew	3	53	10.3	44–64
470 (female)	Skipper	4	44	3.8	39–47
470 (female)	Crew	4	41	8.5	29–47
49er	Skipper and crew	8	57	4.2	50–62
Europe (female)	Skipper	4	36	5.5	29–41
Finn	Skipper	2	68	3.5	65–70
Laser	Skipper	35	64	5.7	59–76
Mistral	Skipper	15	56	3.4	50–62
Mistral (female)	Skipper	6	44	3.4	39–48
Soling	Skipper	3	62	3.5	58–65
Soling	Middleman	3	62	3.5	59–66
Soling	Forward hand	3	63	2.6	61–66
Tornado	Skipper	6	65	4.6	59–70
Tornado	Crew	6	59	4.7	56–68
All classes	Female	18	39.5	3.4	31–47
	Male	87	59.8	6.0	50–77

All sailors are male unless indicated otherwise. Data derived from tests on international- or Olympic-level Australian sailors.

Correct form should be strictly maintained. The following should attract a warning: jerky movements (throwing the head, trunk, or arms forward or letting the trunk fall to the mat), not sitting up far enough, not completing the repetitions at the prescribed rate, or lifting the feet off the ground. If the subject does not correct the deviation, the test should be ended.

An alternative or additional test of abdominal muscle function is the abdominal stage test as described in chapter 9. It can become part of physiotherapy screening, or can be used in conjunction with the test just described in cases in which a weakness in this area has been identified.

Grip Strength Test Procedure:

- The subject holds the dynamometer comfortably at his or her side, arm approximately straight, and then squeezes as hard as possible for 3 s.
- After one squeeze of the right hand, test the left grip strength before the sailor changes back to the right hand again for a second round of squeezes to determine the maximum grip strength (kilograms).
- Wipe sweat away with a cloth to keep the subject's hands from slipping.

Ergometer Tests of Sailing

Rationale: Pulling on ropes controlling the sail(s) ("sheeting") is common to all classes of sailing boat and most often involves one-handed isometric and isotonic contractions. In strong winds, the ability to rapidly adjust the sails will dramatically influence the boat's speed through the water, so a specific test of sheeting power is included here. Blackburn and Hubinger (1996) reported a correlation of 0.77 between sheeting on this test and subjects' sailing performance, albeit with a group of sailors heterogeneous in national ranking.

Intermittent isometric contractions dominate the sport in the form of hiking. Unique physiological demands are created because of the duration for which these contractions must be held during a race (Blackburn 1994; Vogiatzis et al. 1995). The test of hiking endurance presented here is incremental, ensuring that a specific estimate of the sailor's hiking ability is recorded; it is not easy to obtain such an estimate during competition (e.g., Mackie and Legg 1999) or when sailors hike off a replica of their boat on dry land. In their study, Blackburn and Hubinger (1996) found a strong correlation between hiking endurance on this test and subjects' sailing performance (r = 0.82).

In sailboarding, the sheeting of the sail is achieved more directly, with the sailor gripping the boom. Sailboard racing rules allow unlimited pumping of the sail, that is, repeated pulling on the boom to increase the velocity of air across the sail and increase board speed. This pumping is especially vigorous as a race starts and at other specific points during a race when it is important to gain a tactical advantage. At other times, pumping is less frequent but must be carried out to maintain a high average speed. Hence, both alactic and aerobic tests of pumping capacity can be used to evaluate the sailor's fitness.

To date, few measurements of sailboard pumping power and endurance have been performed using the protocol presented here, or in fact with any other procedure. Therefore, appropriate normative data cannot be presented, and the validity of this test in evaluating on-water performance cannot be documented as yet. However, the tests were developed in consultation with leading sailors and coaches and are believed to measure important and specific capacities in the event.

Test Procedure for Sheeting Power (All Except Sailboarders)

See table 24.5 (page 355) for normative data. Figure 24.1 illustrates the sheeting power test.

With use of a Concept II rowing ergometer, the test involves 2 min of maximal effort.

- On the Concept IIc ergometer, select a resistance of 10; on the IIb ergometer, ensure that the chain is on the small cog and that the vent is fully open.
- The subject sits on the seat of the rowing ergometer, and the feet are placed in the straps. No force may be exerted with the legs, and they are kept nearly straight.
- The sailor is to complete a 60 s warm-up consisting of light to moderate (males: under 110 W; females: under 70 W) one-arm pulling on the rowing handle, at the same time practicing changing hands.
- During the test, the sailor pulls on the rowing handle with one arm for six pulls, then changes arms and completes six more pulls, and so on for 2 min. The tester should point out to the subject that the elapsed time is displayed on the work monitor and that the criterion for the test is mean watts, also displayed on the monitor.
- Subjects should be instructed to aim to pull as hard as they can for the 2 min. Stroke rate is self-selected.
- In terms of hand placement on the rowing handle, the fingers of the sheeting hand can be wrapped around the handle on one side of the chain, or some of the fingers can be wrapped around the handle on either side of the chain, according to individual preference.
- The subject should lean the trunk in slightly so that the pulling arm is more in line with the chain as it comes out of the ergometer and the chain does not grate on the chain guard.

Figure 24.1 Sheeting power test.

- Before the test begins, the flywheel is stopped and the work monitor zeroed. The tester must ensure that correct form is maintained during the test: the trunk is to remain still with only the arm and shoulder contributing to the action.
- The average power (watts) with which the subject can sheet over 1 min (record this figure during the test) and over 2 min is recorded.

Test Procedure for Hiking Endurance (Hiking Sailors Only)

See table 24.5 for normative data. See figure 24.2 for an illustration of this test.

This is a maximal incremental test of hiking endurance, with no specific warm-up.

- The sailor sits on an elevated padded bench (e.g., an Acromat [approximately 35 mm thick padding] attached to a bench or an elevated weight-lifting bench with integral padding). The bench must be stable and of sufficient height (more than approximately 900 mm high) to allow a tall person to adopt the prescribed joint angle throughout the test. The subject sits with the back of the knees touching the edge of the bench. Place a towel, cloth, or paper towel under the subject's legs to soak up sweat.
- The test uses a steel bucket with a padded hiking strap attached in place of the handle. Testing begins with three 5 kg disks (approximately a 16 kg load including the bucket). The subject places his or her feet (with shoes) under the hiking strap (with the strap across the instep of the feet), lifting the bucket off the floor.
- The sailor is to maintain a knee angle of approximately 130° (the angle between the surface of the bench and the tibia), and the joint angle should be checked regularly using a goniometer or a fixed guide. Subjects can brace themselves by gripping the bench (hence the dimensions of the bench must be able to accommodate this) and can move around (e.g., shift the load from one leg to another).
- Five-kilogram disks are added to the bucket every minute until the subject can no longer (with encouragement) hold the bucket at the prescribed angle. Record the subject's heart rate each minute and the final endurance time (min:s).

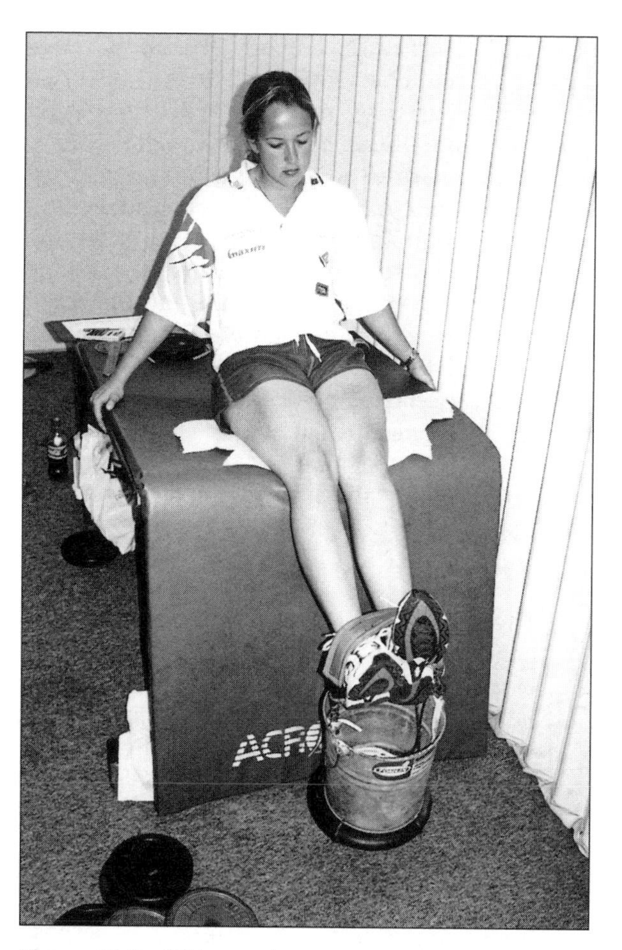

Figure 24.2 Hiking endurance test.

Equipment note: Five-kilo-weight disks are approximately 195-210 mm in diameter and 25-30 mm high. A conventional bucket sturdy enough for the job is approximately 240 mm deep. Three 5 kg disks, or a 10 kg and a 5 kg disk, should be wired or tied to the bottom of the bucket, since the test will always start with 15 kg. Thus, there is enough room for seven to nine 5 kg disks, that is, 7-9 min of the test. However, the longest period a subject has endured this test is more than 14 min. A solution is to tie one or two 5 kg and/or 10 kg disks to the sides of the bucket toward the end of the test. Therefore, in total, 75 kg is required, including one or two 10 kg disks.

Sailboard Pumping Test

Figure 24.3 illustrates the sailboard pumping test.

Preparation: A Concept II rowing ergometer is used to complete two separate tests, alactic and aerobic. The ergometer is placed on a solid table or bench(es) (approximate height: 730-900 mm) with the seat end near the edge of one end of the bench. This allows the subject to grasp the rowing handle near chest level. The front of the ergometer is secured by rope either to the bench (where it cannot be shifted) or to a solid fixture to keep the apparatus still during pumping. Care should be taken to minimize damage to the ergometer during pumping that could result from the use of these unusual tie-down points. The rowing handle is removed from the chain, and the chain is lengthened by approximately 1.5 m by tying a double length of light rope (e.g., 4 mm pre-stretch) between the end of the chain and the rowing handle. The highest resistance is selected on the ergometer (i.e., resistance of 10 on the Concept IIc ergometer; use the smaller cog with the vent fully open on the Concept IIb ergometer).

Test Procedure:

- The sailboarder takes hold of the ergometer handle with both hands and stands at the end of the bench. Ensure that the subject cannot pull the ergometer's chain to its limit while "pumping" in this position (otherwise, lengthen the rope between the handle and the chain). The subject may stand with the feet parallel or with one foot in front of the other. In the latter case, it is permissible to alternate the forward foot between right and left during the test.
- Except when the subject is alternating between left or right foot forward, the feet should remain in the one position during the test so that the sailor uses a technique similar to that for on-water sailing. If the subject loses balance and then regains a static foot position, the test may be continued. It is important that subjects not "creep" backward, away from the ergometer during the test, or they may reach the chain's limit. Subjects wear rubber-soled shoes.
- The tester should point out to the subject that elapsed time is displayed on the work monitor and that the criteria for the first test are peak power and mean watts, also displayed on the monitor.
- Set the units on the ergometer to watts. The subject undertakes a 60 s warm-up consisting of light to moderate pumping (males: under 170 W; females: under 130 W). Then the flywheel is allowed to stop.
- A 10 s alactic test is completed first. The sailor is instructed to complete 10 s of maximal pumping using an "upwind" pumping technique. The tester must watch the ergometer display closely to record the peak power that will occur at some stage during the test. Mean watts at 10 s must also be recorded.
- After 2 min of rest, the sailor starts the aerobic test (again, with the flywheel stopped). Set the ergometer units to meters (min/500 m) and point out to the subject the components of this display (i.e., distance rowed and 500 m pace). The sailor completes 1000 m of pumping, again using an "upwind" pumping technique. At each 100 m the elapsed time, min/500 m pace, and heart rate are recorded.

Figure 24.3 Sailboard pumping test.

Table 24.5 Mean Sheeting and Hiking Scores for Olympic-Level Male and Female Sailors

		Sheeting (W)				Hiking (min:s)			
		n	Mean	SD	Range	n	Mean	SD	Range
470	Skipper	3	129.6	15.6	113–144	3	5:37	0:53	4:37–6:21
470	Crew	3	126.3	15.3	109–138	–	–	–	–
470 (female)	Skipper	3	82.1	21.6	64–106	3	3:52	0:37	3:15–4:30
470 (female)	Crew	3	77.3	19.2	64–99	–	–	–	–
49er	Skipper and crew	8	169.4	38.9	103–220	–	–	–	–
Europe (female)	Skipper	4	115.7	12.1	99–128	2	6:01	0:12	5:31–6:50
Finn	Skipper	2	240.3	11.7	232–248	2	9:44	1:00	9:01–10:27
Laser	Skipper	34	250.7	34.4	179–285	34	10:50	2:11	7:37–13:30
Soling	Skipper	3	183.1	26.0	158–210	–	–	–	–
Soling	Middleman	3	170.4	30.5	140–201	3	4:47	0:44	4:03–5:32
Soling	Forward hand	3	167.9	28.0	144–199	3	5:10	0:56	4:18–6:10
Tornado	Skipper	6	194.1	18.6	169–216	6	5:58	0:29	5:33–6:31
Tornado	Crew	6	163.7	42.0	87–208	–	–	–	–
All classes	Women	10	88.6	32.9	47–119	7	5:08	1:18	3:20–6:10
	Men	71	175.3	49.5	88–287	51	7:28	2:54	4:18–14:00

All sailors are male unless indicated otherwise. Data derived from tests on international- or Olympic-level Australian sailors.

chapter 25

Protocols for the Physiological Assessment of Male and Female Soccer Players

Douglas Tumilty

Soccer is one of the most widely played games in the world, certainly among males, and the women's game is also increasing in popularity. The men's World Cup attracts a bigger television audience than the Olympic Games; the third World Cup in the female game was held in 1999, and the best teams from this event will appear in the Olympic Games in Sydney in 2000.

A game lasts 90 min, of which about 60 min are actual playing time. The latter figure can vary considerably depending on tactics, injuries, and other stoppages, while there may be an extra 30 min of play in order to produce a conclusive outcome. Styles of play vary from country to country and have also changed over the years, so it is now generally agreed that the pace of the game is much higher than formerly.

Although in the past many coaches may have felt that fitness played a minor role in soccer compared to skill, there is now no disagreement that soccer players need a wide range of fitness attributes to achieve success.

Internationally, there are no standard protocols for testing soccer players. However, international results do provide information on the physiological attributes that appear to distinguish the best players from the less successful, while the substantial testing of Australian players has also begun to produce indications of the values necessary for success at high levels in the game.

Most of the results cited in this chapter are from tests performed at the Australian Institute of Sport (AIS) in Canberra. In addition to a resident male youth squad, a number of male and female national-level squads have been tested at the AIS. Testing has typically been restricted to basic field tests, since the coaches are usually reluctant to release the players for more intensive evaluation, preferring to use the time at the AIS for skill development and team play. However, it has been shown that results on the anthropometric and performance tests, particularly the latter, are very good indicators of the likelihood of selection for national squads (Hugg 1994). Hugg's thesis demonstrates the need for a high level of athletic ability to achieve success in the sport.

In accord with protocols set out in chapter 9, leg power should be assessed using a countermovement jump on a Yardstick device or similar product; sprint times will be measured over distances of 5, 10, and 20 m, starting level with the first timing gate. Aerobic power is measured by the multistage fitness test. There are also fitness components in soccer for which as yet no agreed-upon tests exist; these include agility and anaerobic capacity.

Although there is considerable scientific literature for the men's game, this is not the case for the women's. However, the evidence indicates that the demands of the female game are similar to those of the men's game in terms of the stress placed on both the aerobic and the anaerobic energy systems.

Test Environment and Subject Preparation

A key factor in obtaining accurate test results is consistency—in the test environment, in subject preparation, and in test protocols. Changes in any one of these can have an effect, often unquantifiable, on test scores and can raise doubts about comparisons between individuals and groups.

• **Test environment.** Although the tests to be discussed are sometimes called field tests, there are advantages in performing as many of them as possible in a controlled environment—in a laboratory or hall. Climate and floor surface will then remain reasonably constant, and distances can perhaps be permanently marked. Some argue that since the game is played on grass with the players in soccer boots, that is how a test should be conducted. However, the tests usually performed are designed to assess physiological qualities, not specifically game-related capabilities. Therefore it is probably better to attempt to obtain consistency in as many aspects of the test performance as possible, and to allow players to obtain their maximum possible scores for such characteristics as speed or endurance, by conducting the tests under optimal conditions of surface, footwear, and the like.

If tests such as sprints and the multistage fitness test must be performed outside, it is essential that test conditions be as controlled as possible. If tests are done at the same time of day, there is a greater chance that temperature, humidity, wind, and radiation will be less variable. Details of these conditions should, in any case, be recorded. The surface should be smooth but not slippery. If the surface is grass, it should be consistently short.

• **Subject preparation.** Changes in the fitness of the subject should be the only variable that produces a change in the score on any test. Therefore the condition in which a subject performs a test should be as similar as possible each time the test is done. Chapters 2 and 9 present details of subject preparation.

• **Equipment and testers.** The standards for equipment and testers are set out in chapters 2 and 9.

• **Scheduling tests.** Apart from the details of test scheduling already mentioned, there is the more general aspect of when in the training year one should conduct tests. In general, tests should be done as infrequently as possible commensurate with giving the coach the information needed to plan training, or to give the sport scientist the information necessary for assisting the coach in this task. If tests are done too frequently, players become jaded and find it difficult to give their best efforts. This is particularly the case if players are not told why tests are being done, if they do not feel that test information will enable them to perform better, or if test results are not returned to them—preferably as soon as possible after the tests are performed. Because soccer follows periodized training programs, appropriate testing times are at the beginning and end of phases of training, in order to determine the positive and sometimes negative effects of that phase. This information enables the coach to better plan the next phase. Enough is known about the physiological requirements of soccer to enable the coach to determine not only whether the team as a whole is strong or weak in important components of the game, but also whether groups of players (e.g., the midfielders or center backs, or even individual players) need additional work in specific areas. Another useful time to test is before and after a tournament, to help determine which fitness qualities tend to drop off during this period when the usual training pattern will likely be disrupted.

Equipment Checklist

1. Anthropometry

Appropriate equipment is described in chapter 5.

2. Vertical jump
   - ☐ Yardstick jumping device as described on page 130.

3. 20 m sprint

Refer to page 129.

4. Multistage fitness test

Refer to page 129 for a description of relevant equipment.

PROTOCOLS: Soccer

Anthropometry

See table 25.1 for normative data.

Rationale: Soccer is a game that requires continual changes in direction and speed. Since nobody can escape Newton's second law, most relevantly expressed as Acceleration = Force/Mass, it is reasonable to believe that soccer players should carry as little fat mass as is consistent with good health in order to reduce their unproductive mass.

In addition, because fat is a good insulator, an overfat player may overheat since most of the energy consumed by the body appears as heat. This is especially a problem in the hot and humid conditions in which the game is often played.

Skinfolds are used to assess the body fat of soccer players both male and female, and despite the apparent importance of low values, soccer players are in general not the leanest athletes measured. Soccer is a contact, or more correctly, a collision sport that does require significant muscle mass to produce the power necessary to accelerate, horizontally or vertically, and to contest possession. So players of either sex should attempt fat loss carefully and should combine diet with aerobic and anaerobic exercise in order to ensure that weight loss is mostly fat while muscle loss is kept to a minimum.

The goalkeeper is invariably the tallest player on a team, followed by the defenders, especially the center backs. Midfielders and forwards tend

Table 25.1 Anthropometry Data on Australian Male and Female Soccer Squads Tested at the Australian Institute of Sport

		Height (cm)	Mass (kg)	Skinfold sum 7 sites* (mm)
Male				
AIS youth, February 1999	Mean	178.7	74.9	52.5
(n=19)	SD	8.2	7.2	11.2
	Range	164.3–191.6	62.5–88.9	32.4–76.1
Nat U17, July 1999	Mean	177.1	74.3	57.1
(n=21)	SD	7.0	6.8	10.5
	Range	164.1–188.9	59.7–86.4	42.1–86.1
Olympic, May 1998	Mean	181.8	79.9	57.8
(n=31)	SD	6.3	7.4	14.1
	Range	173.1–190.2	68.5–93.4	38.4–99.6
Female				
National, March 1999	Mean	167.4	62.5	81.4
(n=27)	SD	6.9	6.1	19.1
	Range	155.9–184.0	54.0–77.8	57.5–147.1
Junior, October 1995	Mean	162.1	57.0	97.1
(n=31)	SD	4.9	6.5	20.6
	Range	152.8–171.4	41.5–68.7	61.0–156.0
Youth, December 1995	Mean	165.5	59.8	97.2
(n=35)	SD	5.7	5.7	23.3
	Range	153.2–175.7	49.2–70.3	57.4–176.2

*Skinfold sum of biceps, triceps, subscapular, supraspinale, abdomen, front thigh, medial calf.

SD = standard deviation.

to be smaller. Goalkeepers tend to be fatter than outfield players.

Test Procedure: All measurements are taken in accord with the recommendations developed by the Laboratory Standards Assistance Scheme as described in chapter 5. A more detailed description of all 42 anthropometric measures is presented by Norton et al. (1996).

The usual measures taken for both sexes have been stretch height, weight, and the seven skinfold sites of biceps, triceps, subscapular, supraspinale, abdomen, front thigh, and medial calf.

Vertical Jump

See table 25.2 for normative data.

Rationale: Studies usually indicate that soccer players of both sexes have only moderate vertical jump heights in comparison with many other groups of athletes. However, comparisons between studies are difficult because the exact protocol used for this test can make a significant difference to the results, and the protocol is seldom detailed in reports.

One might imagine that the results for the vertical jump and the times for short sprints would be highly correlated, but testing at the AIS has not always shown this to be the case. Technique plays a part in both these tests, and it is believed that a good vertical jump can indicate the potential, at least, to develop sprint speed—even though the latter may be quite moderate because of lack of coordination, especially in growing youths. Figures cited by Hugg (1994) indicate that players who were successful in being selected for a variety of Australian squads attained better vertical jump scores than unsuccessful players, supporting the results of studies in other nations.

Test Procedure: The standard test requires a vertical jump using a Yardstick or similar device. The exact protocol is set out in page 141. Note that this jump permits the subject to employ both a dip (countermovement) and an arm swing with the subject's arm vertically overhead at the highest point of the jump.

Sprints

See table 25.3 for normative data.

Rationale: Acceleration and speed are crucial requirements in soccer, and many studies have shown that these variables differentiate between players at different levels of the game.

Current Test Procedure: The standard test measures the time to cover distances of 5 m, 10 m, and 20 m, starting with the front foot level with the light gate. The test procedure is described on page 130.

Aerobic Power

Table 25.4 gives the results obtained on the multistage fitness test, reported in levels and shuttles and in predicted maximum oxygen uptake. The score with the levels and shuttles expressed as a decimal is also given, since this allows a more valid calculation of the mean and standard deviation for a group. The figures for level/shuttle in the table were obtained by translating the corresponding decimal values.

Rationale: Measurement of maximal oxygen consumption remains the single most commonly

Table 25.2 Vertical Jump Scores Using Yardstick Device

	Vertical jump (cm)			
Squad	**Mean**	**SD**	**Minimum**	**Maximum**
Males				
AIS youth, February 1999 (n=16)	64	5	52	71
National U17, July 1999 (n=21)	61	5	53	70
Olympic, May 1998 (n=31)	60	7	47	70
Females				
NSWIS state, January 1998 (n=19)	48	5	40	58
National, March 1999 (n=20)	51	5	42	59

SD = standard deviation.

AIS = Australian Institute of Sport, NSWIS = New South Wales Institute of Sport.

AIS and NSWIS are state-level players, one level below national-level players.

Table 25.3 Sprint Times Over 5 m, 10 m, and 20 m (Starting at the Zero Point)

Squad		Sprint times (s)		
		5 m	10 m	20 m
Males				
AIS Youth, February 1999	Mean	1.07	1.78	3.01
(n=15)	SD	0.03	0.05	0.06
	Range	1.02–1.11	1.70–1.86	2.89–3.12
National U17, July 1999 (n=21)	Mean	1.11	1.85	3.12
	SD	0.03	0.04	0.07
	Range	1.03–1.16	1.74–1.91	2.96–3.19
Olympic, May 1998 (n=31)	Mean	1.10	1.81	3.04
	SD	0.05	0.06	0.07
	Range	1.00–1.18	1.69–1.90	2.92–3.19
Females				
NSWIS state, January 1998	Mean	1.07	1.85	3.24
(n=19)	SD	0.05	0.07	0.11
	Range	1.00–1.14	1.75–1.96	3.10–3.46
National, March 1999 (n=20)	Mean	1.14	1.91	3.26
	SD	0.04	0.04	0.06
	Range	1.06–1.22	1.84–1.99	3.15–3.34

SD = standard deviation.

AIS = Australian Institute of Sport, NSWIS = New South Wales Institute of Sport.

AIS and NSWIS are state-level players, one level below national-level players.

quoted figure in the determination of aerobic or endurance capability, though it is known that other factors such as the lactate threshold or running economy will also affect this capability. The figure for maximum oxygen consumption ($\dot{V}O_2max$) is widely quoted in the soccer literature, and is one of the few figures that allows comparisons across a number of studies. Good endurance is very important in soccer, particularly as the workload demanded of players has tended to increase over the years.

Values given in almost all studies of male players cluster between 55 and 65 ml · kg^{-1} · min^{-1} for outfield players. A value of 60 ml · kg^{-1} · min^{-1} is common even at high levels of the game, though midfielders and fullbacks may have to have higher values to comfortably cope with the greater workload expected of them, while center backs and strikers can get away with values slightly less. The optimal values are not known.

Less information exists for females, though values of about 50 ml · kg^{-1} · min^{-1} are quoted. It is likely that the demands of the women's game will increase much more rapidly than those of the male game because of its early stage of development in combination with its rapid increase in international popularity. Values of 48-50 ml · kg^{-1} · min^{-1} may become a base value for top-level players, with levels of 55 ml · kg^{-1} · min^{-1} and upward becoming necessary for some positions.

Some scores are obtained from treadmill running with gas analysis, and various field tests such as the 12 min run are still popular; but the test beginning to see the most widespread use as a field test to assess endurance capability and to estimate maximum oxygen uptake is the multistage fitness test. This allows a number of players to be tested at the same time and requires little judgment of pacing. Readers should note that this test provides nothing more than an estimate of maximum oxygen uptake. Although comparisons with accurate gas analysis tests invariably produce a significant correlation between the two methods for groups of subjects, there may be a few players who obtain results differing by several points. This may be a consequence of several

Table 25.4 Results on the Multistage Fitness Test for Australian Male and Female Squads

		Multistage fitness test score		
Squad		**Level/Shuttle**	**Level/Shuttle as a decimal value**	**Predicted $\dot{V}O_2max$ ($ml \cdot kg^{-1} \cdot min^{-1}$)**
Males				
AIS Youth, February 1999	Mean	13;05	13.4	58.6
(n=15)	SD		1.1	3.7
	Range	11;01–14;11	11.1–14.9	50.5–63.5
Nat U17, July 1999	Mean	12;12	13.0	56.9
(n=21)	SD		1.1	3.8
	Range	11;08–15;07	11.1-15.5	
Olympic, May 1998	Mean	13;09	13.7	59.7
(n=22)	SD		1.1	3.7
	Range	11;12–15;10	12.0–15.8	53.7–66.7
Females				
National, March 1999	Mean	11;01	11.0	50.3
(n=17)	SD		1.5	5.1
	Range	8;05–13;03	8.5–13.2	41.5–57.9
Junior, October 1995	Mean	10;01	10.1	47.3
(n=31)	SD		1.1	3.8
	Range	8;02–12;02	8.2–12.2	40.5–54.2
Youth, December 1995	Mean	9;06	9.6	45.2
(n=15)	SD		1.4	4.8
	Range	6;07–11;07	6.7–11.6	35.4–52.2

SD = standard deviation.

factors: the differing actions required (the treadmill requires running in a straight line with intermittent increases in pace, while the multistage test requires constant changes in direction with regular acceleration and deceleration), differing durations of the tests, and the protocol used in the treadmill test. However, the multistage test does provide a very cost-effective method for estimating maximum oxygen uptake, though some testing centers prefer to report the result only in levels and shuttles, without making the step to predicted $\dot{V}O_2max$.

Test Procedure: Refer to page 130 for a comprehensive description of the test procedure. To obtain the most reliable results for the multistage fitness test, ensure that athletes adhere to the instructions for subject preparation detailed in chapters 2 and 9. Additional guidelines are as follows:

- Stress to subjects that this is a test requiring maximum effort and that they must push themselves hard. It should go without saying that no one for whom this effort might pose a danger should do this test—persons experiencing severe viral or respiratory infections, those with a heart condition, or those with severe asthma.
- If possible, perform the test inside on a nonslippery floor surface and under constant environmental conditions.
- Before the test begins, warn that subjects who fall behind the pace will be eliminated if they fail to catch up within two shuttles.
- During the test, ensure that subjects turn in time with the beeps. Stress to them that there is no advantage in getting ahead of the signal.
- Begin the test at Level 1, even though this is quite slow for fit subjects.
- Encourage subjects during the test, especially in the later stages.

Other Performance Tests

Although other components of fitness are important in soccer, no standard tests have yet been agreed upon.

The constant changes in direction in the sport mean that agility is essential. Several tests have been used by various testers, including the Illinois agility test and the 505 agility test.

Attempts are being made in several team sports to devise appropriate agility tests. However, debate continues over aspects such as what degree of sport specificity should be required, whether agile movements should be in response to a game-related stimulus, and whether the test should incorporate a ball or other sport-specific implement.

Debate exists over the degree of involvement of the lactic acid system in soccer. The evidence indicates that the higher the level at which the game is played, the higher the levels of lactate generated. Though values as high as 12-15 mmol/L have been recorded, readings of 6-8 mmol/L seem more common, indicating a moderate level of involvement. Tests aimed at assessing the capacity of the lactic acid system include 400 m run, 30 s cycle ergometer test, and high-speed treadmill runs such as the Cunningham test. However, all such tests also engage both the phosphagen and the aerobic energy systems to varying degrees and have a high motivational component, so their results are not as precise as one would desire.

Many sports, including soccer, require the player to repeat high-intensity efforts at irregular intervals interspersed with periods of lower-intensity recovery. The aim in tests of repeatability is to measure the ability to repeat high-intensity efforts with minimum fatigue between individual efforts. Many are variations of the so-called phosphagen recovery test, in which the player either sprints as far as possible in a specific time or is timed over a set distance. The efforts are repeated at fixed intervals; and the variables of time, distance, and number of repetitions are chosen to be representative of occurrence.

There is not yet agreement on the values of these variables that are appropriate to soccer. A further point for consideration is that the tests used are of short duration; just as important in soccer is the ability to repeat high-intensity efforts throughout the whole 90 min of the game. No agreed-upon test yet exists to measure this ability.

Protocols for the Physiological Assessment of Softball Players

Lidnsay Ellis, Paul Smith, David Aitken, Lachlan Penfold, and Bob Crudgington

This chapter contains guidelines for assessing physiological abilities of softball players in the field or training venue. The tests are those currently in use by the Australian National Elite Training Program and consist of field tests that can be performed at the softball training venue, as well as strength tests that require access to a weight-training facility. Most of the current tests have been found to be sensitive to training while being cost and time effective.

The core of the test battery is based on the physiological factors that have an important relation to softball performance. Because of the need in softball to be fast at base running as well as agile in the field, the speed and agility components have the strongest emphasis in the test battery. The other tests selected measure physical abilities related to flexibility, "explosive" leg power, and endurance.

Resistance training is an important component of the softball player's physical preparation, and the selected strength tests enable basic diagnosis of strength and of changes related to training. Readers should note that the field tests and strength tests are two separate procedures and should be conducted on different occasions.

Equipment Checklist

1. Field test equipment

In addition to the equipment listed in chapter 9, the following items are required:

- ☐ Home plate and two bases with associated fasteners to secure them to the ground
- ☐ Recording sheets

2. Strength test equipment
   - ☐ Power rack
   - ☐ Olympic bar
   - ☐ Weights (20 kg disks down to 1.25 kg disks)
   - ☐ Weight belt
   - ☐ Blocks or disks that can be used for heel lift
   - ☐ Bench press rack

General Administration of Tests

- **Frequency of tests.** In general, both field tests and strength tests should take place at the end of each appropriate macrocycle to help determine the effectiveness of the program completed and structure of the next program to be implemented. Thus, the majority of the field tests will be carried out four or five times per year.
- **Athlete preparation and test conditions.** For general guidelines on athlete preparation and testing and environmental conditions, refer to chapter 2. The following guidelines are more specific to softball:

1. The tests should be carried out at the same time of day for each testing session.
2. Testing should not be conducted in extreme environmental conditions, particularly hot and humid conditions. While it is sometimes impossible to control the environmental conditions for field testing, it is possible to test at times of the day that are the least extreme.

3. The time of day, temperature (°C), humidity (%), the type of surface used for testing, and the ground surface condition should be recorded.

• **Role of the coach.** Coaches are encouraged to supervise, help organize the athletes during the test session, and ensure that athletes perform the tests correctly. If possible, coaches should avoid being involved as a tester or scorer. Supervision of the athletes includes ensuring that they perform the correct warm-up effectively; ensuring that they use correct and effective starting techniques for sprint and agility tests; and taking note of cornering technique used by athletes during the two-base sprint, which may be helpful for future coaching of this skill.

• **Field test order.** It is important that the field tests be completed in a standardized order to help reduce interference between tests. This order also allows valid comparison of different test occasions. The recommended order is as follows:

1. Height, mass, and skinfolds
2. Warm-up (see recommended warm-up described further on)
3. One-base sprint
4. Agility tests
5. Two-base sprint
6. Sit and reach
7. Vertical jump
8. Multistage fitness tests

Skinfold measurements have been included in the test battery, but it is not imperative to do them at the same test session. Skinfolds done on the day of field testing will greatly extend the length of the test session. If more than one set of timing lights is available for use, the test order might be changed slightly to facilitate concurrent testing of different athletes. To allow this, the agility and one-base sprint tests may be conducted at the same time, and the two-base sprint may be administered at the same time as the sit-and-reach test. It is, however, essential that the multistage fitness test be the last test completed.

• **Strength test order.** Strength testing should proceed in the following order:

1. Squat
2. Bench press
3. Abdominal strength test

• **Warm-up procedure for field tests.** Athletes should observe the following guidelines for warming up before field tests:

1. Maximum 20 min
2. 3-5 min continuous running and activity: for example, jog, squats, skip, lunges, crossover skips, hip twists, ice skates, half burpees, backward run, side skips, squats
3. 5-10 min static and dynamic stretching/flexibility
4. Run-throughs over 25 m:

1 × 60%	jog return
1 × 70%	jog return
1 × 80%	jog return

(1-3 min final stretching)

1 × 80%	walk return
1 × 90%	walk return
1 × 95%	walk return
1 × 100%	10 m (emphasizing maximum effort at the start)

5. 2 × 5 m shuttle runs:

1 × 95%

1 × 100% (emphasizing maximum effort in acceleration)

• **Variations in administration of field tests.** Readers should note that not all tests need to be carried out at all test sessions:

1. Height does not need to be measured more than once per year on senior athletes who have stopped growing.
2. Skinfolds might be measured less frequently, particularly if the body mass of the athletes is not fluctuating.
3. The multistage fitness test should be administered to all athletes at the beginning and end of the early preparation phases of training. This test can be used as a screening measure, with only those players who do not achieve a desirable level having to repeat the test at test sessions performed in the later parts of the season (for details see section on endurance later in this chapter).

Anthropometry

See table 26.1 for normative data.

Rationale: The rationale for and importance of measures of stature, body mass, and sum of skinfolds are the same as for netball (page 303). Excess fat has no functional role in activities such as sprinting between bases and can be regarded as dead weight that will decrease acceleration (Norton et al. 1994). When recommending a particular skinfold level, one should consider the playing position or role of the player, as well as the extent to which excess body fat may affect performance.

Test Procedure: The equipment and techniques for measuring height, body mass, and seven skinfold thicknesses are covered in chapter 5. The seven sites are triceps, subscapular, biceps, supraspinale, abdominal, front thigh, and medial calf. Issues related to reliability are discussed on page 303.

The individual skinfold measures, as well as the sum of the seven, should be reported.

Flexibility

See table 26.2 for normative data.

Rationale: The sit-and-reach test requires a combined joint action movement that provides a crude approximation of flexibility around the hip joint. In spite of its limitations as a test of specific flexibility (Jackson and Baker 1986), the test does have some specific relation to the physical requirements in softball for bending and reaching to field ground balls. The test is extremely easy to administer and requires little equipment, and substantial sport-specific normative data are available. A more comprehensive rationale for selection of flexibility measures is given on pages 99 and 102.

Test Procedure—Sit and Reach: The equipment and test procedure are described on pages 129 and 138.

Table 26.1 Anthropometric Target Scores of National- and State-Level Players

Squad	n	Mean	SD	Range
Height (cm)				
AIS/QAS	14	170.0	5.6	154–185
Mass (kg)				
AIS/QAS	14	71.57	10.7	55.7–99.0
Skinfolds Σ7* (mm)				
AIS/QAS	14	110.2	41.8	76.5–227.5

*Σ7 = triceps, subscapular, biceps, supraspinale, mid-abdominal, front thigh, and medial calf.

AIS = Australian Institute of Sport, QAS = Queensland Academy of Sport. Data compiled from best test scores collected during 1997.

Table 26.2 Flexibility Target Scores of National-Level Players

Squad	n	Mean	SD	Range	75th percentile	Desirable	Optimal
Sit and reach (cm)							
AIS	19	17.1	7.4	1–28	26.0	15–20	15–20

AIS = Australian Institute of Sport.

Mean, SD, and range collected from several tests conducted during 1997.

Desirable and optimal levels are based on the best average and best 75th percentile scores, respectively, from tests conducted between 1992 and 1997. Note that very high scores are not necessarily optimal, as hyper-mobility may predispose toward injury.

Power

See table 26.3 for normative data.

Rationale: The vertical jump test was selected to evaluate explosive leg power. The test is easy to perform in the field and requires limited equipment, and substantial sport-specific normative data are available.

Vertical jump height can be used to monitor the effects of strength-training programs that target the legs and to make comparisons between players.

Vertical Jump Test Procedure: The equipment and test procedure are described on pages 130 and 141.

The method of jump measurement (wall-mounted board or Yardstick) should be recorded; however, all measurements on high-performance players will be made using the Yardstick or similar equipment.

Speed and Acceleration

See table 26.4 for normative data.

Rationale: An obvious performance requirement of the game of softball is the ability to run quickly between bases. The tests include electronic timing of the softball player's performance for the one-base and two-base sprints. The one-base sprint includes the time from 0 to 10 m, which is used to indicate the ability to accelerate in a straight line from a stationary position. The performance of the two-base sprint includes cornering ability as well as acceleration. Measurement of these components of the performance is not part of the two-base sprint; however, coaches who require further information may observe or film the player's cornering technique for qualitative analysis.

General Procedures: The speed tests should be set up on a firm, level, and even ground surface with a pegged home plate and securely fixed first and second bases. A standard playing field can be used. Figures 26.1 and 26.2 illustrate the test setup dimensions (Softball Australia 1993). It is important to ensure that the condition of the ground is consistent for each test. Therefore, select a test site that will serve during wet weather and still remain firm under foot. Athletes must complete tests in their competition shoes.

Athletes must perform a rigorous warm-up in preparation for the speed and agility tests; a poor warm-up may result in unreliable results and increase the chance of muscle strain. Athletes should follow the procedure for the standardized warm-up described earlier in this chapter, and they should have a minimum of 2 min recovery between each sprint test trial.

One-Base Sprint Test

Equipment:

- Three electronic timing gates, leads, and related timing equipment
- Home plate
- One base
- Measuring tape

The timing gates should be set up at home plate (at the point of the plate), at 10 m, and at the front of first base (17.90 m) as illustrated in figure 26.1. The timing gate positioned at 0 m (the start) should be set at a height of approximately 1 m to the top of the gate.

Procedure: This test consists of three trials of a 17.90 m sprint (home plate to first base) timed electronically. The athlete starts behind the first timing gate in a stationary position (no rocking or step toward the starting line should be allowed). The front foot is to be positioned 10 cm back from the starting line (the point of home plate), and the athlete should have the upper and lower body facing first base. The athlete then does a maximal

Table 26.3 Vertical Jump Target Scores of National-Level Players

Squad	n	Mean	SD	Range	75th percentile	Desirable	Optimal
				Vertical jump (cm)			
AIS	19	45.7	4.0	36–50	50.0	50	58

AIS = Australian Institute of Sport.

Mean, SD, and range collected from several tests conducted during 1997.

Desirable and optimal levels are based on the best average and best 75th percentile scores, respectively, from tests conducted between 1992 and 1997.

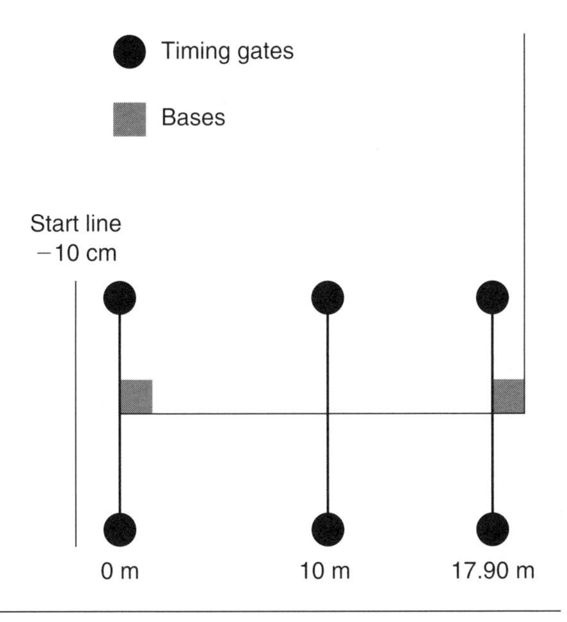

Figure 26.1 Equipment setup for one-base sprint.

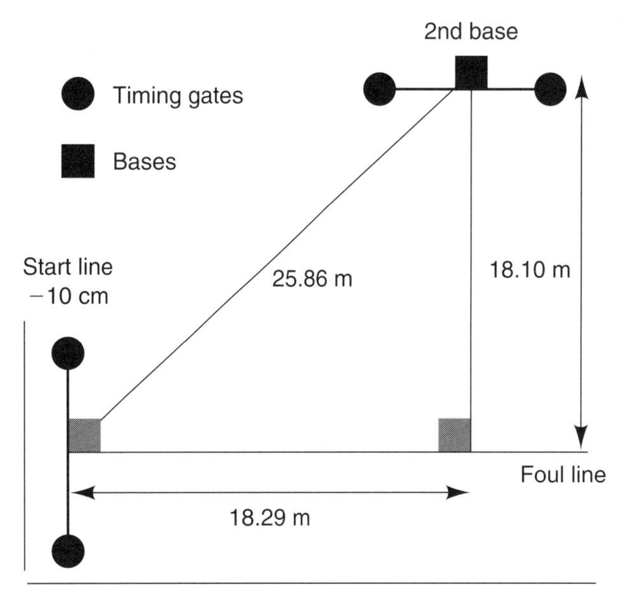

Figure 26.2 Equipment setup for two-base sprint.

Dimensions: 18.29 m = point of home base to back of first base; 18.10 m = foul line to center of second base; 25.86 m = center of second base to point of home base.

sprint effort through the timing gates. In order for the test to be valid, it is essential that the athlete touch first base. The best time for each segment—0 (home) to 10 m and 0 to 17.90 m—is recorded to a hundredth of a second (i.e., two decimal places).

Some athletes have a tendency to slow down before reaching first base. Athletes should be instructed to continue to sprint past first base.

Two-Base Sprint Test (With Cornering)

Equipment:

- Two electronic timing gates, leads, and related timing equipment
- Home plate
- Two bases
- 30 m measuring tape

In this test, two timing gates are set up as illustrated in figure 26.2. The gates are positioned at home plate (at the rear point of the plate) and at the front of second base (18.10 m from the foul line and positioned in line with the front of second base).

Procedure: This test consists of two trials of a maximal two-base run in which the athlete must touch first and second base. The finish of the run is straight over second base with no stopping. Athletes should be instructed to continue to sprint past second base. The starting position is the same as for the one-base sprint. The best time to second base should be recorded to a hundredth of a second.

Agility

See table 26.5 for normative data.

Rationale: Softball play entails instances in which the athlete must have a high level of agility requiring sudden changes of direction and acceleration. These situations are apparent in fielding, for example, when a player must move quickly to cover a hard base drive or in base running when a player must quickly recover or return to base.

A modification of the 505 test, described on page 132, has been selected as the measure of softball agility or footwork. The 505 test is a relatively simple test that measures the time for a single, rapid change of direction over a short up-and-back course with a running start. The test has been shown to be a valid and reliable measure of agility (Draper and Lancaster 1985).

The test has been modified in several ways to make it more relevant to softball. When executing the change of direction, the player must also touch the ground with the "glove hand" to simulate fielding a ground ball. (Note that "glove hand" refers to the hand only; no glove should be worn on that hand.) Included in the test is the time from 0 to 10 m using a "side-on" starting action, which further assesses the player's ability to accelerate from a

Table 26.4 Sprint and Base Run Target Times of National-Level Players

Squad	n	Mean	SD	Range	75th percentile	Desirable	Optimal
One-base sprint to 10 m mark (s)							
AIS	19	1.94	0.08	1.82–2.05	1.84	1.95	1.85
One-base sprint time (s)							
AIS	19	3.06	0.12	2.90–3.27	2.92	3.05	2.90
Two-base sprint time (s)							
AIS	19	6.09	0.18	5.88–6.43	5.88	6.10	5.80

AIS = Australian Institute of Sport.

Mean, SD, and range collected from several tests conducted during 1997.

Desirable and optimal levels are based on the best average and best 75th percentile scores, respectively, from tests conducted between 1992 and 1997.

stationary position and posture as typically used in softball during fielding.

Modified 505 Test

Equipment:

- Three sets of electronic timing gates, leads, and related timing equipment
- Measuring tape

Procedure: Timing gates are set up as illustrated in figure 26.3. Two trials take place on the backhand side and two on the forehand side.

Right-Side Trial: Two trials are conducted (backhand for right-handers and forehand for left-handers).

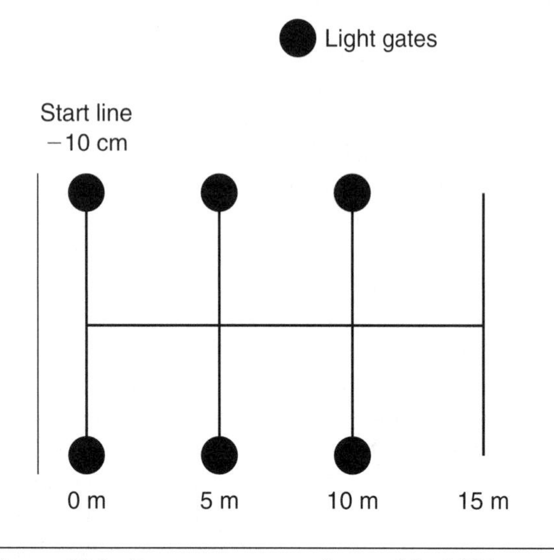

Figure 26.3 Equipment setup for modified 505 agility test.

1. The athlete starts side on (in fielding position) with the outside of the right foot touching the start line.

2. The athlete then drives off from this starting position and sprints through the timing gates toward the 15 m line.

3. For the right-hander, the athlete must touch the 15 m line with the right foot and must reach across the body to touch the ground on the far side of the line with the left hand (i.e., backhand play). For the left-hander, the right foot touches the line and the right hand is used to touch the ground on the far side of the line (i.e., to reach out from the body as in a forehand play).

Note that both left- and right-handers must have the right foot closer to the 15 m line than the left foot, and the chest roughly facing the left side of the line, immediately prior to touching the ground with the hand. The idea here is to simulate the action used in moving to the right and fielding a ground ball.

4. Having touched the ground with the hand, the athlete sprints back through the timing gates positioned on the 10 m line. Note that athletes must touch or pass over the 15 m line with their foot and must touch ground over this line with the required hand in order for the test to be valid.

5. The best time for each segment (0 to 5 m gates, and 10 m to 15 m line and back to 10 m gates) is recorded to a hundredth of a second (i.e., two decimal places).

Left-Side Trial: This test is the reverse of the right-side trial. Two trials are conducted (forehand for right-handers and backhand for left-handers).

1. The athlete starts side on (in fielding position), but this time with the left foot touching the start line. The athlete drives off from this starting position and sprints through the timing gates toward the 15 m line.

2. The instructions from this point are simply the reverse of the right-side trial (steps 3 through 5 of the right-side trial).

After the athlete has performed both trials, the athlete's dominant hand must be recorded.

Endurance

Rationale: Endurance is a physiological factor that does not play a major role in the energy requirements of softball—a game in which most efforts are of very short duration and in which there is substantial recovery between the efforts. However, a good level of endurance will aid in preparation of the softball player for other aspects of the game. For example, a good level of endurance will assist in the immediate recovery from more intense anaerobic efforts completed in training and therefore will help sustain the training intensity required.

The multistage fitness test has been selected because it has been found to be a valid and reliable estimate of endurance (aerobic power) and is a simple and efficient test to administer. The test can be used as a screening measure of a softball player's aerobic power rather than a permanent item in the field test battery. It is helpful to perform the test at the beginning and end of the early preparation phase of training.

Multistage Fitness Test Procedure: The equipment and test procedure are described on pages 129 and 130.

Normative Data: Players who do not achieve level 9 and shuttle 1 (~43 $ml \cdot kg^{-1} \cdot min^{-1}$) at the end of the first phase of physical preparation may be required to repeat the test at the next test session. For athletes who achieve this level, there is no further testing of aerobic power.

Strength

Rationale: Softball play involves explosive and dynamic action exhibited in batting, pitching, and fielding. These explosive actions require high levels of force to be exerted over a relatively short period of time. Maximum force and rate of force development have been demonstrated to be important to the performance of such activities (Hakkinen et al. 1986; Viitasalo et al. 1984; Schmidtbliecher 1992). Furthermore, resistance and plyometric training effectively develop the strength qualities of maximum force and rate of force development (Hakkinen 1985; Schmidtbliecher 1992). Resistance training, therefore, should be an important component of the softball player's physical preparation; and the selected strength tests enable basic diagnosis of strength levels and changes related to training. The bench press and squat lifts have been employed because

Table 26.5 Modified 505 Agility Run Target Times of National-Level Players

Squad	n	Mean	SD	Range	75th percentile	Desirable	Optimal
				Right-side time 0–5 m (s)			
AIS	19	1.18	0.06	1.08–1.33	1.10	1.15	1.10
				Right-side time 505 (s)			
AIS	19	2.53	0.08	2.40–2.67	2.43	2.55	2.40
				Left-side time 0–5 m (s)			
AIS	19	1.19	0.05	1.12–1.33	1.12	1.15	1.10
				Left-side time 505 (s)			
AIS	19	2.52	0.09	2.37–2.68	2.42	2.55	2.40

AIS = Australian Institute of Sport.

Mean, SD, and range collected from one test only conducted during 1997.

Desirable and optimal levels are based on the average and 75th percentile scores, respectively, from one administration of this test.

they both test multijoint function, use the major muscle groups of the upper and lower body, and are common in training exercises familiar to the high-level player.

General Procedures: Refer to procedures described for netball (chapter 20).

Squat

For normative data, see table 26.6. The equipment and test procedures are as described for netball in chapter 20.

Split Squat

The equipment and test procedures are as described in chapter 20 for netball.

Bench Press

For normative data, see table 26.7. The equipment and test procedures are described in chapter 20.

Chin-Up Test

Athletes perform a three-repetition maximum (3RM) test using an overhand grip (palms facing away from the body); if this test is successful, they may make further attempts using masses added to the belt attached to the waist. If this test is unsuccessful, the athlete attempts a 3RM test using an underhand grip (palms facing toward the body); and if successful with this technique, he or she may make further attempts using masses added to the belt attached to the waist. Athletes who are un-

Table 26.6 Three-Repetition Maximum Squat Results of Australian International Squad Players

		Body mass (kg)			Mass lifted (kg)			Normalized mass lifted* (kg/kg)		
Position	n	Mean	SD	Range	Mean	SD	Range	Mean	SD	Range
Catcher	7	74.5	5.1	69.0–81.1	87.5	8.3	80–100	1.756	0.144	1.537–1.919
Corner	5	78.7	11.7	63.0–97.2	85.0	13.8	65–100	1.653	0.406	1.305–2.369
Outfield	10	67.7	7.9	55.0–77.5	73.8	12.0	45–85	1.627	0.213	1.140–1.866
Pitcher	10	80.9	14.4	66.6–111.5	79.2	8.2	70–95	1.494	0.237	1.004–1.896
Pivot	6	66.1	3.8	58.2–70.0	84.2	7.9	75–95	1.907	0.188	1.599–2.113

*Normalized mass lifted = mass lifted/(body mass × 0.67) (Challis 1999).

Position key: Corners = first base and third base, pivots = second base and shortstop.

Mean, SD, and range collected from several tests conducted during December 1997.

Table 26.7 Three-Repetition Maximum Bench Press Results of Australian International Squad Players

		Body mass (kg)			Mass lifted (kg)			Normalized mass lifted* (kg/kg)		
Position	n	Mean	SD	Range	Mean	SD	Range	Mean	SD	Range
Catcher	6	75.6	4.7	69.0–80.0	54.2	4.2	50–60	1.079	0.155	0.960–1.298
Corner	5	78.7	11.7	63.0–97.2	55.0	7.6	42.5–62.5	1.067	0.220	0.771–1.421
Outfield	11	68.3	7.5	55.0–77.5	46.6	7.2	35–55	1.019	0.126	0.835–1.244
Pitcher	11	79.4	14.3	66.6–111.5	50.0	4.0	45–55	0.961	0.143	0.736–1.200
Pivot	6	66.0	3.8	58.2–70.0	49.6	7.6	40–62.5	1.119	0.146	0.967–1.390

*Normalized mass lifted = mass lifted/(body mass × 0.67) (Challis 1999).

Position key: Corners = first base and third base, pivots = second base and shortstop.

Mean, SD, and range collected from several tests conducted during December 1997.

successful at either overhand or underhand technique for chin-ups perform a 3RM test on a lat pull-down machine using an overhand grip. Normative data for the chin-up test are not available.

Equipment:

- Chin-up bar
- Belt with fittings to attach masses
- Disk weights (from 10 kg disks down to 2.5 kg)

Procedure—Chin-Up

1. The starting position is with the hands approximately shoulder-width apart, with arms fully extended.

2. Athletes pull themselves up so that the chin is above the bar; they then lower themselves to the start position with the arms fully extended.

3. Two additional repetitions of these steps must be completed successfully for the performance to be counted as a 3RM.

4. If attempts are made with added masses on the waist belt, increments are 2.5 kg.

5. The test does not count if any of the following occur: the athlete fails to elevate the chin above the bar; the athlete fails to return to the start position under full control; the athlete swings the legs or body.

Lat Pull-Down

1. The athlete grips the bar in an overhand grip with the hands approximately shoulder-width apart. With the athlete sitting in the apparatus, the starting position is with the arms fully extended.

2. Athletes pull the bar down so that it is lower than the chin; they then return the bar to the start position with the arms fully extended.

3. The test does not count if any of the following occur: the athlete fails to pull the bar below chin level; the athlete fails to return to the start position under full control; the athlete rocks back using body weight and momentum to help pull the bar down.

Abdominal Strength: Seven-Level Sit-Up Test

The problems and issues of strength assessment are described in chapter 10. The equipment and test procedure are described on pages 130 and 141. For normative data, see table 26.8.

Table 26.8 Abdominal Strength Scores of State- and National-Level Players

Squad	n	Mean	SD	Range
		Score on 7-stage test		
AIS/QAS	11	5.6	0.8	4–7

AIS = Australian Institute of Sport, QAS = Queensland Academy of Sport.

Data compiled from best test scores collected during 1997.

chapter 27

Protocols for the Physiological Assessment of Swimmers

David Pyne, Graeme Maw, and Wayne Goldsmith

Swimming presents significant opportunities to apply information gained from physiological testing (Prins 1988; Treffene 1979; Troup 1984, 1986). A well-planned testing program can yield a number of benefits, as discussed in chapter 17. Although national-level swimmers in some countries (notably Australia) have a formal requirement to undertake these tests, a cooperative approach between the swimmer, coach, and scientist is the best way to obtain the benefits.

The tests outlined in this chapter are largely pool based for reasons of specificity and practicality. Few nations possess the purpose-built swimming flume necessary for routine measurement of oxygen uptake during swimming. Although it is possible to measure oxygen uptake during swimming in a conventional pool, or to estimate it using techniques such as backward extrapolation, the practical difficulties associated with these approaches have limited their use in testing large numbers of swimmers. For this reason, the swimming tests outlined here are centered on performance measures, stroke characteristics, and physiological parameters (such as heart rate and blood lactate) that are easily measured poolside.

The underlying energy systems and the characteristics of swimmers and competitive events form the basis of the training program (Roberts 1991). In simple terms, the two main physiological characteristics of highly trained swimmers are anaerobic power and endurance. Endurance is related to the power and capacity of the aerobic energy system and is assessed indirectly with the graded incremental swimming test (the 7 × 200 m step test) or a continuous swim (2000 or 3000 m time trial). The stroke mechanics necessary for efficient utilization of energy is assessed in the pool with the 7 × 50 m incremental test. This test facilitates the assessment of stroke mechanics from submaximal to maximal speeds. The relative importance of a given test will depend on the swimmer's main competitive event and the phase of the training season.

Physiological Factors in Swimming Performance

Pool swimming events sanctioned by Fédération Internationale De Natation Amateur (FINA), the international governing body of swimming, range between 50 m (approximately 22-35 s) and 1500 m (approximately 15-17 min). Open-water or long-distance events may range between 1 km (approximately 10-12 min) and 25 km (5-6 h). The three systems that supply energy for skeletal muscle contraction (adenosine triphosphate-phosphocreatine [ATP-PC], anaerobic glycolysis, and the aerobic system) are all activated simultaneously during exercise. The intensity and duration of exercise determine the relative contribution of each system to the resupply of adenosine triphosphate (ATP). In the shortest swimming event, the 50 m sprint, the relative contribution for each system is (approximately) ATP-PC, 65%; anaerobic glycolysis, 30%; and aerobic, 5%. For a 200 m event the contributions are ATP-PC, 10%; anaerobic glycolysis, 50%; and aerobic, 40%. For a 1500 m the breakdown is ATP-PC, 2-5%; anaerobic glycolysis, 20%; and aerobic, 75-80%

(Maglischo 1987; Roberts 1991; Sharp 1992; Troup 1984). Open-water or long-distance events rely almost exclusively on the aerobic energy system.

As already stated, the training program is based on the energy systems along with the physiological requirements of competitive events (Roberts 1991). Although indirect assessment of the underlying physiological capacities is important, swimming is a technically demanding sport; and it is prudent to assess stroke characteristics in parallel with physiological responses.

Testing Environment and Subject Preparation

As outlined in chapter 2, the pretest preparation must be standardized if one is to obtain reliable and valid physiological data. This is essential for both laboratory and pool testing. The following guidelines relate more specifically to swimming.

- **Training.** Athletes should not perform any highly stressful swimming training within the preceding 24 h. For blood testing, it is essential for swimmers not to have trained within the preceding 6 h, as the effects of exercise on the distribution of blood cells persist for several hours. It is also advisable for swimmers to refrain from heavy weight training or unaccustomed exercise on both the preceding day and the day of testing.
- **Diet.** Swimmers should follow a normal high-carbohydrate, low-fat diet on the days before and on the day of testing; they should abstain from food and beverages containing caffeine or alcohol in the 2 h prior to testing. Encourage adequate hydration with either water or a sports drink. This is particularly important for testing in conditions of high temperature and humidity.
- **Pool conditions.** Most pools have closely monitored temperature (usually 27.0 ± 1.0° C) and water-quality levels. These conditions should be checked with pool staff if there is any doubt about the suitability of the pool for testing. Results may not be valid if the water temperature is outside comfortable limits. Pool testing can be completed either indoors or outdoors. If testing is outdoors, ambient air temperature and wind may affect swimmers waiting to be tested. Close cooperation with coaches and swimmers is recommended to ensure appropriate timing of warm-up and testing. Finally, the choice of pool length (25 or 50 m) is critical, as short-course and long-course swimming yield different results (Telford et al. 1988; Richardson et al. 1996).
- **Warm-up.** Prior to pool testing, swimmers should complete a standardized warm-up of 1000-1500 m, consisting primarily of aerobic swimming of low-to-moderate intensity. Swimmers normally undertake the warm-up in freestyle with some elements of pull, kick, and the main stroke they will use for the pool tests (if this is not freestyle). Some swimming at the end of the warm-up at the speed of the first swim in the test set is recommended. A swimmer should complete the same warm-up prior to each test.
- **Swimmer condition.** Test results should be valid if the swimmer is reasonably well rested and free of illness and injury as indicated in the checklists in chapter 2. If there is any doubt whether the swimmer is ready to undertake testing, it may be more appropriate to postpone testing to another day. Results of testing with swimmers who are not adequately prepared can be difficult to interpret.

Equipment Checklist

1. Blood lactate testing
   - ☐ Blood lactate analyzer(s)
   - ☐ Autolet sampler, lancets, and platforms
   - ☐ Sterile alcohol swabs
   - ☐ Tissues
   - ☐ Heparinized capillary tubes
   - ☐ Pipette
   - ☐ Sample tray or rack
   - ☐ Biohazard bags
   - ☐ Sharps container
   - ☐ Disposable surgical gloves
   - ☐ Surgical tape
2. Anthropometry
   - ☐ Stadiometer (wall mounted)
   - ☐ Balance scales (accurate to ±0.05 kg)
   - ☐ Skinfold calipers (Harpenden, dynamically calibrated)
   - ☐ Marker pen
   - ☐ Steel anthropometric measuring tape (Lufkin, model W606PM)
3. Aerobic test (7 × 200 m)
   - ☐ 50 m pool (long course)
   - ☐ Stopwatch with stroke-rate facility
   - ☐ Electronic heart rate meter (e.g., Polar Sport Tester or Precision Heart Rate Meter)

- ☐ Clipboard with data sheets and pens
- ☐ Blood lactate testing equipment (as listed above)

4. Stroke mechanics test (7 × 50 m)
   - ☐ 50 m pool
   - ☐ Stopwatch with stroke-rate facility

Administration of Tests

It is recommended that physiological testing of national- and international-level swimmers take place approximately three times per competitive cycle, based on the preparation for the major championship in each calendar year. Testing dates should be organized at least one full competitive cycle in advance.

Based on an average 12- to 16-week preparation, testing should take place at the following points:

- Early in the preparation (2 weeks into preparation: 10-14 weeks from competition)
- Midpreparation (6-8 weeks from competition)
- Pretaper (3-4 weeks from competition)

To avoid the interference of residual fatigue, do not schedule testing during weeks of high training volume or intensity. Standardize testing so that it is on the same day of the week (e.g., Monday, Tuesday) each time. Table 27.1 illustrates a typical testing schedule.

Table 27.1 A Suggested Schedule for Physiological Testing of Swimmers

Day	Saturday	Sunday	Monday
A.M.	Normal training	Off	1. Blood testing 2. Anthropometry Normal training
P.M.	Normal training	Off	3. 7 × 200 m aerobic test 4. 7 × 50 m stroke mechanics test

PROTOCOLS: Swimming

Anthropometry

Rationale: Measurement of height, body mass, and sum of skinfolds in highly trained swimmers should be routine. In younger swimmers, measurement of height and body mass is useful to monitor growth and development. In older swimmers, measurement of body mass and estimation of body fat using the sum of skinfolds technique provide useful feedback on body composition and the cumulative effects of training and diet. In practice, most use a two-compartment model of body composition that divides the body into components of fat mass and lean body mass. Dual measurement of body mass and sum of skinfolds (body fat) gives an indirect estimation of the relative proportion of these components.

In swimming, the relationship between body fat and body drag is important. Body drag—the amount of resistance that the body encounters while moving through the water—is influenced by body size, the speed of swimming, and other mechanical factors. The issue here is the effect of body size and buoyancy on performance. As with many factors, there is considerable interindividual variation in the relationship between body fat and performance. Above a certain individual level, an increase in body fat will be deleterious to performance because of increased body drag. Although increased body fat is likely to enhance buoyancy, the increase in body drag will offset any advantage resulting from improved buoyancy. The other two key factors that influence this relationship are gender and event distance. Clearly, females carry more fat than males as a biological requirement, and distance swimmers generally carry more than sprint swimmers. The higher body fat levels of ultra-endurance long-distance swimmers confer some advantage in buoyancy and thermoregulation.

Procedures: For the measurement of height, body mass, and skinfolds (sum of seven sites for males and females), refer to chapter 5. More advanced anthropometric assessment (somatotype and limb lengths, breadths, and girths), as described by Norton et al. (1996), will provide even more detail if needed.

Interpretation: Height and body mass are evaluated in terms of the chronological and biological age of the swimmer. Although comparison to normal ranges and group means is of interest, comparison with each swimmer's own previous results has the most value.

Basic Anthropometric Assessment

For the measurement of height, body mass, and skinfolds (sum of seven sites for males and females), refer to chapter 5.

Advanced Anthropometric Assessment

It may be helpful to perform more advanced anthropometric assessment (fractionation, somatotype, limb lengths, breadths, and girths) as described by Norton et al. (1996). These measurements are helpful for two reasons. With age-group or junior swimmers they are useful for monitoring growth and physical maturation. With senior swimmers, these measurements may permit more detailed assessment of changes in lean body mass and fat mass associated with specific training programs and/or dietary regimes.

Presentation of Results: Results are presented in both tabular and graphical form. Table 27.2 illustrates the recommended presentation for more advanced assessment of the results of fractionation (computer-generated estimates of muscle mass, fat mass, skeletal mass, and residual mass).

Even elite swimmers show a large degree of individual variation; interpretation of results must take this into account. Mean results for national team swimmers are presented in table 27.3. It is advisable for swimmers to consult with a sport dietitian to review dietary practices and develop appropriate management strategies.

Aerobic Test (7 × 200 Meters)

Rationale: The aim of this test is to provide objective information on the aerobic or endurance fitness of the swimmer. The protocol involves a graded incremental test for measurement of cardiovascular (heart rate) and metabolic (blood lactate) responses to increasing swimming speeds. These data are processed using standard computer software for the generation of heart rate-velocity and lactate-velocity curves. The basic

Table 27.2 Longitudinal Changes in Body Composition of a 19-Year-Old Male Swimmer Using the Process of Fractionation*

Date	February	June	August	October
Height (cm)	192.9	193.0	192.7	193.5
Mass (kg)	77.9	80.7	83.0	83.3
Skinfolds (mm)	43.9	40.4	42.5	46.2
Fat (kg)	6.8	6.4	6.7	7.2
Muscle (kg)	37.0	38.6	39.1	38.7
Skeletal (kg)	13.6	13.5	13.6	13.8
Residual (kg)	21.6	21.9	22.5	21.8

*For more on the process of fractionation, see Norton et al. (1996).

Estimates of skeletal muscle mass, fat mass, skeletal mass, and residual mass are shown. Muscle mass increased by 2.1 kg, but fat mass only by 0.1 kg in the six months between February and August.

Table 27.3 Height, Weight, and Sum of Skinfolds Values of 1999 Australian National Team Swimmers (Mean ± Standard Deviation)

	Females (n = 25)	Males (n = 29)
Height (cm)	171.8 ± 5.0	185.7 ± 5.4
Weight (kg)	64.8 ± 5.6	81.2 ± 7.1
Σ skinfolds (mm)	64.3 ± 8.4	48.1 ± 9.1

premise is that heart rate and blood lactate responses to submaximal exercise are sensitive indicators of endurance fitness, and that blood lactate responses relate to training-induced adaptations occurring within skeletal muscle (Weltman 1995). In practice, improvements in fitness are indicated by characteristic changes in the heart rate-velocity and lactate-velocity relationships.

These curves are used for two main purposes: to prescribe training speeds and to monitor longitudinal changes in aerobic fitness with training. Within the scientific and coaching literature, numerous protocols have been devised and debated (Counsilman and Counsilman 1993): many have evolved because of the great variety of information that one can gain from testing. In swimming, the basic interval in the various aerobic tests normally ranges from 100 to 400 m. The longer intervals such as 300 or 400 m are more likely to produce a physiological steady state; this is important for assessing the relative metabolic cost of a given swimming speed. These longer intervals ensure that heart rate, blood lactate, and oxygen uptake are truly representative of that particular swimming speed.

Notwithstanding the need to achieve a physiological steady state, the shorter 100 or 200 m intervals are more specific to the training and competitive requirements of swimmers. The 200 m interval chosen here represents a compromise between achieving a steady state in metabolism and using swimming speeds more specific to competition levels. Recent work has demonstrated the stability of blood lactate-heart rate relationships in competitive athletes, indicating that a single well-conducted test permits evaluation of longitudinally stable training markers (Foster et al. 1998). These tests may be used either in isolation or in conjunction with other tests, depending on the requirements of the scientist and coach.

Procedures: Swimmers undertake a series of seven 200 m swims in their specialist stroke (i.e., freestyle, backstroke, breaststroke). Given the difficulties in "holding stroke" over a series of 200 m efforts, butterfly swimmers should swim this test freestyle. Individual-medley swimmers also swim the test using freestyle. The basic protocol is:

7 × 200 m swim on 5 min cycle

Swimmers perform seven evenly paced swims graded from easy to maximal on a 5 min cycle.

1. Preparation of swimmers should follow the guidelines presented earlier in this chapter.

2. Calculate target times for each swimmer before the session and then discuss these with the swimmer and the coaching and testing staff. Calculate times according to the guidelines in the list that follows. As an example, table 27.4 shows the times for an elite male distance swimmer with a 200 m freestyle best time of 1:50 (min:s). In everyday terms, one could summarize the protocol by telling the swimmer, "Add 30 s to your predicted 200 m time and then descend by 5 s."

- Ascertain the swimmer's 200 m personal best (pb) (e.g., 1:50 for a male 200 m freestyle swimmer).
- Add a 5 s differential to account for push start and training situation to estimate the time for the final swim (no. 7) (e.g., 1:50 + 0:05 = 1:55).
- Working in reverse order from the seventh and final swim, add 5 s for each subsequent interval to establish the full test protocol, for example:

Swim no. 7	= pb + 0:05	= 1:55
Swim no. 6	= 1:55 + 0:05	= 2:00
Swim no. 5	= 2:00 + 0:05	= 2:05
Swim no. 4	= 2:05 + 0:05	= 2:10
Swim no. 3	= 2:10 + 0:05	= 2:15
Swim no. 2	= 2:15 + 0:05	= 2:20
Swim no. 1	= 2:20 + 0:05	= 2:25

3. Prepare a results sheet on which to record all pertinent information for each swimmer, including recent and/or current illness, recent and/or current injury, recent training absences, the number of weeks completed in the current training cycle, and any other information relevant to the interpretation of test performances and results.

4. Swimmers should perform the test in small groups of three to four at a time (each group takes approximately 45 min). Three to four testing staff will be needed to collect all the necessary data. The warm-up should finish with one or two 50 or 100 m swims at the same pace as required for the first 200 m swim of the test.

5. All swims utilize a push start. Before each swim, emphasize "even pace" or "even splits," and give the target time. A common mistake is to go too fast on the first swim.

6. Record the time for each 100 m. With manual timing, the first movement is used as the starting time. It is important to note that for breaststroke, hand touch is used to time splits—not feet off the wall. If required, the stroke rate (third 50 m) and

Table 27.4 An Example of Individualized Protocol for the 7 × 200 m Step Test

Swim no.	Approximate % of target time	Approximate heart rate below max (beats/min)	Seconds slower than final target time	Example (pb 1:50) (min:s)
1	70%	–70	–30	2:25
2	75%	–60	–25	2:20
3	80%	–40	–20	2:15
4	85%	–30	–15	2:10
5	90%	–20	–10	2:05
6	95%	–10	–5	2:00
7	100 %	10 to 0	0	1:55

stroke count (fourth 50 m) are also recorded as outlined in the section on stroke mechanics later in this chapter.

7. Measurement of heart rate: Immediately upon completion of each swim, heart rate is measured with an electronic heart rate meter (Precision Heart Rate Meter [Heart Rate Industries, Brisbane, Australia] or a waterproof Sport Tester PE [Kempele Oy, Finland]). The rating of perceived exertion (RPE) can also be recorded at this time.

8. Measurement of blood lactate: The swimmer exits the pool after each swim and has a blood sample taken as soon as possible from the earlobe (or fingertip). Samples need to be drawn rapidly to ensure that the swimmer adheres to the 5 min cycle. See chapter 6 for details of the procedures for blood lactate testing.

9. The swimmer then has a short break (depending on the time taken to extract the blood sample) before commencing the next swim exactly 5 min after the preceding swim. Swimmers should enter the pool at least 15 s prior to the push start.

10. Continue this cycle until all seven swims have been completed.

11. Record all results on the data sheet and forward them for processing.

Data Presentation: A typical data set is presented in table 27.5. The swimming velocity is represented directly as velocity (m/s) or indirectly as time per 100 m of swimming (s). Coaches prefer the measure of time per 100 m.

Using the data collected, graphs are derived to establish the heart rate- and lactate-velocity relationship. In simple terms, heart rate (beats/min) and lactate (mmol/L) for each swim are plotted on the y-axis against swimming speed on the x-axis. Heart rate (beats/min) is plotted against the time for the final 100 m, while lactate (mmol/L) is plotted against the time for the average 100 m. The scientist will make the choice of a simple line chart, exponential, or second- or third-order polynomial plot (depending on the analysis used). Additional information such as stroke rate, stroke count, stroke efficiency index, and RPE can also be graphed in a similar manner if necessary. Figure 27.1 shows a typical graph for heart rate.

Interpretation: The most important principle in the interpretation of heart rate and blood lactate testing of swimmers is that of individual analysis for each swimmer (Pyne and Telford 1988). Individual analysis is indicated by the twin effects of gender and event specificity. It is well documented that the physiological demands of swimming vary considerably between each of the four strokes (freestyle, backstroke, butterfly, and breaststroke) and the individual medley. Apart from these basic differences, responses to swimming tests may be affected by age, training background, immediate training history, injury, and motivation. The usual practice is to analyze and interpret each swimmer's results individually. This does not, of course, preclude some group comparisons if these are of interest.

In practice, consideration of the following points helps one interpret test results for a given swimmer:

- Have the heart rate and lactate-velocity curves moved, and if so in which direction?
- Was the final time (i.e., the seventh 200 m) faster or slower than before, and how does it relate to the swimmer's personal best time for the 200 m?
- What were the increments in speed between each of the seven swims?
- Were the swims completed with even or appropriate splits?

Table 27.5 A Typical Set of Results From the 7 × 200 m Step Test

Swim no.	200 m (min:s)	1st 100 m (s)	2nd 100 m (s)	Av 100 m (s)	Heart rate (beats/min)	Lactate (mmol/L)	Stroke rate (st/min)	Stroke count (st/50 m)	RPE
1	2:21.40	71.0	70.4	70.7	120	1.8	35.5	38	8
2	2:18.10	69.3	68.9	69.1	133	1.9	35.4	37	11
3	2:11.15	65.1	66.1	65.6	142	2.2	38.9	38	12
4	2:05.13	62.5	62.6	62.6	156	3.3	39.5	38	13
5	1:59.47	59.5	60.0	59.7	172	5.7	42.5	38	14
6	1:55.99	57.5	58.3	57.9	178	8.2	43.1	38	16
7	1:51.91	56.1	55.9	56.0	181	11.2	43.7	38	18

HR = heart rate, RPE = rating of perceived exertion.

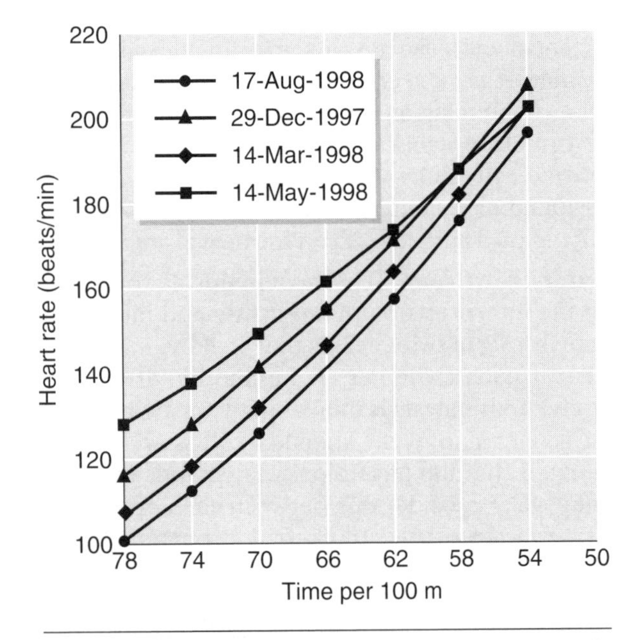

Figure 27.1 Heart rate-velocity relationship derived from the 7 × 200 m step test. A downward and rightward shift in the curve is evidence of improvement in aerobic fitness.

- Were trends in stroke measurements (e.g., stroke rate and stroke count) consistent with the physiological data?
- Was there any biomechanical and/or subjective evaluation of the swimmer's technique?
- What was perception of the swimmer regarding the 7 × 200 m test?
- What was the recent training history, including quality sets within the last week?

Prescription of Training Velocities: Prescription of swimming training intensities is largely regulated by physiological and metabolic parameters (see pages 61-62). The various training intensities can be arranged into a system of training zones: for example, low-intensity aerobic, moderate-intensity aerobic, anaerobic threshold, maximal aerobic, and sprint (race pace). Accurate prescription of these zones depends upon the identification of physiological transitions between the various aerobic and anaerobic energy systems. These transitions largely reflect changes in the dynamics of blood lactate accumulation and muscle metabolism. The initial rise of blood lactate concentration above baseline has been variously termed the aerobic threshold, the lactate threshold, or the onset of blood lactate accumulation. Its determination relies on identification of the baseline lactate concentration (the first three steps of the 7 × 200 m step test) coupled with reliable determination of the first significant increase in lactate concentration. With further increases in velocity, most swimmers will exhibit a secondary and more substantial rise in the blood lactate concentration: this point is generally referred to as the anaerobic threshold (AT) and usually occurs between the fourth and fifth step.

A number of investigators have proposed methods to analyze heart rate and lactate curves for the prescription of training speeds (Anderson and Rhodes 1989; Bishop and Martino 1993; Maglischo et al. 1987; Maglischo et al. 1984; Roberts 1990). The aerobic training zone is sometimes divided in half to differentiate low-intensity (A1) and moderate-intensity (A2) aerobic efforts. The midpoint, where the rate of blood lactate accumulation escalates significantly above the baseline, is commonly termed the aerobic threshold (AeT). The anaerobic threshold is usually defined as the swimming speed at which lactate production exceeds removal, leading to a significant elevation in muscle and blood lactate concentrations. However, it is not correct to think that this point simply represents a change from aerobic to anaerobic metabolism or that the escalating lactate concentration is necessarily debilitating; rather, having crossed this point, the swimmer will not be in steady state and cannot continue swimming at this pace for an extended period of time. Above the AT is a predominance of energy supply from hypoaerobic as compared to hyperaerobic fibers (Antonutto and Di Prampero 1995), with ensuing fatigue relating to the accumulation of intramuscular acid (hydrogen ion, H^+) (Hultman and Sahlin 1980).

The theory underlying the concept of the AT is the subject of much scientific debate (see chapter 4); but in practice, attention focuses on determining the swimming velocity associated with the AT. Among the various ways of estimating swimming velocity at AT (VAT), a popular method is the so-called Dmax protocol (Cheng et al. 1992). The lactate tolerance rating (dVL5-L10), defined as the differential velocity between lactate concentrations of 5.0 and 10.0 mmol/L, is a useful measure indicating the efficiency of the swimmer in the area of maximal oxygen uptake (Holroyd and Swanwick 1993; Pyne et al. 1999). This method, which identifies an acceleration in the rate of blood lactate accumulation, can easily be computerized for automatic threshold determination (figure 27.2). Training at the AT will improve the muscles' ability to tolerate or buffer acid and remove lactate from the intracellular environment.

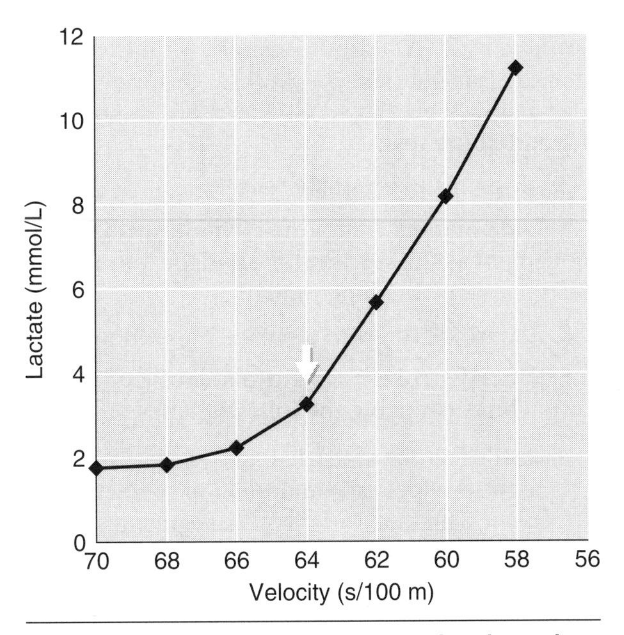

Figure 27.2 Lactate-velocity relationship derived from the 7 × 200 m step test. Computer-based analysis is used to indicate the speed at which the anaerobic threshold (64.0 s/100 m) occurs.

Aerobic training can also be regulated using Treffene's model of critical velocity (Vcrit) (1978). This velocity is determined at the speed at which maximum heart rate is first reached. Training intensity then approaches and is sustained close to maximum heart rate, eliciting high blood lactate concentrations to stimulate the rate of lactate removal from the cell. In practice, the Vcrit relates to the maximal sustainable work rate over typical training sets of 2000 to 3000 m. Changes in critical velocity are thought to indicate the swimmer's level of aerobic conditioning or residual fatigue and maladaptation. In addition, the lactate response might provide an indication of a swimmer's current sprint (< 8 mmol/L) or endurance (> 12 mmol/L) potential (Treffene et al. 1979).

Assessing Changes in Fitness: The parameters measured during the aerobic step test (i.e., AeT, AT, dVL5-L10, and Vcrit) all provide distinct indicators of a swimmer's training status. In practice, people "eyeball" the heart and lactate curves and evaluate them subjectively for indications of improvement, plateauing, or degradation of performance. In graphical terms, the heart rate-velocity and lactate-velocity curves will shift upward and/or leftward if fitness has deteriorated. In contrast, a shift downward and/or rightward is evidence of improved aerobic fitness (Madsen and Lohberg 1987; Pyne 1989; Treffene 1979; Weltman 1995) (see figure 27.1). Taking this one step further, an increase in velocity at AT (a shift to the right in the lower part of the curve) is thought to indicate an improvement in basal aerobic fitness (see chapter 4). Lactate production has presumably been delayed due to increased fat oxidation. A rightward shift (to a higher velocity) in AT might indicate an improvement in aerobic endurance—of particular importance to middle- and long-distance swimmers—possibly caused by improved lactate removal or acid buffering.

One should always interpret the lactate and heart rate curves cautiously, as other factors may influence the test result (Pyne 1995). For example, a swimmer who is glycogen depleted might show an attenuated lactate response due to a lack of available substrate (Tegtbur et al. 1993; Reilly and Woodbridge 1999). This will generally be accompanied by slower times during the final stages of the test and a decrease in the ratio of blood lactate concentration to the swimmer's perceived level of exertion (Snyder et al. 1993); that is, the lactate response will be low even though the swimmer is apparently trying very hard. It is useful to record RPEs in order to preempt development of fatigue or overtraining. Lactate and heart rate may also be high during the test without a true decrease in aerobic condition if the swimmer is ill or is struggling to adapt to the current training load (Pyne 1995); again, one needs to make judgments about the cause of deterioration in indicators of aerobic fitness.

Apart from physiological factors, the primary cause of changes in lactate and heart rate curves is a change in swimming efficiency. For example, when a swimmer's technique is poor, the heart rate may be elevated without there having been a loss in fitness; this simply reflects the extra effort required to compensate for a loss in efficiency. Wakayoshi et al. (1995) showed that a change in the slope of the stroke rate-velocity relationship—as compared to a parallel displacement—distinguished between technical and physiological swimming changes. It might be simpler to measure the stroke rate needed to achieve a given speed. For example, if an increase in heart rate is accompanied by a significant increase in stroke rate between corresponding tests, one might deduce that the swimmer's technique has suffered rather than the underlying physiology. Of course, this does not discount the possibility that the technical changes are themselves caused by changes in fitness, and further investigation may be necessary.

In summary, the 7 × 200 m step test can provide useful information for the prescription, regulation, and monitoring of swimmers' training. In simple terms, gross shifts in the heart rate-velocity or lactate-velocity curves, or both, suggest improvements or regression in aerobic condition.

Stroke Mechanics (7 × 50 Meters)

Rationale: Swimming is a technically demanding sport, and a substantial proportion of training time addresses the refinement of a swimmer's technique. A series of 7 × 50 m swims of progressively increasing speed serves to establish the relationship between swimming velocity (V), stroke rate (SR), and distance per stroke (DPS). These relationships can be summarized (Costill et al. 1985; Craig and Pendergast 1979; Maw and Volkers 1996) as:

$$V = SR \times DPS$$

$$V\ (m/s) = SR\ (strokes/s) \times DPS\ (m/stroke)$$

$$DPS\ (m) = (V\ [m/s] \times 60)/SR\ (strokes/m)$$

In practice, distance per stroke is difficult to measure in the pool without sophisticated biomechanical analysis. Simply counting strokes per lap may be inaccurate, as this does not account for how much distance was traveled underwater and whether the lap finished on a complete stroke. One can calculate distance per stroke from velocity and stroke rate by recording stroke characteristics over a known distance in the pool (Maw and Volkers 1996) as outlined in step 6 of the procedures described next.

Procedures:

1. The protocol for this test is 7 × 50 m swims on a 2 min cycle. All swimmers use their main stroke. For individual-medley swimmers, the usual practice is to undertake the test using butterfly, which is the leadoff stroke. A 50 m pool is mandatory for this test. Swimmers complete the same warm-up as for the 7 × 200 m step test. They should finish the warm-up with one or two 50 m efforts at the same pace as for the first swim in the 7 × 50 m test.

2. Determine the target times for the test starting with the slowest swim. The slowest swim (i.e., swim no. 1) is performed approximately 12 s slower than the predicted best time on the day. Each of the subsequent swims is then undertaken approximately 2 s faster than the preceding swim, until the seventh and final (and maximal effort) swim is completed. In everyday terms, one could summarize the protocol by telling the swimmer, "Add 12 s to your predicted 50 m time and then descend by 2." A common mistake is for the swimmer to start too fast on the first swim or to descend slowly and have a big increase in speed at the end of the test.

3. All swims use a push start.

4. With manual timing, use the first observed movement as the starting time and the hand touch at 50 m as the finishing time.

5. Record all times to a tenth of a second.

6. Record stroke rate and distance per stroke for each repeat using the following procedures:

   - Record stroke data (distance per stroke) from between flags at 5 m and 45 m of each 50 m swim. The time to swim this segment is taken with a stopwatch. At approximately the 15 m mark (i.e., within the first 25 m segment) and the 35 m mark (i.e., within the second 25 m segment), three complete stroke cycles are recorded. Taking these measurements accurately requires that swimmers be surfaced from their push start by the 5 m point. Timing the head as it goes through the 5 and 45 m points is recommended as the best method to record the time taken. An alternative method involves measurements taken at the 15 m mark (the false-start line) and the 45 m point.

- Stroke rate (strokes/min) is measured using the base three stroke-rate facility on the stopwatch. The stopwatch is started as the swimmer's hand enters the water to commence a stroke. At the end of three complete stroke cycles, the stopwatch is stopped as the same hand enters the water for the fourth time. Alternatively, the stroke cycles can be timed and the stroke rate calculated using the equation (Maw and Volkers 1996):

$$SR = (60 \times 3)/\text{Time for Three Strokes (s)}$$

For example, if three consecutive strokes take 4.08 s, then SR = $(60 \times 3)/4.08 = 44.1$ strokes/min; for three strokes in 3.90 s, the SR = $(60 \times 3)/3.90 = 46.2$ strokes/min.

- For breaststroke, it is often easier to use the point where the head comes up rather than the hand entry. The average of the two stroke rates (i.e., at the 15 and 35 m points) is used to represent the stroke rate for that swim.
- Distance per stroke is calculated using the equation DPS = $(V \times 60)/SR$ (strokes/min), which converts the units from strokes per second to strokes per minute.

7. Plot on separate graphs the stroke rate and distance per stroke (y-axis) against swimming speed (x-axis).

Data Presentation: The following example is for a female 100 m breaststroke swimmer (table 27.6 and figures 27.3 and 27.4).

Interpretation: The aim of this test is to provide a qualitative analysis of stroke mechanics during a series of progressively faster swims (Craig and Pendergast 1979; Maw and Volkers 1996). This information should be used in conjunction with the coach's subjective assessment of the technical quality of the stroke (Wakayoshi et al. 1995). Most importantly, each swimmer will have a different combination of distance per stroke and stroke rate for his or her particular stroke.

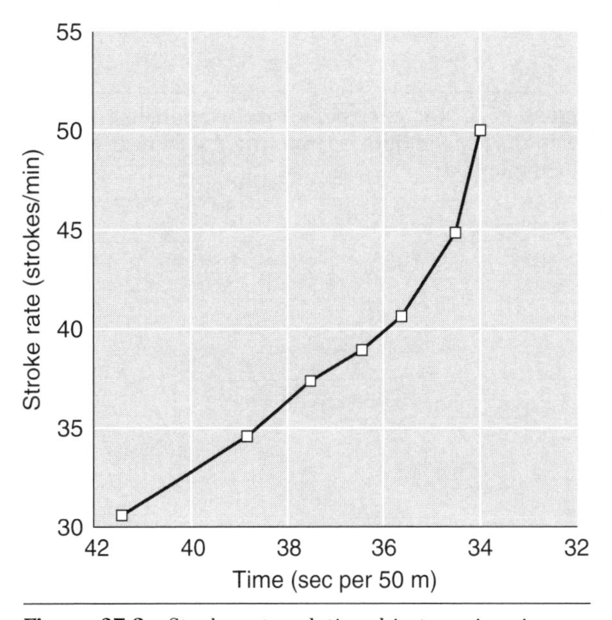

Figure 27.3 Stroke rate relationship to swimming speed derived from the 7 × 50 m stroke mechanics test.

Table 27.6 A Typical Example of Results From the 7 × 50 m Stroke Mechanics Test for a World-Class Female Breaststroke Swimmer

Swim no.	Time (s)	Stroke rate (cycles/min)	Stroke count (strokes/50 m)	Distance per stroke (m)
1	41.4	30.6	19	2.28
2	38.8	34.6	20	2.18
3	37.5	37.4	21	2.14
4	36.4	38.9	21	2.07
5	35.6	40.6	22	2.03
6	34.5	44.8	23	1.91
7	34.0	50.0	23	1.74

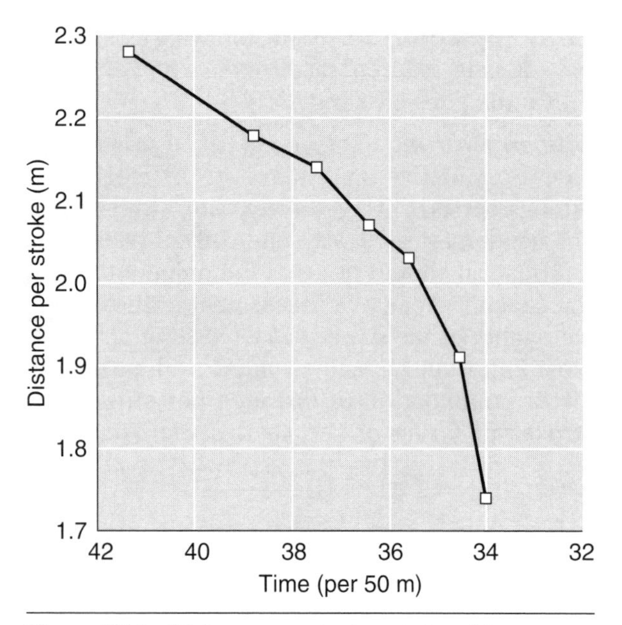

Figure 27.4 Distance per stroke relationship to swimming speed derived from the 7 × 50 m stroke mechanics test.

It is desirable for a swimmer to maintain good technique from the slowest to the fastest swim. Better swimmers are able to "hold their stroke together" at the fastest speeds while less skilled performers lose control as evidenced by non-linear changes in stroke rate, distance per stroke, or both (Wakayoshi et al. 1993). Inspection of the graph should indicate the speed at which control of stroke mechanics starts to deteriorate.

Protocols for the Physiological Assessment of High-Performance Tennis Players

Andrea Buckeridge, Damian Farrow, Paul Gastin, Mark McGrath, Peter Morrow, Ann Quinn, and Warren Young

Tennis is a unique game requiring at one instant explosive power to hit a serve or get to a short ball, tempered with the fine motor control utilized in net play. Similarly, pure leg speed around the court is traded off with the ability to change footwork patterns and movement direction instantaneously.

Developing a Testing Protocol for Tennis

In developing a testing protocol for the physiological assessment of high-performance tennis players, one needs to consider the nature of the game at the elite level, in addition to identifying the most relevant physiological parameters that influence performance.

The Nature of the Game

At the elite level (i.e., full-time players on the professional circuit), the competition phase is approximately 11 months per year, making periodization of training a difficult task. Players must travel continuously to compete, adapting to a variety of climatic conditions and court surfaces (i.e., clay, grass, and hardcourt). Matches may be between 1 to 4 h in duration, with players required to play either the best-of-three or -five set matches.

Physiological parameters significantly influence a player's performance, but owing to the open nature of the sport, psychological and technique factors also interact to affect performance. However, when players of equal skill level meet in competition, the player with the better physical characteristics (whether through training or endowment) has an advantage over the opponent (Reilly 1990). By using standardized testing to identify the physiological performance characteristics advantageous to successful performance in tennis, a coach can design individualized training programs to optimize improvements in the desired quality.

Kibler et al. (1988) stated that determining fitness levels should be as sport specific as possible, mimicking the actual coordination and movement of the game. Over the past four years, the authors have developed a field testing battery that addresses these concerns. By administering the tests at regular intervals throughout the year, we have been able to determine the strengths and weaknesses of players and subsequently gauge the effectiveness of their training programs and adjust the programs accordingly.

What Physiological Components Are Important?

Roetert et al. (1996) examined a number of physiological parameters of tennis in order to determine their relationship to tennis performance. In summary, a strong, significant relationship between the elements of speed, strength, flexibility, agility, and tennis performance was identified.

- **Strength and power.** A vital aspect of racquet sports is the ability to exert muscular force at a high speed. Hitting with power means that the tennis player must optimize the action between the larger body parts to hit at high velocity and still allow the upper limb to maintain good control over the racket (Groppel 1986). Koziris et al. (1991) evaluated unilateral and bilateral strength profiles in 38 collegiate women tennis players. Greatest strength differentials occurred in the shoulder and elbow. These joints are highly involved in tennis skills and the demands of a unilateral sport that places stress on the dominant shoulder and arm.
- **Agility and speed.** Agility is crucial to good court movement and correct positioning on the court. To strike the ball effectively in tennis, the body must be properly positioned, which requires the use of eyes, legs, and feet (Groppel et al. 1989; Leach 1988; Loehr 1988). Elliott et al. (1990) found that agility separated high-performance junior male players from an intermediate-level group and a control group.
- **Energy system contribution.** Performance of singles tennis involves both aerobic and anaerobic metabolism, although the contribution of both systems remains controversial and poorly understood. The alactic system is most likely to be the predominant mechanism of adenosine triphosphate formation during a point. However as the length of rallies may vary depending on the surface and the opponent, the contribution from the lactic acid system would also increase as the length of the point increases. Rallies during men's singles matches at Roland Garros (clay) average 10 s, compared to only 4 s at Wimbledon (grass). Docherty (1982), Friedman et al. (1984), and Elliott et al. (1985) found heart rates exceeding 60% of predicted maximum in singles play. These findings suggest that tennis singles competition has an aerobic as well as an anaerobic component. Bergeron et al. (1991) monitored metabolic function during singles tennis and a treadmill test with Division 1 collegiate tennis players. They found that although tennis is characterized by periods of high-intensity exercise, the overall metabolic response resembled that for prolonged moderate-intensity exercise.
- **Body composition and aerobic capacity.** Few studies have examined the body structure and characteristics of tennis players. Buono et al. (1980) found relatively high maximal oxygen uptakes (mean 58.4 $ml \cdot kg^{-1} \cdot min^{-1}$) and low body fat percentage (mean 10.4), indicating that skilled tennis players have excellent cardiovascular endurance. Elliott et al. (1990) conducted a prospective study of physiological and kinanthropometric indicators of junior tennis performance. They found no significant differences between elite junior tennis players and noncompetitive performers (who played no more than once per week) with respect to any aerobic capacity or lung function variables tested. The authors did find that higher-level performers, both male and female, were more linear and carried less body fat than the less skilled participants, although this did not discriminate performance.

From this brief review of the literature in relation to the physiological assessment of high-performance tennis players, two factors become clear:

- Speed, strength, flexibility, and agility are important physiological components of tennis performance. However, the exact contribution of aerobic capacity to tennis performance—while intuitively important—is equivocal.
- There has been a dearth of research on the very best professional tennis players (i.e., top 100 ranked players) in relation to the physiological components of tennis performance. Hence the normative data provided in this chapter are drawn from the results of the Australian Institute of Sport tennis scholarship holders, the Victorian Institute of Sport tennis squads, and the top male and female 12-, 14-, and 16-year-olds in Australia.

The following protocols are offered in an attempt to standardize test procedures in tennis. The relevance and specificity of the tests, in addition to the ease of administration and practical application, were considered in the selection process. Not all tests are used in every testing session. For example, the repeated-effort agility test is typically used once a year. The use of a standard protocol will aid in developing national databases.

Athletes are also encouraged to complete tennis-specific medical and musculoskeletal screening protocols (available from Australian Sports Commission; see appendix for details) as an adjunct to the physiological protocols outlined in this chapter. It is important to note that this chapter does not include tests to assess flexibility because a comprehensive assessment of this parameter is included in the musculoskeletal screening protocol.

All checklists and forms from chapter 2 should be filled out before any testing is undertaken. This is extremely important to ensure standardization of test conditions and hence the reliability of results. A reproducible form for recording test data is provided on pages 401-403.

Equipment Checklist

1. Anthropometry

A description of appropriate equipment is given in chapter 5.

2. Aerobic power—multistage fitness test

Refer to page 129 for a description of relevant equipment.

3. Speed—sprint tests

Refer to page 129 for a description of relevant equipment. Note that the test is conducted over 10 m only.

4. Agility—backward movement agility test

See page 129 for a description of relevant equipment. The following additional items are required:

- ☐ Tennis racquet
- ☐ Recommended surface—tennis court (hardcourt)
- ☐ Recommended footwear—tennis shoes

5. Agility—sideways movement; change-of-direction/acceleration sideways; and change-of-direction/acceleration forward agility tests

Refer to page 129 for a description of relevant equipment. Other required items are the following:

- ☐ Masking tape
- ☐ Measuring tape
- ☐ Tennis racquet
- ☐ Recommended surface—tennis court (hardcourt)
- ☐ Recommended footwear—tennis shoes

6. Agility—repeated-effort test
   - ☐ Masking tape
   - ☐ Measuring tape
   - ☐ Recommended surface—tennis court (hardcourt)
   - ☐ Recommended footwear—tennis shoes
   - ☐ 3 cones

7. Leg power—vertical jump (dynamic, left leg, right leg)

See page 130 for a description of relevant equipment.

8. Power—medicine ball throw (overhead; side-arm, left and right)
   - ☐ 2 kg medicine ball
   - ☐ Measuring tape (20-30 m)
   - ☐ Recommended surface—tennis court (hardcourt)
   - ☐ Recommended footwear—tennis shoes

9. Power—tennis serve speed
   - ☐ Radar gun
   - ☐ New tennis balls
   - ☐ Tennis racquet
   - ☐ Measuring tape
   - ☐ Recommended surface—tennis court (hardcourt)
   - ☐ Recommended footwear—tennis shoes

PROTOCOLS: Tennis

Anthropometry

For normative data, see table 28.1.

Rationale: Tennis is a game that requires great agility and speed. Furthermore, tennis is often played in hot conditions; and since fat is a good insulator, a player with excess adipose tissue may be more prone than others to overheat. In order to have optimal acceleration and good capacity for thermoregulation, both of which can affect performance, tennis players should carry as little adipose tissue as possible but not so little that health is compromised.

Test Procedure: The anthropometric measurements of stretch height, mass, and sum of seven skinfolds are taken in accordance with the recommendations developed by the Laboratory Standards Assistance Scheme as described in chapter 5. The seven skinfold measurement sites are biceps, triceps, subscapular, supraspinale, mid-abdominal, front thigh, and medial calf.

Table 28.1 Anthropometric Data for Female and Male High-Performance Tennis Players

Group	n	Age (yr)	Height (cm)	Mass (kg)	Sum of 7 skinfolds* (mm)
Female AIS scholarship holders	12				
Mean		17.8	170.0	64.9	93.5
SD		1.2	5.5	6.1	25.2
Range		16–19	160.7–181.3	52.5–71.6	39.8–136.1
Male AIS scholarship holders	7				
Mean		17.8	184.6	79.2	64.3
SD		0.8	6.8	6.9	14.4
Range		17–19	178.0–196.0	69.9–87.0	50.4–92.2
Female VIS scholarship holders	13				
Mean		15.1	166.4	59.1	90.4
SD		0.8	6.9	8.9	15.9
Range		14–16	158.5–177.0	47.2–70.1	62.4–110.7
Male VIS scholarship holders	13				
Mean		16.5	179.0	71.5	62.9
SD		1.7	7.0	7.3	20.8
Range		14–19	167.5–192.5	63.2–89.6	41.1–123.2
Female 16s National Camp	8				
Mean		15.6	164.9	58.0	82.2
SD		0.6	4.3	5.2	15.5
Range		15.0–16.3	159.7–173.1	49.4–67.3	61.2–100.5
Male 16s National Camp	8				
Mean		15.5	176.7	69.3	61.6
SD		0.6	7.3	7.1	19.6
Range		14.7–16.3	167.2–186.8	60.1–78.2	37.8–93.6
Female 14s National Camp	8				
Mean		13.7	162.1	51.2	69.6
SD		0.4	4.5	7.0	21.8
Range		13.1–14.1	154.5–168.4	39.8–61.3	46.7–98.7
Male 14s National Camp	8				
Mean		13.8	167.5	58.9	67.8
SD		0.6	7.9	11.3	25.8
Range		12.7–14.3	154.7–177.2	46.6–76.6	37.8–93.6
Female 12s National Camp	7				
Mean		12.0	152.8	41.8	65.1
SD		0.3	6.7	5.8	13.2
Range		11.5–12.3	142.8–162.3	34.2–50.8	51.9–89.1
Male 12s National Camp	8				
Mean		11.6	154.0	46.4	87.6
SD		0.7	7.9	7.1	38.7
Range		10.1–12.4	143.7–164.0	34.0–55.3	36.1–152.1

*Sum of seven skinfold sites (triceps, subscapular, biceps, supraspinale, mid-abdominal, front thigh, and medial calf).

AIS = Australian Institute of Sport/national-level players, VIS = Victorian Institute of Sport/state-level players, 16s National Camp = national-junior players/16 years and under, 14s National Camp = national-junior players/14 years and under, 12s National Camp = national-junior players/12 years and under.

Aerobic Power: Multistage Fitness Test

For normative data, see table 28.2.

Rationale: The multistage fitness test has been selected as the test of aerobic power for a number of reasons. It is easily administered both in the lab and on the road, is inexpensive, and is similar to tennis with respect to the stop, start, and change-of-direction movement patterns. This 20 m progressive shuttle run test has been found to be a sufficiently accurate estimate of aerobic power (Brewer et al. 1988; Leger and Lambert 1982).

Test Procedure: Refer to page 130 for a comprehensive description of the test procedure.

Speed: Sprint Tests (5 and 10 Meter)

For normative data, see table 28.3.

Rationale: The 5 and 10 m sprint tests may be used as tests of an athlete's explosive ability and rate of force development. Tennis players rarely run more than 3 m in a straight line during a match. Thus, testing the speed of tennis players over distances greater than 10 m is not relevant as players do not have the opportunity to achieve such stride patterns during match play. Additionally, the standing start used in the sprint tests is specific to tennis, in that players must run forward, backward, and sideways over short distances from a standing start.

Test Procedure:

1. Set timing gates at 0, 5, and 10 m intervals.
2. Mark a starting line (0 m) and a finishing line (10 m) with masking tape.
3. Starting position is with front foot up to starting line.
4. Subject may start when ready (thus there is no assessment of reaction time).
5. Subject sprints as fast as possible through to finish line, making sure not to slow down before the finish gate.
6. Split times (5 m and 10 m) are recorded.
7. Subject completes three trials. Each trial is recorded.
8. The best time for 5 m and 10 m is used even if these times come from different trials.

Leg Power

Rationale: A number of field tests designed to measure sport-specific jumping ability (vertical jump: dynamic leg jump, right leg jump, left leg jump) have been included.

For normative data for all three of the jumps described in this section, see table 28.4.

Vertical Jump (Dynamic Leg Jump) Test Procedure:

1. The subject should stand side-on to the Yardstick jumping device.
2. Keeping the heels on the floor, the subject should reach upward as high as possible, fully elevating the shoulder to displace the corresponding plastic vane.
3. Subject then takes one step backward from the Yardstick, still remaining side-on.
4. Subject is then instructed to take one step forward and jump for maximum height by executing a dip or countermovement immediately before upward propulsion.
5. As the subject jumps, the arms are permitted to swing, displacing the vane at the height of the jump. The takeoff must be from two feet.
6. Record the distance between the reach and jump vanes to the nearest 1 cm.
7. Subject performs a minimum of three trials but may continue as long as there are improvements. The best of the trials is recorded.

Vertical Jump (Right Leg Jump) Test Procedure:

1. It is very important that the subject use correct technique when completing the right and left leg jumps.
2. The subject should stand side-on to the Yardstick jumping device.
3. Keeping the heels on the floor, the subject should reach upward as high as possible to displace the corresponding plastic vane.
4. Subjects should lift their left leg off the ground and stand on the right leg only. The left leg may swing back and forward in order to provide momentum to assist with coordination and height of jump.
5. Subject then jumps off the single leg and reaches upward as high as possible to displace the plastic vane at the height of the jump.
6. Record the distance between the reach and jump vanes to the nearest 1 cm.
7. Subject performs a minimum of three trials, but may continue as long as improvements are being made. The best of the trials is recorded.

Table 28.2 Multistage Fitness Test Data for Female and Male High-Performance Tennis Players

Group	n	Mass (kg)	Shuttle level (decimal)	Estimated $\dot{V}O_2max$ ($ml \cdot kg^{-1} \cdot min^{-1}$)
Female AIS scholarship holders	12			
Mean		64.9	11.0	49.9
SD		6.0	1.1	3.8
Range		52.5–71.6	8.9–13.2	42.7–57.6
Male AIS scholarship holders	7			
Mean		79.2	12.9	56.9
SD		6.9	1.4	4.8
Range		69.9–87.0	11.4–15.1	51.3–64.2
Female VIS scholarship holders	8			
Mean		59.0	10.0	47.5
SD		8.5	1.3	4.2
Range		47.2–70.1	8.0–13.2	40.2–57.6
Male VIS scholarship holders	8			
Mean		71.5	12.6	55.4
SD		7.3	1.3	4.3
Range		69.0–75.5	10.2–14.0	47.4–60.8
Female 16s National Camp	8			
Mean		58.0	11.4	51.5
SD		5.2	1.3	4.3
Range		49.4–67.3	10.1–13.4	47.1–58.4
Male 16s National Camp	8			
Mean		69.3	12.6	55.7
SD		7.1	0.9	3.2
Range		60.1–78.2	11.3–14.3	51.3–61.6
Female 14s National Camp	8			
Mean		51.2	10.0	46.6
SD		7.0	0.6	1.9
Range		39.8–61.3	9.1–10.7	43.6–49.3
Male 14s National Camp	8			
Mean		58.9	10.0	46.8
SD		11.3	1.7	5.9
Range		46.6–76.6	7.2–12.1	37.1–53.9
Female 12s National Camp	7			
Mean		65.1	8.5	41.5
SD		13.2	1.2	4.0
Range		51.9–89.1	6.3–10.0	34.0–46.8
Male 12s National Camp	8			
Mean		46.4	9.3	44.3
SD		7.1	1.5	5.2
Range		34.0–55.3	7.0–11.5	36.4–51.9

AIS = Australian Institute of Sport/national-level players, VIS = Victorian Institute of Sport/state-level players, 16s National Camp = national-junior players/16 years and under, 14s National Camp = national-junior players/14 years and under, 12s National Camp = national-junior players/12 years and under.

Table 28.3 Speed Data for Female and Male High-Performance Tennis Players

Group	n	Sprint test 5 m (s)	Sprint test 10 m (s)
Female AIS scholarship holders	12		
Mean		1.20	2.02
SD		0.08	0.10
Range		1.10–1.34	1.90–2.23
Male AIS scholarship holders	8		
Mean		1.08	1.83
SD		0.08	0.10
Range		0.97–1.19	1.68–1.96
Female VIS scholarship holders	15		
Mean		1.15	1.98
SD		0.06	0.08
Range		1.06–1.24	1.80–2.09
Male VIS scholarship holders	16		
Mean		1.08	1.83
SD		0.05	0.09
Range		0.99–1.17	1.68–2.00
Female 16s National Camp	8		
Mean		1.14	1.96
SD		0.05	0.05
Range		1.10–1.22	1.91–2.06
Male 16s National Camp	8		
Mean		1.08	1.83
SD		0.04	0.05
Range		1.03–1.13	1.77–1.93
Female 14s National Camp	8		
Mean		1.13	1.96
SD		0.02	0.03
Range		1.09–1.16	1.91–1.99
Male 14s National Camp	8		
Mean		1.11	1.92
SD		0.05	0.07
Range		1.03–1.19	1.81–2.04
Female 12s National Camp	8		
Mean		1.18	2.02
SD		0.05	0.08
Range		1.10–1.25	1.87–2.11
Male 12s National Camp	8		
Mean		1.15	2.00
SD		0.06	0.11
Range		1.08–1.24	1.85–2.19

AIS = Australian Institute of Sport/national-level players, VIS = Victorian Institute of Sport/state-level players, 16s National Camp = national-junior players/16 years and under, 14s National Camp = national-junior players/14 years and under, 12s National Camp = national-junior players/12 years and under.

Table 28.4 Vertical Jump Data for Female and Male High-Performance Tennis Players

Group	n	Jump height (cm) Dynamic leg	Jump height (cm) Left leg	Jump height (cm) Right leg
Female AIS scholarship holders	12			
Mean		50.1	36.5	36.3
SD		5.2	6.2	3.9
Range		43–58	25–47	31–45
Male AIS scholarship holders	7			
Mean		66.4	47.7	46.6
SD		9.2	5.1	6.8
Range		57–83	57–83	41–61
Female VIS scholarship holders	15			
Mean		48.5	38.7	36.7
SD		5.9	5.5	6.2
Range		37–58	26–48	21–46
Male VIS scholarship holders	16			
Mean		64.1	47.1	46.2
SD		5.8	8.2	7.1
Range		54–76	28–62	30–58
Female 16s National Camp	8			
Mean		43	32	32
SD		4	6	4
Range		37–50	24–42	26–37
Male 16s National Camp	8			
Mean		49	35	36
SD		6	5	6
Range		38–59	26–42	25–42
Female 14s National Camp	8			
Mean		38	27	29
SD		4	3	3
Range		32–44	22–31	24–31
Male 14s National Camp	8			
Mean		44	33	29
SD		8	5	8
Range		36–57	25–39	18–38
Female 12s National Camp	7			
Mean		34	25	25
SD		4	4	5
Range		28–41	20–33	19–30
Male 12s National Camp	8			
Mean		49	25	22
SD		6	9	8
Mean		38–59	9–36	11–35

AIS = Australian Institute of Sport/national-level players, VIS = Victorian Institute of Sport/state-level players, 16s National Camp = national-junior players/16 years and under, 14s National Camp = national-junior players/14 years and under, 12s National Camp = national-junior players/12 years and under.

Vertical Jump (Left Leg Jump) Test Procedure: Repeat the procedure as explained for the other jumps, with the subject using the left leg to jump.

Power: Overhead Medicine Ball Throw

Rationale: The ability to exert muscular force at a high speed is an important aspect of tennis. The following tests have been selected for reasons of specificity and practical application.

See table 28.5 for normative data.

Test Procedure:

- A 2 kg medicine ball is used.
- Run a measuring tape (20-30 m tape) across two adjacent tennis courts.
- Instruct the subject to stand behind the designated start line in the service stance position.
- Subject holds the medicine ball in two hands, elevating and flexing arms so that ball is held behind the subject's head.
- The subject then propels the ball maximally from this position. Subjects are permitted to use leg flexion/extension and trunk/shoulder rotation. They are not permitted to step.
- Encourage a release of approximately 45° to maximize distance achievable.
- Record the distance to the nearest centimeter.
- Subject performs a total of three throws. The best of the three trials is recorded.

Power: Sidearm Medicine Ball Throw

See table 28.5 for normative data.

Test Procedure:

- A 2 kg medicine ball is used.
- Run a measuring tape (20-30 m tape) across two adjacent tennis courts.
- Instruct the subject to stand sideways behind the designated start line.
- Subject holds the medicine ball in two hands, right hand at the back of the ball and left hand under the ball for a right-side throw (vice versa for the left-side throw).
- Subject's arms are nearly straightened in front of the body and are held in a near-horizontal plane.
- Subject then propels the ball maximally from this position. Subjects are permitted to use trunk/shoulder rotation (take a backswing) and attempt to sling the medicine ball. They are not permitted to step or to bend the arms significantly. The arms should move in a horizontal plane.
- Encourage a release of approximately 45° to maximize distance achievable.
- Record the distance to the nearest centimeter.
- Subject performs a total of three throws. The best of the three trials is recorded.
- Repeat the test on the left hand side.

Power: Tennis Serve Speed

See table 28.6 for normative data.

Test Procedure:

Court Setup

Measure the following:

- 90 cm either side of center mark (points A and B)
- 1.25 m inside baseline (point C)
- 2.20 m for the height of the radar gun (point C)

Point radar gun at net height toward the center of the court.

1. Subjects are allowed to use their own racquet and assume their service stance 90 cm from center mark on the right side of court (this distance is marked on court).
2. New tennis balls are used.
3. Subject is instructed to serve a flat first serve for speed and accuracy down the middle of the court.
4. Subject performs a maximum of 12 serves. Each trial is recorded with a notation whether the serve was legal (i.e., whether the serve was in or out).
5. The best five legal serves are then averaged for the final score.

Agility

Rationale: Agility is a very important physiological parameter for tennis players. The following tests have been selected for reasons of specificity and practical application.

Table 28.5 Medicine Ball Throw Data for Female and Male High-Performance Tennis Players

Group	n	Distance (m) overhead	Distance (m) sidearm (left)	Distance (m) sidearm (right)
Female AIS scholarship holders	11			
Mean		9.48	12.06	12.28
SD		1.44	1.28	1.24
Range		7.00–12.60	9.85–14.07	10.90–15.60
Male AIS scholarship holders	6			
Mean		11.88	16.25	16.68
SD		1.16	1.70	1.40
Range		11.00–14.10	13.30–18.20	14.05–18.20
Female VIS scholarship holders	11			
Mean		7.96	10.80	10.64
SD		0.82	1.37	1.46
Range		6.80–9.20	9.00–13.10	8.10–13.20
Male VIS scholarship holders	11			
Mean		11.34	15.40	15.49
SD		0.90	1.49	1.80
Range		10.00–12.70	13.10–17.80	13.60–18.10
Female 16s National Camp	8			
Mean		9.14	10.24	10.76
SD		1.29	0.93	0.66
Range		7.16–10.75	8.30–11.20	9.72–11.66
Male 16s National Camp	8			
Mean		10.72	12.43	12.16
SD		1.39	0.56	0.67
Range		8.80–12.61	11.52–13.18	11.30–13.23
Female 14s National Camp	8			
Mean		7.83	7.88	8.25
SD		0.85	0.87	0.38
Range		6.39–8.85	6.54–9.35	7.65–8.85
Male 14s National Camp	8			
Mean		9.01	8.92	9.54
SD		2.01	1.62	1.52
Range		6.95–12.80	7.10–11.60	7.50–12.35
Female 12s National Camp	7			
Mean		6.47	7.24	7.15
SD		0.72	1.67	1.42
Range		5.11–7.17	4.96–9.02	5.41–9.74
Male 12s National Camp	8			
Mean		6.54	7.31	8.01
SD		0.72	1.03	1.28
Range		5.65–7.95	5.86–8.74	6.58–10.63

AIS = Australian Institute of Sport/national-level players, VIS = Victorian Institute of Sport/state-level players, 16s National Camp = national-junior players/16 years and under, 14s National Camp = national-junior players/14 years and under, 12s National Camp = national-junior players/12 years and under.

Table 28.6 Tennis Serve Speed Data for Female and Male High-Performance Tennis Players

Group	n	Speed (km/h)
Female AIS scholarship holders	12	
Mean		137.9
SD		6.1
Range		127–148
Male AIS scholarship holders	7	
Mean		155.6
SD		6.1
Range		146–165
Female VIS scholarship holders	13	
Mean		126.5
SD		8.1
Range		115–142
Male VIS scholarship holders	13	
Mean		146.8
SD		6.5
Range		140–159

AIS = Australian Institute of Sport/national-level players, VIS = Victorian Institute of Sport/state-level players.

Backward Movement Agility Test

For normative data, see table 28.7.

Preparation:

- Using the measuring tape and masking tape, mark out the points of the course as shown in figure 28.1.
- Set up the timing lights ~2 m apart to form a gate for the subjects.
- Light gates are positioned for the start at a distance 5 m from the service line (toward the net).
- For the finish, light gates are positioned on the service line. The starting tape is placed 30 cm in front of the light gates (toward the net).

Test Procedure:

1. Describe the test pattern to subject and allow the subject to run through a number of times to ensure completion of the correct pattern. Ensure that the subject performs each requirement correctly, and give correctional advice during this brief familiarization period.
2. The subject, holding a racquet in the playing hand, should assume the starting position, on the starting mark (point A) facing the net. The racquet should be held in the smash ready position.

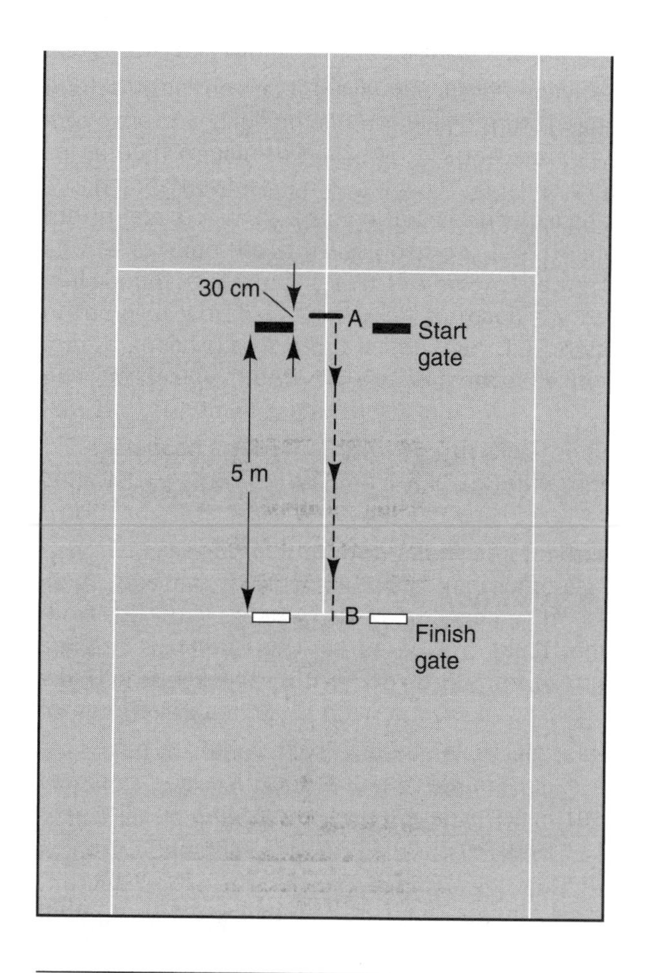

Figure 28.1 Setup for backward movement agility test.

Table 28.7 Backward Movement Agility Test Data for Female and Male High-Performance Tennis Players

Group	n	Time (s)
Female AIS scholarship holders	9	
Mean		1.43
SD		0.12
Range		1.26–1.63
Male AIS scholarship holders	3	
Mean		1.34
SD		0.12
Range		1.21–1.44
Female VIS scholarship holders	5	
Mean		1.49
SD		0.07
Range		1.42–1.58
Male VIS scholarship holders	8	
Mean		1.40
SD		0.11
Range		1.25–1.55

AIS = Australian Institute of Sport/national-level players, VIS = Victorian Institute of Sport/state-level players.

3. The subject should start when ready, as light gate timing will start automatically.

4. Subjects should move in a backward direction to the service line. Subjects may use either a sideways or a crossover step to move back to hit a smash through the finish gate (point B).

5. Subjects complete three trials. Each trial is recorded.

6. The best time is used as the final score.

Sideways Movement Agility Test (Forehand and Backhand)

For normative data, see table 28.8.

Preparation:

- Using the measuring tape and masking tape, mark out the points of the course as in figure 28.2.
- The starting tape is placed 30 cm to the left of the center mark of the baseline (for forehand test for right-handed subjects and backhand test for left-handed subjects).
- Set up the timing lights in line with the center mark of the baseline ~2 m apart to form a gate for the players.
- The tee-ball is placed so that it hangs on the singles sideline, 70 cm inside the baseline,

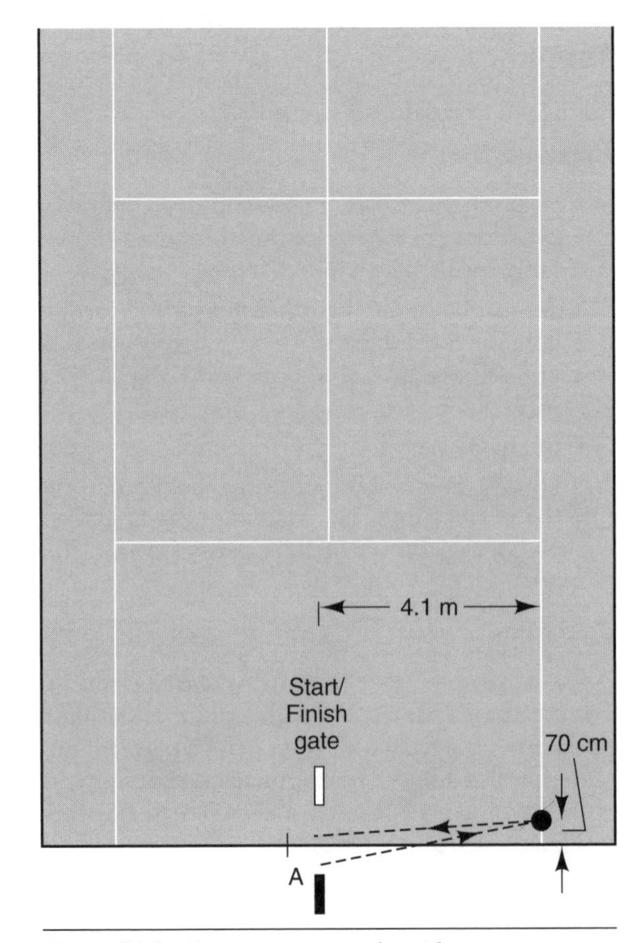

Figure 28.2 Equipment setup for sideways movement agility test (forehand).

exactly 4.1 m from the center mark. The height of the hanging tee-ball for contact is 1 m from the ground. The ball is attached to Velcro on the string hanging down from the tee; this allows the ball to be hit from the tee into the court.

Test Procedure:

1. The subject, holding a racquet in the playing hand, should assume the starting position, on the starting mark (point A). The subject should be in the tennis ready position (feet shoulder-width apart, knees slightly bent).

2. The subject should start when ready, as light gate timing will start automatically.

3. The subject turns and runs naturally sideways and slightly forward to the singles sideline.

4. The subject plays a forehand off the tee-ball and returns with a sideways movement back through the light gates.

5. The ball must go over the net and into the singles court to be a successful attempt.

6. Three successful turns are allowed. The best of these three scores is recorded as the score.

7. Repeat the test for the backhand side by moving the tee to the left singles sideline.

Change-of-Direction/ Acceleration Sideways Agility Test (Left and Right):

For normative data, see table 28.9.

Preparation:

- Using the measuring tape and masking tape, mark out the points of the course as in figure 28.3.
- The starting tape is placed 30 cm to the right of the singles sideline, level with the baseline (for the right-side test).
- Set up the timing lights in line with the singles sideline ~2 m apart to form a gate for the players.

Test Procedure:

1. Subjects should assume the starting position, on the starting mark (point A) level with the baseline. The subject should be in the tennis ready position (without a racquet) facing the net.

2. The subject should start when ready, as light gate timing will start automatically.

3. The subject uses sideways movement (side-step, not a crossover step) along the baseline to touch the center mark on the baseline.

4. Subject turns and runs naturally back along baseline until he or she crosses the singles sideline.

5. The subject's foot must touch the center mark in order for the trial to be valid.

6. Three trials are allowed. The best of these three trials is recorded as the score.

7. Repeat the test on the opposite singles sideline.

Change-of-Direction/ Acceleration Forward Agility Test (Left and Right):

For normative data, see table 28.10.

Preparation:

- Using the measuring tape and masking tape, mark out the points of the course as in figure 28.4.
- The starting tape is placed 30 cm to the right of the singles sideline, level with the baseline (for the right test).
- Set up the timing lights in line with the singles sideline ~2 m apart to form a gate for the players.
- The second set of timing lights are placed 4.1 m inside the baseline (toward the net) to record the finishing time. The gates are 2 m apart, with right gate (for a right-side trial) 50 cm from the center line (opposite for left-side start).
- A 1 m line (starting at the center mark) is marked 30 cm behind and parallel to the baseline. The subject must stay within boundary in order for trial to be valid.

Test Procedure:

1. The subject should assume the starting position, on the starting mark (point A) level with the baseline. The subject should be in the tennis ready position (without a racquet) facing the net.

2. The subject should start when ready, as light gate timing will start automatically.

3. The subject uses sideways movement (side-step, not a crossover step) along the baseline to touch the center mark on the baseline.

4. Subject then runs naturally forward toward the net through the timing gates (4.1 m).

5. Subject's foot must touch the center mark in order for the trial to be valid.

Table 28.8 Sideways Movement Agility Test (Forehand and Backhand) Data for Female and Male High-Performance Tennis Players

Group	n	Forehand (s)	Backhand (s)
Female AIS scholarship holders	9		
Mean		2.36	2.65
SD		0.14	0.15
Range		2.08–2.61	2.46–2.90
Male AIS scholarship holders	3		
Mean		2.37	2.39
SD		0.10	0.08
Range		2.27–2.46	2.32–2.47
Female VIS scholarship holders	5		
Mean		2.56	2.60
SD		0.08	0.12
Range		2.45–2.64	2.46–2.76
Male VIS scholarship holders	8		
Mean		2.49	2.59
SD		0.15	0.11
Range		2.23–2.70	2.44–2.80
Female 16s National Camp	8		
Mean		2.42	2.58
SD		0.10	0.11
Range		2.24–2.54	2.36–2.71
Male 16s National Camp	8		
Mean		2.47	2.52
SD		0.12	0.15
Range		2.26–2.62	2.31–2.71
Female 14s National Camp	8		
Mean		2.77	2.83
SD		0.10	0.14
Range		2.67–2.90	2.68–3.08
Male 14s National Camp	8		
Mean		2.66	2.75
SD		0.19	0.17
Range		2.41–2.96	2.51–2.98
Female 12s National Camp	7		
Mean		2.63	2.83
SD		0.13	0.14
Range		2.44–2.78	2.66–3.04
Male 12s National Camp	8		
Mean		2.70	2.76
SD		0.11	0.09
Range		2.54–2.89	2.60–2.88

AIS = Australian Institute of Sport/national-level players, VIS = Victorian Institute of Sport/state-level players, 16s National Camp = national-junior players/16 years and under, 14s National Camp = national-junior players/14 years and under, 12s National Camp = national-junior players/12 years and under.

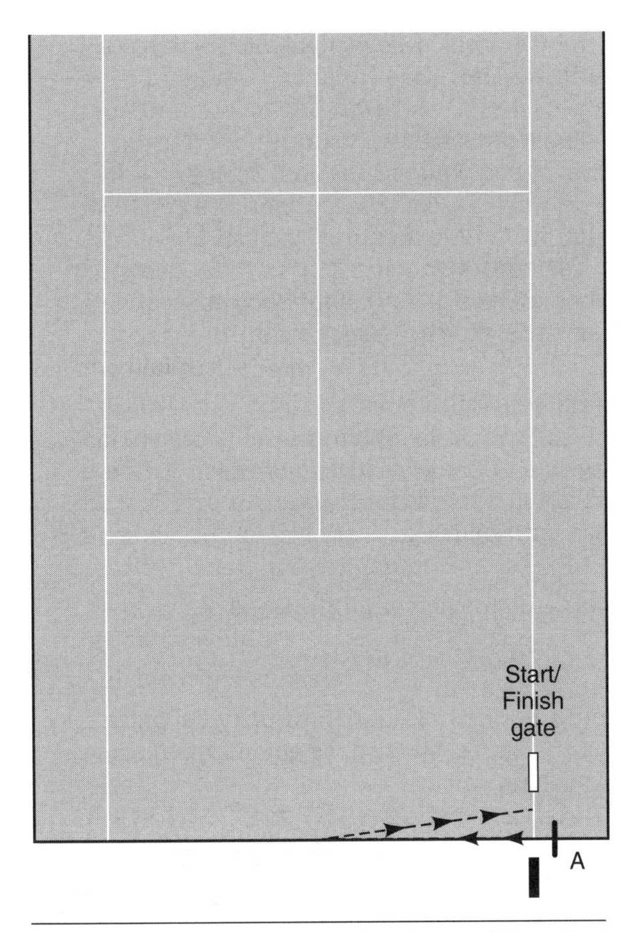

Figure 28.3 Equipment setup for change-of-direction/acceleration sideways agility test (right).

6. Three trials are allowed. The best of these three trials is recorded as the score.

7. Repeat the test on the opposite singles sideline.

Repeated-Effort Agility Test

For normative data, see table 28.11.

Preparation:

- Using the measuring tape, masking tape, and cones, mark out the points of the course as in figure 28.5.
- One cone is placed 5 cm from the net on the center line; one on the intersection of the right singles sideline and baseline; one on the intersection of the right singles sideline and the service line; and one on the intersection of the left singles sideline and service line.
- Set up the timing lights 1 m behind and parallel with the baseline, equidistant from the center mark and ~2 m apart to form a gate for the players.
- The starting tape is placed 30 cm behind the start timing lights, directly behind the baseline center mark.
- The second set of timing lights are placed in line with the singles sideline to record the finish time. The gates are 2 m apart, with one gate 1 m inside the court on the singles

Table 28.9 Change-of-Direction/Acceleration Sideways Agility Test (Left and Right) Data for Female and Male High-Performance Tennis Players

Group	n	Left (s)	Right (s)
Female AIS scholarship holders	9		
Mean		2.39	2.45
SD		0.14	0.16
Range		2.22–2.71	2.28–2.78
Male AIS scholarship holders	3		
Mean		2.25	2.16
SD		0.17	0.14
Range		2.06–2.40	2.08–2.33
Female VIS scholarship holders	5		
Mean		2.51	2.49
SD		0.13	0.12
Range		2.33–2.68	2.35–2.64
Male VIS scholarship holders	8		
Mean		2.31	2.34
SD		0.16	0.16
Range		1.94–2.45	2.02–2.54

AIS = Australian Institute of Sport/national-level players, VIS = Victorian Institute of Sport/state-level players.

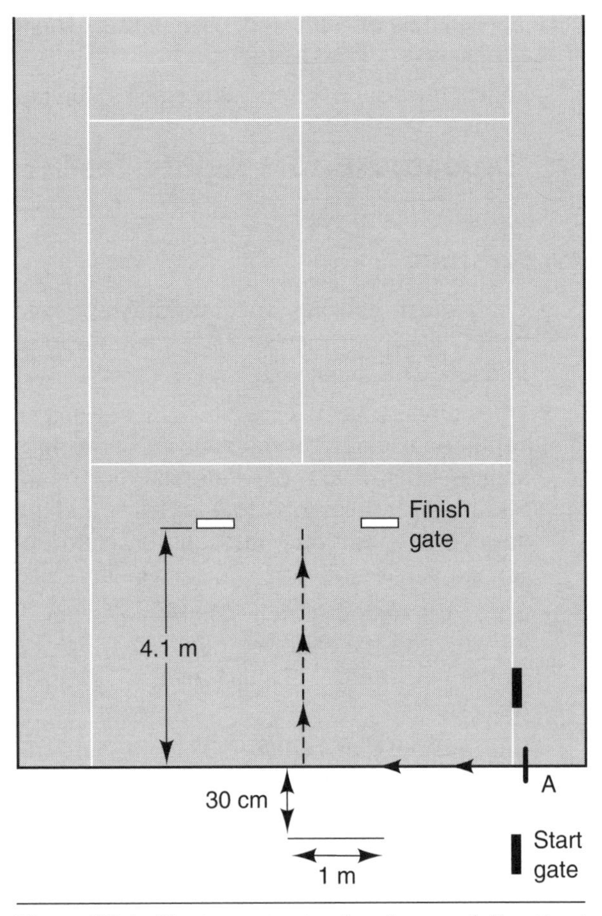

Figure 28.4 Equipment setup for change-of-direction/acceleration forward agility test (right).

sideline and the other 1 m behind the baseline in line with the singles sideline.

Test Procedure:

1. The subject should assume the starting position, on the starting mark (point A) behind the timing gates.
2. The subject should start when ready, as light gate timing will start automatically.
3. Subject sprints forward to the net to touch the cone with the hand.
4. Subject returns to center mark while facing the net (i.e., backward movement).
5. Subject's foot must touch the center mark in order for the trial to be valid.
6. Subject sprints along baseline to touch cone on right singles sideline and baseline.
7. Subject returns to starting position while facing the net.
8. Subject sprints on the short diagonal to the right side of the court to touch cone on service line and sideline.
9. Subject returns to center mark facing the net.
10. Subject sprints on the short diagonal to the left side of the court to touch cone on service line and sideline.
11. Subject returns to center mark facing the net.

Table 28.10 Change-of-Direction/Acceleration Forward Agility Test (Left and Right) Data for Female and Male High-Performance Tennis Players

Group	n	Left (s)	Right (s)
Female AIS scholarship holders	9		
Mean		2.28	2.25
SD		0.16	0.20
Range		2.03–2.55	2.04–2.69
Male AIS scholarship holders	3		
Mean		1.96	2.02
SD		0.04	0.05
Range		1.93–2.00	1.99–2.08
Female VIS scholarship holders	5		
Mean		2.21	2.24
SD		0.10	0.10
Range		2.08–2.34	2.15–2.39
Male VIS scholarship holders	8		
Mean		2.06	2.12
SD		0.14	0.17
Range		1.87–2.26	1.83–2.30

AIS = Australian Institute of Sport/national-level players, VIS = Victorian Institute of Sport/state-level players.

Table 28.11 Repeated-Effort Agility Test Data for Female and Male High-Performance Tennis Players

Group	n	Trial 1 (s)	Trial 2 (s)	Trial 3 (s)	Trial 4 (s)	Trial 5 (s)	Decrement (%)	Average (s)
Female 16s National Camp	8							
Mean		17.36	17.64	17.75	17.98	17.66	5.0	17.64
SD		0.57	0.55	0.64	0.37	0.50	2.2	0.55
Range		16.51–18.17	16.84–18.35	16.97–18.62	17.56–18.72	16.97–18.57	1.0–7.8	16.84–18.35
Male 16s National Camp	8							
Mean		16.78	17.30	17.13	17.49	17.29	5.8	17.20
SD		1.10	1.33	0.99	0.97	0.93	2.2	1.02
Range		15.45–18.48	15.95–19.53	15.96—18.82	16.22–19.16	16.17–18.73	2.4–9.1	16.13–18.94
Female 14s National Camp	8							
Mean		17.64	18.11	18.19	18.59	18.38	6.0	18.17
SD		0.77	0.62	0.55	0.84	0.82	3.1	0.63
Range		16.82–19.24	17.41–19.28	17.35–18.75	17.66–19.80	17.39–19.94	2.6–9.9	17.55–19.37
Male 14s National Camp	8							
Mean		17.24	17.64	17.70	18.19	17.61	5.3	17.67
SD		1.25	1.79	1.48	1.72	1.11	2.5	1.45
Range		16.08–19.82	16.08–21.79	16.49–21.10	16.93–22.17	16.61–20.02	3.1–10.6	16.67–20.98
Female 12s National Camp	7							
Mean		18.29	18.65	18.68	19.03	18.51	5.0	18.58
SD		0.84	1.60	1.36	1.21	1.53	2.6	1.29
Range		17.10–19.90	17.13–22.09	17.13–21.32	18.14–21.30	17.24–21.83	1.6–9.9	17.27–21.29
Male 12s National Camp	8							
Mean		18.65	19.03	19.32	19.30	18.96	5.3	19.05
SD		0.82	0.71	0.69	0.76	0.76	2.6	0.67
Range		17.62–20.15	18.13–20.42	18.44–20.37	18.32–20.52	17.64–20.07	2.0–10.3	18.26–20.31

16s National Camp = national-junior players/16 years and under, 14s National Camp = national-junior players/14 years and under, 12s National Camp = national-junior players/12 years and under.

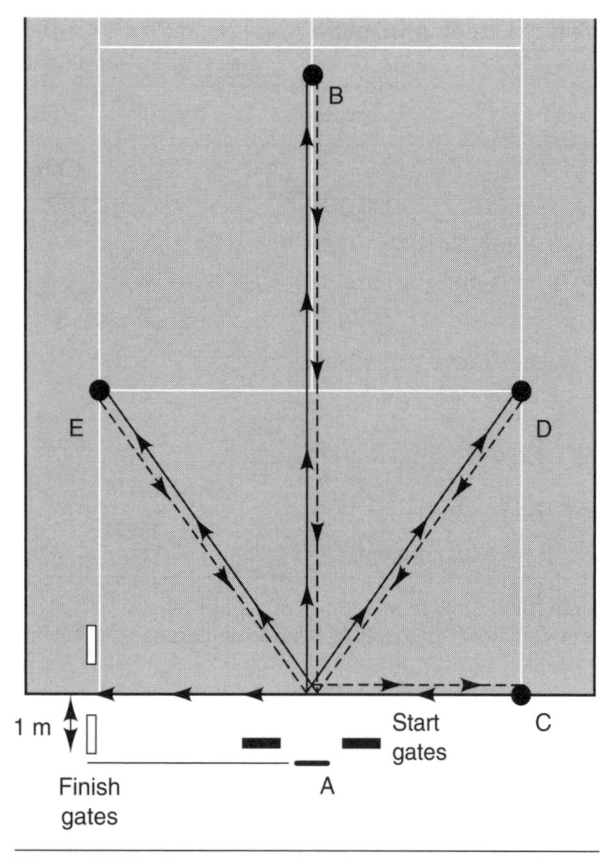

Figure 28.5 Equipment setup for the repeated-effort agility test.

12. Subject sprints through the light gates on left singles sideline.

13. Subject rests for 25 s.

14. This test is completed a total of five times, each with a recovery of 25 s (similar to the time between points).

15. Five trials are recorded. The difference between the fastest and the slowest trials is expressed as a percentage decrement.

Test Data Sheet

Name: ______________________ Date of testing: __________

Place of testing: ______________ Tester: ______________

Date of birth: ________ Age: __________ Playing standard (e.g., AIS): __________

Anthropometry

Height (cm) _____ Supraspinale (mm) _____
Weight (kg) _____ Mid-abdominal (mm) _____
Skinfolds—biceps (mm) _____ Front thigh (mm) _____
Triceps (mm) _____ Medial calf (mm) _____
Subscapular (mm) _____ Sum of skinfolds (mm) _____

Aerobic power

Multistage Fitness Test

Level (decimal) _____
$\dot{V}O_2max$ (ml · kg^{-1} · min^{-1}) _____

Speed (sprint tests)

Surface ______________ Footwear ______________

5 m—Trial 1 (s) _____ 10 m—Trial 1 (s) _____
Trial 2 (s) _____ Trial 2 (s) _____
Trial 3 (s) _____ Trial 3 (s) _____
Best (s) _____ Best (s) _____

Agility

Backward Agility Test

Surface ______________
Footwear ______________
Trial 1 (s) _______
Trial 2 (s) _______
Best (s) _______

Sideways Movement Agility Test

Forehand:

Trial 1 (s) _____
Trial 2 (s) _____
Trial 3 (s) _____
Best (s) _____

Backhand:

Trial 1 (s) _____
Trial 2 (s) _____
Trial 3 (s) _____
Best (s) _____

(continued)

Change-of-Direction/Acceleration Sideways Agility Test

Surface: ______________________ Footwear: ______________________

Left		**Right**	
Trial 1 (s)	_____	Trial 1 (s)	_____
Trial 2 (s)	_____	Trial 2 (s)	_____
Trial 3 (s)	_____	Trial 3 (s)	_____
Best (s)	_____	Best (s)	_____

Change-of-Direction/Acceleration Forward Agility Test

Surface: ______________________ Footwear: ______________________

Left		**Right**	
Trial 1 (s)	_____	Trial 1 (s)	_____
Trial 2 (s)	_____	Trial 2 (s)	_____
Trial 3 (s)	_____	Trial 3 (s)	_____
Best (s)	_____	Best (s)	_____

Repeated-Effort Agility Test

Surface: ______________________ Footwear: ______________________

Trial 1 (s)	_____	Fastest	_____
Trial 2 (s)	_____	Slowest	_____
Trial 3 (s)	_____	Decrement (%)	_____
Trial 4 (s)	_____	Mean (s)	_____
Trial 5 (s)	_____		

Leg Power

Vertical Jump (Dynamic)

Trial 1 (cm) _____
Trial 2 (cm) _____
Trial 3 (cm) _____
Best (cm) _____

Vertical Jump (Right Leg)

Trial 1 (cm) _____
Trial 2 (cm) _____
Trial 3 (cm) _____
Best (cm) _____

Vertical Jump (Left Leg)

Trial 1 (cm) ______

Trial 2 (cm) ______

Trial 3 (cm) ______

Best (cm) ______

Power

Overhead Medicine Ball Throw

Trial 1 (m) ______

Trial 2 (m) ______

Trial 3 (m) ______

Best (m) ______

Sidearm Medicine Ball Throw (Left Side)

Trial 1 (m) ______

Trial 2 (m) ______

Trial 3 (m) ______

Best (m) ______

Sidearm Medicine Ball Throw (Right Side)

Trial 1 (m) ______

Trial 2 (m) ______

Trial 3 (m) ______

Best (m) ______

Tennis Serve Speed

Trial 1 (km/h) ______

Trial 2 (km/h) ______

Trial 3 (km/h) ______

Trial 4 (km/h) ______

Trial 5 (km/h) ______

Best (km/h) ______

chapter 29

Protocols for the Physiological Assessment of High-Performance Triathletes

Darren Smith and Rob Pickard

Competitive elite triathlon is a physiologically demanding sport requiring swimming, cycling, running, and transitioning skills. The Olympic distance triathlon of 1.5 km swim, 40 km cycle, and 10 km run takes between 1:45 and 1:55 h:min and 1:55 and 2:05 h:min for elite males and females, respectively. Although the metabolic determinants of triathlon performance have been reasonably well documented for the traditional Olympic distance triathlon (Slievert and Rowlands 1996), new regulations allowing drafting within the cycle leg for elite competitors have in practice increased the emphasis of the swim and run legs in relation to overall performance. No scientific data are currently available to confirm this.

In triathlon, as with most other endurance-based sports, it is necessary to describe training intensity when prescribing training. As with other endurance sports that have adopted training zones specific to their sport, the Triathlon Sport Science Committee has developed triathlon-specific zones (table 29.1). These zones are based on one discernible point on the blood lactate (or pH) versus workload curve—the anaerobic threshold (AT) where blood lactate begins to increase disproportionately with additional workload.

The present battery of tests includes blood testing, anthropometry, and submaximal and maximal aerobic power tests for the three disciplines of swimming, cycling, and running. A combined cycle and run test is currently being developed. The contents of this chapter have been heavily influenced by the testing protocols prepared for the individual sports of cycling (chapter 17), distance running (chapter 23), and swimming (chapter 27), in keeping with the aim of promoting uniform sport science testing of triathletes.

Triathlon Training Zones

Training zones (e.g., heart rate [HR], cycling wattage, and running, cycling, and swimming velocities, if appropriate) should be prepared in order to show the athlete and coach the information in table 29.1.

Laboratory Environment and Athlete Preparation

The discussion of running (chapter 23) includes a thorough description of necessary preparation.

Acknowledgements: The authors would like to sincerely thank the following people for their contributions to this chapter: Bronwyn Clift, members of the Triathlon Australia Sports Science Committee, Kate Cameron, and Kenneth Graham.

Table 29.1 Triathlon Training Zones

Name	Description	HR zones	Perceived exertion	Hla (mmol/L)
T1	Recovery	Below 75% AT	Easy	< 1.5
T2	Aerobic endurance	76–85% AT	Comfortable	1.5–3.0
T3	Intense aerobic	86–95% AT	Uncomfortable	3.5–5.0
T4	Threshold	96–102% AT	Stressful	4.0–7.0
T5	Lactate tolerance	103% AT-Max	Very stressful	> 7.0
T6	$\dot{V}O_2max$	Max	Maximal	> 7.0

AT = Anaerobic threshold (Dmax), Hla = blood lactate concentration.

Note that in order to help the athlete and testing coordinator confirm adherence to the pretesting conditions, athletes should fill out the pretest questionnaire in chapter 2 (pages 16-18).

Equipment Checklist

1. Routine hematology

Refer to page 312 for a complete list.

2. Anthropometry

Refer to chapter 5 for a comprehensive list.

3. Blood sampling

Refer to page 92 for a complete list.

4. Progressive run and cycle tests
   - ☐ Treadmill (calibrated for velocity and elevation)
   - ☐ Cycle ergometer (dynamically calibrated wind-braked or electromagnetically braked types)
   - ☐ Gas analysis system
   - ☐ Lactate and blood gas analyzers
   - ☐ Heart rate monitoring system (electrocardiogram or other, plus associated consumables, e.g., electrode gel, razors)
   - ☐ Electric fan
   - ☐ Stopwatch
   - ☐ Blood sampling equipment (as listed above)
5. Swim testing
   - ☐ 50 m swimming pool
   - ☐ Lactate and blood gas analyzers
   - ☐ Waterproof HR monitoring system
   - ☐ Stopwatch
   - ☐ Blood sampling equipment (as listed above)

PROTOCOLS: Triathlon

Routine Hematology

Blood Testing: A full blood profile and assessment of iron status should be conducted prior to the submaximal and maximal aerobic power tests. Athletes must be in a rested state (no exercise in the previous 12 h) before the blood sample is drawn. A 5 ml sample is drawn from an antecubital vein with the athlete in a supine position. The sample is analyzed as described in chapter 23 (page 336).

If possible, blood samples from each athlete should be obtained at the same time of day, preferably in the morning after a 12 h fast, every three months throughout the training year.

Normative Data: See chapter 23 for information on comparison to normative data.

Anthropometry

Testing: Anthropomorphic measurements should be conducted as described for running (page 336).

More advanced assessments may be necessary if the athletes are part of a talent identification program. These may include somatotype, limb lengths and girths, or bone density scans. See Norton et al. (1996) for descriptions of procedures for these additional measures.

Normative Data: Guidelines for the evaluation of normative data are the same as described in chapter 23 (page 336).

Progressive Swimming Test

Rationale: The aim of this test is to provide information on the swimming economy and swimming fitness of the athlete. The protocol is for a graded incremental test measuring cardiovascular (HR) and metabolic (blood lactate) responses to various intensities of swimming. This test should be conducted using small groups of swimmers (e.g., less than four). Swimmers perform seven evenly paced 200 m swims graded from easy to maximal, on a 5 min cycle, using a push start. Table 29.2 outlines the target times for swims.

Test Procedure and Data Collection: Athletes should complete a standard warm-up (>500 m) prior to testing. The test comprises seven evenly paced 200 m freestyle swims beginning with a push start from the wall. Each swim should be paced relative to the athlete's personal best time for 200 m as outlined in table 29.2.

1. Split times and total time are recorded. Stroke rate can also be recorded during the final 50 m of each 200 m repetition.

2. Record peak HR after each 200 m interval as soon as the athlete reaches the end of the pool. Athletes do not need to wear an HR monitor during each swim, but instead can reach for the monitor immediately after completion of each swim interval.

Table 29.2 Target Times for Progressive 200 m Swim Intervals

200 m interval (number)	Total time (s)
1	30 s slower than personal best
2	25 s slower than personal best
3	20 s slower than personal best
4	15 s slower than personal best
5	10 s slower than personal best
6	5 s slower than personal best
7	Maximal effort

3. Obtain a preswim lactate sample from either an earlobe or finger. Thereafter, collect blood lactates from the same sampling site at the end of each swim. It is necessary to collect samples swiftly, as the swimmer must start the next 200 m at the end of the 5 min cycle (i.e., there is not a 5 min rest between swims).

4. Swimmers should enter the pool 15 s before the start of the next interval to begin swimming using a push start.

5. Swimmers repeat these steps until they have completed all seven swims.

6. Sample blood immediately after the final swim and then 2 and 4 min later.

Data Analysis and Presentation: Data should be entered into the ADAPT computer package (see appendix for details) to determine the following values:

- Swimming velocity (m/s)
- Heart rate (beats/min)
- Lactate (mmol/L) at lactate threshold (LT), at AT, and at maximum

A table prepared for each swim should have the following headings:

- Interval number
- Swim time (min:s)
- Swim velocity (m/s)
- RPE (Borg 1962)
- Heart rate (beats/min)
- Blood lactate (mmol/L)
- Stroke count (strokes taken to swim the final 50 m of each interval)
- Stroke rating (strokes/min from the last 50 m of each interval)

Graphs should show the following relationships:

- Heart rate (y-axis) versus velocity (x-axis)
- Lactate (y-axis) versus velocity (x-axis)
- Stroke count (y-axis) versus velocity (x-axis)
- Stroke rate (y-axis) versus velocity (x-axis)

Swim Performance Test

This test takes place in a 50 m pool and represents an Olympic distance triathlon-specific performance test of 1500 m. Average HR (if possible), 100 m split times, and final time plus pre- and postswim blood lactates should be obtained. This test may also be used to check the training HR zones set previously.

Test Procedure: Warm-up should be decided by the coach and athlete but should be standardized for each test session. At least a 500 m slow-medium swim is advisable.

1. Explain the test to the athlete.
2. If possible, set the HR monitor to record HR every 5 s.
3. Obtain resting, midpoint (optional), and immediate postswim lactates.
4. Record times for every 100 m split.
5. Athlete will have a push start.
6. Stroke count and stroke rate, along with video footage, can also be obtained during this test to further evaluate if training adaptations have been transferred to competition-like scenarios (i.e., does the triathlete maintain efficient swimming form while fatigued?).

Data Presentation: A table should show distance, split times, stroke counts, HR, and lactates.

Cycling Progressive Test

Rationale: The format of the 40 km cycle leg within the Olympic distance triathlon race—which in the past was a time-trial event—has recently changed for elite competitors. The inclusion of drafting rules has changed the energy requirements somewhat and perhaps reduced emphasis on this section of the triathlon. Nevertheless, success in the cycling leg of the triathlon race and the ability to maintain contact with the bunch will still require a high $\dot{V}O_2max$ and the ability to sustain a high percentage of work output for extended periods of time. Hence, measurement of $\dot{V}O_2max$ and blood LTs can be useful in predicting performance, monitoring adaptations to training, and prescribing cycle training.

Equipment Preparation: Prepare the equipment for the test according to the following guidelines:

- Check that the cycle ergometer is clean and that it has been recently dynamically calibrated for cadence and power output (PO).
- Ensure that the lactate and blood gas meters are sufficiently warmed up by switching them on at least 45 min before the first test. Calibrate the analyzers against reference standards just prior to the first test.
- Prior to the first test, ensure that the gas analysis system is also adequately warmed up and calibrated for ventilation and gas composition (using three alpha-standard reference gases spanning the physiological range).
- Have athletes set up the ergometer with their own pedals, seat height, and handlebar position. Have athletes select gearing to suit themselves, in order to optimize work output (cadence should be between 85 and 115 rpm). Record all these settings for future testing.

Cycle Protocol: This is a continuous test of 5 min (males) or 3 min (females, juniors) intervals.

- Senior males: Begin at 100 W and increase by 50 W each 5 min interval.
- Junior males: Begin at 100 W and increase by 50 W each 3 min interval.
- Senior females: Begin at 100 W and increase by 25 W each 3 min interval.
- Junior females: Begin at 100 W and increase by 25 W each 3 min interval.

Test Procedure: When the athlete arrives, verify that he or she has met the guidelines for athlete preparation, and then have the athlete complete the pretest questionnaire presented in chapter 2. Administration of the test should then proceed as follows:

1. Weigh the athlete (±0.05 kg) in minimal clothing. Also measure the mass of one of the athlete's cycling shoes and record the value on the pretest questionnaire (chapter 2, pages 16-18).
2. Prepare the athlete's skin for placement of the HR monitoring equipment by shaving any excess body hair, cleaning with alcohol swabs, and drying with a tissue. Electrode gel will help with electrical conduction.
3. Place Finalgon on the athlete's right ear, or on some fingertips, to ensure hyperemia prior to sampling. Apply Finalgon at least 5 min before the test starts and remove it just before the start.
4. Place a piece of Micropore tape across the bridge of the athlete's nose. This will help keep the noseclip from slipping during the test.
5. Instruct the athlete to take position on the ergometer and allow a brief warm-up and familiarization period. During this time, outline the test procedures to the athlete (i.e., 5 min at each workload with each interval increasing by 50 W; timing of blood samples).
6. After the warm-up, obtain a pre-exercise blood sample.

7. Position the headset and mouthpiece so as to ensure maximum comfort. Instruct the athlete to take hold of the noseclip and position it comfortably to block nasal airflow.

8. When ready, instruct the athlete to begin the test, and simultaneously start the gas analysis and timing devices. During each workload, indicate the time remaining and check with the athlete about his or her ability to continue.

9. At the end of each work stage, a blood sample should be drawn from the earlobe (or fingertip) and analyzed for lactate, pH, and bicarbonate levels.

10. Athletes should continue exercising until exhaustion. With a few seconds remaining in each 30 s period throughout the expected final workload, the tester should ask athletes if they are capable of cycling for at least a further 30 s. Athletes should indicate their decision with hand signals (i.e., thumb up is yes; thumb down is no).

11. The test ends when the athlete stops voluntarily or when the athlete indicates to the tester that he or she cannot complete the next workload.

12. At the end of the test, remove the noseclip and headset and immediately obtain the blood sample for the last workload. Collect blood samples also at 2 and 5 min posttest with the subject seated and not in an active recovery.

13. If no more tests are scheduled, ensure that the gas analysis system, the blood analyzers, and all other equipment are shut down and packed away in accordance with normal procedures. Ensure that the respiratory apparatus is thoroughly sterilized, rinsed, and hung up to dry and that all analyzed blood samples are disposed of into a biohazard bag.

Data Collection: For each minute, the following should be recorded:

- $\dot{V}_E$, $\dot{V}O_2$, $\dot{V}CO_2$, respiratory exchange ratio (RER), $\dot{V}_E/\dot{V}O_2$, HR, and PO for each 30 s of the test
- Resting blood variables (lactate [Hla], pH, and bicarbonate, if possible) prior to athlete warm-up, after each completed workload, at the end of the test, and 2 and 5 min posttest
- The $\dot{V}O_2$max: the average of the highest two consecutive 30 s readings

Data Analysis: Data should be transferred to the ADAPT computer package for computation of LT, AT, and peak values by the Dmax method. The following indexes should be reported for each threshold: $\dot{V}O_2$, %$\dot{V}O_2$max, HR, %HRmax, PO, lactate, and pH (if measured). $\dot{V}O_2$max should be presented as L/min and $ml \cdot kg^{-1} \cdot min^{-1}$.

Note that for purposes of data analysis, $\dot{V}O_2$ is taken as the average of the data collected during the last 2 min of each interval; the PO is calculated as the average power during each interval; and HR is averaged over the last 15 s of each interval.

Cycle Field Tests

30-Minute Time Trial: In this performance test the athlete performs the maximal amount of work in a 30 min time period. The cycle course should be traffic free, flat, and looped (e.g., velodrome or flat, smooth road course 2-5 km in length). During this test HR should be monitored (at least each minute), as should lactate where possible. Environmental and road conditions can affect the HR and lactate response; so unless conditions are ideal, a laboratory-based test on an ergometer, as described next, is a more controlled alternative—although less preferred by athletes and coaches.

Ergometer Time Trial: A test similar to the laboratory incremental test can also be conducted on an ergometer (either wind-braked or magnetically braked ergometers in good working order are suitable). See the cycling progressive $\dot{V}O_2$max test and follow the guidelines for the ergometer setup for gears, cadence, and resistance; record HR and lactate as described for that test.

Junior Testing: A 10 km cycle over a 5 km out-and-back course uses total time as the performance score.

Running Progressive Test

Rationale: The progressive running test is designed to allow determination of the running velocity, HR, lactate, and pH corresponding to the blood lactate transitions (LT and AT) and $\dot{V}O_2$max. It also allows estimates of running economy (i.e., the $\dot{V}O_2$ at a given running speed) and construction of HR/lactate/pH versus running velocity curves. Previous training and diet will affect economy and HR and lactate curves; therefore adherence to pretest conditions is essential.

Equipment Preparation: Prepare the equipment for this test according to the following guidelines:

- Check that the treadmill is clean and has been recently calibrated for velocity and elevation.

To check for proper operation, run the treadmill through the range of velocities required in the test before the athlete arrives.

- To ensure that the lactate and blood gas meters are sufficiently warmed, switch them on at least 45 min prior to the first test. Calibrate the analyzers against reference standards just before the first test.
- Prior to the first test, ensure that the gas analysis system is also adequately warmed up and calibrated for ventilation and gas composition (using three alpha-standard reference gases spanning the physiological range).
- Adjust the slope of the treadmill to 1% elevation in preparation for the test.

Treadmill Protocol:

- The athlete exercises for 3 min intervals at predetermined running velocities.
- Each workload is separated by a 1 min rest interval.
- The initial workload is typically 10 km/h (females and juniors) or 12 km/h (male athletes), with successive workloads 2 km/h faster than the previously completed stage.
- Prior to the test, all athletes should complete a standardized warm-up (e.g., similar to one preceding a running race).

Test Procedure: On the athlete's arrival, check to ensure that he or she has followed the guidelines for athlete preparation, and have the individual complete the pretest questionnaire (see chapter 2). Administration of the test should then proceed as follows:

1. Measure the athlete's body mass (±0.05 kg) using the balance scales in the anthropometry room. The mass of one of the athlete's running shoes should also be measured and recorded on the pretest questionnaire.

2. Prepare the athlete's skin for placement of the HR monitoring equipment as described for the cycling progressive test earlier in this chapter.

3. Apply Finalgon as described for the cycling progressive test.

4. Apply Micropore tape as described for the cycling progressive test.

5. Have the athlete take position at the front of the treadmill; when ready, allow a brief warm-up and familiarization period. During this time, outline the test procedures to the athlete (i.e., 3 min at each workload with 1 min rest with each workload increasing in velocity by 2 km/h; frequency of blood samples).

6. After the warm-up, obtain a pre-exercise blood sample.

7. Position the headset and mouthpiece and have the athlete position the noseclip as described for cycling progressive test.

8. Before beginning the test, have the athlete hold the side rails and straddle the treadmill. Start the treadmill; when it is operating at the desired speed, have the athlete start running, and at the same time start the gas analysis and time devices. During each workload, indicate the time remaining and check with the athlete about his or her ability to complete the workload. Stop the treadmill at the end of the 3 min, giving the athlete prior warning with a countdown over the last 5 s.

9. Immediately on cessation of exercise, a blood sample should be drawn from the earlobe (or fingertip) and analyzed for lactate, pH, and bicarbonate levels. Have the athlete identify on the rating of perceived exertion (RPE) chart (Borg 1962) his or her subjective assessment of exertion, and record this value on the data sheet (page 339).

10. During the rest interval, the athlete may remove the noseclip and mouthpiece, but the headset should remain in place. With approximately 30 s before the start of the next workload, instruct the athlete to replace the mouthpiece and noseclip and to assume the starting position. The treadmill should be started at least 15 s before the start of the next workload to allow it to reach the required speed (i.e., 2 km/h faster than the previous workload). After the 60 s of rest, have the athlete start running for a further 3 min.

11. Proceed as outlined above, with the athlete completing at least five (preferably six) different workloads.

12. To assess the appropriateness of a further stage, observe athletes' HR, RPE, and some cardiorespiratory values (i.e., $\dot{V}O_2$, R value) in response to the previous level, and ask them whether they feel that they could attempt to run for another 3 min at a higher speed.

13. The athlete should continue exercising until exhaustion. With a few seconds left in each 30 s period throughout the expected final interval, the tester should ask athletes if they are capable of running for at least a further 30 s. Athletes should indicate their decision with hand signals (i.e., thumb up is yes; thumb down is no).

14. The test ends when athletes voluntarily press the emergency "Stop" button or when they

indicate to the tester that they cannot complete the next workload.

15. At the end of the test, remove the noseclip and headset and immediately obtain the blood sample for the last workload. Blood samples are also collected at 2 and 5 min posttest. Note that the athlete must be standing but must not be in an active recovery.

16. If this is the last scheduled test, ensure that the gas analysis system, the blood analyzers, and all other equipment are shut down and packed away in accordance with normal procedures. Ensure that the respiratory apparatus is thoroughly sterilized, rinsed, and hung up to dry and that all analyzed blood samples are disposed of into a biohazard bag.

Data Collection: For each minute the following should be recorded:

- $\dot{V}_E$, $\dot{V}O_2$, $\dot{V}CO_2$, RER, $\dot{V}_E/\dot{V}O_2$, HR, and treadmill speed for each 30 s of the test
- Resting blood variables (Hla, pH, and bicarbonate, if possible) before athlete's warm-up, after each completed workload, at the end of the test, and 2 and 5 min posttest.
- The $\dot{V}O_2$max: the average of the highest two consecutive 30 s readings
- RPE at the end of each workload

Data Analysis: Data should be transferred to the ADAPT computer package for computation of LT, AT, and peak values by the Dmax method. The following indexes should be reported for each threshold: $\dot{V}O_2$, %$\dot{V}O_2$max, HR, %HRmax, PO, lactate, and pH (if measured). $\dot{V}O_2$max should be presented as L/min and ml · kg^{-1} · min^{-1}, and is the average of the two highest consecutive 30 s values.

Note that for purposes of data analysis, $\dot{V}O_2$ is taken as the average of the data collected during the last minute of each 3 min workload, and HR is averaged over the last 15 s of each 3 min workload.

Run Field Tests

Seniors: This performance test requires the athlete to run for 30 min on a 400 m track. Heart rate should be recorded throughout. One-kilometer split times should be recorded, and an optional blood lactate should be obtained at the immediate completion of the trial.

Juniors: This could take the form of a 5 km run on a 400 m track with total time taken as the performance score. Heart rate and blood lactate measures should be obtained.

Triathlon Field Tests for Talent Identification

It is important to adhere to preparation procedures for field tests as strictly as with normal testing. Please read the sections on athlete preparation and general test conditions in this chapter and in chapter 2.

Swim Testing

100-Meter Time Trial Speed Test:

- This test should be performed in a 50 m pool.
- The trial begins with a push start.
- Possible factors to record: HR, stroke count and rate (last 50 m), split times, final time.

400-Meter Time Trial Speed-Endurance Test:

- This test should be completed in a 50 m pool.
- The start is a push start.
- Possible factors to record are listed for the 100 m time trial.

Endurance:

- Please see the swim tests described earlier in this chapter, which are field tests (i.e., 7 × 200 m or 1500 m time trial).

Cycle Testing

Cycle testing can consist of a 10 km cycle over a 5 km out-and-back course with total time used as the performance score.

Run Testing

This could take the form of a 5 km run on a 400 m track with total time taken as the performance score.

chapter 30

Protocols for the Physiological Assessment of Elite Water Polo Players

Douglas Tumilty, Peter Logan, Wayde Clews, and Don Cameron

Elite performance in the complex sport of water polo relies upon developing all the energy systems, as well as upon high levels of muscular strength, power, and flexibility. The sport involves numerous maximal swimming efforts over distances ranging from a few to as many as 25 m. These efforts may be interspersed with slower swimming, treading water, grappling with opponents, and lunging in offense and defense, as well as the skills of passing, receiving, and shooting the ball.

The men's and women's games are played within areas of 30 m × 20 m and 25 m × 20 m, respectively. Both the male and female games are played over four quarters; these are 7 min in duration, but because of stoppages the actual duration is 5-10 min longer. Two-minute intervals occur between quarters. Players can be substituted from the pool only when a goal is scored. Therefore, during low-scoring games, the energy costs to players can be extreme.

Equipment Checklist

- ☐ Stopwatch
- ☐ Waterproof heart rate monitor (e.g., Polar PE 3000T)
- ☐ Autolet platforms and sterile lancets
- ☐ Sterile alcohol swabs
- ☐ Heparinized capillary tubes (100 μl)
- ☐ Yellow Springs Instruments 25 μl pipette (positive displacement)
- ☐ Yellow Springs Instruments lysing buffer
- ☐ Eppendorf tubes
- ☐ Tissues
- ☐ Sample tray or rack
- ☐ Latex medical examination gloves
- ☐ Biohazard bag
- ☐ Sharps container
- ☐ Wall-mounted stadiometer
- ☐ Scales (accurate to ±0.05 kg)
- ☐ Harpenden skinfold calipers (dynamically calibrated)
- ☐ Marker pen
- ☐ Steel anthropometric measuring tape
- ☐ Recording proforma
- ☐ Bench

Pool Test Procedures

Standardization of testing of male and female water polo players has proved difficult because of limitations in resources relating to testing time, access to pools of the same lengths in different locations, and test staff and equipment. The tests described are the best compromise currently available between the existing limitations and the information desired.

The tests were chosen to give an indication of the capacity of all the energy systems. However, limitations on these tests should be appreciated:

- The protocol of 3 × 200 m for the males and 4 × 200 m for females is insufficient to give a really accurate assessment of the intensities (times or heart rates) at which threshold training paces occur for individual players. Even so, changes in the shape and position of the curve within the chosen axes are good indicators of changes in each player's aerobic fitness, and threshold training pace accurate to within a few seconds can be determined. Performing this test with five steps as described here is preferable if determination of an accurate threshold is desired.
- Tests of acceleration over a few meters would be desirable, since this is obviously an important requirement in the sport. However, the timing of such tests using only stopwatches, and the inability to control body position, make their accuracy too uncertain. Current development of a pool dynamometer at the Australian Institute of Sport (AIS) promises to allow a range of new pool-based strength, power, and acceleration tests to be conducted in the future for players in some locations.
- As the repetition 200 m test is usually performed, even the fastest of the 200 m swims is slightly less than a maximal effort. Even if the swim is done maximally, the duration is somewhat short to give a good assessment of maximal aerobic power. Currently therefore there is no in-pool test for estimating this capacity. (A swim test equivalent to the 20 m multistage fitness running, developed at the West Australian Institute of Sport but not yet adopted as a national test protocol, may remedy this deficiency. The modified, in-water, multistage fitness test is still being validated for water polo players. It may also be structured to give information on technique on the turn and acceleration.)
- Though the sport does require the ability to repeat high-intensity efforts from one end of the pool to the other, no standard test of this ability has yet been agreed on. The validity of the maximal-effort 100 m freestyle, considered a speed-endurance test, to describe this ability has yet to be demonstrated.

PROTOCOLS: Water Polo

Musculoskeletal Screening

Rationale: A comprehensive musculoskeletal screening procedure not only will identify existing injuries and predisposition to injury, but also is useful for monitoring the efficacy of flexibility and combined strength and flexibility programs. Screening may also identify findings known to inhibit performance, thus improving awareness of these limitations.

Test Procedure: Generalized musculoskeletal screening protocols have been published in Screening Test Protocols (see appendix for details).

Pool Tests

Rationale: Water polo is a sport that engages all energy systems:

- The aerobic system—because of the duration and continuous nature of a game, and because of the need for recovery from the frequent high-intensity efforts.
- The phosphagen system—because of the high-intensity bursts of swimming required to gain position, the physical struggles with opponents for possession, and the explosive movements used to rise from the water and shoot the ball.
- The lactic acid system—because situations of possession of the ball by one side, or turnover to the opposing side, can result in prolonged periods of high-intensity effort of a minute or more with very little chance of recovery. Levels of lactic acid of up to 12 mmol/L have been found even in games somewhat below world standard.

High levels of strength and power are also required to cope with frequent high-intensity acceleration, vertical jumps, and struggles with opponents.

Test Procedure:

- The first test (preferably the 200 m repeat swims) should be conducted after a day of only light activity. The players should follow their normal high-carbohydrate diet and should be well hydrated on the days of the tests.
- Athletes should perform a standard warm-up. For the males this has been 20 × 50 m freestyle and for the females 500-800 m, swum at low to moderate intensity.
- Ideally, the tests should take place on three consecutive days, since measurements have

shown that quite high levels of lactic acid remain for some time after each test (i.e., the repetition 200 m, the 100 m, and the 4-5 × 25 m swims), and this may influence the results of the later test. If more than one type of test must be performed in the same session, players should have a period of at least 10 min between each test. During this time they should swim continuously at an easy to moderate pace in order to speed up recovery from the first test.

- Access to pools of the same length in different locations has been difficult to arrange. If possible, each group of players should always be tested in the same pool, whether it be 25 m or 50 m. Pool lengths are given for the test results quoted in this chapter.
- All tests are begun with the player in the water and feet under water.
- Tumble turns are used.
- Timing is by stopwatch, preferably with the same tester timing each session.

Aerobic Test (Repetition 200-Meter Freestyle for Males and Females)

Normative data on national-level male and female players are provided in tables 30.1 and 30.2, respectively.

The primary information gained from this test is from graphs of heart rate against time and lactate against time. An improved aerobic capability would be indicated by a shift in these curves downward—that is, achievement of the same time for each 200 m at lower heart rates and lower lactates. An indication of training paces can be obtained from the results, though if possible these should be checked by taking heart rate and blood lactate readings during training.

It is important for subjects to swim at a constant pace throughout each of the efforts so that the heart rate and lactate values are reliable. Even the final 200 m is not a maximal effort. However, if the interest is in a pure performance test, and if a maximum (or very near maximum) heart rate is required, a final maximal 200 m swim can be performed.

Test Procedure:

- Each 200 m freestyle swim is performed on a 6 min cycle.
- Timing is by stopwatch, and intermediate times can be recorded if desired. Timing begins when the player's feet leave the wall on the initial push-off.
- Heart rate is recorded immediately after each swim with the subject in an upright position in the water. The preferable method is by Polar heart rate monitor. Note that there is no need to have male players wear the monitor; the receiver and transmitter can be kept at the edge of the pool, and the transmitter can be placed on the player's chest as soon as he finishes the swim. Be sure to keep the transmitter in place long enough to ensure a correct reading. Occasionally the first reading displayed is spurious.
- The player leaves the pool immediately after heart rate recording, and an earlobe or fingertip blood sample is taken as soon as practicable, with the player seated.
- Players return to the water as soon as the blood draw is complete. They may swim a few meters easily if there is time, or may simply rest and set themselves up for the next step.
- With the use of five steps, these are the intensities for the steps (pb, personal best):
    - Step 1: approximately 70% of 200 m pb speed (pb + 30 s)
    - Step 2: approximately 80% of 200 m pb speed (pb + 20 s)
    - Step 3: approximately 85% of 200 m pb speed (pb + 15 s)
    - Step 4: approximately 90% of 200 m pb speed (pb + 10 s)
    - Step 5: approximately 100% of 200 m pb speed (pb + 5 s) (15 min rest including a 500 m easy swim)
- Swim is 200 m at maximum intensity.
- Currently, male and female goalkeepers perform a test of 3 × 200 m breaststroke at intensities of 75%, 85%, and 95% of their pb time for a 200 m breaststroke swim.

Speed-Endurance Swim—100-Meter Freestyle

Normative data on national-level male and female players are provided in table 30.3.

This is a maximal 100 m swim. The duration of the effort is approximately 1 min, thereby stressing the lactic acid energy system maximally. However, one recognizes that this energy system cannot be isolated by this test, which also has significant phosphagen and aerobic systems involvement.

Table 30.1 Aerobic Test Results for Australian Institute of Sport National Senior Male Players

Stage/Intensity	Group	Measure	Mean	Range	SD
Step 1—~80% best 200 m speed (~200 m pb + 25 s)	Freestyle	Time (m:s)	2:30	2:21–2:45	0:07
		Heart rate (bpm)	132	104–151	14
		Lactate (mmol/L)	1.8	1.0–3.2	0.7
Step 1—~75% best 200 m speed (~200 m pb + 30 s)	Breaststroke	Time (m:s)	3:11	3:04–3:18	0:07
		Heart rate (bpm)	143	137–151	6
		Lactate (mmol/L)	2.4	1.9–3.0	0.5
Step 2—~88% best 200 m speed (~200 m pb + 15 s)	Freestyle	Time (m:s)	2:19	2:09–2:28	0:06
		Heart rate (bpm)	150	135–167	12
		Lactate (mmol/L)	3.7	2.3–4.7	0.9
Step 2—~85% best 200 m speed (~200 m pb + 15 s)	Breaststroke	Time (m:s)	3:02	2:57–3:08	0:06
		Heart rate (bpm)	167	156–181	10
		Lactate (mmol/L)	4.8	4.1–5.3	0.5
Step 3—~96% best 200 m speed (~200 m pb + 5 s)	Freestyle	Time (m:s)	2:08	2:00–2:15	0:05
		Heart rate (bpm)	168	153–190	11
		Lactate (mmol/L)	7.7	5.7–9.4	1
Step 3—~95% best 200 m speed (~200 m pb + 5 s)	Breastroke	Time (m:s)	2:56	2:48–3:01	0:07
		Heart rate (bpm)	176	167–187	9
		Lactate (mmol/L)	7.5	5.2–9.3	1.7

Incremental-intensity 200 m tests completed in a 25 m pool (n = 12 field players who completed freestyle tests, n = 4 goalkeepers who completed breaststroke tests), pb = personal best.

Results from this test may be difficult to interpret, since all three variables measured (time, heart rate, lactate) will be influenced by the training phase previously undertaken. One hopes at least for maintenance of the 100 m time; a deterioration in time, especially if accompanied by lower heart rate and lactate, might follow an aerobic phase in which high-intensity work has been neglected. A faster time, even though accompanied by a higher heart rate and lactate, would follow a high-intensity phase, or one that stressed technique, and would be considered desirable especially if the results of the aerobic test had not deteriorated.

Test Procedure:

- The swimmer performs a single maximal 100 m freestyle swim (except for goalkeepers, who perform 100 m breaststroke).
- The subject starts in his or her own time, and the stopwatch is started as soon as the feet leave the wall.
- Heart rate is measured as soon as the swim is finished and the swimmer has assumed an upright position.
- The swimmer leaves the pool, and an earlobe or fingertip blood sample is taken 3 min after the end of the swim, with the subject seated.

Sprint Tests

Normative sprint data on national-level male and female players are provided in table 30.4; sprint as well as 100 and 400 m results for junior male and female players are presented in table 30.5.

A total of four (five for goalkeepers) 25 m swims are performed. A 3 min interval should be allowed between each swim. Except for the freestyle, which begins with a push from the wall, the tests begin with the subject away from the wall with the aim of ensuring that the swimmer uses the correct action from the beginning of the test.

Test Procedure:

1. 25 m freestyle

- Two attempts at maximum effort.
- Swimmer starts in his or her own time.
- Swimmer pushes off wall, and timing starts when the feet leave the wall.

Table 30.2 Aerobic Test Results for Australian Institute of Sport National Senior Female Players

Stage/Intensity	Measure	Mean	Range	SD
Four-stage incremental test				
Step 1—~70% best 200 m speed (~200 m pb + 30 s)	Time (s)	166.8	153.9–173.8	5.9
	Heart rate (bpm)	148	132–165	13
	Lactate (mmol/L)	2.9	1.6–4.5	0.8
Step 2—~80% best 200 m speed (~200 m pb + 20 s)	Time (s)	161.9	150.6–166.3	5
	Heart rate (bpm)	161	146–174	10
	Lactate (mmol/L)	3.5	2.0–9.2	2.1
Step 3—~90% best 200 m speed (~200 m pb + 10 s)	Time (s)	154.1	148.0–158.9	3.3
	Heart rate (bpm)	173	156–186	11
	Lactate (mmol/L)	4.9	3.6–8.8	1.8
Step 4—~95% best 200 m speed (~200 m pb + 5 s)	Time (s)	148.2	141.4–152.6	3.8
	Heart rate (bpm)	182	170–194	7
	Lactate (mmol/L)	6.6	5.2–10.3	1.6
Five-stage incremental test				
Step 1—~80% best 200 m speed (~200 m pb + 25 s)	Time (s)	179.5	165–195	11.1
	Heart rate (bpm)	115	90–129	16
	Lactate (mmol/L)	1.5	0.8–3.0	0.8
Step 2—~84% best 200 m speed (~200 m pb + 20 s)	Time (s)	172.7	156–182	10.3
	Heart rate (bpm)	124	110–142	14
	Lactate (mmol/L)	1.8	0.1–3.7	1.3
Step 3—~88% best 200 m speed (~200 m pb + 15 s)	Time (s)	163.8	154–174	7.3
	Heart rate (bpm)	139	130–147	7
	Lactate (mmol/L)	4.5	2.0–7.7	1.9
Step 4—~92% best 200 m speed (~200 m pb + 10 s)	Time (s)	158.2	152–169	6.4
	Heart rate (bpm)	153	151–158	3
	Lactate (mmol/L)	7.1	5.0–8.7	1.4
Step 5 — ~96% best 200 m speed (~200 m pb + 5 s)	Time (s)	151.3	145–164	7.1
	Heart rate (bpm)	166	160–175	7
	Lactate (mmol/L)	9	5.3–11.7	2.3

Incremental-intensity 200 m tests completed in a 25 m pool. Both four-stage and five-stage tests were completed (n = 10 field players who completed freestyle tests); pb = personal best.

2. 25 m freestyle—arms only

- One attempt at maximum effort.
- Legs are held together with a rubber band at the ankles—legs supported by holding pull-buoy between legs superior to knees.
- Swimmer starts with the forehead level with the front of the transverse black lane line. (Note: It is useful to have a teammate place a hand lightly on the swimmer's forehead to help the swimmer obtain the exact starting position.)
- The swimmer should be in a position as close to horizontal as possible.
- The time keeper calls "Go" and at the same time starts the stopwatch.
- Starting and timing are performed by the same person.

3. 25 m freestyle kick

- One attempt at maximum effort.
- Water polo ball is pushed along the surface—males use male ball, females use female ball.

Table 30.3 Speed-Endurance Swim Results for Australian Institute of Sport National Senior Male and Female Players

Test	Measure	Mean	Range	SD
Male				
Freestyle	Time (s)	56.4	53.3–61.1	2.2
	Heart rate (beats/min)	157	137–184	13.9
	Lactate (mmol/L)	8.7	6.1–13.1	2
Breaststroke	Time (s)	77.41	73.4–81.5	4.56
	Heart rate (beats/min)	158	132–174	19
	Lactate (mmol/L)	7.4	5.3–9.2	1.7
Female				
Freestyle	Time (s)	68.7	65.6–75.4	2.4
	Heart rate (beats/min)	176	160–192	11
	Lactate (mmol/L)	6.9	4.1–10.3	2

All tests completed in a 25 m pool. Men were field players (n = 12) who completed freestyle tests and goalkeepers (n = 4) who completed breaststroke tests. Women were field players (n = 16) who completed freestyle tests.

- Hands are held over the ball with thumbs touching.
- Elbows are locked and head is out of the water.
- Swimmer starts with the forehead level with the front of the transverse black lane line. (Note: A teammate may place a hand lightly on the swimmer's forehead to help the swimmer obtain the exact starting position.)
- The swimmer should be in a position as close to horizontal as possible.
- The time keeper calls "Go" and at the same time starts the stopwatch.
- Starting and timing are performed by the same person.

4. 25 m eggbeater kick

- One attempt at maximum effort
- Protocol as described for the preceding tests

5. 25 m breaststroke kick (goalkeepers only)

- One attempt at maximum effort
- Protocol as described for the preceding tests

Laboratory Tests

Rationale: Laboratory tests provide additional information on the characteristics of water polo players, which—though important to the ability to play the game—are difficult or impossible to measure in the water. The tests cover aspects such as the aerobic capacity, strength, and bodily dimensions of players.

Maximum Oxygen Consumption

No standard test of maximum oxygen uptake exists for water polo players. Tests that have been done occasionally have utilized either an arm-and-leg ergometer in the laboratory or tethered swimming in the pool. The advantage of such tests is that they give a well-understood final value that is of a physiological capability only, whereas the results of pool tests in terms of time for a distance swum are influenced by the subject's technique. However, the nonspecific nature of laboratory tests and the difficulty of setting up tethered swim tests have limited their use.

Normative data on national-level players are shown in table 30.6.

Strength

Considerable research demonstrates that strength adaptation is context specific. That is, strength changes that result from a given conditioning intervention may be discernible only if the characteristics of muscle contraction(s) are similar between the conditioning intervention and the strength-testing protocol (see Abernethy et al. 1995). For this reason strength testing of elite Australian water polo players is conducted in two ways.

Table 30.4 Sprint Results for AIS National Senior Male and Female Players

Test	Group	Mean	Range time (s)	SD
Male				
Freestyle	Field players	12.08	11.64–12.87	0.34
	Keepers	12.30	11.64–12.86	0.52
Freestyle—arms	Field players	14.29	13.48–15.02	0.43
	Keepers	14.74	14.16–15.43	0.53
Freestyle—legs	Field players	23.61	20.49–28.14	2.13
	Keepers	26.71	23.45–31.34	3.32
Eggbeater—legs	Field players	26.05	23.31–29.67	2.17
	Keepers	25.68	24.10–27.41	1.55
Breastroke	Field players	–	not done	–
	Keepers	23.25	21.30–24.83	1.49
Female				
Freestyle	Field players	13.99	13.61–14.42	0.41
Freestyle—arms	Field players	25.94	25.49–26.73	0.69
Freestyle—legs	Field players	26.91	26.23–27.57	0.67
Eggbeater—legs	Field players	23.85	23.04–24.74	0.85
Breaststroke	Field players	–	not done	–

Tests completed in a 25 m pool. Men were field players (n = 12) and goalkeepers (n = 4); women were field players (n = 3). AIS = Australian Institute of Sport.

Table 30.5 Mean Sprint, 100 m, and 400 m Results for Junior Male and Female Players

		Sprint				Speed-endurance	
Group	n	Freestyle time (s)	Freestyle—legs time (s)	Eggbeater—legs time (s)	Breaststroke—legs time (s)	100 m time (s)	400 m time (m:s)
14 years male	11	15.28	26.68	29.62	26.67	74.9	05:50.1
14 years female	7	15.71	24.77	30.74	27.92	67.0	05:59.0
15 years male	22	13.94	25.48	29.07	26.30	69.6	05:27.0
15 years female	11	15.15	26.34	30.59	26.37	72.4	05:34.8
16 years male	24	13.39	25.22	28.21	25.30	64.6	05:11.8
16 years female	12	14.53	28.33	30.88	26.48	68.7	05:09.1
17 years male	18	13.03	23.83	27.41	24.63	64.5	05:13.2
17 years female	8	15.56	26.99	29.92	25.52	73.9	05:41.9
18 years male	8	12.69	23.14	27.85	24.07	60.6	04:52.9

All tests were freestyle unless indicated otherwise. Tests completed in both 25 m and 50 m pools (but no time corrections applied).

Table 30.6 Maximal Oxygen Consumption for Testing Completed on an Arm-and-Leg Ergometer

	Mean	Range	SD
Maximum oxygen uptake			
(L/min)	4.95	4.17–5.29	0.39
($ml \cdot kg^{-1} \cdot min^{-1}$)	57.1	45.6–65.6	5.5
Peak heart rate (beats/min)	193	175–210	10
Peak power (W)	447	395–488	32

n = 16 national-level players; none were goalkeepers.

One type of isoinertial strength testing takes place on dry land using conventional, resistance-training apparatuses and exercises. This complements the significant amount of strength training that occurs in the weight room with use of similar apparatuses and similar muscle activity. Secondly, much resistance work is performed in the water in an attempt to improve water polo-specific strength. To this end, isoinertial strength testing also takes place in the water, invoking movements that are more specific to water polo demands.

Dry-Land Maximum Strength Test: Three-repetition maximum (3RM) protocols are used to assess the athlete's maximal strength for chin-up, bench press, and leg press tasks. These movements, performed with heavy weights for few repetitions, are meant to stress those muscle groups used in an explosive or strength-oriented manner during water polo performance.

1. 3RM chin-up test

- Warm-up: Athlete performs lat pull-down exercise with a load not in excess of 90% of body weight.
- Athlete uses overhand (pronated) grip at shoulder-width.
- Body is to be kept straight throughout movement (legs may be crossed).
- Arms must be fully extended at the bottom of each repetition.
- A 3 s count is applied if the athlete is stalled at any stage of movement.
- The repetition is counted only if chin is clearly above the bar at the top of the movement.
- Resistance is progressively added (via a "weights belt" fixed around the athlete's waist) until the athlete can complete no greater than three repetitions for a given load. The additional load is recorded as the 3RM score.
- In the event that the athlete attempts a load and does not complete three repetitions, the previous load for which the three repetitions were completed is recorded as the 3RM score.

2. Bench press (3RM)

- Warm-up: Athlete performs bench press with a load not in excess of 90% of previous 3RM score (or estimated 3RM score).
- Shoulders, back, and buttocks must touch bench at all times.
- Athlete must touch chest with the bar and pause at bottom of movement.
- Full elbow extension must be attained at the completion of each repetition.
- Resistance is progressively added to the bar until athlete can complete no greater than three repetitions for a given load. The absolute load (bar plus additional resistance) is recorded as the 3RM score.
- If the athlete attempts a load and does not complete three repetitions, the previous load for which three repetitions were completed is recorded as the 3RM score.

3. Leg press (3RM)

- Warm-up: Athlete performs leg press with a load not in excess of 90% of previous 3RM score.
- Athlete performs leg press on 45° sled.
- Lower back and buttocks must touch back and seat supports at all times.
- Athlete must lower sled to 90° knee flexion and pause at bottom of movement.
- Full knee extension must be attained at the completion of each repetition.
- Resistance is progressively added to the sled until the athlete can complete no greater than three repetitions for a given load. The total

load that has been added to the sled is recorded as the 3RM score.

- In the event that the athlete attempts a load and does not complete three repetitions, the previous load for which three repetitions were completed is recorded as the 3RM score.

Dry Land Strength Endurance Tests: The aim of maximal-repetition (MR) protocols is to assess an athlete's strength endurance for the bench press and leg press tasks. These protocols require athletes to complete as many repetitions as they possibly can at a load of 70% of body weight for the bench press, and two times body weight for the leg press. These movements, performed with a moderate resistance for several repetitions, are meant to stress those muscle groups that are repetitively used in an explosive or strength-oriented manner throughout a water polo match.

1. Bench press (3RM)

- Warm-up: Athlete will have already completed the 3RM protocol for bench press and will need no additional warm-up.
- Shoulders, back, and buttocks must touch bench at all times.
- Athlete must touch chest with the bar and pause at bottom of movement.
- Full elbow extension must be attained at completion of each repetition.
- Against an absolute load of 70% of body weight, the athlete performs as many repetitions as possible until the arms can no longer be straightened.
- Throughout the test the athlete must maintain an even tempo.
- The number of completed repetitions is recorded as the MR score.

2. Leg press (MR)

- Warm-up: Athlete will have already completed 3RM protocol for leg press and will need no additional warm-up.
- Leg press is performed on 45° sled.
- Lower back and buttocks must touch back and seat supports at all times.
- Athlete must lower sled to 90° knee flexion and pause at the bottom of the movement.
- Full knee extension must be attained at completion of each repetition.
- Against an absolute load of two times body weight, the athlete performs as many repetitions as possible until the legs can no longer be straightened.
- Throughout the test the athlete must maintain an even tempo.
- The number of completed repetitions is recorded as the MR.

In-Water Strength Tests: The in-water component of the strength-testing protocol is designed to assess the athlete's ability to explode out of the water in a vertical plane. This capacity is vital in order for field players to overcome the downward resistance of opponents and also for the act of shooting over the top of defenders. Moreover, the ability to explode into the air is critical for successful goalkeeping.

For assessment of the cumulative strength qualities of those muscles responsible for vertical explosiveness in the water, the athlete must perform a vertical lunge while tethered to an electronically braked dynamometer. A load of 10 N applied to the dynamometer acts to partially resist the vertical movement of the athlete. A light load of this magnitude allows near-maximum power output and promotes the use of normal movement skill.

The in-water strength test protocol differs between field players and goalkeepers in order to reflect the differing physiological demands of the roles. On the one hand, research has shown that goalkeepers need to perform few maximal vertical efforts throughout a 30 s defensive phase (Smith 1991). There are usually none of these efforts from the goalkeeper during an offensive play. On the other hand, field players must perform several maximal efforts in both offensive and defensive plays. Therefore the view is that testing for one-off or singular explosiveness is applicable to both goalkeepers and field players, and that the capacity for repeated, explosive efforts should also be tested in field players.

1. Maximal explosiveness

- This test is for field players and goalkeepers.
- Athletes must complete three trials in which they perform four consecutive vertical lunges.
- Prior to the commencement of each lunge, the athlete must be treading water at a resting intensity.
- On "Go," the athlete attempts to lunge into the air as high as possible, leading with the dominant throwing arm.
- Minimum rest intervals of 2 min and 3 s are required between trials and individual repetitions, respectively.

- REEL-IN software interfaced with a flux-vector drive dynamometer is used to assess the force versus time functions associated with each vertical lunge (figure 30.1). (For details about this software, see the appendix.)
- The peak force and peak power scores across all repetitions are recorded as the athlete's maximal explosiveness scores.

2. Repeat explosiveness

- This test is for field players only.
- Athletes must complete one 30 s trial in which they perform 15 consecutive vertical lunges.
- Lunges are started at 2 s intervals throughout the entire 30 s trial period.
- Prior to the commencement of each lunge, the athlete must attempt to return to resting water-treading position.
- On "Go," given at the start and at 2 s intervals thereafter, the athlete attempts to lunge into the air as high as possible, leading with the dominant throwing arm.
- REEL-IN software interfaced with a flux-vector drive dynamometer is used to assess the force versus time functions associated with each vertical lunge.
- The peak force across all repetitions is recorded. A fatigue index is then determined based on the decrement in force recorded by the athlete from the first to last repetition. This fatigue index is then given as the athlete's repeat explosiveness score.

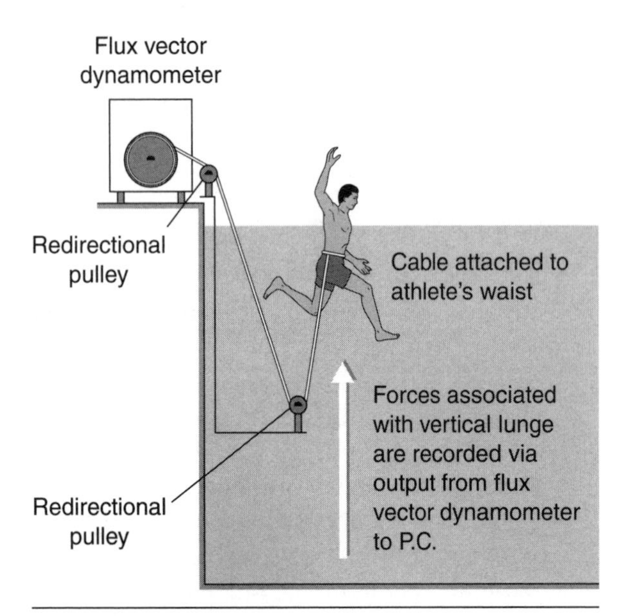

Figure 30.1 The interface of software with dynamometer ensures accurate data collection for maximal explosiveness.

Test Schedules: Dry-land and in-water strength-testing protocols take place on separate days, with a minimum intervening period of 24 hours. Within each testing session, all maximum strength and maximal explosiveness trials should precede strength endurance or repeat explosiveness trials. Furthermore, during dry-land maximum strength testing, performance of the lower limb test (leg press) should separate that of the two upper limb tests (i.e., the bench press and chin-up tests). Table 30.7 shows the test schedule used for the national men's team based at the AIS.

Technical Error of Measurement and Normative Values: The technical error of measurement (TEM; see page 83; Pederson and Gore 1996) for the various strength tests was established using 10 experienced water polo players. The TEM of 3RM and 1RM bench press was 1.6% and 3.1%, respectively; the corresponding values for the leg press were 5.2% and 6.1%. The TEM for maximal explosiveness (in water) was 10.1%, but a recent upgrade to the flux-vector dynamometer that was mentioned earlier would be expected to improve test precision.

Normative data pertaining to the national open men's team (AIS squad) are reported in table 30.8. These data are taken from a representative test occasion and include results for all strength test protocols detailed in this chapter.

To date, routine testing with all recommended strength protocols has not been conducted for national-level open (international level) females or for Australian Intensive Training-level males and females. For these athletes, only dry-land maximum strength data have been collected. Again, the table shows representative scores for these tests.

Anthropometry

Table 30.9 shows normative data for anthropometry results for AIS national-level male water polo players.

Rationale: Results of the Kinanthropometry in Aquatic Sports study (Carter and Ackland 1994) indicate that water polo players of both sexes tend to be tall and muscular, though not extremely so in comparison to shorter-distance swimmers. Differences do exist between playing positions, goalkeepers tending to be more ectomorphic and center backs and center forwards larger and taller than other players. Body fat levels of water polo players are somewhat higher than those of swimmers (sum of six skinfold sites 62.5 vs. 44.9 mm for males, 89.8 vs.

Table 30.7 Recommended Schedule for Routine Strength Testing of Elite Water Polo Players

	Timing	Test order
In-water strength tests	Monday P.M.	1. Maximal explosiveness 2. Repeat explosiveness
Dry-land strength tests	Wednesday A.M.	1. 3RM bench press 2. 3RM leg press 3. 3RM chin-ups 4. MR bench press 5. MR leg press

3RM = three-repetition maximum, MR = maximal repetition.

Table 30.8 Normative Strength Values (Mean ± SD) for Australian Water Polo Players at ITC (U/17) and Open, International Level as Determined From National Strength-Testing Protocols

Test instrument	Open females national squad	Open males national squad
Maximal explosivenes (N)	–	110 (75 to 325)
Repeat explosiveness (%)	–	–8 (+1 to –43)
3RM bench press (kg)	54.4 (42.5 to 65)	97.08 (75 to 120)
3RM leg press (kg)	191.1 (160 to 220)	242.5 (175 to 300)
3RM chin-ups (kg)	–	21.25 (10 to 30)
MR bench press	19.3 (15 to 30)	16.7 (14 to 20)
MR leg press	17.2 (8 to 30)	18.6 (9 to 31)

ITC = Intensive Training Centre, 3RM = three-repetition maximum, MR = maximal repetition.

Table 30.9 Anthropometry Results for AIS National Senior Male (n = 18) and Female (n = 17) Players

Measure	Group	Mean	Range	SD
Mass (kg)	Men	90.1	77.2–101.8	6.9
Height (cm)	Men	190.3	175.9–199.1	5.4
Arm span (cm)	Men	198.7	185.5–210.5	6.4
Σ skinfolds (mm)	Men	66.5	242.3–125.8	22.2
Mass (kg)	Women	72.2	64.0–90.0	6.8
Height (cm)	Women	173.9	165.7–178.5	4.1
Σ skinfolds (mm)	Women	94.2	53.6–175.5	28.0

Σ skinfolds = sum of seven sites: triceps, subscapular, biceps, supraspinale, abdominal, front thigh, and medial calf. AIS = Australian Institute of Sport.

70.6 mm for females, Kinanthropometry in Aquatic Sports figures). There may be some advantage in terms of buoyancy for water polo players to be slightly fatter, given the amount of time spent in the vertical position—although this is likely to be largely offset by the greater moment of inertia to be overcome during the frequent accelerations necessary. A large arm span would also be advantageous in reaching for the ball and in gaining extra height for shots.

An annual full anthropometric profile is useful, especially to monitor the progress of developing players. Somatotype and O-scale readings have sometimes been used to follow changes in male players undergoing a hypertrophy weight-training program. However, as with many other sports, the most frequently measured dimensions have been height, weight, and skinfold sum, used in conjunction with performance test results as an indicator of overall fitness and condition.

Appendix

This guide is included in the book because similar information is not readily available elsewhere. No attempt was made to include or exclude listings based on product quality, nor to offer an exhaustive list. Possible omissions were inadvertent. Thus, inclusion in this guide does not represent endorsement by the Australian Sports Commission or the authors.

Information was correct at the time of publication.

Measurement Apparatus and Equipment

Blood collection

Vacuette (product name)
Greiner laortechnik, CA Greiner & Söhne GmbH
A-4550 Kremsmünster, Bad Haller Strasse 32
Germany
Tel: +43 7583 6791
Fax: +43 7583 6348
Web site: **www.greiner-lab.com**

Calibration, dynamometer/ergometers

Bio-med Electronic Services
P.O. Box 46
Brooklyn Park, South Australia 5032
Australia
e-mail: **biomed_electronics@hotmail.com**

South Australian Sports Institute
P.O. Box 219
Brooklyn Park, South Australia 5032
Australia
Contact: Tom Stanef
e-mail: **stanef.tom@saugov.sa.gov.au**

VacuMed
4483 McGrath St. #102
Ventura, California 93003
United States
Tel: 805-644-7461, Toll-Free: 800-235-3333
Fax: 805-654-8579
e-mail: **info@vacumed.com**
Web site: **www.zest.net/vacumed**

Calibration, gases

BOC Gases Australia Ltd
127 Newton Rd.
Wetherill Park
New South Wales 2164, Australia
Tel: +61 2 9616 3300
Fax: +61 2 9616 3450
Web site: **www.boc.com.au**

Calibration, syringes—sinusoidal and motor driven

JV Precision Engineering
3 Raw Pl.
Farrer
Australian Capital Territory 2607, Australia
Tel: +61 2 6290 1930
Fax: +61 2 6286 6252

Dynamometers, grip-strength

Fabrication Enterprises Inc. (manufacturer)
Trent Building
South Buckout St.
Irvington, New York 10533
United States
Tel: 914-591-9300
Fax: 914-591-4083

Mentone Educational Centre
(distributes in Australia)
24 Woorayl St.
Carnegie, Victoria 3163, Australia
Tel: +61 3 9563 3488

Ergometers

K1 Kayak Ergometer *(kayak)*
Roger Cargill
59 Gilmore Crescent
Garran, Australian Capital Territory 2605
Australia
Tel: +61 2 6281 5660
Fax: +61 2 6285 2763
e-mail: **k1ergo@dynamite.com.au**

Repco Cycle Company *(modified kayak)*
88 Peters Ave.
Mulgrave, Victoria 3170, Australia
Tel: +61 3 9574 7000
Fax: +61 3 9574 7012

Concept II Inc. *(rowing)*
105 Industrial Park Dr.
Morrisville, Vermont 05661-9727
United States
Tel: 802-888-7971
Fax: 802-888-4791
Web site: **www.conceptii.com**

Kingcycle *(cycle)*
Lane End Rd.
Sands, High Wycombe
Bucks HP124JQ
United Kingdom
Tel: +44 1494 524004
Fax: +44 1494 437591

SRM High Performance Ergometer *(cycle)*
Ingenieurbüro Schoberer
Fuchsend 24, 52428 Julich, Welldorf
Germany
Tel: +49 2463 3156
Fax: +49 2463 3090
Web site: **www.srm.de**

Repco *(air-braked cycle and Exertech work meter)*
P.O. Box 225
Mulgrave
Victoria 3170, Australia
Tel: +61 3 9574 7000
Fax: +61 3 9574 7012

AusTredEx *(treadmill)*
5a Junction St.
Preston, Victoria 3072, Australia
Tel: +61 3 9480 5222
Fax: +61 3 9480 0385
Web site: **www.ausbusiness.com/au/m/austredex**

Quinton Instrument Co. *(treadmill)*
3303 Monte Villa Pkwy.
Bothell, Washington 98021-8906
United States
Tel: United States and Canada
(sales/information)—800-426-0337 ext. 2440
(technical service)—800-426-0538
Outside North America
(sales/information)—425-402-2440
(technical service)—425-402-2485
Fax: 425-402 2440 (sales/information)
Web site: **www.quinton.com**

Stanton Engineering Pty Ltd *(treadmill)*
Factory 2/40 Forge St.
Blacktown, New South Wales 2148, Australia
Tel: +61 2 9831 5833
Fax: +61 2 9831 1012

Gas analyzers

AEI Technologies Inc. (formerly Ametek, Applied Electrochemistry)
300 William Pitt Way
Pittsburgh, Pennsylvania 15238
United States
Tel: 412-826-3280
Fax: 412-826-3281
Web site: **www.inhouse.ca/aei/index.html**

Jump systems

Swift Performance Equipment *(countermovement)*
P.O. Box 639
Lismore
New South Wales 2480, Australia
Tel: +61 2 6628 7778
Fax: +61 2 6628 7779
e-mail: **info@spe.com.au**
Contact: Robert Baglin or Mark Fisher

Sports Imports *(Vertec jumping device)*
P.O. Box 21040
Columbus, Ohio 43221
United States
Tel: 614-771-0246
Fax: 614-771-0750

Swift Performance Equipment *(Yardstick jumping device)*
P.O. Box 639
Lismore
New South Wales 2480, Australia
Tel: +61 2 6628 7778
Fax: +61 2 6628 7779
e-mail: **info@spe.com.au**
Web site: **www.spe.com.au**
Contact: Robert Baglin or Mark Fisher

Plyopower Smith machine and vertical jump system
Contact: Dr. Robert Newton, PhD
Senior Lecturer
Director of Postgraduate Studies and Research
School of Exercise Science and Sport Management
Southern Cross University
P.O. Box 157
Lismore
New South Wales 2480, Australia
Tel: + 61 2 6620 3234
Fax: + 61 02 6620 3880
e-mail: **rnewton@scu.edu.au**

Swift Performance Equipment *(timing mat)*
P.O. Box 639
Lismore
New South Wales 2480, Australia
Tel: +61 2 6628 7778
Fax: +61 2 6628 7779
e-mail: **info@spe.com.au**
Web site: **www.spe.com.au**
Contact: Robert Baglin or Mark Fisher

Leak detection

Swagelok
29500 Solon Rd.
Solon, Ohio 44139
United States
Tel: 216-248-4600
Fax: 216-349-5970
Web site: **www.swagelok.com**

Light gates

Swift Performance Equipment
P.O. Box 639
Lismore, New South Wales 2480, Australia
Tel: +61 2 6628 7778
Fax: +61 2 6628 7779
e-mail: **info@spe.com.au**
Web site: **www.spe.com.au**
Contact: Robert Baglin or Mark Fisher

AST Stopwatches *(TAG Heuer systems)*
The Old Vicarage Rd.
Ashby-de-la Zouch, Leicestershire
United Kingdom
Tel: +44 1530 411321
Web site: **www.astopwatch.co.uk**

Cambrian Timing *(Alge Timing systems)*
Garreg Lwyd, Berthddu, Rhosesmor, nr Mold
Flintshire CH7 6PS
United Kingdom
Tel: +44 1352 78130
Web site: **www.cambrian-timing.co.uk**

Manometers

Pegler Beacon Pty Ltd
(supply Dwyer manometers)
Unit 3, 8 South Street
Rydalmere, New South Wales 2116
Australia
Tel: +61 2 9841 2345
Fax: +61 2 9684 2067

Pressure Biofeedback Stabilizer

Pacific Health Pty Ltd
31 Ethel Street
Yeerongpilly, Queensland 4105
Australia
Tel: +61 7 3255 8600
Fax: +61 7 3255 8611
e-mail: **sales@pacifichealth.com.au**
Web site: **www.pacifichealth.com.au/contact_us/contact_us.html**

Respiratory valves, tubing, and handheld volumetric calibration syringes

Hans Rudolph Inc.
7205 Wyandotte
Kansas City, Missouri 64114
United States
Tel: 816-363-5522
Fax: 816-822-1414
Web site: **www.rudolphkc.com**

VacuMed
4483 McGrath St. #102
Ventura, California 93003
United States
Tel: 805-644-7461
Fax: 805-654-8759
Web site: **www.vacumed.com**

Warren E Collins Inc.
220 Wood Rd.
Braintree, Massachusetts 02184
United States
Tel: 617-843-0610
Fax: 617-843-4024

$\dot{V}O_2$ systems

Erich Jaeger Gmb H
P.O. Box 5846
D-97008 Wuerzburg, Germany
Tel: +49 931 4972-0
Fax: +49 931 4972-46

Medical Graphics Corporation
350 Oak Grove Pkwy.
St. Paul, Minnesota 55127-8599
United States
Tel: 612-484-4874
Fax: 612-484-8941
Web site: **www.medgraph.com**

PK Morgan Ltd
4 Bloors Lane
Rainham, Gillingham, Kent ME8 7ED
United Kingdom
Tel: +44 634 37 3865
Fax: +44 634 37 1681
Web site: **www.morganmedical.co.uk**

Quinton Instrument Co.
3303 Monte Villa Pkwy.
Bothell, Washington 98021-8906
United States
Tel: United States and Canada
(sales/information)—800-426-0337 ext. 2440
(technical service)—800-426-0538
Outside North America
(sales/information)—425-402-2440
(technical service)—425-402-2485
Fax: 425-402 2440 (sales/information)
Web site: **www.quinton.com**

Sensor Medics Corporation
22705 Savi Ranch Pkwy.
Yorba Linda, California 92687
United States
Tel: 714-283-2228
Fax: 714-283-8439
Web site: **www.sensormedics.com**

Volume measurement

American Meter Company *(dry gas meters)*
(does not sell direct, but uses eight distributors throughout America)
Contact: **www.americanmeter.com/ContactUs.htm**

Measurement Control Systems *(dry gas meters)*
1331 South Lyon St.
Santa Ana, California 92705
United States
Tel: 714-835-0995
Fax: 714-835-1103

PK Morgan Ltd *(turbine ventilometer)*
4 Bloors Lane
Rainham, Gillingham, Kent ME8 7ED
United Kingdom
Tel: +44 634 37 3865
Fax: +44 634 37 1681
Web site: **www.morganmedical.co.uk**

Weighing scales

A & D Mercury Pty Ltd
32 Dew St.
Thebarton, South Australia 5031, Australia
Tel: +61 8 8352 3033
Fax: +61 8 8352 7409
Web site: **www.andmercury.com.au**

Salter Weigh-Tronix Pty Ltd
20 Terracotta Dr.
Blackburn, Victoria 3130, Australia
Tel: +61 3 9894 2444
Fax: +61 3 9894 2882
Web site: **www.weigh.tronix.com**

Wedderburn Scales
P.O. Box 180
Summerhill, New South Wales 2130, Australia
Tel: +61 2 9797 0111
Fax: +61 2 9799 2013
Web site: **www.wedderburn.com.au**

Software and Video/Audio

ADAPT software program for lactate threshold

Sport Sciences, Australian Institute of Sport
P.O. Box 176
Belconnen, Australian Capital Territory 2616
Australia
E-mail: robyn.power@ausport.gov.au

Audiocassette for *Multistage Fitness Test*

Australian Sports Commission
P.O. Box 176
Belconnen, Australian Capital Territory 2616
Australia
Tel: +61 2 6214 1915
Fax: +61 2 6214 1995
e-mail: **pubs@ausport.gov.au**

Compact disk for *20m Shuttle Run Test/Multistage Fitness Test*

Australian Sports Commission
P.O. Box 176
Belconnen, Australian Capital Territory 2616
Australia
Tel: +61 2 6214 1915
Fax: +61 2 6214 1995
e-mail: **pubs@ausport.gov.au**

Data collection software for cycle ergometers

Developed and written at the:
South Australian Sports Institute
P.O. Box 219
Brooklyn Park, South Australia 5032
Australia
Tel: +61 8 8416 6677
E-mail: **stanef.tom@saugov.sa.gov.au**
Contact: Tom Stanef

LifeSize software for calculating technical error of measurement
Human Kinetics
P.O. Box 5076
Champaign, IL 61825-5076
Tel: 800-747-4457
Web site: **www.humankinetics.com**

REEL-IN software for assessing force versus time functions in water polo
Australian Institute of Sport
P.O. Box 176
Belconnen, Australian Capital Territory 2616
Australia
e-mail: **cmackintosh@ausport.gov.au**
Contact: Colin Mackintosh

Sport-Specific Musculoskeletal Screening Protocols

Golf

Melbourne's Golf Injury Clinic
925 Dandenong Rd.
East Malvern, Victoria 3145, Australia
Tel: +61 3 9572 3033

Tennis

Tennis Australia
Private Bag 6060
Richmond South, Victoria 3121, Australia
Tel: +61 3 9286 1177
Fax: +61 3 9650 2743
Web site: **www.tennisaustralia.com.au**

References

Chapter 1

Carlyon RG, Gore CJ, Woolford SM and Bryant RW (1996) Calibrating Harpenden skinfold calipers. In: Norton KI and Olds T (eds) *Anthropometrica.* Sydney: University of New South Wales Press, pp. 97-118.

Gore CJ, Catcheside PG, French SN, Bennett JM and Laforgia J (1997) Automated $\dot{V}O_2max$ calibrator for open-circuit indirect calorimetry systems. *Medicine and Science in Sports and Exercise* 29:1095-1103.

International Organization for Standardization (ISO) 9001/2/3 series (1994) *Quality Systems-Model for quality assurance in Design/Development, Production, Installation and Servicing.* Switzerland: International Organization for Standardization.

International Organization for Standardization (ISO) Guide 25 (1990) *General Requirements for the Competence of Calibration and Testing Laboratories.* Switzerland: International Organization for Standardization.

Journal of Applied Physiology (1996) Guiding principles for research involving animals and human beings. Information for authors. *Journal of Applied Physiology* 80(6).

National Association of Testing Authorities (NATA) (1989) *Guide to Assessment of Laboratories.* Chatswood, New South Wales, Australia: National Association of Testing Authorities, p. 1-2.

Pederson DG and Gore CJ (1996) Anthropometry measurement error. In: Norton KI and Olds T (eds) *Anthropometrica.* Sydney: University of New South Wales Press, pp. 77-96.

Chapter 3

Abler P, Foster C, Thompson NN, Crowe M, Alt K, Brophy A and Palin WD (1986) Determinants of anaerobic muscular performance. *Medicine and Science in Sports and Exercise* 18:2, S1.

Åstrand PO, Hultman E, Juhlin-Dannfelt A and Reynolds G (1986) Disposal of lactate during and after strenuous exercise in humans. *Journal of Applied Physiology* 61:338-343.

Bangsbo J (1998) Quantification of anaerobic energy production during intense exercise. *Medicine and Science in Sports and Exercise* 30:47-52.

Bangsbo J, Gollnick PD, Graham TE, Juel C, Kiens B, Mizuno M and Saltin B (1990) Anaerobic energy production and O_2 deficit-debt relationship during exhaustive exercise in humans. *Journal of Physiology* (London) 422:539-559.

Bangsbo J, Michalsik L and Peterson A (1993) Accumulated O_2 deficit during intense exercise and muscle characteristics of elite athletes. *International Journal of Sports Medicine* 14:207-213.

Bar-Or O, Dotan R and Inbar O (1977) A 30 s all-out ergometer test: Its reliability and validity for anaerobic capacity. *Israel Journal of Medical Sciences* 13:326-327.

Bouchard C (1985) Specificity of aerobic and anaerobic work capacities and powers. *International Journal of Sports Medicine* 6:325-328.

Bouchard C, Taylor AW, Simoneau JA and Dulac S (1991) Testing anaerobic power and capacity. In: MacDougall JD, Wenger HA and Green HJ (eds) *Physiological Testing of the High Performance Athlete* (2nd ed). Champaign, Illinois: Human Kinetics, pp. 175-221.

Boulay MR, Lortie G, Simoneau JA, Hamel P, Leblanc C and Bouchard C (1985) Specificity of aerobic and anaerobic work capacities and powers. *International Journal of Sports Medicine* 6:325-328.

Cheetham ME, Williams C and Lakomy HKA (1985) A laboratory running test: Metabolic responses of sprint and endurance trained athletes. *British Journal of Sports Medicine* 19:81-84.

Craig NP, Norton KI, Bourdon PC, Woolford SM, Stanef T and Conyers RAJ (1995) Influence of test duration when assessing the maximal accumulated oxygen deficit in high-performance track cyclists. *International Journal of Sports Medicine* 16:534-540.

Di Prampero PE, Boutellier U and Pietsch P (1983) Oxygen deficit and stores at onset of muscular exercise in humans. *Journal of Applied Physiology* 55:146-153.

Finn JP, Sainsbury DA and Withers RT (1996) A macro-driven Excel template for determining the anaerobic capacity using an air-braked ergometer. *International Journal of Clinical Monitoring and Computing* 13:179-189.

Foley MJ, McDonald KS, Green MA, Schrager M, Synder AC and Foster C (1991) Comparison of methods for estimation of anaerobic capacity. *Medicine and Science in Sports and Exercise* 23:202.

Foster C, Hector LL, McDonald KS and Synder AC (1995) Measurement of anaerobic power and capacity. In: Maud PJ and Foster C (eds) *Physiological Assessment of Human Fitness.* Champaign, Illinois: Human Kinetics.

Gastin PB (1994) Quantification of anaerobic capacity. *Scandinavian Journal of Medicine and Science in Sports* 4:91-112.

Gastin PB, Costill DL, Lawson DL, Krzeminski K and McConell GK (1995) Accumulated oxygen deficit during supramaximal all-out and constant intensity exercise. *Medicine and Science in Sports and Exercise* 27:255-263.

Gastin PB and Lawson DL (1994a) Influence of training status on maximal accumulated oxygen deficit during all-out cycle exercise. *European Journal of Applied Physiology* 69:321-330.

Gastin PB and Lawson DL (1994b) Variable resistance all-out test to generate accumulated oxygen deficit and predict anaerobic capacity. *European Journal of Applied Physiology* 69:331-336.

Gastin P, Lawson D, Hargreaves M, Carey M and Fairweather I (1991) Variable resistance loadings in anaerobic power testing. *International Journal of Sports Medicine* 12:513-518.

Gladden LB and Welch HG (1978) Efficiency of anaerobic work. *Journal of Applied Physiology* 44:564-578.

Goldspink G (1978) Energy turnover during contraction of different types of muscle. In: Asmussen E and Jorgensen K (eds) *Biomechanics VI-A.* Baltimore: University Park Press, pp. 27-39.

Graham KS and McLellan TM (1989) Variability of time to exhaustion and oxygen deficit in supramaximal exercise. *Australian Journal of Science and Medicine in Sport* 21:11-14.

Green S and Dawson B (1993) Measurement of anaerobic capacities in humans. Definitions, limitations and unsolved problems. *Sports Medicine* 15:312-327.

Green S and Dawson B (1995) The oxygen uptake-power regression in cyclists and untrained men: Implications for the accumulated oxygen deficit. *European Journal of Applied Physiology* 70:351-359.

Green S and Dawson BT (1996) Methodological effects on the $\dot{V}O_2$-power regression and the accumulated O_2 deficit. *Medicine and Science in Sports and Exercise* 28:392-397.

Green S, Dawson B, Goodman C and Carey MF (1996) Anaerobic ATP production and accumulated O_2 deficit in cyclists. *Medicine and Science in Sports and Exercise* 28:315-321.

Hermansen L and Medbø JI (1984) The relative significance of aerobic and anaerobic processes during maximal exercise of short duration. In: Marconnet P, Poortmans J and Hermansen L (eds) *Medicine and Sport Science vol 17: Physiological Chemistry of Training and Detraining.* Basel: Karger, pp. 56-67.

Inbar O, Bar-Or O and Skinner JS (1996) *The Wingate Anaerobic Test.* Champaign, Illinois: Human Kinetics.

Lakomy HKA (1984) An ergometer for measuring the power generated during sprinting. *Journal of Physiology* (London) 354:33P.

Lakomy HKA (1987) The use of a non-motorized treadmill for analysing sprint performance. *Ergonomics* 30:627-637.

McLean BD (1993) The relationship between frontal surface area and anthropometric parameters in racing cyclists. In: *Abstracts of the International Society of Biomechanics,* 14th Congress, Paris, pp. 856-857.

Medbø JI and Burgers S (1990) Effect of training on the anaerobic capacity. *Medicine and Science in Sports and Exercise* 22:501-507.

Medbø JI, Mohn AC, Tabata I, Bahr R, Vaage O and Sejersted OM (1988) Anaerobic capacity determined by maximal accumulated O_2 deficit. *Journal of Applied Physiology* 64:50-60.

Medbø JL and Sejersted OM (1985) Acid-base and electrolyte balance after exhausting exercise in endurance-trained and sprint-trained subjects. *Acta Physiologica Scandinavica* 125:97-109.

Medbø JL and Tabata I (1989) Relative importance of aerobic and anaerobic energy release during short-lasting bicycle exercise. *Journal of Applied Physiology* 67:1881-1886.

Morton DP and Gastin PB (1997) Effect of high intensity board training on upper body anaerobic capacity and short-lasting exercise performance. *Australian Journal of Medicine and Science in Sport* 29:17-21.

Nummela A and Rusko H (1995) Time course of anaerobic and aerobic energy expenditure during short-term exhaustive run in athletes. *International Journal of Sports Medicine* 16:522-527.

Olesen HL (1992) Accumulated oxygen deficit increases with inclination of uphill running. *Journal of Applied Physiology* 73:1130-1134.

Olesen HL, Raabo E, Bansbo J and Secher NH (1994) Maximal oxygen deficit of sprint and middle distance runners. *European Journal of Applied Physiology* 69:140-146.

Perez-Landaluce J, Fernandez B, Gonzalez V, Montolic MA, Gorostiaga E and Terrados N (1992) Anaerobic capacity in professional cyclists. *Second IOC World Congress on Sports Sciences.* Institut Nacional d'Educacio Fiscia de Catalunga, p. 231.

Poole PC, Ward SA and Whipp BJ (1990) The effects of training on the metabolic and respiratory profile of high intensity cycle ergometer exercise. *European Journal of Applied Physiology* 59:421-429.

Ramsbottom R, Nevill AM, Nevill ME, Newport S and Williams C (1994) Accumulated oxygen deficit and short-distance running performance. *Journal of Sports Sciences* 12:447-453.

Saltin B (1990) Anaerobic capacity: Past, present and prospective. In: Taylor AW, Gollnick PD, Green HJ, Ianuzzo CD, Noble EG, Métivier G and Sutton JR (eds) *Biochemistry of Exercise VII.* International Series on Sport Sciences vol 21. Champaign, Illinois: Human Kinetics, pp. 387-412.

Scott CB, Roby FB, Lohman TG and Bunt JC (1991) The maximally accumulated oxygen deficit as an indicator of anaerobic capacity. *Medicine and Science in Sports and Exercise* 23:618-624.

Serresse O, Ama PFM, Simoneau JA et al. (1989) Anaerobic performances of sedentary and trained subjects. *Canadian Journal of Sports Sciences* 4:46-52.

Spencer MR, Gastin PB and Payne WR (1996) Energy system contribution during 400 to 1500 metres running. *New Studies in Athletics* 11:59-65.

Telford RD (1982) Specific performance analysis with air-braked ergometers. *Journal of Sports Medicine and Physical Fitness* 22:349-357.

Vandewalle H, Pérès G and Monod H (1987) Standard anaerobic exercise tests. *Sports Medicine* 4:268-289.

Weyland PG, Cureton KJ, Conley DS, Sloniger MA and Liu YL (1994) Peak oxygen deficit predicts sprint and middle-distance track performance. *Medicine and Science in Sports and Exercise* 26:1174-1180.

Whipp BJ and Wasserman K (1969) Efficiency of muscular work. *Journal of Applied Physiology* 26:644-648.

Withers RT, Sherman WM, Clark DG, Esselbach PC, Nolan SR, Mackay MH and Brinkman M (1991) Muscle metabolism during 30, 60 and 90 s of maximal cycling on an air-braked ergometer. *European Journal of Applied Physiology* 63:354-362.

Withers RT, Van Der Ploeg G and Finn JP (1993) Oxygen deficits incurred during 45, 60, 75 and 90-s maximal cycling on an air-braked ergometer. *European Journal of Applied Physiology* 67:185-191.

Woods GF, Day L, Withers RT, Ilsley AH and Maxwell BF (1994) The dynamic calibration of cycle ergometers. *International Journal of Sports Medicine* 15:168-171.

Woolford SM, Withers RT, Craig NP, Bourdon PC, Stanef T and McKenzie, I (1999) Effect of pedal cadence on the accumulated oxygen deficit, maximal aerobic power and blood lactate transition thresholds of high-performance junior endurance cyclists. *European Journal of Applied Physiology* 80: 285-91.

Chapter 4

ADAPT (1995) [software program] Sport Sciences Division, Australian Institute of Sport.

Allen WK, Seals DR, Hurley BF, Ehsani AA, and Hagberg JM (1985) Lactate threshold and distance running performance in young and older endurance athletes. *Journal of Applied Physiology* 58:1281-1284.

Aunola S and Rusko H (1986) Aerobic and anaerobic thresholds determined from venous lactate or from ventilation and gas exchange in relation to muscle fibre composition. *International Journal of Sports Medicine* 7:161-166.

Beaver WL, Wasserman KJ and Whipp BJ (1985) Improved detection of lactate threshold during exercise using a log-log transformation. *Journal of Applied Physiology* 59:1936-1940.

Beneke R (1995) Anaerobic threshold, individual anaerobic threshold, and maximal lactate steady state in rowing. *Medicine and Science in Sports and Exercise* 27:863-867.

Beneke R and Petelin von Duvillard S (1996) Determination of maximal lactate steady state response in selected sports events. *Medicine and Science in Sports and Exercise* 28:241-246.

Bishop D, Jenkins DG and MacKinnon LT (1998) The relationship between plasma lactate parameters, Wpeak and 1-h cycling performance in women. *Medicine and Science in Sports and Exercise* 30: 1270-1275.

Blomstrand E, Bergh U, Essen-Gustavsson B et al. (1984) Influence of low muscle temperature on muscle metabolism during intense dynamic exercise. *Acta Physiologica Scandinavica* 120:229-236.

Borch KW, Ingjer F, Larsen S et al. (1993) Rate of accumulation of blood lactate during graded exercise as a predictor of anaerobic threshold. *Journal of Sports Sciences* 11:49-55.

Brettoni M, Alessandri F, Cupelli V et al. (1989) Anaerobic threshold in runners and cyclists. *Journal of Sports Medicine and Physical Fitness* 29:230-233.

Celsing F and Ekblom B (1986) Anaemia causes a relative decrease in blood lactate concentration during exercise. *European Journal of Applied Physiology* 55:74-78.

Cheng B, Kuipers H, Snyder A et al. (1992) A new approach for the determination of ventilatory and lactate thresholds. *International Journal of Sports Medicine* 13:518-522.

Coen B, Schwartz L, Urhausen A et al. (1991) Control of training in middle and long distance running by means of the individual anaerobic threshold. *International Journal of Sports Medicine* 12:519-524.

Conconi F, Ferrari M, Ziglio PG et al. (1982) Determination of the anaerobic threshold by a noninvasive field test in runners. *Journal of Applied Physiology* 52:869-873.

Coyle EF, Coggan AR, Hemmert MK, and Walters TJ (1984) Glycogen usage and performance relative to lactate threshold. *Medicine and Science in Sports and Exercise* 16:120-121.

Coyle EF, Hemmer MC and Coggan AR (1986) Effects of detraining on cardiovascular responses to exercise: Role of blood volume. *Journal of Applied Physiology* 60:95-99.

Coyle EF, Martin WH, Ehsani AA et al. (1983) Blood lactate threshold in some well-trained ischemic heart disease patients. *Journal of Applied Physiology* 54:18-23.

Craig NP (1987) The measurement of blood lactate to monitor and prescribe aerobic and anaerobic training: Concepts and controversies. *Proceedings: The Australian Sports Medicine Federation National Scientific Conference.* Adelaide: Australian Sports Medicine Federation, 118-135.

Craig NP, Norton KI, Bourdon PC et al. (1993) Aerobic and anaerobic indices contributing to track endurance cycling performance. *European Journal of Applied Physiology* 67:150-158.

Craig NP, Norton KI, Conyers RAJ et al. (1995) Influence of test duration and event specificity on maximal accumulated oxygen deficit on high performance track cyclists. *International Journal of Sports Medicine* 16:534-540.

Daniels JT, Yarborough RA and Foster C (1978) Changes in $\dot{V}O_2$max and running performance with training. *European Journal of Applied Physiology* 39:249-254.

Davies SC, Iber C, Keene SA et al. (1986) Effect of respiratory alkalosis during exercise on blood lactate. *Journal of Applied Physiology* 61:948-952.

Davis HA, Bassett J, Hughes P et al. (1983) Anaerobic threshold and lactate turnpoint. *European Journal of Applied Physiology* 50:383-392.

Davis JA, Frank MH, Whipp BJ et al. (1979) Anaerobic threshold alterations caused by endurance training in middle aged men. *Journal of Applied Physiology* 46:1039-1046.

Denis C, Dormis D and Lacour JR (1984) Endurance training, $\dot{V}O_2$max and OBLA: A longitudinal study of two different age groups. *International Journal of Sports Medicine* 5:167-173.

Denis C, Fouquet R, Poty P et al. (1982) Effect of 40 weeks of endurance training on the anaerobic threshold. *International Journal of Sports Medicine* 3:208-214.

Dotan R, Rotstein A and Grodjinovsky A (1989) Effect of training load on OBLA determination. *International Journal of Sports Medicine* 10:346-351.

Farrell PA, Wilmore JH, Coyle EF et al. (1979) Plasma lactate accumulation and distance running performance. *Medicine and Science in Sports* 11:338-344.

Fink WJ, Costill DL and Van Handel PJ (1975) Leg muscle metabolism during exercise in the heat and cold. *European Journal of Applied Physiology* 34:183-190.

Flore P, Therminarias A, Oddou-Chirpaz MF et al. (1992) Influence of moderate cold exposure on blood lactate during incremental exercise. *European Journal of Applied Physiology* 64:213-217.

Fohrenbach R, Mader A and Hollmann W (1987) Determination of endurance capacity and prediction of exercise intensities for training and competition in marathon runners. *International Journal of Sports Medicine* 8:11-18.

Foster C, Cohen J, Donovan K et al. (1993) Fixed time versus fixed distance protocols for the blood lactate profile in athletes. *International Journal of Sports Medicine* 14:264-268.

Foster C, Pollock ML, Farrell PA et al. (1982) Training responses of speed skaters during a competitive season. *Research Quarterly for Exercise and Sport* 53:243-246.

Foster C, Schrager M and Snyder AC (1995) Blood lactate and respiratory measurement of the capacity for sustained exercise. In: Maud PJ and Foster C (eds) *Physiological Assessment of Human Fitness.* Champaign, Illinois: Human Kinetics.

Foxdal P, Sjodin A, Ostman B et al. (1991) The effect of different blood sampling sites and analyses on the relationship between exercise intensity and 4.0 mmol/l blood lactate concentration. *European Journal of Applied Physiology* 63:52-54.

Foxdal P, Sjodin B, Rudstam H et al. (1990) Lactate concentration differences in plasma, whole blood, capillary finger blood and erythrocytes during submaximal graded exercise in humans. *European Journal of Applied Physiology* 61:218-222.

Foxdal P, Sjodin A and Sjodin B (1996) Comparison of blood lactate concentrations obtained during incremental and constant intensity exercise. *International Journal of Sports Medicine* 17:360-365.

Foxdal P, Sjodin B, Sjodin A et al. (1994) The validity and accuracy of blood lactate measurements for prediction of maximal endurance running capacity. Dependency of analysed blood media in combination with different designs of the exercise test. *International Journal of Sports Medicine* 15:89-95.

Freund H, Oyono-Euguelle S, Heitz A et al. (1989) Effect of exercise duration on lactate kinetics after short muscular exercise. *European Journal of Applied Physiology* 58:534-542.

Frohlich L, Urhausen A, Seul U et al. (1989) The influence of low-carbohydrate and high-carbohydrate diets on the individual anaerobic threshold. *Leistungssport* 19:18-20.

Gaesser GA and Rich RG (1985) Influence of caffeine on blood lactate response during incremental exercise. *International Journal of Sports Medicine* 6:207-211.

Genovely H and Stanford BA (1982) Effects of prolonged warm-up exercise above and below anaerobic threshold on maximal performance. *European Journal of Applied Physiology* 48:323-330.

Gollnick P (1982) Peripheral factors as limitations to exercise capacity. *Canadian Journal of Applied Sports Science* 7:14-21.

Gollnick PD, Bayly WM and Hodgson DR (1986) Exercise intensity, training, diet, and lactate concentration in muscle and blood. *Medicine and Science in Sports and Exercise* 18:334-340.

Gollnick PD and Saltin B (1982) Significance of skeletal muscle oxidative enzyme enhancement with endurance training. *Clinical Physiology* 2:1-12.

Green HJ, Hughson RL, Orr GW et al. (1983) Anaerobic threshold, blood lactate and muscle metabolites in progressive exercise. *Journal of Applied Physiology* 54:1032-1038.

Greenhaff PL, Gleeson M and Maughan RL (1987) The effects of dietary manipulation on blood acid-base status and the performance of high intensity exercise. *European Journal of Applied Physiology* 56:331-337.

Hagberg JM (1986) Physiological implications of the lactate threshold. *International Journal of Sports Medicine* 5 (suppl):106-109.

Hagberg JM and Coyle EF (1983) Physiological determinants of endurance performance such as studied in competitive racewalkers. *Medicine and Science in Sports and Exercise* 15:287-289.

Heck H, Mader A, Hess G et al. (1985) Justification of the 4 mmol/l lactate threshold. *International Journal of Sports Medicine* 6:117-130.

Heitkamp HC, Holdt M and Scheib K (1991) The reproducibility of the 4 mmol/l lactate threshold in trained and untrained women. *International Journal of Sports Medicine* 12:363-369.

Hughes EF, Turner SC and Brooks GA (1982) Effects of glycogen depletion and pedaling speed on anaerobic threshold. *Journal of Applied Physiology* 58:534-542.

Hurley BF, Hagberg JM, Allen WK et al. (1984) Effect of training on blood lactate levels during submaximal exercise. *Journal of Applied Physiology* 56:1260-1264.

Ivy JL, Costill DL, van Handel PJ et al. (1981) Alteration in the lactate threshold with changes in substrate availability. *International Journal of Sports Medicine* 2:139-142.

Ivy JL, Withers RT, Van Handel PJ et al. (1980) Muscle respiratory capacity and fibre type as determinants of the lactate threshold. *Journal of Applied Physiology* 48:523-527.

Jacobs I (1981) Lactate, muscle glycogen and exercise performance in man. *Acta Physiologica Scandinavica* (suppl 495):1-35.

Jacobs I (1986) Blood lactate: Implications for training and sports performance. *Sports Medicine* 3:10-25.

Jones NL and Ehrsour RE (1982) The anaerobic threshold. *Exercise and Sports Sciences Review* 10:49-53.

Jorfeldt L, Juhein-Dannfeldt A and Karlsson J (1978) Lactate release in relation to tissue lactate in human skeletal muscle during exercise. *Journal of Applied Physiology* 44:350-352.

Karlsson J and Jacobs I (1982) Onset of blood lactate accumulation during muscular exercise as a threshold concept: Theoretical considerations. *International Journal of Sports Medicine* 3:190-201.

Karlsson J, Nordesjo LO, Jorfeldt L and Saltin B (1972) Muscle lactate, ATP and CP levels during exercise after physical training in man. *Journal of Applied Physiology* 33:199-203.

Katch V, Weltman A, Sady S and Freedson P (1978) Validity of the relative percent concept for equating training intensity. *European Journal of Applied Physiology* 39:219-227.

Keith SP, Jacobs I and McLellan TM (1992) Adaptations to training at the individual anaerobic threshold. *European Journal of Applied Physiology* 65:316-323.

Keul J, Suison G, Berg A et al. (1979) Determination of the individual anaerobic threshold in the assessment of efficiency and in the designing of training. *Deutsche Zeitschrift für Sportmedizin* 7:212-218.

Kindermann W, Simon G and Keul J (1979) The significance of the aerobic-anaerobic transition for the determination of work load intensities during endurance training. *European Journal of Applied Physiology* 42:25-34.

Kowalchuk JM, Heigenhauser GJF and Jones NL (1984) Effect of pH on metabolic and cardiorespiratory responses during progressive exercise. *Journal of Applied Physiology* 57:1558-1563.

Kumagai S, Tanaka K, Matsuura Y et al. (1982) Relationships of the anaerobic threshold with the 5km, 10km and 10 mile races. *European Journal of Applied Physiology* 49:13-23.

LaFontaine TP, Londeree BR and Spath WK (1981) The maximal steady state versus selected running events. *Medicine and Science in Sports and Exercise* 13:190-192.

Londeree BR and Ames SA (1975) Maximal steady state versus state of conditioning. *European Journal of Applied Physiology* 34:269-278.

Lormes W, Steinmacker JM, Michalsky R et al. (1987) Comparison of the multi-stage test on a rowing ergometer in a racing shell (abstract). *International Journal of Sports Medicine* 8:165.

Maassen N and Busse MW (1989) The relationship between lactic acid and work load: A measure for endurance capacity or an indicator of carbohydrate deficiency? *European Journal of Applied Physiology* 58:728-737.

MacDougall JD, Reddan WG, Layton CR et al. (1974) Effects of metabolic hyperthermia on performance during heavy prolonged exercise. *Journal of Applied Physiology* 36:538-544.

MacRae HSH, Dennis SC, Bosch AN et al. (1992) Effects of training on lactate production and removal during progressive exercise in humans. *Journal of Applied Physiology* 72:1649-1656.

Mader A and Heck H (1986) A theory of the metabolic origin of "anaerobic threshold." *International Journal of Sports Medicine* 7:45-65.

Mader A, Heck H and Hollmann W (1976) Evaluation of lactic acid anaerobic energy contribution by determination of post-exercise lactic acid concentration in ear capillary blood in middle distance runners and swimmers. In: Landing F and Orban W (eds) *Exercise Physiology.* Florida: Symposia Specialists Inc.

Madsen O and Lohberg M (1987) The lowdown on lactates. *Swimming Technique* May-July:21-28.

McLellan TM (1987) The anaerobic threshold: Concept and controversy. *Australian Journal of Science and Medicine in Sport* 19:3-8.

McLellan TM, Cheung KSY and Jacobs I (1991) Incremental test protocol, recovery mode and the individual anaerobic threshold. *International Journal of Sports Medicine* 12:190-195.

Olbrecht J, Madsen O, Mader A et al. (1985) Relationship between swimming velocity and lactate concentration during continuous and intermittent training exercises. *International Journal of Sports Medicine* 6:74-77.

Oyono-Enguelle S, Heitz A, Marbach J et al. (1990) Blood lactate during constant-load exercise at aerobic and anaerobic thresholds. *European Journal of Applied Physiology* 60:321-330.

Poole DC and Gaesser GA (1985) Response of ventilatory and lactate thresholds to continuous and interval training. *Journal of Applied Physiology* 58:1115-1121.

Poole DC, Ward SA and Whipp BJ (1990) The effects of training on the metabolic and respiratory profile of high-intensity cycle ergometer exercise. *European Journal of Applied Physiology* 59:421-429.

Pyne DB (1989) The use and interpretation of blood lactate testing in swimming. *Excel* 5(4): 23-26.

Rieu M, Miladi J, Ferry A and Duvallet A (1989) Blood lactate during submaximal exercises. Comparison between intermittent incremental exercises and isolated exercises. *European Journal of Applied Physiology* 59:73-79.

Rusko H, Luhdtanen P, Rahkila P et al. (1986) Muscle metabolism, blood lactate and oxygen uptake in steady-state exercise at aerobic and anaerobic thresholds. *European Journal of Applied Physiology* 55:181-186.

Saltin B, Nazer K, Costill DL et al. (1976) The nature of training response: Peripheral and central adaptations to one-legged exercise. *Acta Physiologica Scandinavica* 96:289-305.

Saltin B and Rowell LB (1980) Functional adaptations to physical activity and inactivity. *Federation Proceedings* 39:1506-1513.

Schmidt P, Jacob E, Huber G et al. (1984) The behaviour of heart rate and blood levels in laboratory tests and field tests at cross-country skiing (abstract). *International Journal of Sports Medicine* 5:297.

Schnabel A, Kindermann W, Schmitt WM et al. (1982) Hormonal and metabolic consequences of prolonged running at the individual anaerobic threshold. *International Journal of Sports Medicine* 3: 163-168.

Sjodin B and Jacobs I (1981) Onset of blood lactate accumulation and marathon running performance. *International Journal of Sports Medicine* 2:23-26.

Sjodin B, Jacobs I and Karlsson J (1981) Onset of blood lactate accumulation and enzyme activities in m vastus lateralis in man. *International Journal of Sports Medicine* 2:166-170.

Sjodin B, Jacobs I and Svedenhag J (1982) Changes in onset of blood lactate accumulation (OBLA) and muscle enzymes after training at OBLA. *European Journal of Applied Physiology* 49:45-57.

Sjodin B and Svendenhag J (1985) Applied physiology of marathon running. *Sports Medicine* 2:83-99.

Skinner JS and McLellan TM (1980) The transition from aerobic to anaerobic metabolism. *Research Quarterly for Exercise and Sport* 51:234-248.

Stegmann H and Kindermann W (1982) Comparison of prolonged exercise tests at the individual anaerobic threshold and the fixed anaerobic threshold of 4 mmol/l lactate. *International Journal of Sports Medicine* 3:105-110.

Stegmann H, Kindermann W and Schnabel A (1981) Lactate kinetics and individual anaerobic threshold. *International Journal of Sports Medicine* 2:160-165.

Svendenhag J and Sjodin B (1985) Physiological characteristics of elite male runners in and off season. *Canadian Journal of Applied Sports Science* 10:127-133.

Tanaka K, Nakagawa T, Hazana T et al. (1985) A prediction equation for indirect assessment of anaerobic threshold in male distance runners. *European Journal of Applied Physiology* 54:386-390.

Tanaka K, Watanabe H, Konishi Y et al. (1986) Longitudinal associations between anaerobic threshold and distance running performance. *European Journal of Applied Physiology* 55:248-252.

Telford R (1984) Lactic acid measurements—are they useful? *Sports Science and Medicine Quarterly* 1:2-7.

Thoden JS (1991) Testing aerobic power. In: MacDougall JD, Wenger HA and Green HJ *Physiological Testing of the High-Performance Athlete.* Champaign, Illinois: Human Kinetics.

Urhausen A and Kindermann W (1992) Exercise physiology: Performance diagnostics and training control. In: Haag H, Grupe O and Kirsch A. *Sports Science in Germany.* Berlin: Springer-Verlag.

Welsman J (1992) Methodological problems of lactate testing. *Coaching Focus* 21:14-15.

Weltman A (1995) *The Blood Lactate Response to Exercise.* Champaign, Illinois: Human Kinetics.

Weltman A, Seip RL, Snead D et al. (1992) Exercise training at and above the lactate threshold in previously untrained women. *International Journal of Sports Medicine* 13:257-263.

Weltman A, Snead D, Seip R et al. (1987) Prediction of lactate threshold and fixed blood lactate concentrations from 3200m running performance in male runners. *International Journal of Sports Medicine* 8:401-406.

Weltman J, Snead D, Stein P et al. (1990) Reliability and validity of a continuous incremental treadmill protocol for the determination of lactate threshold, fixed blood lactate concentrations, and $\dot{V}O_2$max. *International Journal of Sports Medicine* 11:26-32.

Withers RT, Sherman WM, Miller JM et al. (1981) Specificity and anaerobic threshold in endurance trained cyclists and runners. *European Journal of Applied Physiology* 47:93-104.

Woolford SM, Withers RT, Craig NP, Bourdon PC, Stanef T, and McKenzie I (1999) Effect of pedal cadence on the accumulated oxygen deficit, maximal aerobic power and blood lactate transition thresholds of high performance junior endurance cyclists. *European Journal of Applied Physiology* 80:285-291.

Yoshida T (1984a) Effect of dietary modifications on lactate threshold and onset of blood lactate accumulation during incremental exercise. *European Journal of Applied Physiology* 53:200-205.

Yoshida T (1984b) Effect of exercise duration during incremental exercise on the determination of anaerobic threshold and the onset of blood lactate accumulation. *European Journal of Applied Physiology* 53:196-199.

Yoshida T (1986) Effect of dietary manipulation on anaerobic threshold. *Sports Medicine* 3:4-9.

Yoshida T, Childa M, Ichioka M et al. (1987) Blood lactate parameters related to aerobic capacity and endurance performance. *European Journal of Applied Physiology* 56:7-11.

Yoshida T, Suda Y and Takeuchi N (1982) Endurance training regimen based upon arterial blood lactate: Effects on anaerobic threshold. *European Journal of Applied Physiology* 49:223-230.

Yoshida T, Takeuchi N and Suda T (1982) Arterial versus venous blood lactate increase in the forearm during incremental bicycle exercise. *European Journal of Applied Physiology* 50:87-93.

Chapter 5

Baumgartner TA (1989) Norm-referenced measurement: Reliability. In: Safrit MJ and Woods TM (eds) *Measurement Concepts in Physical Education and Exercise Science.* Champaign, Illinois: Human Kinetics, pp. 45-72.

Dahlberg G (1940) Errors of estimation. In: Dahlberg G (ed) *Statistical Methods for Medical and Biological Students.* London: Allen & Unwin, pp. 122-132.

Knapp TR (1992) Technical error of measurement: A methodological critique. *American Journal of Physical Anthropology* 87:235-236.

Norton K, Whittingham N, Carter L, Kerr D and Gore C (1996) Measurement techniques in anthropometry. In Norton K and Olds T (eds) *Anthropometrica.* Sydney: University of New South Wales Press, pp. 25-75.

Pederson DG and Gore CJ (1996) Anthropometry measurement error. In Norton K and Olds T (eds) *Anthropometrica.* Sydney: University of New South Wales Press, pp. 77-96.

Utermohle CJ, Zegura SL and Heathcote GM (1983) Multiple observers, humidity and choice of precision statistics. Factors influencing craniometric data quality. *American Journal of Physical Anthropology* 61: 85-95.

Verducci FM (1980) Intraclass correlation coefficient (analysis of variance). In: Verducci FM (ed) *Measurement Concepts in Physical Education.* St Louis: Mosby, pp. 85-97.

Withers RT, Craig NP, Bourdon PC and Norton KI (1987) Relative body fat and anthropometric prediction of body density of male athletes. *European Journal of Applied Physiology* 56:191-200.

Woolford S, Bourdon P, Craig N and Stanef T (1993) Body composition and its effects on athletic performance. *Sports Coach* 16(4):24-30.

Chapter 6

Andersen R (1995) Practical consequences of blood sampling. *New Studies in Athletics* 10(3):31-34.

Australian National Council on AIDS (ANCA) (1990) *Laboratory safety guidelines that take account of HIV and other blood-borne agents.* Bulletin no. 3. Australian National Council on AIDS. Canberra, Australia.

Australian National Council on AIDS and Australian Sports Medicine Federation (1994) HIV and Sports. *Joint bulletin of the Australian National Council on AIDS and Australian Sports Medicine Federation.* Canberra: Australian Sports Medicine Federation.

Castle M (1980) *Hospital Infection Control: Principles and Practice.* New York: Wiley.

Dejonghe P and Parkinson B (1992) Benefits and costs of vaccination. *Vaccine* 10:936-939.

Ekblom B, Goldberg AN and Gulbring B (1972) Response to exercise after venesection and reinfusion of red blood cells. *Journal of Applied Physiology* 33:175-180.

El-Sayed MS, George KP, Wilkinson D, Mulan N, Fenoglio R, Flannigan J (1993) Fingertip and venous blood lactate concentration in response to graded treadmill exercise. *Journal of Sports Sciences* 11: 139-143.

Garza D and Becan-McBride K (1989) *Phlebotomy Handbook.* Norwalk, CT: Appleton & Lange.

Harrison MH (1985) Effects of thermal stress and exercise on blood volume in humans. *Physiological Reviews* 65(1):149-209.

Hill AV and Lupton H (1924) Muscular exercise, lactic acid, and the supply and utilization of oxygen. *Quarterly Journal of Medicine* 135-171.

Mader A, Liesen H, Heck H, Philippi H, Rost R, Schürch P, Hollman W (1976) Zur beurteilung der sportartspezifischen ausdauerleistungs fähigkeit im labor. *Sportarzt Sportmed* 27:80-112.

Mast ST, Woolwine JD and Gerberding JL (1993) Efficacy of gloves in reducing blood volumes transferred during simulated needlestick injury. *Journal of Infectious Diseases* 168:1589-1592.

Maw GJ, Mackenzie IL and Taylor NAS (1995) Redistribution of body fluids during postural manipulations. *Acta Physiologica Scandinavica* 155:157-163.

Munster DC (1993) Drugs in sport: Towards the use of blood samples in doping control? *Deutsche Zeitshrift für Sportmedizin* 44(19):18-21.

National Committee for Clinical Laboratory Standards (NCCLS) (1991) Protection of laboratory workers from infectious disease transmitted by blood, body fluids and tissues. National Committee for Clinical Laboratory Standards 11(14): 1-38.

National Health and Medical Research Council (NHMRC) and Australian National Council on AIDS (1996) *Infection Control in the Health Care Setting.* Canberra: AGPS.

New South Wales Department of Health (1987) *Code of Safe Practice in Clinical Laboratories.* Sydney: New South Wales Department of Health.

Pendergraph GE (1988) *Handbook of Phlebotomy.* Philadelphia: Lea & Febiger.

Todd JC (1908) *Manual of Clinical Diagnosis.* Philadelphia: Saunders.

Tokars JI, Marcus R, and Culver DH (1993) Surveillance of human immunodeficiency virus (HIV) infection and zidovudine use among health care workers with occupational exposure to HIV-infected blood. *Annals of Internal Medicine* 118:913-919.

Urhausen A and Kindermann W (1992) Biochemical monitoring of training. *Clinical Journal of Sport Medicine* 2:52-61.

World Health Organization (WHO) (1976) Biomedical research: A revised code of ethics (The Declaration of Helsinki). *World Health Organization Chronicle* 30:360-362.

World Health Organization (WHO) (1993) *Laboratory Biosafety Manual* (2nd ed). Geneva: World Health Organization.

Chapter 7

Agre JC and Baxter TL (1987) Musculoskeletal profile of male collegiate soccer players. *Archives of Physical Medicine and Rehabilitation* 68:147-150.

Bergh U (1980) Human power at subnormal body temperatures. *Acta Physiologica Scandinavica* 478 (suppl):1-39.

Blanch P (1997) *The Swimming Machine: Make the Most of Every Stroke by Being Flexible and Strong.* Dickson, Australian Capital Territory: Australian Swimming Inc.

Cameron D and Bohannon R (1993) Relationship between active knee extension and active straight leg raise test measurements. *Journal of Orthopaedic and Sports Physical Therapy* 17(5):257-260.

Cavagna G (1977) Storage and utilisation of elastic energy in skeletal muscle. *Exercise and Sports Sciences Reviews* 5:89-129.

Ciullo J and Zarins B (1983) Biomechanics of the musculotendinous unit: Relation to athletic performance and injury. *Clinical Journal of Sport Medicine* 2(1):71-86.

DeVries H (1963) The "looseness" factor in speed and O_2 consumption of an anaerobic 100 yard dash. *Research Quarterly for Exercise and Sport* 34:305-313.

DeVries H (1980) *Physiology of Exercise for Physical Education and Athletics.* Dubuque, Iowa: Brown.

Dickinson S (1929) The efficiency of bicycle pedalling as affected by speed and load. *Journal of Physiology* (London) 67:242-255.

Finch C (1995) *Preliminary Statistical Report of All Injuries Treated During the VicHealth 5th Australian Masters Games October 1995.* Monash University Accident Research Centre report.

Gajdosik R and Lusin G (1983) Hamstring muscle tightness: Reliability of an active-knee-extension test. *Physical Therapy* 63:1085-1090.

Garrett WE Jr (1996) Muscle strain injuries. *American Journal of Sports Medicine* 24(6):S2-S8.

Garrett WE Jr, Rich FR, Nikolaou PK, and Vogler JB (1989) Computed tomography of hamstring muscle strains. *Medicine and Science in Sports and Exercise* 21:506-514.

Gleim G, Stachenfeld N and Nicholas J (1990) The influence of flexibility on the economy of walking and jogging. *Journal of Orthopaedic Research* 8:814-823.

Godges J, MacRae H, Longdon C, Tinberg C and MacRae P (1989) The effects of two stretching procedures on hip range of motion and gait economy. *Journal of Orthopaedic and Sports Physical Therapy* 10(9):350-357.

Hennessy L and Watson AWS (1993) Flexibility and posture assessment in relation to hamstring injury. *British Journal of Sports Medicine* 27(4):243-246.

Hoare D (1997) Testing Data on Female Junior Netball Players. Unpublished.

Hortobagyi T, Faludi J, Tihanyi J and Merkley B (1985) Effects of intense "stretching"-flexibility training on the mechanical profile of the knee extensors and on the range of motion of the hip joint. *International Journal of Sports Medicine* 6(6):317-321.

Hutton RS (1992) Neuromuscular basis of stretching exercises. In: Komi PV (ed) *The Encyclopedia of Sports Medicine: Strength and Power in Sport.* London: Blackwell Scientific, 2C:29-39.

Jackson AW and Baker AA (1986) The relationship of the sit and reach test to criterion measures of hamstring and back flexibility in young females. *Research Quarterly for Exercise and Sport* 57(3): 183-186.

Jones BH (1997) The role of medical surveillance and research in army injury prevention. American College of Sports Medicine Conference abstract, Denver.

Kane Y and Bernasconi J (1992) Analysis of a modified active knee extension test. *Journal of Orthopaedic and Sports Physical Therapy* 15(3):141-146.

Kapandji IA (1970) *The Physiology of the Joints* (2nd ed). New York: Churchill Livingstone, pp. 66-68.

Knapik JJ, Jones BH and Harris JM (1992) Strength, flexibility and athletic injuries. *Sports Medicine* 14(5):277-288.

Krivickas LS and Feinberg JH (1996) Lower extremity injuries in college athletes: Relation between ligamentous laxity and lower extremity muscle tightness. *Archives of Physical Medicine and Rehabilitation* 77(11):1139-43.

Loudon JK, Jenkins W and Loudon KL (1996) The relationship between posture and ACL injury in female athletes. *Journal of Orthopaedic and Sports Physical Therapy* 24(2):91-97.

Lysens RJ, Ostyn MS, Auweele YU, Lefevre J, Vuylsteke M, and Renson L (1989) The accident-prone and overuse prone profiles of the young athlete. *American Journal of Sports Medicine* 17(5):612-619.

Norkin C and Levangie P (1992) *Joint Structure and Function: A Comprehensive Analysis* (2nd ed). Philadelphia: Davis, p. 105.

Pieper HG and Schulte A (1996) Muscular imbalances in elite swimmers and their relation to typical sports lesions. *Sports Exercise and Injury* 2(2):96-99.

Reid DC, Burnham RS, Saboe LA and Kushner SF (1987) Lower extremity flexibility patterns in classical ballet dancers and their correlation to lateral hip and knee injuries. *American Journal of Sports Medicine* 15(4):347-352.

Riegger-Krugh C and Keysor JJ (1996) Skeletal malalignments of the lower quarter: Correlated and compensatory motions and postures. *Journal of Orthopaedic and Sports Physical Therapy* 23(2): 164-170.

Safran MR, Seaber AU and Garrett NE (1989) Warm up and muscular injury prevention. An update. *Sports Medicine* 8(4):239-249.

Shellock F and Prentice W (1985) Warming-up and stretching for improved physical performance and prevention of sports-related injuries. *Sports Medicine* 2:267-278.

Shorten M (1987) Muscle elasticity and human performance. In: Gheluwe B van and Atha J (eds) *Current Research in Sports Biomechanics.* Basel: Karger, pp. 1-18.

Sinclair A and Tester G (1993) The sit and reach test—What does it actually measure? *ACHPER National Journal* 2:8-13.

Stanish WD, Curwin SL and Bryson G (1990) The use of flexibility exercises in preventing and treating sports injuries. In: *Sports Induced Inflammation: Clinical and Basic Science Concepts.* Park Ridge, Illinois: American College of Orthopaedic Surgeons, pp. 731-745.

van Mechelen W, Hlobil H and Kemper HGG (1992) Incidence, severity, aetiology and prevention of sports related injuries. *Sports Medicine* 14(2):82-89.

Wilkinson A (1992) Stretching the truth. A review of the literature on muscle stretching. *Australian Journal of Physiotherapy* 38(4):283-287.

Wilson G, Elliott B and Wood G (1991a) Performance benefits through flexibility training. *Sports Coach* 14(2):7-10.

Wilson G, Elliott B and Wood G (1992) Stretch shorten cycle performance enhancement through flexibility training. *Medicine and Science in Sports and Exercise* 24(1):116-123.

Wilson G, Wood G and Elliott B (1991b) The relationship between stiffness of the musculature and static flexibility: An alternative explanation for the occurrence of muscular injury. *International Journal of Sports Medicine* 12(4):403-407.

Worrell T, Smith T and Winegardner J (1994) Effects of hamstring stretching on hamstring muscle performance. *Journal of Orthopaedic and Sports Physical Therapy* 20(3):154-159.

Yamashita T, Ishii S and Oota I (1993) Effect of muscle stretching on the activity of neuromuscular transmission. *Medicine and Science in Sports and Exercise* 24:80-84.

Chapter 8

American College of Sports Medicine Position Stand (1990) The recommended quantity and quality of exercise for developing and maintaining cardiorespiratory and muscular fitness in healthy adults. *Medicine and Science in Sports and Exercise* 22: 265-274.

Åstrand PO and Rodahl K (1986) *Textbook of Work Physiology* (3rd ed). New York: McGraw-Hill, p. 366.

Beck, HV (1970) *Displacement Gas Meters: Operating, Testing and Repairing (Handbook E-4).* Philadelphia: American Meter Company.

Bouchard C, Lesage R, Lortie G et al. (1986) Aerobic performance in brothers, dizygotic and monozygotic twins. *Medicine and Science in Sports and Exercise* 18:639-646.

Buchfuhrer MJ, Hansen JE, Robinson TE et al. (1983) Optimizing the exercise protocol for cardiopulmonary assessment. *Journal of Applied Physiology: Respiratory, Environmental and Exercise Physiology* 55:1558-1564.

Cumming GR and Alexander WD (1968) The calibration of bicycle ergometers. *Canadian Journal of Physiology and Pharmacology* 46:917-919.

Dahlberg G (1940) *Statistical Methods for Medical and Biological Students.* London: Allen Unwin, pp. 122-132.

Diem K and Lentner C (eds) (1974) *Scientific Tables* (7th ed). Basel: Ciba-Geigy Ltd.

Douglas CG (1911) A method for determining the total respiratory exchange in man. *Journal of Physiology* (London) 42:xvii-xviii.

Gardner RM (1979) American Thoracic Society Statement—Snowbird workshop on standardisation of spirometry. *American Review of Respiratory Disease* 119:831-838.

Geppert J and Zuntz N (1888) Ueber die Regulation der Athmung. *Pflüegers Archives* 42:189-245.

Gore CJ, Catcheside PG, French SN, Bennett JM and Laforgia J (1997) Automated $\dot{V}O_2$max calibrator for open-circuit indirect calorimetry systems. *Medicine and Science in Sports and Exercise* 29(8):1095-1103.

Haldane JS (1912) *Methods of Air Analysis*. London: Charles Griffin, p. 56.

Hart JD and Withers RT (1996) The calibration of gas volume measuring devices at continuous and pulsatile flows. *Australian Journal of Science and Medicine in Sport* 28:61-65.

Hart JD, Withers RT and Ilsley AH (1992) The accuracy of dry gas meters at continuous and sinusoidal flows. *European Respiratory Journal* 5:1146-1149.

Hart JD, Withers RT and Tucker RC (1994) Precision and accuracy of Morgan ventilometers at continuous and sinusoidal flows. *European Respiratory Journal* 7:813-816.

Howley ET, Bassett Jr DR and Welch HG (1995) Criteria for maximal oxygen uptake: Review and commentary. *Medicine and Science in Sports and Exercise* 27:1292-1301.

Huszczuk A, Whipp BJ and Wasserman K (1990) A respiratory gas exchange simulator for routine calibration in metabolic studies. *European Respiratory Journal* 3:465-468.

Jones NL (1997) *Clinical Exercise Testing* (4th ed). Sydney: Saunders, p. 226.

Katch VL, Sady SS and Freedson P (1982) Biological variability in maximum aerobic power. *Medicine and Science in Sports and Exercise* 14:21-25.

National Association of Testing Authorities (1988) *The In-situ Calibration of Barometers*. Technical Note 8. Sydney: NATA Public Affairs.

Nevill AM, Ramsbottom R and Williams C (1992) Scaling physiological measurements for individuals of different body size. *European Journal of Applied Physiology and Occupational Physiology* 65:110-117.

Poole DC and Whipp BJ (1988) [Letter to the editor]. *Medicine and Science in Sports and Exercise* 20:420-421.

Porszasz J, Barstow TJ and Wasserman K (1994) Evaluation of a symmetrically disposed Pitot tube flowmeter for measuring gas flow during exercise. *Journal of Applied Physiology* 77:2659-2665.

Rowell LB (1974) Human cardiovascular adjustments to exercise and thermal stress. *Physiological Reviews* 54:75-159.

Russell JC and Dale JD (1986) Dynamic torquemeter calibration of bicycle ergometers. *Journal of Applied Physiology* 61:1217-1220.

Sainsbury DA, Gore CJ, Withers RT and Ilsley AH (1988) An on-line microcomputer program for the monitoring of physiological variables during rest and exercise. *Computers in Biology and Medicine* 18:17-24.

Taylor HL, Buskirk E and Henschel A (1955) Maximal oxygen intake as an objective measure of cardiorespiratory performance. *Journal of Applied Physiology* 8:73-80.

Telford RD, Hooper LA and Chennells MHD (1980) Calibration and comparison of air-braked and mechanically-braked bicycle ergometers. *Australian Journal of Sports Medicine* 12:40-46.

Wasserman K, Hansen JE, Sue DY, Whipp BJ and Casaburi R (1994) *Principles of Exercise Testing and Interpretation* (2nd ed). Philadelphia: Lea & Febiger.

Wilmore JH and Costill DL (1974) Semiautomated systems approach to the assessment of oxygen uptake during exercise. *Journal of Applied Physiology* 36:618-620.

Woods GF, Day L, Withers RT, Ilsley AH and Maxwell BF (1994) The dynamic calibration of cycle ergometers. *International Journal of Sports Medicine* 15:168-171.

Chapter 9

Australian Coaching Council (1998) *20m Shuttle Run Test*. Belconnen, Australian Capital Territory: Australian Coaching Council.

Brewer J, Ramsbottom R and Williams C (1988) *Multistage Fitness Test*. Belconnen, Australian Capital Territory: Australian Coaching Council.

Draper J and Pyke F (1988) Turning speed: A valuable asset in cricket run making. *Sports Coach* April-June:30-31.

Draper JA and Lancaster MG (1985) The 505 test: A test for agility in the horizontal plane. *Australian Journal of Science and Medicine in Sport* 17(1):15-18.

Fitzsimons M, Dawson B, Ward D and Wilkinson A (1993) Cycling and running tests of repeated sprint ability. *Australian Journal of Science and Medicine in Sport* 25(4):82-87.

Hubley-Kozey CL (1991) Testing flexibility. In: MacDougall JD, Wenger HA and Green HJ (eds) *Physiological Testing of the High Performance Athlete*. Champaign, Illinois: Human Kinetics, p. 309.

Leger LA and Lambert J (1982) A maximal multistage 20 m shuttle run test to predict $\dot{V}O_2$max. *European Journal of Applied Physiology* 49:1-5.

Luhtanen P and Komi PV (1978) Segmental contribution to forces in vertical jumping. *European Journal of Applied Physiology* 38:181-188.

Narita S and Anderson T (1992) Effects of upper body strength training on vertical jumping ability of high school volleyball players. *Sports Medicine, Training and Rehabilitation* 3:34.

Telford RD, Minikin BR, Hahn AG and Hopper LA (1989) A simple method for assessment of general fitness: The tri-level profile. *Australian Journal of Science and Medicine in Sport* 21(3):6-9.

Ward (1991) Laboratory Test of Repeated Effort Ability and Its Relationship to Aerobic Power, Anaerobic Power and Anaerobic Capacity. Honors thesis. University of Western Australia, Department of Human Movement.

Young W (1994) A simple method for evaluating the strength qualities of the leg extensor muscles and jumping abilities. *Strength and Conditioning Coach* 2:5-8.

Chapter 10

Abe T, Kawakami Y, Ikegawa S, Kanehisa H and Fukunaga T (1992) Isometric and isokinetic knee joint performance in Japanese alpine ski racers. *Journal of Sports Medicine and Physical Fitness* 31:353-357.

Abernethy P, Wilson G and Logan P (1995) Strength and power assessment: Issues, controversies and challenges. *Sports Medicine* 19:401-417.

Abernethy PJ, Howard A and Quigley BM (1996) Isokinetic torque and instantaneous power data do not necessarily mirror one another. *Journal of Strength and Conditioning Research* 10:220-223.

Abernethy PJ and Jürimäe J (1996) Cross-sectional and longitudinal uses of isoinertial, isometric, and isokinetic dynamometry. *Medicine and Science in Sports and Exercise* 28:1180-1187.

Abernethy PJ, Jürimäe J, Logan PA, Taylor AW and Thayer RE (1994) Acute and chronic responses of skeletal muscle to resistance exercise. *Sports Medicine* 17:22-38.

Aura O and Viitasalo JT (1989) Biomechanical characteristics of jumping. *International Journal of Sport Biomechanics* 5:89-98.

Baker D, Wilson G and Carlyon B (1994) Generality versus specificity: A comparison of dynamic and isometric measures of strength and speed-strength. *European Journal of Applied Physiology* 68:350-355.

Fry AC, Kraemer WJ, Weseman CA et al. (1991) The effects of an offseason strength and conditioning program on starters and non-starters in women's intercollegiate volleyball. *Journal of Applied Sports Science Research* 5:174-181.

Hortobagyi T, Katch FI and LaChance PF (1989) Interrelations among various measures of upper body strength assessed by different contraction modes. Evidence for a general strength development. *European Journal of Applied Physiology* 58:749-755.

Huijing PA (1992) Elastic potential of muscle. In: Komi PV (ed) *Strength and Power in Sport*. Oxford: Blackwell Scientific.

Jaric S, Ristanovic D and Corcos M (1989) The relationship between muscle kinetic parameters and kinematic variables in a complex movement. *European Journal of Applied Physiology* 59:370-376.

Komi PV, Suominen H, Heikkinen E, Karlsson J and Tesch P (1982) Effects of heavy resistance training and explosive type strength training methods on mechanical, functional and metabolic aspects of performance. In: Komi PV (ed) *Exercise and Sports Biology*. Champaign, Illinois: Human Kinetics, pp. 90-102.

Marshall RN, Mazur SM and Taylor NAS (1990) Three-dimensional surfaces for human muscle kinetics. *European Journal of Applied Physiology* 61:263-270.

Marshall RN and Taylor NAS (1990) The skeletal muscle force-velocity relationships: 1. Its significance and its measurement. *New Zealand Journal of Sports Medicine* 18(1):8-10.

Murphy AJ, Wilson GJ and Pryor JF (1994) The use of the isoinertial force mass relationship in the prediction of dynamic human performance. *European Journal of Applied Physiology* 69:250-257.

Pearson DR and Costill DL (1988) The effects of constant external resistance exercise and isokinetic training on work-induced hypertrophy. *Journal of Applied Sports Science Research* 2:39-41.

Pryor JF (1995) Rate of Force Development Assessment of the Trained Athlete: A Comparison of Testing Modalities and Force-Time Quantification Methods. Unpublished Masters of Health Science thesis, Southern Cross University.

Sale DG (1991) Testing strength and power. In: MacDougall JD, Wenger HA and Green HJ (eds) *Physiological Testing of the High-performance Athlete*. Champaign, Illinois: Human Kinetics, pp. 21-103.

Sale DG (1992) Neural adaptation to strength training. In: Komi PV (ed) *Strength and Power in Sport*. Oxford: Blackwell Scientific.

Schmidtbleicher D (1992) Training for power events. In: Komi PV (ed) *Strength and Power in Sport*. Oxford: Blackwell Scientific.

Taylor NAS, Sanders RH, Howick EI and Stanley SN (1991) Static and dynamic assessment of the biodex dynamometer. *European Journal of Applied Physiology* 62:3:180-188.

Thomas JR and Nelson JK (1990) *Research Methods in Physical Activity* (2nd ed). Champaign, Illinois: Human Kinetics.

Young WB and Bilby GE (1993) The effect of voluntary effort to influence speed of contraction on strength, muscular power and hypertrophy development. *Journal of Strength and Conditioning Research* 7:172-178.

Chapter 11

Abe T, Kawakami Y, Ikegawa S, Kanehisa H and Fukunaga T (1992) Isometric and isokinetic knee joint performance in Japanese alpine ski racers. *Journal of Sports Medicine and Physical Fitness* 32:353-357.

Asmussen E and Bonde-Petersen F (1974) Storage of elastic energy in skeletal muscles in man. *Acta Physiologica Scandinavica* 91:385-392.

Atha J (1981) Strengthening muscle. In: Miller DI (ed) *Exercise and Sport Science Reviews* vol 9. Philadelphia: Franklin Institute Press, pp. 1-73.

Baker D, Wilson G and Carlyon B (1994) Generality versus specificity: A comparison of dynamic and isometric measures of strength and speed-strength. *European Journal of Applied Physiology* 68(4):350-355.

Bemben MG, Massey BH, Boileau RA and Misner JE (1992) Reliability of isometric force-time curve parameters for men aged 20 to 79 years. *Journal of Applied Sports Science Research* 6(3):158-164.

Bloomfield J, Blanksby BA, Ackland TR and Allison GT (1990) The influence of strength training on overhead throwing velocity of elite water polo players. *Australian Journal of Science and Medicine in Sport* 22:63-67.

Christ CB, Slaughter MH, Stillman RJ, Cameron J and Boileau RA (1994) Reliability of selected parameters of isometric muscle function associated with testing 3 days × 3 trials in women. *Journal of Strength and Conditioning Research* 8(2):65-71.

Considine W and Sullivan W (1973) Relationship of selected tests of leg strength and leg power on college men. *Research Quarterly* 44:404-415.

Costill D, Miller S, Myers W, Kehoc F and Hoffman W (1968) Relationship among selected tests of explosive strength and power. *Research Quarterly* 39:785-787.

Fry AC, Kraemer WJ, van Borselen F et al. (1994) Performance decrements with high-intensity resistance exercise overtraining. *Medicine and Science in Sports and Exercise* 26(9):1165-1173.

Fry AC, Kraemer WJ, Weseman CA et al. (1991) Effects of an off-season strength and conditioning program on starters and non-starters in women's collegiate volleyball. *Journal of Applied Sports Science Research* 5:174-181.

Hakkinen K, Alen M and Komi PV (1985) Electromyographic and muscle fibre characteristics of human skeletal muscle during strength training and detraining. *Acta Physiologica Scandinavica* 125:573-585.

Hakkinen K and Komi PV (1986) Training-induced changes in neuromuscular performance under voluntary and reflex conditions. *European Journal of Applied Physiology* 55:147-155.

Hakkinen K, Komi PV and Kauhanen H (1986) Electromyographic and force production characteristics of leg extensor muscles of elite weightlifters during isometric, concentric and various stretch-shortening cycle exercises. *International Journal of Sports Medicine* 7:144-151.

Hortobagyi T and Lambert NJ (1992) Influence of electrical stimulation on dynamic forces of the arm flexors in strength-trained and untrained men. *Scandinavian Journal of Medicine and Science in Sport* 2:70-75.

Jaric S, Ristanovic D and Corcos DM (1989) The relationship between muscle kinematic parameters and kinematic variables in a complex movement. *European Journal of Applied Physiology* 59:370-376.

Komi PV and Bosco C (1978) Utilization of stored elastic energy in leg extensor muscles by men and women. *Medicine and Science in Sports* 10:261-265.

Mero A, Luhtanen P, Viitasalo JT and Komi PV (1981) Relationship between maximal running velocity, muscle fibre characteristics, force production and force relaxation of sprinters. *Scandinavian Journal of Sports Science* 3:16-22.

Murphy AJ and Wilson GJ (1996) Poor correlations between isometric test and dynamic performance: Relation to muscle activation. *European Journal of Applied Physiology* 73:353-357.

Murphy AJ, Wilson GJ and Pryor JF (1994) The use of the isoinertial force mass relationship in the prediction of dynamic human performance. *European Journal of Applied Physiology* 69(3):250-257.

Nakazawa K, Kawakami Y, Fukunaga T, Yano H and Miyashita M (1993) Differences in activation patterns in elbow flexors during isometric, concentric and eccentric contractions. *European Journal of Applied Physiology* 66:214-220.

Pryor JF, Wilson GJ and Murphy AJ (1994) The effectiveness of eccentric, concentric and isometric rate of force development tests. *Journal of Human Movement Studies* 27:153-172.

Rasch P (1957) Relationship between maximum isometric tension and maximum isotonic elbow flexion. *Research Quarterly* 28:85.

Sale D and Norman R (1982) Testing strength and power. In: MacDougall J, Wenger H and Green H (eds) *Physiological Testing of the Elite Athlete*. New York: Mouvement, pp. 7-34.

Schmidtbleicher D and Buehrle M (1987) Neuronal adaptations and increase of cross-sectional area studying different strength training methods. In: Johnson GB (ed) *Biomechanics X-B vol 6-B*. Champaign, Illinois: Human Kinetics, pp. 615-620.

Strauss D (1991) Force-time and electromyographical characteristics of arm shoulder muscles in explosive type force production in sprint swimmers. *Journal of Swimming Research* 7(1):19-27.

Tax AAM, Denier van der Gon JJ and Erkelens CJ (1990) Differences in coordination of elbow flexor muscles in force tasks and in movement tasks. *Experimental Brain Research* 81:567-572.

Ter Haar Romeny B, Denier van der Gon J and Gielen C (1982) Changes in recruitment order of motor units in the human biceps muscle. *Experimental Neurology* 78:360-368.

Ter Haar Romeny BM, Denier van der Gon JJ and Gielen CAM (1984) Relationship between location of a motor unit in the human biceps brachii and its critical firing levels for different tasks. *Experimental Neurology* 85:631-650.

Tidow G (1990) Aspects of strength training in athletics. *New Studies in Athletics* 1:93-110.

Viitasalo JT, Hakkinen K and Komi PV (1981) Isometric and dynamic force production and muscle fibre composition in man. *Journal of Human Movement Studies* 7:199-209.

Viitasalo JT, Saukkonen S and Komi PV (1980) Reproducibility of measurements of selected neuromuscular performance variables in man. *Electromyography and Clinical Neurophysiology* 20:487-501.

Wilson GJ, Lyttle AD, Ostrowski KJ and Murphy AJ (1995) Assessing dynamic performance: A comparison of rate of force development tests. *Journal of Strength and Conditioning Research* 9(3):176-181.

Wilson GJ and Murphy AJ (1995) The efficacy of isokinetic, isometric and vertical jump tests in exercise science. *Australian Journal of Science and Medicine in Sport* 27(1):62-66.

Wilson GJ, Murphy AJ and Pryor JF (1994) Musculo-tendinous stiffness: Its relationship to eccentric, isometric and concentric performance. *Journal of Applied Physiology* 76(6):2714-2719.

Wilson GJ, Newton RU, Murphy AJ and Humphries BJ (1993) The optimal training load for the development of dynamic athletic performance. *Medicine and Science in Sports and Exercise* 25: 1279-1286.

Wilson GJ, Wood GA and Elliott BC (1991) Optimal stiffness of the series elastic component in a stretch shorten cycle activity. *Journal of Applied Physiology* 70:825-833.

Young WB and Bilby GE (1993) The effect of voluntary effort to influence speed of contraction on strength, muscular power and hypertrophy development. *Journal of Strength and Conditioning Research* 7:172-178.

Chapter 12

Aagaard P, Simonsen EB, Trolle M, Bangsbo J and Klausen K (1994) Moment and power generation during maximal knee extensions performed at low and high speeds. *European Journal of Applied Physiology* 69:376-381.

Aagaard P, Simonsen EB, Trolle M, Bangsbo J and Klausen K (1995) Isokinetic hamstring/quadriceps strength ratio: Influence from joint angular velocity, gravity correction and contraction mode. *Acta Physiologica Scandinavica* 154:421-427.

Abe T, Kawakami Y, Ikegawa S, Kanehisa H and Fukunaga T (1992) Isometric and isokinetic knee joint performance in Japanese alpine ski racers. *Journal of Sports Medicine and Physical Fitness* 32(4):353-357.

Agre JC and Baxter TL (1987) Musculoskeletal profile of male collegiate soccer players. *Archives of Physical Medicine and Rehabilitation* 68:147-150.

Alderink GJ and Kuck DJ (1986) Isokinetic shoulder strength of high school and college-aged pitchers. *Journal of Orthopaedic and Sports Physical Therapy* 7:163-172.

Appen L and Duncan PW (1986) Strength relationship of the knee musculature: Effects of gravity and sport. *Journal of Orthopaedic and Sports Physical Therapy* 7(5):232-235.

Appling SA and Weiss LW (1993) The association of jump performance with quadriceps muscle function and body composition in women (abstract). *Physical Therapy* 73(6):S13.

Armstrong N and Welsman J (1994) Assessment and interpretation of aerobic fitness in children and adolescents. In: Holloszy JO (ed) *Exercise and Sports Sciences Reviews* vol 22. Indianapolis: American College of Sports Medicine, pp. 435-476.

Arnold BL and Perrin DH (1993) The reliability of four different methods of calculating quadriceps peak torque and angle-specific torques at 30°, 60° and 75°. *Journal of Sport Rehabilitation* 2:243-250.

Arrigo CA, Wilk KE and Andrews JR (1994) Peak torque and maximum work repetition during isokinetic testing of the shoulder internal and external rotators. *Isokinetics and Exercise Science* 4(4):171-175.

Ashley CD and Weiss LW (1994) Vertical jump performance and selected physiological characteristics of women. *Journal of Strength and Conditioning Research* 8:5-11.

Åstrand P-O and Rodahl K (1977) Body dimensions and muscular work. In: *Textbook of Work Physiology*. New York: McGraw-Hill, pp. 369-388.

Baltzopoulos V, Williams JG and Brodie DE (1991) Sources of error in isokinetic dynamometry: Effects of visual feedback on maximum torque measurements. *Journal of Orthopaedic and Sports Physical Therapy* 13:138-142.

Bandy WD and McLaughlin S (1993) Intramachine and intermachine reliability for selected dynamic muscle performance tests. *Journal of Orthopaedic and Sports Physical Therapy* 18(5):609-613.

Barbee J and Landis D (1984) Reliability of Cybex computer measures (abstract). *Physical Therapy* 64(5):737.

Barnes WS (1981) Isokinetic fatigue curves at different contractile velocities. *Archives of Physical Medicine and Rehabilitation* 62:66-69.

Bartonietz K (1994) Das Drehmoment—der grundlegende Parameter bei isokinetischen Tests: Hintergrunwissen für die kranken-gymnastische Praxis. *Krankengymnastik* 46:1646-1660.

Baumgartner TA (1989) Norm-referenced measurement: Reliability. In: Safrit MJ and Wood TM (eds) *Measurement Concepts in Physical Education and Exercise Science*. Champaign, Illinois: Human Kinetics, pp. 45-72.

Baumgartner TA and Jackson AS (1991) Validity. In: *Measurement for Evaluation in Physical Education and Exercise Science*. Dubuque, Iowa: Brown, pp. 149-162.

Beimborn DS and Morrissey MC (1988) A review of the literature related to trunk muscle performance. *Spine* 13(6):655-660.

Berg K, Miller M and Stephens L (1986) Determinants of 30 meter sprint time in pubescent males. *Journal of Sports Medicine and Physical Fitness* 26:225-231.

Bobbert MF and Harlaar J (1992) Evaluation of moment-angle curves in isokinetic knee extension. *Medicine and Science in Sports and Exercise* 25(2): 251-259.

Bond V, Gresham K, McRae J and Tearney RJ (1986) Caffeine ingestion and isokinetic strength. *British Journal of Sports Medicine* 20(3):135-137.

Brady EC, O'Regan M and McCormack B (1993) Isokinetic assessment of uninjured soccer players. In: Reilly T, Clarys J and Stibbe A (eds) *Science and Football II.* London: Spon, pp. 351-356.

Brett S (1992) Comparisons of peak torque and angular data derived from Cybex II and Kin-Com dynamometers. *PELOPS* 7:32-38 (Dept of Human Movement, Royal Melbourne Institute of Technology, Victoria, Australia).

Brown LE, Whitehurst M, Findley BW, Gilbert R and Buchalter DN (1995) Isokinetic load range during shoulder rotation exercise in elite male junior tennis players. *Journal of Strength and Conditioning Research* 9(3):160-164.

Brown LP, Niehues SL, Harrah A, Yavorksy P and Hirshman HP (1988) Upper extremity range of motion and isokinetic strength of the internal and external shoulder rotators in major league baseball pitchers. *American Journal of Sports Medicine* 16:577-585.

Brown SL and Wilkinson JG (1983) Characteristics of national, divisional and club male alpine ski racers. *Medicine and Science in Sports and Exercise* 15(6):491-495.

Burdett RG and Van Swearingen J (1987) Reliability of isokinetic muscle endurance tests. *Journal of Orthopaedic and Sports Physical Therapy* 8(10):484-488.

Byl NN and Sadowski SH (1993) Intersite reliability of repeated isokinetic measurements: Cybex back systems including trunk rotation, trunk extension-flexion and Liftask. *Isokinetics and Exercise Science* 3:139-147.

Byl NN, Wells L, Grady D, Friedlander A and Sadowski S (1991) Consistency of repeated isokinetic testing: Effect of different examiners, sites and protocol. *Isokinetics and Exercise Science* 1:122-130.

Cabri J, DeProft E, Dufour W and Clarys JP (1988) The relation between muscular strength and kick performance. In: Reilly T, Lees A, Davids K and Murphy WJ (eds) *Science and Football.* London: Spon, pp. 186-193.

Caiozzo VJ, Barnes WS, Prietto CA, McMaster CA and McMaster WC (1981) The effect of isometric precontractions on the slow velocity-high force region of the in vivo force-velocity relationship (abstract). *Medicine and Science in Sports and Exercise* 13:128.

Caiozzo VJ, Laird T, Chow K, Prietto CA and McMaster WC (1982) The use of precontractions to enhance the in vivo force-velocity relationship (abstract). *Medicine and Science in Sports and Exercise* 14: 162.

Callister R, Callister RJ, Fleck SJ and Dudley GA (1990) Physiological and performance responses to overtraining in elite judo athletes. *Medicine and Science in Sports and Exercise* 22:816-824.

Capranica L, Cama G, Fanton F, Tessitore A and Figura F (1992) Force and power of preferred and non-preferred leg in young soccer players. *Journal of Sports Medicine and Physical Fitness* 32(4):358-363.

Carlson AJ, Bennett G and Metcalf J (1992) The effect of visual feedback in isokinetic testing. *Isokinetics and Exercise Science* 2(2):60-64.

Chandler TJ, Kibler WB, Stracener EC, Ziegler AK and Pace B (1992) Shoulder strength, power and endurance in college tennis players. *American Journal of Sports Medicine* 20:455-458.

Chin M-K, Lo YSA, Li CT and So CH (1992) Physiological profiles of Hong Kong elite soccer players. *British Journal of Sports Medicine* 26(4):262-266.

Chin M-K, So CH, Yuan YWY, Li CT and Wong ASK (1994) Cardiorespiratory fitness and isokinetic muscle strength of elite Asian junior soccer players. *Journal of Sports Medicine and Physical Fitness* 34(3):250-257.

Chook KK, Teh KC and Giam CK (1986) The isokinetic strength of dominant quadriceps and hamstring muscles of 47 Singapore national sportsmen. In: Giam CK and Teh KC (eds) *Proceedings of 2nd International Sports Science Conference.* Singapore: Singapore Sports Council, pp. 127-133.

Clarke DH and Manning JM (1985) Properties of isokinetic fatigue at various movement speeds in adult males. *Research Quarterly for Exercise and Sport* 56(3):221-226.

Clarke DH and Wysochanski PM (1986) Torque-velocity curves in high-strength and low-strength men. In: Dotson CO and Humphrey JH (eds) *Exercise Physiology: Current Selected Research* vol 2. New York: AMS Press, pp. 123-131.

Cook EE, Gray VL, Savinor-Nogue E and Medeiros J (1987) Shoulder antagonistic strength ratios: A comparison between college-level baseball pitchers and nonpitchers. *Journal of Orthopaedic and Sports Physical Therapy* 8:451-461.

Coombes J and McNaughton LR (1993) Effects of bicarbonate ingestion on leg strength and power during isokinetic knee flexion and extension. *Journal of Strength and Conditioning Research* 7(4):241-249.

Costain R and Williams AK (1984) Isokinetic quadriceps and hamstring torque levels of adolescent, female soccer players. *Journal of Orthopaedic and Sports Physical Therapy* 5(4):196-200.

Cox PD (1995) Isokinetic strength testing of the ankle: A review. *Physiotherapy Canada* 47(2):97-119.

Cress EM, Johnson J and Agre JC (1991) Isokinetic strength testing in older women: A comparison of two systems. *Journal of Orthopaedic and Sports Physical Therapy* 13(4):199-202.

Daniel D, Malcolm L, Stone ML, Perth H, Morgan J and Riehl B (1982) Quantification of knee stability and function. *Contemporary Orthopaedics* 5:83-91.

Davis JA, Brewer J and Atkin D (1992) Pre-season physiological characteristics of English first and second division soccer players. *Journal of Sports Sciences* 10:541-547.

DeKoning FL, Vos JA, Binkhorst RA and Vissers ACA (1984) Influence of training on the force-velocity relationship of the arm flexors of active sportsmen. *International Journal of Sports Medicine* 5:43-46.

Delitto A, Crandell CE and Rose SJ (1989) Peak torque-to-body weight ratios in the trunk: A critical analysis. *Physical Therapy* 69(2):138-143.

Dempster P (1987) Limb length measurement. *Lido Tech Report* 1(1):1-3.

DeNuccio DK, Davies GJ and Rowinski MJ (1991) Comparison of quadriceps isokinetic eccentric and isokinetic concentric data using a standard fatigue protocol. *Isokinetics and Exercise Science* 1(2):81-86.

DeProft E, Cabri J, Dufour W and Clarys JP (1988) Strength training and kick performance in soccer players. In: Reilly T, Lees A, Davids K and Murphy WJ (eds) *Science and Football.* London: Spon, pp. 108-113.

DiBrezzo R, Fort IN and Brown B (1988) Dynamic strength and work variations during three stages of the menstrual cycle. *Journal of Orthopaedic and Sports Physical Therapy* 10(4):113-116.

DiBrezzo R, Fort IN and Brown B (1991) Relationships among strength, endurance, weight and body fat during three phases of the menstrual cycle. *Journal of Sports Medicine and Physical Fitness* 31(1):89-94.

Duncan P (1987) The effect of a prior quadriceps contraction on knee flexor torque in normal subjects and multiple sclerosis patients with spastic paresis. *Physiotherapy Practice* 3:11-17.

Dvir Z (1991) Clinical applicability of isokinetics: A review. *Clinical Biomechanics* 6:133-144.

Dvir Z (1995) *Isokinetics. Muscle Testing, Interpretation and Clinical Applications.* Edinburgh: Churchill Livingstone.

Dvir Z, Eger G, Halperin N and Shklar A (1989) Thigh muscle activity and anterior cruciate ligament insufficiency. *Clinical Biomechanics* 4:87-91.

Eckerson JM, Housh DJ, Housh TJ and Johnson GO (1994) Seasonal changes in body composition, strength, and muscular power in high school wrestlers. *Pediatric Exercise Science* 6:39-52.

Ekstrand J and Gillquist J (1983) The avoidability of soccer injuries. *International Journal of Sports Medicine* 4(2):124-128.

Ellenbecker TS (1991) A total arm strength isokinetic profile of highly skilled tennis players. *Isokinetics and Exercise Science* 1(1):9-21.

Ellenbecker TS (1992) Shoulder internal and external rotation strength and range of motion of highly skilled junior tennis players. *Isokinetics and Exercise Science* 2(2):65-72.

Emery L, Sitler M and Ryan J (1994) Mode of action and angular velocity fatigue response of the hamstrings and quadriceps. *Isokinetics and Exercise Science* 4(3):91-95.

Ericsson M, Johansson K, Nordgren B, Nordesjo L-O and Borges O (1982) Evaluation of a dynamometer for measurement of isometric and isokinetic torques. *Upsala Journal of Medical Science* 87:223-233.

Farrar M and Thorland W (1987) Relationship between isokinetic strength and sprint times in college-age men. *Journal of Sports Medicine and Physical Fitness* 27:368-372.

Farrell M and Richards JG (1986) Analysis of the reliability and validity of the Kinetic Communicator exercise device. *Medicine and Science in Sports and Exercise* 18:44-49.

Figoni SF and Morris AF (1984) Effects of knowledge of results on reciprocal isokinetic strength and fatigue. *Journal of Orthopaedic and Sports Physical Therapy* 6:190-197.

Fillyaw M, Bevins T and Fernandez L (1986) Importance of correcting isokinetic peak torque for the effect of gravity when calculating knee flexor to extensor muscle ratios. *Physical Therapy* 66(1):23-31.

Finucane SDG, Mayhew TP and Rothstein JM (1994) Evaluation of the gravity-correction feature of a Kin-Com isokinetic dynamometer. *Physical Therapy* 74(12):1125-1133.

Fleshman SA and Keppler MV (1992) A biomechanical comparison study of the adapted Orthotron KT-11, with stress indicators, to the Cybex II for the purpose of isokinetic testing of the knee. *Isokinetics and Exercise Science* 2(4):195-203.

Francis K and Hoobler T (1987) Comparison of peak torque values of the knee flexor and extensor muscle groups using the Cybex II and Lido 2.0 isokinetic dynamometers. *Journal of Orthopaedic and Sports Physical Therapy* 8(10):480-483.

Froese EA and Houston ME (1985) Torque-velocity characteristics and muscle fibre type in human vastus lateralis. *Journal of Applied Physiology* 59(2): 309-314.

Fry AC, Kraemer WJ, van Borselen F, Lynch JM, Marsit JL, Roy EP, Triplett NT and Knuttgen HG (1994) Performance decrements with high-intensity resistance exercise overtraining. *Medicine and Science in Sports and Exercise* 26(9):1165-1173.

Fry RW and Morton AR (1991) Physiological and kinanthropometric attributes of elite flatwater kayakists. *Medicine and Science in Sports and Exercise* 23(11):1297-1301.

Fugl-Meyer AR (1981) Maximum isokinetic ankle plantar and dorsal flexion torques in trained subjects. *European Journal of Applied Physiology* 47:393-404.

Gauffin H, Ekstrand J and Tropp H (1988) Improvement of vertical jump performance in soccer players after specific training. *Journal of Human Movement Studies* 15:185-190.

Genuario SE and Dolgener FA (1980) The relationship of isokinetic torque at two speeds to the vertical jump. *Research Quarterly for Exercise and Sport* 51(4):593-598.

Gibala MJ and MacDougall JD (1993) The effects of tapering on strength performance in trained athletes (abstract). *Medicine and Science in Sports and Exercise* 25(5, suppl):S47.

Gleeson N and Mercer T (1991) Intra-subject variability in isokinetic knee extension and flexion strength characteristics of adult males: A comparative examination of gravity-corrected and uncorrected data (abstract). *Journal of Sports Science* 9(4):415-416.

Gleeson NP and Mercer TH (1996) The utility of isokinetic dynamometry in the assessment of human muscle function. *Sports Medicine* 21(1):18-34.

Gleeson NP, Parry A and Mercer TH (1994) The effect of contraction mode on isokinetic leg strength and associated day-to-day reproducibility in adult males (abstract). *Journal of Sports Science* 12(2):137-138.

Grabiner MD (1994) Maximum rate of force development is increased by antagonist conditioning contraction. *Journal of Applied Physiology* 77(2):807-811.

Grabiner MD and Hawthorne DL (1990) Conditions of isokinetic knee flexion that enhance isokinetic knee extension. *Medicine and Science in Sports and Exercise* 22(2):235-24.

Grace TG, Sweetser ER, Nelson MA, Ydens LR and Skipper BJ (1984) Isokinetic muscle imbalance and knee-joint injuries. A prospective blind study. *Journal of Bone and Joint Surgery* 66A:734-740.

Grana EA and Frontera WR (1993) The use of the work ratio method to measure muscle fatiguability in high performance athletes (abstract). *Medicine and Science in Sports and Exercise* 25(5):S109.

Gransberg L and Knutsson E (1983) Determination of dynamic muscle strength in man with acceleration controlled isokinetic movements. *Acta Physiologica Scandinavica* 119:317-320.

Gravel D, Richards CL and Filion M (1988) Influence of contractile tension development on dynamic strength measurements of the plantarflexors in man. *Journal of Biomechanics* 21(2):89-96.

Gravel D, Richards CL and Filion M (1990) Angle dependency in strength measurements of the ankle plantar flexors. *European Journal of Applied Physiology* 61:182-187.

Gray JC and Chandler JM (1989) Percent decline in peak torque production during repeated concentric and eccentric contractions of the quadriceps femoris muscle. *Journal of Orthopaedic and Sports Physical Therapy* 10(8):309-314.

Greenberger HB, Wilkowski T and Belyea B (1994) Comparison of quadriceps peak torque using three different isokinetic dynamometers. *Isokinetics and Exercise Science* 4(2):70-75.

Greenhaff PL, Casey A, Short AH, Harris R, Soderlund K and Hultman E (1993) Influence of oral creatine supplementation of muscle torque during repeated bouts of maximal voluntary exercise in man. *Clinical Science* 84:565-571.

Gross MT, Huffman GM, Phillips CN and Wray JA (1991) Intramachine and intermachine reliability of the Biodex and Cybex II for knee flexion and extension peak torque and angular work. *Journal of Orthopaedic and Sports Physical Therapy* 13(6):329-335.

Gulch RW (1994) Force-velocity relations in human skeletal muscle. *International Journal of Sports Medicine* 15 (suppl 1):S2-15.

Gur H (1997) Concentric and eccentric isokinetic measurements in knee muscles during the menstrual cycle: A special reference to reciprocal moment ratios. *Archives of Physical Medicine and Rehabilitation* 78(5):501-505.

Hagerman FC and Staron RS (1983) Seasonal variations among physiological variables in elite oarsmen. *Canadian Journal of Applied Sport Science* 8(3):143-148.

Hald RD and Bottjen EJ (1987) Effect of visual feedback on maximal and submaximal isokinetic test measurements of normal quadriceps and hamstrings. *Journal of Orthopaedic and Sports Physical Therapy* 9:86-93.

Handel M, Dickhuth H-H, Mayer F and Gulch RW (1996) Prerequisites and limitations to isokinetic measurements in humans. Investigations on a servomotor-controlled dynamometer. *European Journal of Applied Physiology* 73:225-230.

Harman E, Frykman P, Rosenstein M, Johnson M and Rosenstein R (1990) The relationship of individual torque-velocity curve shapes to sprint running performance (abstract). *Medicine and Science in Sports and Exercise* 22(2):S8.

Harridge SDR and White MJ (1993) Muscle activation and the isokinetic torque-velocity relationship of the human triceps muscle. *European Journal of Applied Physiology* 67:218-221.

Heinrichs KI, Perrin DH, Weltman A, Gieck JH and Ball DW (1995) Effect of protocol and assessment device on isokinetic peak torque of the quadriceps muscle group. *Isokinetics and Exercise Science* 5:5-13.

Hellwig EV and Perrin DH (1991) A comparison of two positions for assessing shoulder rotator peak torque: The traditional frontal plane versus the plane of the scapula. *Isokinetics and Exercise Science* 1(4):202-206.

Hellwig EV and Perrin DH (1995) The mechanical and clinical reliability of the Kinetic Communicator's gravity correction procedure. *Isokinetics and Exercise Science* 5:85-91.

Herzog W (1988) The relation between the resultant moments at a joint and the moments measured by an isokinetic dynamometer. *Journal of Biomechanics* 21(1):5-12.

Hinton RY (1988) Isokinetic evaluation of shoulder rotational strength in high school baseball pitchers. *American Journal of Sports Medicine* 16:274-279.

Hobbel SL and Rose DJ (1993) The relative effectiveness of three forms of visual knowledge of results on peak torque output. *Journal of Orthopaedic and Sports Physical Therapy* 18(5):601-608.

Hoens A and Strauss GR (1994) The effect of deleting nonisokinetic phases of movement from isokinetic strength evaluations. *Isokinetics and Exercise Science* 4(3):96-103.

Hoffman JR, Fry AC, Howard R, Maresh CM and Kraemer WJ (1991) Strength, speed and endurance changes during the course of a Division I basketball season. *Journal of Applied Sport Science Research* 5(3):144-149.

Holm I, Ludvigsen P and Steen H (1994) Isokinetic hamstrings/quadriceps ratios: Normal values and reproducibility in sport students. *Isokinetics and Exercise Science* 4(4):141-145.

Housh TJ, Thorland WG, Tharp GD, Johnson GO and Cisar CJ (1984) Isokinetic leg flexion and extension strength of elite female track and field athletes. *Research Quarterly for Exercise and Sport* 55(4):347-350.

Huang TC, Roberts EM and Youm Y (1982) The biomechanics of kicking. In: Ghista DN (ed) *Human Body Dynamics.* New York: Oxford University Press, pp. 409-443.

Iossifidou AN and Baltzopoulos V (1996) Angular velocity in eccentric isokinetic dynamometry. *Isokinetics and Exercise Science* 6:65-70.

Jacobson BH and Edwards SW (1991) Influence of two levels of caffeine on maximal torque at selected angular velocities. *Journal of Sports Medicine and Physical Fitness* 31(2):147-153.

Jacobson BH, Weber MD, Claypool L and Hunt LE (1992) Effect of caffeine on maximal strength and power in elite male athletes. *British Journal of Sports Medicine* 26(4):276-280.

Jakeman PM, Winter EM and Doust J (1994) A review of research in sports physiology. *Journal of Sports Sciences* 12:33-60.

James C, Sacco P, Hurley MV and Jones DA (1994) An evaluation of different protocols for measuring the force-velocity relationship of the human quadriceps muscles. *European Journal of Applied Physiology* 68:41-47.

Jensen RC, Warren B, Laursen C and Morrissey MC (1991) Static pre-load effect on knee extensor isokinetic concentric and eccentric performance. *Medicine and Science in Sports and Exercise* 23(1):10-14.

Johansson C (1992) Knee extensor performance in runners. Differences between specific athletes and implications for injury prevention. *Sports Medicine* 14(2):75-81.

Johansson C, Lorentzon R, Fagerlund M, Sjostrom M and Fugl-Meyer AR (1987) Sprinters and marathon runners. Does isokinetic knee extensor performance reflect muscle size and structure? *Acta Physiologica Scandinavica* 130:663-669.

Johansson C, Lorentzon R, Rasmuson S, Reiz S, Haggmark S, Nyman H and Fugl-Meyer AR (1988) Peak torque and OBLA running capacity in male orienteers. *Acta Physiologica Scandinavica* 132:525-530.

Johnson R and Meeter D (1977) Estimation of maximum physical performance. *Research Quarterly* 48(1):74-84.

Kannus P (1992) Normality, variability and predictability of work, power and torque acceleration energy with respect to peak torque in isokinetic muscle testing. *International Journal of Sports Medicine* 13:249-256.

Kannus P (1994) Isokinetic evaluation of muscular performance: Implications for muscle testing and rehabilitation. *International Journal of Sports Medicine* 15 (suppl 1):S11-18.

Kannus P and Jarvinen M (1989) Prediction of torque acceleration energy and power of thigh muscles from peak torque. *Medicine and Science in Sports and Exercise* 21:304-307.

Kannus P, Cook L and Alosa D (1992) Absolute and relative endurance parameters in isokinetic tests of muscular performance. *Journal of Sport Rehabilitation* 1:2-12.

Katch FI, McArdle WD, Pechar GS and Perrine JJ (1974) Measuring leg force-output capacity with an isokinetic dynamometer-bicycle ergometer. *Research Quarterly* 45(1):86-91.

Katch VL (1972) Correlational and ratio adjustments of body weight in exercise-oxygen studies. *Ergonomics* 15:671-680.

Katch VL (1973) Use of the oxygen/body weight ratio in correlational analyses: Spurious correlations and statistical considerations. *Medicine and Science in Sports and Exercise* 5:252-257.

Katch VL and Katch FI (1974) Use of weight-adjusted oxygen uptake scores that avoid spurious correlations. *Research Quarterly* 45:447-451.

Keating JL and Matyas TA (1996a) The influence of subject and test design on dynamometric measurements of extremity muscles. *Physical Therapy* 76(8):866-889.

Keating JL and Matyas TA (1996b) Method-related variations in estimates of gravity correction values using electromechanical dynamometry: A knee extension study. *Journal of Orthopaedic and Sports Physical Therapy* 24(3):142-153.

Kellis E and Baltzopoulos V (1995) Isokinetic eccentric exercise. *Sports Medicine* 19(3):202-222.

Kellis E and Baltzopoulos V (1996) Resistive eccentric exercise: Effects of visual feedback on maximum moment of knee extensors and flexors. *Journal of Orthopaedic and Sports Physical Therapy* 23(2):120-124.

Keskula DR and Perrin DH (1994) Effect of test protocol on torque production of the rotators of the shoulder. *Isokinetics and Exercise Science* 4(4):176-181.

Kimura IF, Gulick DT, Alexander DM and Takao SH (1996) Reliability of peak torque values for concentric and eccentric shoulder internal and external rotation on the Biodex, Kinetic Communicator, and Lido dynamometers. *Isokinetics and Exercise Science* 6:95-99.

Kirkendall DT (1979) Comparison of Isokinetic Power-Velocity Profiles in Various Classes of American Athletes. Unpublished doctoral thesis, Ohio State University. Michigan: University Microfilms.

Kirkendall DT (1985) The applied sport science of soccer. *Physician and Sports Medicine* 13(4):53-59.

Klentrou PP and Montpetit RR (1991) Physiologic and physical correlates of swimming performance. *Journal of Swimming Research* 7(1):13-18.

Knapik JJ, Bauman CL, Jones BH, Harris JMcA and Vaughan L (1991) Preseason strength and flexibility imbalances associated with athletic injuries in female collegiate athletes. *American Journal of Sports Medicine* 19(1):76-81.

Knapik JJ, Jones BH, Bauman CL and Harris JMcA (1992) Strength, flexibility and athletic injuries. *Sports Medicine* 14(5):277-288.

Knapp TR (1992) Technical error of measurement: A methodological critique. *American Journal of Physical Anthropology* 87:235-236.

Koutedakis Y, Frischknecht R, Vrbova G, Sharp NC and Budgett R (1995) Maximal voluntary quadriceps strength patterns in Olympic overtrained athletes. *Medicine and Science in Sports and Exercise* 27(4):566-572.

Koutedakis Y, Pacy PJ, Quevedo RM, Millward DJ, Hesp R, Boreham C and Sharp NCC (1994) The effects of two different periods of weight-reduction on selected performance parameters in elite lightweight oarswomen. *International Journal of Sports Medicine* 15(8):472-477.

Koziris LP, Kraemer WJ, Triplett NT, Fry AC, Baver J, Pedro JG, Clemson A and Connors J (1991) Strength imbalances in women tennis players (abstract). *Medicine and Science in Sports and Exercise* 23(4, suppl):S43.

Kramer JF and Balsor B (1990) Lower extremity dominance and knee extensor torques in intercollegiate soccer players. *Canadian Journal of Applied Sport Science* 15:180-184.

Kramer JF, Hill K, Jones IC, Sandrin M and Vyse M (1989) Effect of dynamometer application arm length on concentric and eccentric torques during isokinetic knee extension. *Physiotherapy Canada* 41(2):100-106.

Kramer JF, Ingham-Tupper S, Walters-Stansbury K, Stratford P and MacDermid J (1994) Reliability of absolute and ratio data in assessment of knee extensor and flexor strength. *Isokinetics and Exercise Science* 4(2):51-57.

Kramer JF and Leger A (1991) Oarside and nonoarside torques of the knee extensors and flexors in lightweight and heavy-weight sweep oarsmen. *Physiotherapy Canada* 43(3):23-27.

Kramer JF, Leger A and Morrow A (1991a) Oarside and nonoarside knee extensor strength measures and their relationship to rowing ergometer performance. *Journal of Orthopaedic and Sports Physical Therapy* 14(5):213-219.

Kramer JF, Vaz MD and Hakansson D (1991b) Effect of activation force on knee extensor torques. *Medicine and Science in Sports and Exercise* 23(2):231-237.

Kroemer KHE, Kroemer HJ and Kroemer-Elbert KE (1990) Skeletal muscle. In: *Engineering Physiology. Bases of Human Factors/Ergonomics* (2nd ed). New York: Van Nostrand Reinhold, pp. 51-72.

Kroll PG, Nelson AJ and Nordin M (1996) The effect of previous contraction condition on subsequent eccentric power production in elbow flexor muscle. *Isokinetics and Exercise Science* 6:27-31.

Kulig K, Andrews JG and Hay JG (1984) Human strength curves. In: Terjung RL (ed) *Exercise and Sports Sciences Reviews* vol 12. New York: Macmillan, pp. 417-466.

Lai JS, Wong PL and Lien IN (1986) Isokinetic evaluation of soccer players (abstract). *XXIII FIMS World Congress of Sports Medicine Abstracts.* Australian Sports Medicine Federation, p. 138.

Leatt P, Shephard RJ and Plyley MJ (1987) Specific muscular development in under-18 soccer players. *Journal of Sports Sciences* 5:165-175.

Leslie M, Zachazewski J and Browne P (1990) Reliability of isokinetic torque values for ankle invertors and evertors. *Journal of Orthopaedic and Sports Physical Therapy* 11(12):612-616.

Levene JA, Hart BA, Seeds RH and Fuhrman GA (1991) Reliability of reciprocal isokinetic testing of the knee extensors and flexors. *Journal of Orthopaedic and Sports Physical Therapy* 14(3):121-127.

MacIntosh BR, Herzog W, Suter E, Wiley JP and Sokolosky J (1993) Human skeletal muscle fibre types and force: Velocity properties. *European Journal of Applied Physiology* 67:499-506.

Madsen OR (1996) Torque, total work, power, torque acceleration energy and acceleration time assessed on a dynamometer: Reliability of knee and elbow extensor and flexor strength measurements. *European Journal of Applied Physiology* 74:206-210.

Mahler P, Mora C, Gremion G and Chantraine A (1992) Isotonic muscle evaluation and sprint performance. *Excel* 8:139-145.

Mangine RE, Noyes FR, Mullen MP and Barber SD (1990) A physiological profile of the elite soccer athlete.

Journal of Orthopaedic and Sports Physical Therapy 12(4):147-152.

Martin DT, Scifres JC, Zimmerman SD and Wilkinson JG (1994) Effects of interval training and a taper on cycling performance and isokinetic leg strength. *International Journal of Sports Medicine* 15:485-491.

Mathiassen SE (1989) Influence of angular velocity and movement frequency on development of fatigue in repeated isokinetic knee extensions. *European Journal of Applied Physiology* 59:80-88.

Mawdsley RH (1985) Reciprocal versus non-reciprocal isokinetic testing (abstract). *Physical Therapy* 65(5):730.

Mawdsley RH and Knapik JJ (1982) Comparison of isokinetic measurements with test repetitions. *Physical Therapy* 62(2):169-172.

Mayhew TP and Rothstein JM (1985) Measurement of muscle performance with instruments. In: Rothstein JM (ed) *Measurement in Physical Therapy.* New York: Churchill Livingstone, pp. 57-102.

Mayhew TP, Rothstein JM, Finucane SDG and Lamb RL (1994) Performance characteristics of the Kin-Com dynamometer. *Physical Therapy* 74(11):1047-1054.

McCleary RW and Andersen JC (1992) Test-retest reliability of reciprocal isokinetic knee extension and flexion peak torque measurements. *Athletic Training* 27(4):362-365.

McKay LJ, Dale A, Hochstetler S and Plyley MJ (1987) Physiological profiles of Canadian varsity women soccer players (abstract). *Medicine and Science in Sports and Exercise* 19(2, suppl):S48.

McLean BD and Tumilty DMcA (1993) Left-right asymmetry in two types of soccer kick. *British Journal of Sports Medicine* 27(4):260-262.

Mikesky AE, Edwards JE, Wigglesworth JK and Kunkel S (1995) Eccentric and concentric strength of the shoulder and arm musculature in collegiate baseball pitchers. *American Journal of Sports Medicine* 23(5):638-642.

Moffroid MA, Whipple R, Hofkosh J, Lowman E and Thistle H (1969) A study of isokinetic exercise. *Physical Therapy* 49:735-746.

Mognoni P, Narici MV, Sirtori MD and Lorenzelli F (1994) Isokinetic torques and kicking maximal velocity in young soccer players. *Journal of Sports Medicine and Physical Fitness* 34(4):357-361.

Molczyk L, Thigpen LK, Eickhoff J, Coldgar D and Gallagher JC (1991) Reliability of testing the knee extensors and flexors in healthy adult women using a Cybex II isokinetic dynamometer. *Journal of Orthopaedic and Sports Physical Therapy* 14:37-41.

Montgomery LC, Douglass LW and Deuster PA (1989) Reliability of an isokinetic test of muscle strength and endurance. *Journal of Orthopaedic and Sports Physical Therapy* 10:315-322.

Narici MV, Sirtori MD, Mastore S and Mognoni P (1991) The effect of range of motion and isometric preactivation on isokinetic torques. *European Journal of Applied Physiology* 62:216-220.

Narici MV, Sirtori MD and Mognoni P (1988) Maximal ball velocity and peak torques of hip flexor and knee extensor muscles. In: Reilly T, Lees A, Davids K and Murphy WJ (eds) *Science and Football.* London: Spon, pp. 429-433.

Nevill AM and Holder RL (1995) Scaling, normalizing and per ratio standards: An allometric modelling approach. *Journal of Applied Physiology* 79(3):1027-1031.

Nevill AM, Ramsbottom R and Williams C (1992) Scaling measurements in physiology and medicine for individuals of different size. *European Journal of Applied Physiology* 64:419-425.

Nicholas JA, Strizak AM and Veras G (1976) A study of thigh muscle weakness in different pathological states of the lower extremity. *American Journal of Sports Medicine* 4(6):241-248.

Nitschke JE (1992) Reliability of isokinetic torque measurements: A review of the literature. *Australian Journal of Physiotherapy* 38(2):125-134.

Norman RW (1992) Matching issues in strength measurements. *Canadian Journal of Sports Science* 17:70-71.

Nutter J and Thorland WG (1987) Body composition and anthropometric correlates of isokinetic leg extension strength of young adult males. *Research Quarterly for Exercise and Sport* 58(1):47-51.

Nyland JA, Caborn DNM, Brosky JA, Kneller CL and Freidhoff G (1997) Anthropometric, muscular fitness, and injury history comparisons by gender of youth soccer teams. *Journal of Strength and Conditioning Research* 11(2):92-97.

Oberg B (1993) Evaluation and improvement of strength in competitive athletes. In: Harms-Ringdahl K (ed) *Muscle Strength.* Edinburgh: Churchill Livingstone, pp. 167-185.

Oberg B, Ekstrand J, Moller M and Gillquist J (1984) Muscle strength and flexibility in different positions of soccer players. *International Journal of Sports Medicine* 5:213-216.

Oberg B, Moller M, Ekstrand J and Gillquist J (1985a) Exercises for knee flexors and extensors in uninjured soccer players: Effects of two different programs. *International Journal of Sports Medicine* 6:151-154.

Oberg B, Moller M, Gillquist J and Ekstrand J (1986) Isokinetic torque levels for knee extensors and knee flexors in soccer players. *International Journal of Sports Medicine* 7:50-53.

Oberg B, Odenrick P and Tropp H (1985b) Muscle strength and jump performance in soccer players (abstract). 10th International Congress of Biomechanics, Umea. Abstract book, p. 311.

Oddsson LIE and Westing SH (1991) Jumping height can be accurately predicted from selected measurements of muscle strength and biomechanical parameters. In: Tant C, Patterson P and York S (eds) *Proceedings of 9th ISBS Symposium.* Ames, Iowa: ISBS, pp. 29-33.

Otis JC and Gould JD (1986) The effect of external load on torque production by the knee extensors. *Journal of Bone and Joint Surgery* 68-A(1):65-70.

Pedegana LR, Elsner RC, Roberts D, Lang J and Farewell V (1982) The relationship of upper extremity strength to throwing speed. *American Journal of Sports Medicine* 10(6):352-354.

Perrin DH, Hellwig EV, Tis LL and Shenk BS (1992) Effect of gravity correction on shoulder rotation isokinetic average force and reciprocal muscle group ratios. *Isokinetics and Exercise Science* 2(1):30-33.

Perrin DH, Robertson RJ and Ray RL (1987) Bilateral isokinetic peak torque, torque acceleration energy, power, and work relationships in athletes and nonathletes. *Journal of Orthopaedic and Sports Physical Therapy* 9:184-189.

Perrine JJ and Edgerton VR (1978) Muscle force-velocity and power-velocity relationships under isokinetic loading. *Medicine and Science in Sports* 10(3):159-166.

Piastra G, Capanna R and Greco P (1990) Dinamometria isocinetica e forza esplosiva. Confronto tra due gruppi di pallavoliste di differente livello agonistico (Isokinetic dynamometry and explosive force. Comparison between groups of volleyball players of different level). *Medicina Dello Sport* 43(4):297-301 (in Italian with English abstract).

Pincivero DM, Lephart SM and Karunakara RG (1997) Reliability and precision of isokinetic strength and muscular endurance for the quadriceps and hamstrings. *International Journal of Sports Medicine* 18(2):113-117.

Podolsky A, Kaufman KR, Cahalan TD, Aleshinsky SY and Chao EYS (1990) The relationship of strength and jump height in figure skaters. *American Journal of Sports Medicine* 18(4):400-405.

Posch E, Haglund Y and Eriksson E (1989) Prospective study of concentric and eccentric leg muscle torques, flexibility, physical conditioning, and variation of injury rates during one season of amateur ice hockey. *International Journal of Sports Medicine* 10(2):113-117.

Poulmedis P (1985) Isokinetic maximal torque power of Greek elite soccer players. *Journal of Orthopaedic and Sports Physical Therapy* 6:293-295.

Poulmedis P, Rondoyannis G, Mitsou A and Tsarouchas E (1988) The influence of isokinetic muscle torque exerted in various speeds on soccer ball velocity. *Journal of Orthopaedic and Sports Physical Therapy* 10:93-96.

Pressly SC, Clark RD, Adams R, Lephart SM and Robertson RJ (1991) A comparison of values obtained from the Cybex II and Kin-Com II isokinetic dynamometers for peak torque and muscular endurance (abstract). *Athletic Training* 26(2):150.

Prietto CA and Caiozzo VJ (1989) The in vivo force-velocity relationship of the knee flexors and extensors. *American Journal of Sports Medicine* 17(5):607-611.

Puhl J, Case S, Fleck S and Van Handel P (1982) Physical and physiological characteristics of elite volleyball players. *Research Quarterly for Exercise and Sport* 53(3):257-262.

Pyke FS, Minikin BR, Woodman LR, Roberts AD and Wright TG (1979) Isokinetic strength and maximal oxygen uptake of trained oarsmen. *Canadian Journal of Applied Sports Science* 4(4):277-279.

Rajala GM, Neumann DA, Foster C and Jensen RH (1994) Quadriceps muscle performance in male speed skaters. *Journal of Strength and Conditioning Research* 8:48-52.

Rankin JM and Thompson CB (1983) Isokinetic evaluation of quadriceps and hamstrings function: Normative data concerning body weight and sport. *Athletic Training* 18(Summer):110-114.

Rathfon JA, Matthews KM, Yang AN, Levangie PK and Morrissey MC (1991) Effects of different acceleration and deceleration rates on isokinetic performance of the knee extensors. *Journal of Orthopaedic and Sports Physical Therapy* 14(4):161-168.

Ready AE (1982) Seasonal evaluation of isokinetic strength, power and endurance of middle distance runners (abstract). *Canadian Journal of Applied Sport Sciences* 7(4):238.

Rhodes EC, Mosher RE, McKenzie DC, Franks IM, Potts JE and Wenger HA (1986) Physiological profiles of the Canadian Olympic soccer team. *Canadian Journal of Sport Science* 11(1):31-36.

Roberts ME and Metcalfe A (1968) Mechanical analysis of kicking. In: Wartenweiler J, Jokl E and Hebbelinck M (eds) *Biomechanics I.* Basel: Karger, pp. 315-319.

Roberts ME, Zernicke RF, Youm Y and Huang TC (1974) Kinetic parameters of kicking. In: Nelson T and Morehouse C (eds) *Biomechanics IV.* Baltimore: University Park Press, pp. 157-162.

Robertson DGE and Mosher RE (1983) Work and power of leg muscles in soccer kicking. In: Winter DA, Norman RW, Wells RP, Hayes KC and Patla AE (eds) *Biomechanics IX-B.* Champaign, Illinois: Human Kinetics, pp. 533-542.

Rochcongar P, Morvan R, Jan J, Dassonville J and Beillot J (1988) Isokinetic investigation of knee extensors and knee flexors in young French soccer players. *International Journal of Sports Medicine* 9:448-450.

Roemmich JN and Sinning WE (1996) Sport-seasonal changes in body composition, growth, power and strength of adolescent wrestlers. *International Journal of Sports Medicine* 17(2):92-99.

Rothstein JM (1985) Measurement and clinical practice: Theory and application. In: Rothstein JM (ed) *Measurement in Physical Therapy.* New York: Churchill Livingstone, pp. 1-46.

Rothstein JM, Delitto A, Sinacore DR and Rose SJ (1983) Electromyographic, peak torque, and power relationships during isokinetic movement. *Physical Therapy* 63(6):926-933.

Safrit MJ (1973) *Evaluation in Physical Education.* Englewood Cliffs, New Jersey: Prentice Hall.

Sale DG (1991) Testing strength and power. In: MacDougall JD, Wenger HA and Green HJ (eds) *Physiological Testing of the High-Performance Athlete* (2nd ed). Champaign, Illinois: Human Kinetics, pp. 21-106.

Sale DG and Norman RW (1982) Testing strength and power. In: MacDougall JD, Wenger HA and Green HJ (eds) *Physiological Testing of the Elite Athlete.* Ottawa: Canadian Association of Sport Sciences, pp. 7-37.

Sapega AA (1990) Muscle performance evaluation in orthopaedic practice. *Journal of Bone and Joint Surgery* 72-A(10):1562-1574.

Sapega AA, Nicholas JA, Sokolow D and Saraniti A (1982) The nature of torque "overshoot" in Cybex isokinetic dynamometry. *Medicine and Science in Sports and Exercise* 14(5):368-375.

Schwendner KI, Mikesky AE, Wigglesworth JK and Burr DB (1995) Recovery of dynamic muscle function following isokinetic fatigue testing. *International Journal of Sports Medicine* 16:185-189.

Smith C, Donnelly A, Brewer J and Davis J (1994) An investigation of the specific aspects of fitness in professional and amateur footballers (abstract). *Journal of Sports Science* 12(2):165-166.

Smith DD (1993) Role of body and joint position on isokinetic exercise and testing. *Journal of Sport Rehabilitation* 2:141-149.

So C-H, Siu TO, Chan KM, Chin MK and Li CT (1994) Isokinetic profile of dorsiflexors and plantar flexors of the ankle—a comparative study of elite versus untrained subjects. *British Journal of Sports Medicine* 28(1):25-30.

Soderberg GJ and Blaschak MJ (1987) Shoulder internal and external rotation peak torque production through a velocity spectrum in differing positions. *Journal of Orthopaedic and Sports Physical Therapy* 8(11):518-524.

Stam HJ, Binkhorst RA, van Nieuwenhuyzen JF and Snijders CJ (1993) Influence of the dynamometer driving mechanism on the isokinetic torque angle curve of the knee extensors. *Clinical Biomechanics* 8:91-94.

Stokes IAF, Gookin DM, Ried S and Hazard RG (1990) Effects of axis placement on measurement of isokinetic flexion and extension torque in the lumbar spine. *Journal of Spinal Disorders* 3(2):114-118.

Strauss GR, Allen C, Munt M and Zanoli J (1996) A comparison of continuous and discrete testing approaches on concentric and eccentric production of the knee extensors. *Isokinetics and Exercise Science* 5:135-141.

Svetlize HD (1996) Isokinetic hamstrings and quadriceps evaluation of Argentine elite soccer players (abstract). *Medicine and Science in Sports and Exercise* 28(5, suppl):S9.

Tanner JM (1949) Fallacy of per-weight and per-surface area standards, and their relation to spurious correlation. *Journal of Applied Physiology* 2(1):1-15.

Taylor NAS, Sanders RH, Howick EI and Stanley SN (1991) Static and dynamic assessment of the Biodex dynamometer. *European Journal of Applied Physiology* 62:180-188.

Taylor RL and Casey JJ (1986) Quadriceps torque production on the Cybex II dynamometer as related to changes in lever arm length. *Journal of Orthopaedic and Sports Physical Therapy* 8:148-152.

Tesch PA (1980) Muscle fatigue in man. *Acta Physiologica Scandinavica* suppl 480.

Tesch PA (1995) Aspects of muscle properties and use in competitive Alpine skiing. *Medicine and Science in Sports and Exercise* 27(3):310-314.

Tesch PA, Dudley GA, Duvoisin MR, Hather BM and Harris RT (1990) Force and EMG signal patterns during repeated bouts of concentric or eccentric muscle actions. *Acta Physiologica Scandinavica* 138:263-271.

Tesch PA and Wright JE (1983) Recovery from short term intense exercise: Its relation to capillary supply and blood lactate accumulation. *European Journal of Applied Physiology* 52:98-103.

Tesch PA, Wright JE, Vogel JA, Daniels WL, Sharp DS and Sjodin B (1985) The influence of muscle metabolic characteristics on physical performance. *European Journal of Applied Physiology* 54:237-243.

Thigpen LK, Blanke D and Lang P (1990) The reliability of two different Cybex isokinetic systems. *Journal of Orthopaedic and Sports Physical Therapy* 12(4):157-162.

Thompson MC, Shingleton LG and Kegerreis ST (1989) Comparison of values generated during testing of the knee using the Cybex II Plus and Biodex model B-2000 isokinetic dynamometer. *Journal of Orthopaedic and Sports Physical Therapy* 11(3):108-115.

Thorstensson A (1976) Muscle strength, fibre types and enzyme activities in man. *Acta Physiologica Scandinavica* suppl 443.

Thorstensson A and Karlsson J (1976) Fatiguability and fibre composition of human skeletal muscle. *Acta Physiologica Scandinavica* 98:318-322.

Tihanyi J, Apor P and Fekete G (1982) Force-velocity-power characteristics and fibre composition in human knee extensor muscles. *European Journal of Applied Physiology* 48:31-343.

Timm KE (1989) Comparison of knee extensor and flexor muscle group performances using the Cybex 340 and the Merac isokinetic dynamometers (abstract). *Physical Therapy* 69(5):389.

Timm KE (1994) Comparison of test data from the Cybex TEF and 6000-TMC isokinetic spinal dynamometers. *Isokinetics and Exercise Science* 4(3):112-115.

Timm KE and Fyke D (1993) The effect of test speed sequence on the concentric isokinetic performance of the knee extensor muscle group. *Isokinetics and Exercise Science* 3(2):123-128.

Timm KE, Gennrich P, Burns R and Fyke D (1992) Mechanical and physiological performance reliability of selected isokinetic dynamometers. *Isokinetics and Exercise Science* 2(4):182-190.

Tippett SR (1986) Lower extremity strength and active range of motion in college baseball pitchers: A comparison between stance leg and kick leg. *Journal of Orthopaedic and Sports Physical Therapy* 8(1):10-14.

Tis LL, Perrin DH, Weltman A, Ball DW and Gieck JH (1993) Effect of preload and range of motion on isokinetic torque in women. *Medicine and Science in Sports and Exercise* 25(9):1038-1043.

Togari H, Ohashi J and Ohgushi T (1988) Isokinetic muscle strength of soccer players. In: Reilly T, Lees A, Davids K and Murphy WJ (eds) *Science and Football.* London: Spon, pp. 181-185.

Trolle M, Aagaard P, Simonsen EB, Bangsbo J and Klausen K (1993) Effects of strength training on kicking performance in soccer. In: Reilly T, Clarys J and Stibbe A (eds) *Science and Football II.* London: Spon, pp. 95-97.

Tumilty DMcA, Hahn AG, Telford RD and Smith RA (1988) Is "lactic acid tolerance" an important component of fitness for soccer? In: Reilly T, Lees A, Davids K and Murphy WJ (eds) *Science and Football.* London: Spon, pp. 81-86.

Vagenas G and Hoshizaki B (1991) Functional asymmetries and lateral dominance in the lower limbs of distance runners. *International Journal of Sport Biomechanics* 7:311-329.

van der Leeuw GHF, Stam HJ and Nieuwenhuyzen JF (1989) Correction for gravity in isokinetic dynamometry of knee extensors in below knee amputees. *Scandinavian Journal of Rehabilitation Medicine* 21:141-145.

Vandervoort AA, Sale DG and Moroz J (1984) Comparison of motor unit activation during unilateral and bilateral leg extension. *Journal of Applied Physiology* 56:46-51.

van Ingen Schenau GJ, DeKoning JJ, Bakker FC and DeGroot G (1996) Performance-influencing factors in homogeneous groups of top athletes: A cross-sectional study. *Medicine and Science in Sports and Exercise* 28(1):1305-1310.

Verducci FM (1980) *Measurement Concepts in Physical Education*. St Louis: Mosby.

Vyse WM and Kramer JF (1990) Interaction of concentric and eccentric muscle actions during continuous activation cycles of the elbow flexors. *Physiotherapy Canada* 42(3):123-127.

Walmsley RP (1993) Movement of the axis of rotation of the glenohumeral joint while working on the Cybex II dynamometer. Part 1. Flexion/extension. *Isokinetics and Exercise Science* 3(1):16-20.

Walmsley RP and Dias JM (1995) Intermachine reliability of isokinetic concentric measurements of shoulder internal and external peak torque. *Isokinetics and Exercise Science* 5:75-80.

Walmsley RP and Szybbo C (1987) A comparative study of the torque generated by the shoulder internal and external rotator muscles in different positions and at varying speeds. *Journal of Orthopaedic and Sports Physical Therapy* 9(6):217-222.

Warner MA, Duncan PW, Harned DJ and Garrett WE (1985) A comparison of the Cybex II isokinetic dynamometer and Ariel computerised exerciser (abstract). *Physical Therapy* 65(5):730.

Westblad P and Johansson C (1993) A reliable method for measuring eccentric and concentric knee extensor endurance. *European Journal of Musculoskeletal Research* 2:151-157.

Westblad P, Svedenhag J and Rolf C (1996) The validity of isokinetic knee extensor endurance measurements with reference to treadmill running capacities. *International Journal of Sports Medicine* 17:134-139.

Westing SH, Seger JY and Thorstensson A (1991) Isoacceleration: A new concept of resistive exercise. *Medicine and Science in Sports and Exercise* 23(5):631-635.

Wiklander J and Lysholm J (1987) Simple tests for surveying muscle strength and muscle stiffness in sportsmen. *International Journal of Sports Medicine* 8:50-54.

Wiktorsson-Moller M, Oberg B, Ekstrand J and Gillquist J (1983) Effects of warming up, massage, and stretching on range of motion and muscle strength in the lower extremity. *American Journal of Sports Medicine* 11(4):249-252.

Wilhite MR, Cohen ER and Wilhite SC (1992) Reliability of concentric and eccentric measurements of quadriceps performance using the KIN-COM dynamometer: The effect of testing order for three different speeds. *Journal of Orthopaedic and Sports Physical Therapy* 15(4):175-182.

Wilk KE, Andrews JR and Arrigo CA (1995) The abductor and adductor strength characteristics of professional baseball pitchers. *American Journal of Sports Medicine* 23(3):307-311.

Wilk KE, Arrigo CA and Andrews JR (1992) Isokinetic testing of the shoulder abductors and adductors: Windowed vs nonwindowed data collection. *Journal of Orthopaedic and Sports Physical Therapy* 15(2):107-112.

Wilk KE, Johnson RD and Levine B (1987) A comparison of peak torque values of the knee extension and flexor muscle groups using Biodex, Cybex, and Kin-Com isokinetic dynamometers (abstract). *Physical Therapy* 67(5):789-790.

Wilk KE, Johnson RD and Levine B (1988) Comparison of knee extensor and flexor muscle group strength using the Biodex, Cybex and Lido isokinetic dynamometers (abstract). *Physical Therapy* 68(5): 792.

Wilk KE, Andrews JR, Arrigo CA, Keirns MA and Erber DJ (1993) The strength characteristics of internal and external rotator muscles in professional baseball pitchers. *American Journal of Sports Medicine* 21:61-66.

Wilson GJ and Murphy AJ (1996) Strength diagnosis: The use of test data to determine specific strength training. *Journal of Sports Science* 14:167-173.

Winter DA, Wells RP and Orr GW (1981) Errors in the use of isokinetic dynamometers. *European Journal of Applied Physiology* 46:397-408.

Winter EM (1992) Scaling: Partitioning out differences in size. *Pediatric Exercise Science* 4:296-301.

Winter EM and Nevill AM (1996) Scaling: Adjusting for differences in body size. In: Eston R and Reilly T (eds) *Kinanthropometry and Exercise Physiology Laboratory Manual.* London: Spon, pp. 321-335.

Wood TM (1989) The changing nature of norm-referenced validity. In: Safrit MJ and Wood TM (eds) *Measurement Concepts in Physical Education and Exercise Science.* Champaign, Illinois: Human Kinetics, pp. 23-44.

Woodson C, Bandy WD, Curis D and Baldwin D (1995) Relationship of isokinetic peak torque with work and power for ankle plantar flexion and dorsiflexion. *Journal of Orthopaedic and Sports Physical Therapy* 22(3):113-115.

Wrigley TV (1989) Letter to the editor. *Journal of Orthopaedic and Sports Physical Therapy* 10:340-341.

Wrigley TV (2000) Correlations with athletic performance. In: Brown L (ed) *Isokinetic Performance Enhancement.* Champaign, Illinois: Human Kinetics. In press.

Wrigley TV and Grant M (1995) Isokinetic dynamometry. In: Zuluaga M, Briggs C, Carlisle J, McDonald V, McMeeken J, Nickson W, Oddy P and Wilson D (eds) *Sports Physiotherapy. Applied Science and Practice.* Edinburgh: Churchill Livingstone, pp. 259-287.

Wrigley TV, Vasey A, Watson L and Dalziel R (1995) Reliability of clinical isokinetic dynamometry in pathological athletic shoulders. In: Bauer T (ed) *Proceedings XIII International Symposium for Biomechanics in Sport.* Ontario: International Society of Biomechanics in Sports, pp. 29-33.

Wyse JP, Mercer TH and Gleeson NP (1994) Time-of-day dependence of isokinetic leg strength and associated interday variability. *British Journal of Sports Medicine* 28(3):167-170.

Yates JW and Kamon E (1983) A comparison of peak and constant angle torque-velocity curves in fast and slow-twitch populations. *European Journal of Applied Physiology* 51:67-74.

Zefang W (1993) On the force moment of stretching or flexing the knee and the height of vertical jumping. In: Hamill J, Derrick TR and Elliott EH (eds) *Biomechanics of Sport XI.* Amherst, Massachusetts: ISBS, pp. 56-59.

Chapter 13

Abernethy P, Jurimae J, Logan PA, Taylor AW and Thayer R (1994) Acute and chronic response of skeletal muscle to resistance exercise. *Sports Medicine* 17:22-38.

Abernethy P, Wilson G and Logan PA (1995) Strength and power assessment: Issues, controversies and challenges. *Sports Medicine* 19:401-417.

Abernethy PJ and Jürimäe J (1996) Cross-sectional and longitudinal uses of isoinertial, isometric, and isokinetic dynamometry. *Medicine and Science in Sports and Exercise* 28:1180-1187.

Anderson T and Kearney JT (1983) Effects of three resistance training programs on muscular strength and absolute and relative endurance. *Research Quarterly in Exercise and Sport* 53:1-7.

Asmussen E and Bonde-Petersen F (1974) Apparent efficiency and storage of elastic energy in human muscles during exercise. *Acta Physiologica Scandanavica* 92:537545.

Baker D, Wilson G and Carlyon B (1994) Generality versus specificity: A comparison of dynamic and isometric measures of strength and speed-strength. *European Journal of Applied Physiology* 68:350-355.

Bar-Or O (1987) The Wingate anaerobic test: An update on methodology, reliability and validity. *Sports Medicine* 4:381-391.

Behm DG (1991) An analysis of intermediate speed resistance exercises for velocity-specific strength gains. *Journal of Applied Sports Science Research* 5:1-5.

Bompa T (1983) *Theory and Methodology of Training: The Key to Athletic Performance.* Dubuque, Iowa: Kendall Hunt.

Bosco C, Luhtanen P and Komi PV (1983) A simple method for measurement of mechanical power in jumping. *European Journal of Applied Physiology* 50:273-282.

Brzycki M (1993) Strength testing: Predicting a one-rep max from reps-to-fatigue. *Journal of Health, Physical Education, Recreation and Dance* 64:88-90.

Callister R, Shealy MJ and Fleck SJ (1988) Performance adaptations to sprint, endurance and both modes of training. *Journal of Applied Sports Science Research* 2:46-51.

Cordova ML and Armstrong CW (1996) Reliability of ground reaction forces during a vertical jump: Implications for functional strength assessment. *Journal of Athletic Training* 31(4):342-345.

Dowling JJ and Vamos L (1993) Identification of kinetic and temporal factors related to vertical jump performance. *Journal of Applied Biomechanics* 9:95-110.

Gillespie J and Keenum S (1987) A validity and reliability analysis of the seated shot put as a test of power. *Journal of Human Movement Studies* 13:97-105.

Gullich A and Schmidtbleicher D (1996) MVC-induced short term potentiation of explosive force. *New Studies in Athletics* 11(4):67-81.

Hakkinen K, Parkinen H and Alen M (1987) EMG, muscle fibre and force production characteristics during a one year training period in elite weightlifters. *European Journal of Applied Physiology* 56:419-427.

Hakkinen K, Parkinen H and Alen M (1988) Neuromuscular and hormonal adaptations in athletes to strength training in two years. *Journal of Applied Physiology* 65:2406-2412.

Harman EA, Rosenstein MT, Frykman PN and Rosenstein RM (1990) The effects of arms and countermovement on vertical jumping. *Medicine and Exercise in Sports and Exercise* 22(6):825-833.

Hennessy LC and Watson AWS (1994) The interference effects of training for strength and endurance simultaneously. *Journal of Strength and Conditioning Research* 8(1):12-19.

Hoeger WWK, Hopkins DR and Barette SL (1990) Relationship between repetitions and selected percentages of one repetition maximum: A comparison between trained and untrained males and females. *Journal of Applied Sports Science Research* 4:47-54.

Hortobagyi T and Katch FI (1990) Reliability of muscle mechanical characteristics for isokinetic and isotonic squat and bench press exercise using a multifunction computerised dynamometer. *Research Quarterly in Exercise and Sport* 61:191-195.

Hortobagyi T, Katch F and LaChance P (1989) Interrelationships among various measures of upper body strength assessed by different contraction modes. *European Journal of Applied Physiology* 58:749-755.

Jacobs I, Bell DG and Pope J (1988) Comparison of isokinetic and isoinertial lifting tests as predictors of maximal lifting capacity. *European Journal of Applied Physiology* 57:146-153.

Komi PV and Bosco C (1978) Utilization of stored elastic energy in leg extensor muscles by men and women. *Medicine and Science in Sports and Exercise* 10(4):261-265.

Lander J (1985) Maximums based on reps. *National Strength and Conditioning Association Journal* 6:60-61.

Mayhew JL, Ball TE and Bowen JC (1992) Prediction of bench press lifting ability from submaximal repetitions before and after training. *Sports Medicine of Training and Rehabilitation* 3:195-201.

Mayhew JL, Piper FC and Schweger TM (1989) Contributions of speed, agility and body composition to anaerobic power measurement in college football players. *Journal of Applied Sports Science Research* 3:101-106.

Mero A, Luhtanen P, Viitasalo JT and Komi PV (1981) Relationship between the maximal running velocity, muscle fibre characteristics, force production and force relaxation of sprinters. *Scandinavian Journal of Sports Science* 3:16-22.

Murphy AJ and Wilson GL (1997) The ability of tests of muscular function to reflect training-induced changes in performance. *Journal of Sports Sciences* 15(2):191-200.

Murphy AJ, Wilson GL and Pryor JF (1994) Use of the isoinertial force mass relationship in the prediction of dynamic human performance. *European Journal of Applied Physiology* 69:250-257.

Pearson DR and DL Costill (1988) The effects of constant external resistance exercise and isokinetic exercise training on work induced hypertrophy. *Journal of Applied Sports Science Research* 2:39-41.

Robertson DGE and Fleming D (1987) Kinetics of standing broad and vertical jumping. *Canadian Journal of Sports Science* 12(1):19-23.

Sale DG (1991) Testing strength and power. In: MacDougall J, Wenger H and Green H (eds) *Physiological Testing of the High Performance Athlete.* Champaign, Illinois: Human Kinetics, 21-106.

Schmidtbleicher D (1986) Strength and strength training. *Proceedings: First Elite Coaches Seminar.* Canberra: Australian Coaching Council.

Sewall LP and Lander JE (1991) The effects of rest on maximal efforts in squat and bench press. *Journal of Applied Sports Science Research* 5:96-99.

Shetty AB and Etnrye BR (1989) Contribution of arm movement to the force components of a maximum vertical jump. *Journal of Orthopaedic and Sports Physical Therapy* 11(5):198-201.

Siff MC (1993) Understanding the mechanics of muscle contraction. *National Strength and Conditioning Association Journal* 15(5):30-33.

Tihanyi J, Apor P and Fekete G (1982) Force-velocity-power characteristics and fibre composition in human knee extensor muscles. *European Journal of Applied Physiology* 48:331-343.

Viitasalo JT (1985a) Measurement of force-velocity characteristics for sportsmen in field conditions. In: DA Winter, RW Norman, RP Wells et al. (eds) *Biomechanics IX-A.* Champaign, Illinois: Human Kinetics, 96-101.

Viitasalo JT (1985b) Effects of training on force-velocity characteristics. In: DA Winter, RW Norman, RP Wells et al. (eds) *Biomechanics IX-A.* Champaign, Illinois: Human Kinetics 91-95.

Viitasalo JT and Auro O (1984) Seasonal fluctuations of force production in high jumpers. *Canadian Journal of Applied Sport Sciences* 9(4):209-213.

Vrijens J, Verstuyft J and DeClercq F (1990) Strength training in kayak: A multidimensional concept. *Proceedings: International Seminar on Kayak-Canoe Coaching and Sciences 1989.* Budapest: International Canoe Federation.

Weir JP, Wagner LL and Housh TJ (1994) The effects of rest interval length on repeated maximal bench presses. *Journal of Strength and Conditioning Research* 8:58-60.

White AT and Johnson SC (1991) Physiological comparison of international, national and regional alpine skiers. *International Journal of Sports Medicine* 12:374-378.

Wilson GJ, Lyttle AD, Ostrowski KJ and Murphy AJ (1995) Assessing dynamic performance: A comparison of rate of force development tests. *Journal of Strength and Conditioning Research* 9(3):176-181.

Wilson GJ, Newton RU, Murphy AJ and Humphries BJ (1993) The optimal training load for the development of dynamic athletic performance. *Medicine and Science in Sports and Exercise* 25(11):1279-1286.

Young WB and Bilby GE (1993) The effect of voluntary effort to influence speed of contraction on strength, muscular power, and hypertrophy development. *Journal of Strength and Conditioning Research* 7(3):172-178.

Zamparo P, Antonutto G, Capelli C et al. (1997) Effect of elastic recoil on maximal explosive power of the lower limbs. *European Journal of Applied Physiology* 75:289-297.

Chapter 14

Ackland T, Schreiner A and Kerr D (1994) Anthropometric profiles of world championship female basketball players (abstract). *International Conference of Science and Medicine in Sport*. Brisbane: Sports Medicine Australia.

Fitzsimons M, Dawson B, Ward D and Wilkinson A (1993) Cycling and running tests of repeated sprint ability. *Australian Journal of Science and Medicine in Sport* 25(4):82-87.

Gillam GM (1985) Identification of anthropometric and physiological characteristics relative to participation in college basketball. *National Strength and Conditioning Association Journal* 7(3):34-36.

Leger LA and Lambert J (1982) A maximal multistage 20 m shuttle run test to predict $\dot{V}O_2$max. *European Journal of Applied Physiology* 49:1-12.

Leger LA, Marcier D, Gadoury C and Lambert J (1988) The multistage 20 metre shuttle run test for aerobic fitness. *Journal of Sports Sciences* 6:93-101.

Morrow JR, Hosler WW and Nelson JK (1980) A comparison of women intercollegiate basketball players, volleyball players and non-athletes. *Journal of Sports Medicine and Physical Fitness* 20:435-440.

Norton K, Olds T, Olive S and Craig N (1996a) Anthropometry and sport performance. In: Norton K and Olds T (eds) *Anthropometrica*. Sydney: University of New South Wales Press, pp. 289-364.

Norton K, Whittingham N, Carter L et al. (1996b) Measurement techniques in anthropometry. In: Norton K and Olds T (eds) *Anthropometrica*. Sydney: University of New South Wales Press, pp. 25-76.

Ramsbottom R, Brewer J and Williams C (1987) A progressive shuttle run to estimate maximal oxygen uptake in adults. *British Journal of Sports Medicine* 4:163-165.

Ross WD and Marfell-Jones MJ (1991) Kinanthropometry. In: MacDougall JD, Wenger HA and Green HJ (eds) *Physiological Testing of the High Performance Athlete* (2nd ed). Champaign, Illinois: Human Kinetics, pp. 223-308.

Stone WJ and Kroll WA (1991) *Sports Conditioning and Weight Training* (3rd ed). Dubuque, Iowa: Brown.

Stone WJ and Steingard PM (1993) Year-round conditioning for basketball. *Clinics in Sports Medicine* 12(2):173-191.

Telford RD, Minikin BA, Hahn AG and Hooper LA (1989) A simple method for the assessment of general fitness: The tri-level profile. *Australian Journal of Science and Medicine in Sport* 21(3):6-9.

Withers RT, Craig NP, Bourdon PC and Norton KI (1987) Relative body fat and anthropometric prediction of body density of male athletes. *European Journal of Applied Physiology* 56:191-200.

Young W (1994a) A simple method for evaluating the strength qualities of the leg extensor muscles and jumping abilities. *Strength and Conditioning Coach* 2:5-8.

Young WB (1994b) Specificity of jumping ability and implications for training and testing athletes. *Proceedings of the National Coaching Conference*. Canberra: Australian Sports Commission, pp. 217-221.

Chapter 15

Brewer J, Ramsbottom R and Williams C (1988) *Multistage Fitness Test*. Belconnen, Australian Capital Territory: Australian Coaching Council.

Department of Sport, Recreation and Tourism (1986) Commonwealth Recreation Participation Survey, February 1986. Canberra: Australian Government Publishing Service.

Jackson AW and Baker AA (1986) The relationship of the sit and reach test to criterion measures of hamstring and back flexibility in young females. *Research Quarterly for Exercise and Sport* 57(3):183-186.

Leger LA and Lambert J (1982) A maximal multistage 20 m shuttle run test to predict $\dot{V}O_2$max. *European Journal of Applied Physiology* 49:1-12.

Young W (1994) A simple method for evaluating the strength qualities of the leg extensor muscles and jumping abilities. *Strength and Conditioning Coach* 2: 5-8.

Chapter 16

Bergh U (1982) *Physiology of Cross-Country Ski Racing*. Champaign, Illinois: Human Kinetics.

Bergh U (1987) The influence of body mass in cross-country skiing. *Medicine and Science in Sports and Exercise* 19:324-331.

Bouchard C, Tremblay A, Despres J et al. (1990) The response to long-term overfeeding in identical twins. *New England Journal of Medicine* 322:61-68.

Cheng B, Kuipers H, Snyder A et al. (1992) A new approach for the determination of ventilatory and lactate thresholds. *International Journal of Sports Medicine* 13:518-522.

Clement D, Lloyd-Smith D, Macintyre J et al. (1987) Iron status in winter Olympic sports. *Journal of Sports Sciences* 5:261-271.

Eisenman P, Johnson S, Bainbridge C and Zupan M (1989) Applied physiology of cross-country skiing. *Sports Medicine* 8:67-79.

Hermansen L (1973) Oxygen transport during exercise in human subjects. *Acta Physiologica Scandinavica* 399:19-36.

Jacobs I (1986) Blood lactate: Implications for training and sports performance. *Journal of Sports Medicine* 3:10-25.

Jette M, Thoden J and Spence J (1976) The energy expenditure of a 5-km cross-country ski run. *Journal of Sports Medicine* 16:134-137.

Kindermann W, Simon G and Keul J (1979) The significance of the aerobic-anaerobic transition for the determination of work load intensities during endurance training. *European Journal of Applied Physiology* 20:26-33.

Martin D (1989) Routine testing of U.S. biathletes (unpublished data). Colorado Springs, CO: United States Olympic Training Center.

Millerhagen J, Kelly J and Murphy R (1983) A study of combined arm and leg exercise with application to Nordic skiing. *Canadian Journal of Applied Sports Science* 8:92-97.

Niinimaa V, Shephard R and Dyon M (1979) Determinations of performance and mechanical efficiency in Nordic skiing. *British Journal of Sports Medicine* 13:62-65.

Rundell K (1996) Differences between treadmill running and treadmill roller skiing. *Journal of Strength and Conditioning Research* 10:167-172.

Rundell K and Bacharach D (1995) Physiological characteristics and performance of top US biathletes. *Medicine and Science in Sports and Exercise* 27:281-287.

Sharkey B (1984) *Training for Cross-Country Ski Racing*. Champaign, Illinois: Human Kinetics.

Sharkey B and Heidel B (1981) Physiological tests of cross-country skiers. *Journal of the United States Ski Coaches Association* 5:1-5.

Smith G (1989) Kinetic analysis of the V1 skate in cross-country skiing. *First International World Congress on Sport Sciences*, p. PH-15 (abstract).

Stromme S, Ingjer F and Meen H (1977) Assessment of maximal aerobic power in specifically trained athletes. *Journal of Applied Physiology* 42:833-837.

Zhou S and Weston S (1997) Reliability of using the D-max method to define physiological responses to incremental exercise testing. *Physiological Measurements* 18:145-153.

Chapter 17

Billat V, Renoux JC, Pinoteau J, Petit B and Koralsztein JP (1994) Reproducibility of running time to exhaustion at $\dot{V}O_2$max in subelite runners. *Medicine and Science in Sports and Exercise* 26:254-257.

Bishop D (1997) Reliability of a 1-h endurance performance test in trained female cyclists. *Medicine and Science in Sports and Exercise* 29(4):544-559.

Bouchard C, Taylor AW, Simoneau J and Dulac S (1991) Testing anaerobic power and capacity. In: McDougall JD, Wenger HA and Green HJ (eds) *Physiological Testing of the High Performance Athlete*. Champaign, Illinois: Human Kinetics, pp. 175-221.

Burke ER (1986) The physiology of cycling. In: Burke ER (ed) *Science of Cycling*. Champaign, Illinois: Human Kinetics, pp. 1-19.

Coyle EF, Coggan AR, Hopper MK and Walters TJ (1988) Determinants of endurance in well trained cyclists. *Journal of Applied Physiology* 64:2622-2630.

Coyle EF, Feltner ME, Kautz SA et al. (1991) Physiological and biomechanical factors associated with elite endurance cycling performance. *Medicine and Science in Sports and Exercise* 23:93-107.

Craig NP and Conyers RAJ (1988) Physiological and biochemical correlates of a 40 km individual cycling time trial. In: *Proceedings Australian Sports Medicine Federation National Science Conference*. Sydney, pp. 29-30.

Craig NP, Norton KI, Bourdon PC et al. (1993) Aerobic and anaerobic indices contributing to track endurance cycling performance. *European Journal of Applied Physiology* 67:150-158.

Davies CTM and Sandstrom ER (1989) Maximal mechanical power output and capacity of cyclists and young adults. *European Journal of Applied Physiology* 58:838-844.

Hawley JA and Noakes TD (1992) Peak power output predicts maximal oxygen uptake and performance time in trained cyclists. *European Journal of Applied Physiology* 65:79-83.

Jeukendrup A, Saris WH, Brouns F and Kester AD (1996) A new validated endurance test. *Medicine and Science in Sports and Exercise* 28:266-270.

Kyle CR (1991) Ergogenics of cycling. In: Lamb DR and Williams MH (eds) Ergogenics-Enhancement of Performance in Exercise and Sport. *Perspectives in Exercise Science and Sports Medicine* vol 4. Dubuque, Iowa: Brown, pp. 373-419.

McLean BD (1993) The relationship between frontal surface area and anthropometric parameters in racing cyclists. In: *Proceedings, 14th Congress International Society of Biomechanics*. Paris, pp. 856-857.

McLean BD and Ellis L (1992) Body mass, thigh volume and vertical jumping ability as predictors of short-term cycle ergometer performance in junior cyclists. *Excel* 8:149-153.

Miller FR and Manfredi RG (1987) Physiological and anthropometrical predictors of 15-kilometer time trial cycling performance time. *Research Quarterly* 58:250-254.

Norton KI (1996) Anthropometric estimation of body fat. In: Norton KI and Olds TS (eds) *Anthropometrica*.

Sydney, Australia: University of New South Wales Press, pp. 171-198.

Olds TS, Norton KI and Craig NP (1993) A mathematical model of cycling performance. *Journal of Applied Physiology* 75:730-737.

Olds TS, Norton KI, Craig NP, Olive S and Lowe E (1995a) The limits of the possible: Models of energy supply and demand in cycling. *Australian Journal of Science and Medicine in Sport* 27:29-33.

Olds TS, Norton KI, Lowe ELA et al. (1995b) Modelling road cycling performance. *Journal of Applied Physiology* 78:1596-1611.

Pederson DG and Gore CJ (1996) Anthropometry measurement error. In: Norton KI and Olds TS (eds) *Anthropometrica*. Sydney: University of New South Wales Press, pp. 77-96.

Thoden JS (1991) Testing aerobic power. In: MacDougall JD, Wenger HA and Green HJ (eds) *Physiological Testing of the High Performance Athlete*. Champaign, Illinois: Human Kinetics, pp. 107-173.

White JA and Al-Dawalibi MA (1986) Assessment of the power of racing cyclists. *Journal of Sports Sciences* 4:117-122.

Withers RT, Craig NP, Bourdon PC and Norton KI (1987a) Relative body fat and anthropometric prediction of body density of male athletes. *European Journal of Applied Physiology* 56:191-200.

Withers RT, Whittingham NO, Norton KI et al. (1987b) Relative body fat and anthropometric prediction of body density of female athletes. *European Journal of Applied Physiology* 56:169-180.

Chapter 18

Batt ME (1992) A survey of golf injuries in amateur golfers. *British Journal of Sports Medicine* 26(1):63-65.

Brendecke P (1990) Golf injuries. *Sports Medicine Digest* 12 (4):1.

Cahalan TD, Cooney WP, Tamai K and Chao EYS (1991) Biomechanics of the golf swing in players with pathologic conditions of the forearm, wrist and hand. *American Journal of Sports Medicine* 19:288-293.

Chim J and McMaster R (1996) *Get Fit for Golf Part 1*. Melbourne, Australia: Unlimited Graphics.

Gore C and Edwards D (1992) *Australian Fitness Norms: A Manual for Fitness Assessors*. Adelaide: The Health Development Foundation, South Australia.

Jobe F, Perry WJ and Pink M (1989) Electromyographic shoulder activity in men and women professional golfers. *American Journal of Sports Medicine* 17:782-787.

Jobe F and Yocum L (1988) The dark side to practice. *Golf* 30:22.

Jobe F, Yocum L, Mottram R and Pink M (1994) *Exercise Guide to Better Golf*. Champaign, Illinois: Human Kinetics.

McCarroll JR and Gioe TJ (1982) Professional golfers and the price they pay. *Physician and Sports Medicine* 10:64-70.

Wilmore J and Costill D (1994) *Physiology of Sport and Exercise*. Champaign, Illinois: Human Kinetics.

Chapter 19

Draper J and Lancaster M (1985) The 505 test: A test of agility in the horizontal plane. *Australian Journal of Science and Medicine in Sport* 17(1):15-18.

Fitzsimons M, Dawson B, Ward D and Wilkinson A (1993) Cycling and running tests of repeated sprint ability. *Australian Journal of Science and Medicine in Sport* 25(4):82-87.

Leger LA and Lambert J (1982) A maximal multistage 20 m shuttle run test to predict $\dot{V}O_2$max. *European Journal of Applied Physiology* 49:1-12.

Leger LA, Marcier D, Gadoury C and Lambert J (1988) The multistage 20 metre shuttle run test for aerobic fitness. *Journal of Sports Sciences* 6:93-101.

Morrow JR, Hosler WW and Nelson JK (1980) A comparison of women intercollegiate basketball players, volleyball players and non-athletes. *Journal of Sports Medicine and Physical Fitness* 20:435-440.

Norton K, Olds T, Olive S and Craig N (1996) Anthropometry and sport performance. In: Norton K and Olds T (eds) *Anthropometrica*. Sydney: University of New South Wales Press, pp. 289-364.

Ramsbottom R, Brewer J and Williams C (1987) A progressive shuttle run to estimate maximal oxygen uptake in adults. *British Journal of Sports Medicine* 4:163-165.

Winget CM, DeRoshia CW and Holley DC (1985) Circadian rhythms and athletic performance. *Medicine and Science in Sports and Exercise* 17(5):498-516.

Withers RT, Craig NP, Bourdon PC and Norton KI (1987) Relative body fat and anthropometric prediction of body density of male athletes. *European Journal of Applied Physiology* 56 191-200.

Chapter 20

Baker D (1995) The use of submaximal repetitions to predict maximal squat and bench press strength in trained athletes. *Strength and Conditioning Coach* 3(4):17-19.

Bale P and Hunt S (1986) The physique, body composition and training variables of elite and good netball players in relation to playing position. *Australian Journal of Science and Medicine in Sport* 18(4):16-19.

Chad K and Steele J (1990) *Relationship between Physical Requirements and Physiological Responses to Match Play and Training in Skilled Netball Players: Basis of Tailor-Made Training Programs*. A report presented to the Australian Sports Commission's Applied Sports Research Program. Canberra: Australian Sports Commission.

Draper JA and Lancaster MG (1985) The 505 test: A test for agility in the horizontal plane. *Australian Journal of Science and Medicine in Sport* 17(1):15-18.

Hakkinen K, Komi PV and Alen M (1985) Effect of explosive type strength training on isometric force and relax time, EMG and muscle fibre characteristics of leg extensors muscle. *Acta Physiologica Scandinavica* 125:587-600.

Hakkinen K, Komi PV and Kauhanen H (1986) EMG and force production characteristics of leg extensor muscles of elite weightlifters during isometric, concentric and various stretch shortening cycle exercises. *International Journal of Sports Medicine* 3(7):144-151.

Jackson AW and Baker AA (1986) The relationship of the sit and reach test to criterion measures of hamstring and back flexibility in young females. *Research Quarterly for Exercise and Sport* 57(3):183-186.

Norton KI, Craig NP, Withers RT and Whittingham NO (1994) Assessing the body fat of athletes. *Australian Journal of Science and Medicine in Sport* 26(1&2):6-13.

Otago L (1983) A game analysis of the activity patterns of netball players. *Sports Coach* 7(1):24-28.

Ross WD and Marfell-Jones MJ (1987) Kinanthropometry. In: MacDougall JD, Wenger HA and Green HJ (eds) *Physiological Testing of the Elite Athlete*. Champaign, Illinois: Human Kinetics, pp. 223-308.

Schmidtbleicher D (1992) Training for power events. In: Komi PV (ed) *Strength and Power in Sport*. Oxford: Blackwell Scientific, pp. 381-395.

Steele J and Chad K (1992) An analysis of the movement patterns of netball players during matchplay: Implications for designing training programs. *Sports Coach* 15(1):21-28.

Viitasalo JT and Aura O (1984) Seasonal fluctuations of force production in high jumpers. *Canadian Journal of Applied Sports Science* 9:209-213.

Woolford S and Angove M (1991) A comparison of training techniques and game intensities for national level netball players. *Sports Coach* 14(4):18-21.

Young WB (1994) A simple method for evaluating the strength qualities of the leg extensor muscles and jumping abilities. *Strength and Conditioning Coach* 2(4):5-8.

Chapter 21

Cheng B, Kuipers H, Snyder AC et al. (1992) A new approach for the determination of ventilatory and lactate thresholds. *International Journal of Sports Medicine* 13:518-522.

Gore CJ, Hahn AG, Scroop GC et al. (1996) Increased arterial desaturation in elite athletes during submaximal exercise at 580 m altitude. *Journal of Applied Physiology* 80(6):2204-2210.

Hahn A (1990) Identification and selection of talent in Australian rowing. *Excel* 6(3):5-11.

Medbø JI, Mohn A-C, Tabata I et al. (1988) Anaerobic capacity determined by maximal accumulated O_2 deficit. *Journal of Applied Physiology* 64:50-60.

Norton KI, Whittingham NO, Carter L, Kerr D and Gore CJ (1996) Measurement technique in anthropometry. In: Norton KI and Olds T (eds) *Anthropometrica*. Sydney: University of New South Wales Press, pp. 25-75.

Rodriguez FA (1986) Physical structure of international lightweight rowers. In: Reilly T, Watkins J and Borms J (eds) *Kinanthropometry* III. London: Spon, pp. 255-261.

Ross W, Ward R, Leahy R and Day J (1982) Proportionality of Montreal athletes. In: Carter J (ed) *Physical Structure of Olympic Athletes. Part 1: The Montreal Olympic Games Anthropological Project*. Basel: Karger, pp. 81-106.

Secher N and Vaage O (1983) Rowing performance, a mathematical model based on analysis of body dimensions as exemplified by body weight. *European Journal of Applied Physiology* 52:88-93.

Secher N, Vaage O and Jackson R (1982) Rowing performance and maximal aerobic power of oarsmen. *Scandinavian Journal of Sports Science* 4(1):9-11.

Chapter 22

Dawson B, Ackland T, Roberts C and Lawrence S (1991) The phosphate recovery test revisited. *Sports Coach* 14(3):41.

Deutsch M, Maw G, Jenkins D and Reaburn P (1998). Heart rate, blood lactate and kinematic data of elite colts (under 19) rugby union players during competition. *Journal of Sports Sciences* 16(6): 561-570.

Jenkins D (1993) The importance of aerobic fitness for field-games players. *Sports Coach* 16(2): 22-23.

Jenkins D (ed) (1995) *Rugby Union: Preparing to Play Rugby*. Sydney: Australian Rugby Union.

Norton KI, Whittingham NO, Carter L, Kerr D and Gore CJ (1996) Measurement technique in anthropometry. In: Norton KI and Olds T (eds) *Anthropometrica*. Sydney: University of New South Wales Press, pp. 25-75.

Chapter 23

Borg G (1962) Physical performance and perceived exertion. *Studia Psychogia et Paedagogia*. Geerup vol 11, pp. 1-35.

Cheng B, Kuipers H, Snyder AC et al. (1992) A new approach for the determination of ventilatory and lactate thresholds. *International Journal of Sports Medicine* 13:518-522.

Fohrenbach R, Mader A, Thiele W and W Hollmann (1986) Test procedures and metabolically oriented intensity distribution in sprint training with a submaximal load structure. *Leistungssport* 16:15-24.

Telford RD (1991) Physiological assessment of the runner. In: Draper J, Minikin B and Telford RD (eds) *Test*

Methods Manual. Canberra: National Sports Research Centre, section 3.

Chapter 24

Blackburn M (1994) Physiological responses to 90 minutes of simulated dinghy sailing. *Journal of Sports Sciences* 12:383-390.

Blackburn M and Hubinger L (1996) *Determination of Physiological Profiles and Exercise Training Programs for Competitive Dinghy Sailors*. National Sport Research Centre report. Canberra: Australian Sports Commission.

Gallozzi C, Fanton F, De Angelis M and Dal Monte A (1993) The energetic cost of sailing. *Medical Science Research* 21:851-853.

Mackie HW and Legg SJ (1999) Preliminary assessment of force demands in Laser racing. *Journal of Science and Medicine in Sport* 2:78-85.

Vogiatzis I, Spurway NC, Wilson J and Boreham C (1995) Assessment of aerobic and anaerobic demands of dinghy sailing at different wind velocities. *Journal of Sports Medicine and Physical Fitness* 35:103-107.

Chapter 25

Davis JA and Brewer J (1993) Applied physiology of female soccer players. *Sports Medicine* 16(3):180-189.

Ekblom B (1986) Applied physiology of soccer. *Sports Medicine* 3:50-60.

Hugg PJ (1994) The Selection of Australian Youth Soccer Players Based on Physical and Physiological Characteristics. A thesis for the Degree of Master of Applied Science (Sports Studies), University of Canberra.

Norton KI, Whittingham NO, Carter L, Kerr D and Gore CJ (1996) Measurement technique in anthropometry. In: Norton KI and Olds T (eds) *Anthropometrica*. Sydney: University of New South Wales Press, pp. 25-75.

Tumilty D (1993) Physiological characteristics of elite soccer players. *Sports Medicine* 16(2):80-96.

Chapter 26

Baker D (1995) The use of submaximal repetitions to predict maximal squat and bench press strength in trained athletes. *Strength and Conditioning Coach* 3(4):17-19.

Challis J. Methodological report: the appropriate scaling of weightlifting performance. *Journal of Strength and Conditioning Research* 13(4).

Draper JA and Lancaster MG (1985) The 505 test: A test for agility in the horizontal plane. *Australian Journal of Science and Medicine in Sport* 17(1):15-18.

Hakkinen K, Komi PV and Alen M (1985) Effect of explosive type strength training on isometric force and relax time, EMG and muscle fibre characteristics of leg extensors muscle. *Acta Physiologica Scandinavica* 125:587-600.

Hakkinen K, Komi PV and Kauhanen H (1986) EMG and force production characteristics of leg extensor muscles of elite weightlifters during isometric, concentric and various stretch shortening cycle exercises. *International Journal of Sports Medicine* 3(7):144-151.

Jackson AW and Baker AA (1986) The relationship of the sit and reach test to criterion measures of hamstring and back flexibility in young females. *Research Quarterly for Exercise and Sport* 57(3):183-186.

Norton KI, Craig NP, Withers RT and Whittingham NO (1994) Assessing the body fat of athletes. Australian *Journal of Science and Medicine in Sport* 26(1&2):6-13.

Schmidtbliecher D (1992) Training for power events. In: Komi PV (ed) *Strength and Power in Sport*. Oxford: Blackwell Scientific.

Softball Australia (1993) *Official Playing Rules of the Australian Softball Federation Inc*. International Federation Playing Rules.

Viitasalo JT and Aura O (1984) Seasonal fluctuations of force production in high jumpers. *Canadian Journal of Applied Sports Science* 9:209-213.

Chapter 27

Anderson GS and Rhodes EC (1989) A review of blood lactate and ventilatory methods detecting transition thresholds. *Sports Medicine* 8:43-55.

Antonutto G and Di Prampero PE (1995) The concept of lactate threshold: A short review. *Journal of Sport Medicine and Physical Fitness* 35:6-12.

Bishop P and Martino M (1993) Blood lactate measurement in recovery as an adjunct to training. *Sports Medicine* 16:5-13.

Cheng B, Kuipers H, Snyder AC et al. (1992) A new approach for the determination of ventilatory and lactate thresholds. *International Journal of Sports Medicine* 13:518-522.

Costill DL, Kovaleski J, Porter D et al. (1985) Energy expenditure during front crawl swimming: Predicting success in middle-distance events. *International Journal of Sports Medicine* 6(5):266-270.

Counsilman BE and Counsilman JE (1993) Problems with the physiological classification of endurance loads. *American Swimmer* Dec-Jan:4-20.

Craig AB and Pendergast DR (1979) Relationships of stroke rate, distance per stroke and velocity in competitive swimming. *Medicine and Science in Sports and Exercise* 17:625-634.

Foster C, Fitzgerald DJ and Spatz P (1998) Stability of the blood lactate-heart rate relationship in competitive athletes. *Medicine and Science in Sports and Exercise* 31:578-582.

Holroyd D and Swanwick K (1993) A mathematical model for lactate profiles and a swimming power expenditure formula for use in conjunction with it. *Journal of Swimming Research* 9:25-31.

Hultman E and Sahlin K (1980) Acid-base balance during exercise. *Exercise and Sport Sciences Reviews* 8:41-48.

Madsen O and Lohberg M (1987) The lowdown on lactates. *Swimming Technique* May-July:21-28.

Maglischo EW, Maglischo CW and Bishop RA (1987) Lactate testing for training pace. *Swimming Technique* 19:31-37.

Maglischo EW, Maglischo CW, Smith RE, Bishop RA and Novland PN (1984) Determining the proper training speeds for swimmers. *Journal of Swimming Research* 1:32-38.

Maw GJ and Volkers S (1996) Measurement and application of stroke dynamics during training in your own pool. *Australian Swimming Coach* 12(3):34-38.

Norton K, Whittingham N, Carter L et al. (1996) Measurement techniques in anthropometry. In: Norton K and Olds T (eds) *Anthropometrica*. Sydney: University of New South Wales Press, pp. 25-76.

Prins J (1988) Setting a standard. *Swimming Technique* May-July:13-16.

Pyne DB (1989) The use and interpretation of blood lactate testing in swimming. *Excel* 5(4):23-26.

Pyne DB (1995) Coach, I can't get my heart rate up (or down): The physiology of measuring heart rates. *Australian Swimming Coach* 11(9):19-22.

Pyne DB, Lee H and Swanwick K (1999) Lactate profiling to monitor changes in anaerobic threshold of elite swimmers. *Proceedings of XIIIth FINA Sports Medicine Congress* (abstract), Hong Kong, p. 60.

Pyne DB and Telford RD (1988) Classification of swimming training sessions by blood lactate and heart rate responses. *Excel* 5(2):9-12.

Reilly T and Woodbridge V (1999) Effects of moderate dietary manipulations on swim performance and on blood lactate-swimming velocity curves. *International Journal of Sports Medicine* 20:93-97.

Richardson MT, Zoerink D, Rinehardt CF et al. (1996) Recovery from maximal swimming at the predicted initial onset of blood lactate accumulation. *Journal of Swimming Research* 11:30-35.

Roberts AD (1990) *Heart Rate, Lactic Acid and Swimming Performance*. State of the Art Review no. 25, National Sports Research Centre. Canberra: Australian Sports Commission.

Roberts AD (1991) Physiological capacity for sports performance. In: Pyke FS (ed) *Better Coaching: Advanced Coach's Manual*. Canberra: Australian Coaching Council, pp. 43-54.

Sharp RL (1992) Exercise physiology: Proper conditioning. In: Leonard J (ed) *Science of Coaching Swimming*. Champaign, Illinois: Leisure Press, pp. 71-98.

Snyder AC, Jeukendrup AE, Hesselink MKC, Kuiper H and Foster C (1993) A physiological/psychological indicator of over-reaching during intensive training. *International Journal of Sports Medicine* 14:29-32.

Tegtbur U, Busse MW and Baumann KM (1993) Estimation of an individual equilibrium between lactate production and catabolism during exercise. *Medicine and Science in Sports and Exercise* 25:620-627.

Telford RD, Hahn AG, Catchpole EA, Parker AR and Sweetenham WF (1988) Post-competition blood lactate concentration in highly ranked Australian swimmers. In: Ungerechts BE (ed) *Swimming Science V*. Champaign, Illinois: Human Kinetics, pp. 277-283.

Treffene RJ (1978) Swimming performance test: A method of training and performance time selection. *Australian Journal of Sports Medicine* 10:2-5.

Treffene RJ (1979) Swimming performance test: A method of training and performance time selection. *Swimming Technique* 15:120-124.

Treffene RJ, Craven C, Hobbs K and Wade C (1979) Heart rates and plasma lactate study. *The International Swimmer* June:19-20.

Troup JP (1984) Review, energy systems and training considerations. *Journal of Swimming Research* 1:13-16.

Troup JP (1986) Setting up a season using scientific training. *Swimming Technique* May-July:8-16.

Wakayoshi K, D'Acquisto LJD, Cappaert JM and Troup J (1995) Relationship between oxygen uptake, stroke rate and swimming velocity in competitive swimmers. *International Journal of Sports Medicine* 16:19-23.

Wakayoshi K, Yoshida T, Ikuta Y, Mutoh Y and Miyashita M (1993) Adaptations to six months of aerobic swim training—changes in velocity, stroke rate, stroke length and blood lactate. *International Journal of Sports Medicine* 14:368-372.

Weltman A (1995) *The Blood Lactate Response to Exercise*. Champaign, Illinois: Human Kinetics, pp. 83-92.

Chapter 28

Bergeron MF, Maresh CM, Kraemer WJ, Abreham A, Conroy B and Gabaree C (1991) Tennis: A physiological profile during match play. *International Journal of Sports Medicine* 12:474-479.

Brewer J, Ramsbottom R and Williams C (1988) *Multistage Fitness Test*. Canberra: Australian Coaching Council.

Buono MJ, Constable SH and Stanforth PR (1980) Maximum oxygen uptake and body composition of varsity collegiate tennis players. *Arizona Journal of Health, Physical Education, Recreation and Dance* 23: 6-7.

Docherty D (1982) A comparison of heart rate responses in racquet games. *British Journal of Sports Medicine* 16:96-100.

Elliott BC, Ackland TR, Blanksby BA and Bloomfield J (1990) A prospective study of physiological and kinanthropometric indicators of junior tennis performance. *Australian Journal of Science and Medicine in Sport* 22:87-92.

Elliott B, Dawson B and Pyke F (1985) The energetics of singles tennis. Journal of Human Movement Studies 11:11-20.

Friedman DB, Ramo BW and Gray GJ (1984) Tennis and cardiovascular fitness in middle-aged men. *Physician and Sportsmedicine* 12(7):87-91.

Groppel JL (1986) The biomechanics of tennis: An overview. *International Journal of Sport Biomechanics* 2:141-155.

Groppel JL, Loehr JE, Melville D and Quinn AM (1989) *Science of Coaching Tennis*. Champaign, Illinois: Leisure Press.

Kibler WB, McQueen C and Uhl TL (1988) Fitness evaluations and fitness findings in junior tennis players. *Clinics in Sportsmedicine: Racquet Sports—Injury Treatment and Prevention* 7, 403-416.

Koziris LP, Kraemer WJ, Triplett NT, Fry AC and Bauer J (1991). Strength imbalances in women tennis players. *Medicine and Science in Sports and Exercise* 23 (suppl 4):S43.

Leach RE (1988) Leg and foot injuries in racquet sports. *Clinics in Sportsmedicine: Racquet Sports—Injury Treatment and Prevention* 7, 359-370.

Loehr J (1988) What is periodisation? *Sport Science for Tennis* (summer):1.

Leger LA and Lambert J (1982) A maximal multistage 20 m shuttle run test to predict $\dot{V}O_2$max. *European Journal of Applied Physiology* 49:1-5.

Reilly T (1990) The racquet sports. In: Reilly T, Secher N, Snell P and Williams C (eds) *Physiology of Sports*. London: Spon.

Roetert P, Brown S, Piorkwoski P and Woods R (1996) Fitness comparisons among three different levels of elite tennis players. *Journal of Strength and Conditioning Research* 10:139-143.

Chapter 29

Borg G (1962) Physical performance and perceived exertion. *Studia Psychogia et Paedagogia*. Geerup 11:1-35.

Norton KI, Whittingham NO, Carter L, Kerr D and Gore CJ (1996) Measurement technique in anthropometry. In: Norton KI and Olds T (eds) *Anthropometrica*. Sydney: University of New South Wales Press, pp. 25-75.

Slievert GG and Rowlands DS (1996) Physical and physiological factors associated with success in the triathlon. *Sports Medicine* 22:8-18.

Chapter 30

Abernethy PJ, Wilson G and Logan PA (1995) Strength and power assessment: Issues and controversies. *Sports Medicine* 19(6):401-417.

Carter JEL and Ackland TR (1994) *Kinanthropometry in Aquatic Sports: A Study of World Class Athletes*. Human Kinetics Sports Science Monograph Series vol 5. Champaign, Illinois: Human Kinetics.

Norton KI, Whittingham NO, Carter L, Kerr D and Gore CJ (1996) Measurement technique in anthropometry. In: Norton KI and Olds T (eds) *Anthropometrica*. Sydney: University of New South Wales Press, pp. 25-75.

Pederson DG and Gore CJ (1996) Anthropometry measurement error. In Norton KI and Olds T (eds) *Anthropometrica*. Sydney: University of New South Wales Press, pp. 77-96.

Smith HK (1991) Interpretation and Applications of Time-Motion Data—Senior Men's Goaltending. Personal communication.

Index